Understanding Complex Systems

For further volumes:
http://www.springer.com/series/5394

Understanding Complex Systems

Future scientific and technological developments in many fields will necessarily depend upon coming to grips with complex systems. Such systems are complex in both their composition - typically many different kinds of components interacting simultaneously and nonlinearly with each other and their environments on multiple levels - and in the rich diversity of behavior of which they are capable.

The Springer Series in Understanding Complex Systems series (UCS) promotes new strategies and paradigms for understanding and realizing applications of complex systems research in a wide variety of fields and endeavors. UCS is explicitly transdisciplinary. It has three main goals: First, to elaborate the concepts, methods and tools of complex systems at all levels of description and in all scientific fields, especially newly emerging areas within the life, social, behavioral, economic, neuro and cognitive sciences (and derivatives thereof); second, to encourage novel applications of these ideas in various fields of engineering and computation such as robotics, nano-technology and informatics; third, to provide a single forum within which commonalities and differences in the workings of complex systems may be discerned, hence leading to deeper insight and understanding.

UCS will publish monographs, lecture notes and selected edited contributions aimed at communicating new findings to a large multidisciplinary audience.

Springer Complexity

Springer Complexity is an interdisciplinary program publishing the best research and academic-level teaching on both fundamental and applied aspects of complex systems - cutting across all traditional disciplines of the natural and life sciences, engineering, economics, medicine, neuroscience, social and computer science.

Complex Systems are systems that comprise many interacting parts with the ability to generate a new quality of macroscopic collective behavior the manifestations of which are the spontaneous formation of distinctive temporal, spatial or functional structures. Models of such systems can be successfully mapped onto quite diverse "real-life" situations like the climate, the coherent emission of light from lasers, chemical reaction-diffusion systems, biological cellular networks, the dynamics of stock markets and of the internet, earthquake statistics and prediction, freeway traffic, the human brain, or the formation of opinions in social systems, to name just some of the popular applications.

Although their scope and methodologies overlap somewhat, one can distinguish the following main concepts and tools: self-organization, nonlinear dynamics, synergetics, turbulence, dynamical systems, catastrophes, instabilities, stochastic processes, chaos, graphs and networks, cellular automata, adaptive systems, genetic algorithms and computational intelligence.

The two major book publication platforms of the Springer Complexity program are the monograph series "Understanding Complex Systems" focusing on the various applications of complexity, and the "Springer Series in Synergetics", which is devoted to the quantitative theoretical and methodological foundations. In addition to the books in these two core series, the program also incorporates individual titles ranging from textbooks to major reference works.

Valery I. Klyatskin

Stochastic Equations: Theory and Applications in Acoustics, Hydrodynamics, Magnetohydrodynamics, and Radiophysics, Volume 1

Basic Concepts, Exact Results, and Asymptotic Approximations

Translated from Russian
by A. Vinogradov

Springer

Valery I. Klyatskin
A.M. Obukhov Institute of
Atmospheric Physics
Russian Academy of Sciences
Moscow
Russia

ISSN 1860-0832 ISSN 1860-0840 (electronic)
ISBN 978-3-319-35775-1 ISBN 978-3-319-07587-7 (eBook)
DOI 10.1007/978-3-319-07587-7
Springer Cham Heidelberg New York Dordrecht London

Printed on acid-free paper

Springer is part of Springer Science+Business Media (www.springer.com)

To Sonya Klyatskina

Preface

This monograph is revised and more comprehensive translation of my Russian monograph in 2 volumes [56, 57]. For reasons of usability, the material is divided into two practically independent volumes.

Volume 1 gives the theory of stochastic equations (including ordinary differential equations, partial differential equations, boundary-value problems, and integral equations) in terms of the functional analysis. The developed approach yields exact solutions to stochastic problems for a number of models of fluctuating parameters among which are telegrapher's and generalized telegrapher's processes, Markovian processes with a finite number of states, Gaussian Markovian processes, and functions of the above processes. Asymptotic methods of analyzing stochastic dynamic systems, such as delta-correlated random process (field) approximation and diffusion and higher approximations are also considered. The material is supplemented with the sections concerning dynamical and statistical descriptions of simplest hydrodynamic-type systems. Sections dealing with the stochastic structure formation in random media such as the clustering of parametrically excited random fields, and statistical problems of turbulent dynamo (magnetic field generation in random fields) are also added in this Volume.

Working at this edition, I tried to take into account remarks and wishes of readers about both style of the text and choice of specific problems. Different mistakes and misprints are corrected. The book is destined for scientists dealing with stochastic dynamic systems in different areas, such as acoustics, hydrodynamics, magnetohydrodynamics, radiophysics, theoretical and mathematical physics, and applied mathematics; it can be also useful for senior and postgraduate students. Now, a few words are due on the structure of the text of this volume. It consists of four parts.

The first part may be viewed as an introductory text. It takes up a few typical physical problems to discuss their solutions obtained under random disturbances of parameters affecting the system behavior. More detailed formulations of these problems and relevant statistical analysis can be found in other parts of this Volume and in Volume 2.

The second part deals with the statistical description of random quantities, processes, and fields and includes, in particular, some problems of statistical topography of random fields and stochastic structure formation in random fields.

The third part is devoted to the general theory of statistical analysis of dynamic systems with fluctuating parameters described by differential and integral equations. This theory is illustrated by analyzing specific dynamic systems.

The fourth part treats asymptotic methods of statistical analysis such as the delta-correlated random process (field) approximation, diffusion and higher approximations.

It is worth noting that purely mathematical and physical papers devoted to considered issues run into thousands. It would be physically impossible to give an exhaustive bibliography. Therefore, in this book I confine myself to referencing those papers which are used or discussed in this book and also recent review papers and papers with extensive bibliography on the subject.

Moscow Valery I. Klyatskin

Contents

Mephistopheles (about algebra)
...
I want to warn you:
        careful you should
Be with this science
        which is very tricky.
It can involve your brain into the chaos
        of vain unnecessary transformations.
If you don't manage
        to embrace its basis,
You won't be able indices to differ.
...
What's important –
        deepen into symbols
And, having mastered them,
        you bravely step
Onto the path, which leads into the Kingdom
        of Formulas, eternal and exact.

Kurd Lasswitz, Prost!,
www. gutenberg2000.de/lasswitz/prost/prost.htm

Translated by Alla Parollo.

Introduction

Different areas of physics pose statistical problems in ever-greater numbers. Apart from issues traditionally obtained in statistical physics, many applications call for including fluctuation effects into consideration. While fluctuations may stem from different sources (such as thermal noise, instability, and turbulence), methods used to treat them are very similar. In many cases, the statistical nature of fluctuations may be deemed known (either from physical considerations or from problem formulation) and the physical processes may be modeled by differential, integro-differential or integral equations.

Today the most powerful tools used to tackle complicated statistical problems are the *Markov theory of random processes* and the theory of *diffusion-type processes* evolved from *Brownian motion theory*. Mathematical aspects underlying these theories and their applications have been treated extensively in academic literature and textbooks (see, for example, [17]), and therefore we will not dwell on these issues in this treatise.

We will consider a statistical theory of dynamic and wave systems with fluctuating parameters. These systems can be described by ordinary differential equations, partial differential equations, integro-differential equations and integral equations. A popular way to solve such systems is by obtaining a closed system of equations for statistical characteristics of such systems to study their solutions as comprehensively as possible.

We note that often wave problems are boundary-value problems. When this is the case, one may resort to the imbedding method to reformulate the equations at hand into initial-value problems, thus considerably simplifying the statistical analysis [50].

We shall dwell in depth on dynamic systems whose fluctuating parameters are Gaussian random processes (fields), although what we present in this book is a general theory valid for fluctuating parameters of any nature.

The purpose of this book is to demonstrate how different physical problems described by stochastic equations may be solved on the base of a general approach. This treatment reveals interesting similarities between different physical problems.

Examples of specific physical systems outlined below are mainly borrowed from statistical hydrodynamics, statistical radiophysics and acoustics because of author's research in these fields. However, similar problems and solution techniques occur in such areas as plasma physics, solid-state physics, magnetohydrodynamics to name a few.

In stochastic problems with fluctuating parameters, the variables are functions. It would be natural therefore to resort to functional methods for their analysis. We will use a *functional method* devised by Novikov [114] for Gaussian fluctuations of parameters in a turbulence theory and developed by the author of this book [46, 48–50, 55, 61, 62] for the general case of dynamic systems and fluctuating parameters of arbitrary nature.

However, only a few dynamic systems lend themselves to analysis yielding solutions in a general form. It proved to be more efficient to use an asymptotic method where the statistical characteristics of dynamic problem solutions are expanded in powers of a small parameter which is essentially a ratio of the random impact's correlation time to the time of observation or to other characteristic time scale of the problem (in some cases, these may be spatial rather than temporal scales). This method is essentially a generalization of the theory of Brownian motion. It is termed the *delta-correlated random process (field) approximation.* In Brownian motion theory, this approximation is consistent with a model obtained by neglecting the time between random collisions as compared to all other time scales.

For dynamic systems described by ordinary differential stochastic equations with Gaussian fluctuations of parameters, this method leads to a Markovian problem solving model, and the respective equation for transition probability density has the form of the *Fokker–Planck equation.* In this book, we will consider in depth the methods of analysis available for this equation and its boundary conditions. We will analyze solutions and validity conditions by way of integral transformations. In more complicated problems described by partial differential equations, this method leads to a generalized equation of Fokker–Planck type in which variables are the derivatives of the solution's characteristic functional. For dynamic problems with non-Gaussian fluctuations of parameters, this method also yields Markovian type solutions. Under the circumstances, the probability density of respective dynamic stochastic equations satisfies a closed operator equation. For example, systems with parameters fluctuating in a Poisson profile are converted into the *Kolmogorov–Feller* type of integro-differential equations.

In physical investigations, Fokker–Planck and similar equations are usually set up from rule of thumb considerations, and dynamic equations are invoked only to calculate the coefficients of these equations. This approach is inconsistent, generally speaking. Indeed, the statistical problem is completely defined by dynamic equations and assumptions on the statistics of random impacts. For example, the Fokker–Planck equation must be a logical sequence of the dynamic equations and some assumptions on the character of random impacts. It is clear that not all problems lend themselves for reducing to a Fokker–Planck equation. The functional approach allows one to derive a Fokker–Planck equation from the problem's dynamic equation along with its applicability conditions. In terms of formal mathematics, our approach corresponds to that of R.L. Stratonovich (see, e.g., [126]).

For a certain class of random processes (Markovian telegrapher's processes, Gaussian Markovian process and the like), the developed functional approach also yields closed equations for the solution probability density with allowance for a finite correlation time of random interactions.

For processes with Gaussian fluctuations of parameters, one may construct a better physical approximation than the delta-correlated random process (field) approximation,—the *diffusion approximation* that allows for finiteness of correlation time radius. In this approximation, the solution is Markovian and its applicability condition has transparent physical meaning, namely, the statistical effects should be small within the correlation time of fluctuating parameters. This book treats these issues in depth from a general standpoint and for some specific physical applications.

Part I
Dynamical Description of Stochastic Systems

Chapter 1
Examples, Basic Problems, Peculiar Features of Solutions

In this chapter, we consider several dynamic systems described by differential equations of different types and discuss the features in the behaviors of solutions to these equations under random disturbances of parameters. Here, we content ourselves with problems in the simplest formulation. More complete formulations will be discussed below, in the sections dealing with statistical analysis of the corresponding systems.

1.1 Ordinary Differential Equations: Initial-Value Problems

1.1.1 Particles under Random Velocity Field

In the simplest case, a particle under random velocity field is described by the system of ordinary differential equations of the first order

$$\frac{d}{dt}\boldsymbol{r}(t) = \boldsymbol{U}(\boldsymbol{r},t), \qquad \boldsymbol{r}(t_0) = \boldsymbol{r}_0, \tag{1.1}$$

where $\boldsymbol{U}(\boldsymbol{r},t) = \boldsymbol{u}_0(\boldsymbol{r},t) + \boldsymbol{u}(\boldsymbol{r},t)$, $\boldsymbol{u}_0(\boldsymbol{r},t)$ is the deterministic component of the velocity field (mean flow), and $\boldsymbol{u}(\boldsymbol{r},t)$ is the random component. In the general case, field $\boldsymbol{u}(\boldsymbol{r},t)$ can have both nondivergent (solenoidal, for which $\operatorname{div}\boldsymbol{u}(\boldsymbol{r},t) = 0$) and divergent (for which $\operatorname{div}\boldsymbol{u}(\boldsymbol{r},t) \neq 0$) components.

We dwell on stochastic features of the solution to problem (1.1) for a system of particles in the absence of mean flow ($\boldsymbol{u}_0(\boldsymbol{r},t) = 0$). From Eq. (1.1) formally follows that every particle moves independently of other particles. However, if random field $\boldsymbol{u}(\boldsymbol{r},t)$ has a finite spatial correlation radius l_{cor}, particles spaced by a distance shorter than l_{cor} appear in the common zone

V.I. Klyatskin, *Stochastic Equations: Theory and Applications in Acoustics, Hydrodyn., Magnetohydrodyn., and Radiophys., Vol. 1*, Understanding Complex Systems,
DOI: 10.1007/978-3-319-07587-7_1, © Springer International Publishing Switzerland 2015

of infection of random field $\boldsymbol{u}(\boldsymbol{r},t)$ and the behavior of such a system can show new collective features.

For steady velocity field $\boldsymbol{u}(\boldsymbol{r},t) \equiv \boldsymbol{u}(\boldsymbol{r})$, Eq. (1.1) reduces to

$$\frac{d}{dt}\boldsymbol{r}(t) = \boldsymbol{u}(\boldsymbol{r}), \quad \boldsymbol{r}(0) = \boldsymbol{r}_0. \tag{1.2}$$

This equation clearly shows that steady points $\widetilde{\boldsymbol{r}}$ (at which $\boldsymbol{u}(\widetilde{\boldsymbol{r}}) = 0$) remain the fixed points. Depending on whether these points are stable or unstable, they will attract or repel nearby particles. In view of randomness of function $\boldsymbol{u}(\boldsymbol{r})$, points $\widetilde{\boldsymbol{r}}$ are random too.

It is expected that the similar behavior will be also characteristic of the general case of the space-time random velocity field of $\boldsymbol{u}(\boldsymbol{r},t)$.

If some points $\widetilde{\boldsymbol{r}}$ remain stable during sufficiently long time, then clusters of particles (i.e., compact regions with elevated particle concentration, which occur merely in rarefied zones) must arise around these points in separate realizations of random field $\boldsymbol{u}(\boldsymbol{r},t)$. On the contrary, if the stability of these points alternates with instability sufficiently rapidly and particles have no time for significant rearrangement, no clusters of particles will occur.

Simulations (see [76, 123, 142]) show that the behavior of a system of particles essentially depends on whether the random field of velocities is nondivergent or divergent. By way of example, Fig. 1.1*a* shows a schematic of the evolution of the two-dimensional system of particles uniformly distributed within the circle for a particular realization of the nondivergent steady field $\boldsymbol{u}(\boldsymbol{r})$.

Here, we use the dimensionless time related to statistical parameters of field $\boldsymbol{u}(\boldsymbol{r})$. In this case, the area of surface patch within the contour remains intact and particles relatively uniformly fill the region within the deformed contour. The only feature consists in the fractal-type irregularity of the deformed contour. This phenomenon—called chaotic advection —is under active study now (see, e.g., [6]).

On the contrary, in the case of the divergent velocity field $\boldsymbol{u}(\boldsymbol{r})$, particles uniformly distributed in the square at the initial instant will form clusters during temporal evolution. Results simulated for this case are shown in Fig. 1.1*b*. We emphasize that the formation of clusters is purely the kinematic effect. This feature of particle dynamics disappears on averaging over an ensemble of realizations of the random velocity field.

To demonstrate the process of particle clustering, we consider the simplest problem [70], in which random velocity field $\boldsymbol{u}(\boldsymbol{r},t)$ has the form

$$\boldsymbol{u}(\boldsymbol{r},t) = \boldsymbol{v}(t) f(\boldsymbol{k}\boldsymbol{r}), \tag{1.3}$$

where $\boldsymbol{v}(t)$ is the random vector process and

$$f(\boldsymbol{k}\boldsymbol{r}) = \sin(2\boldsymbol{k}\boldsymbol{r}) \tag{1.4}$$

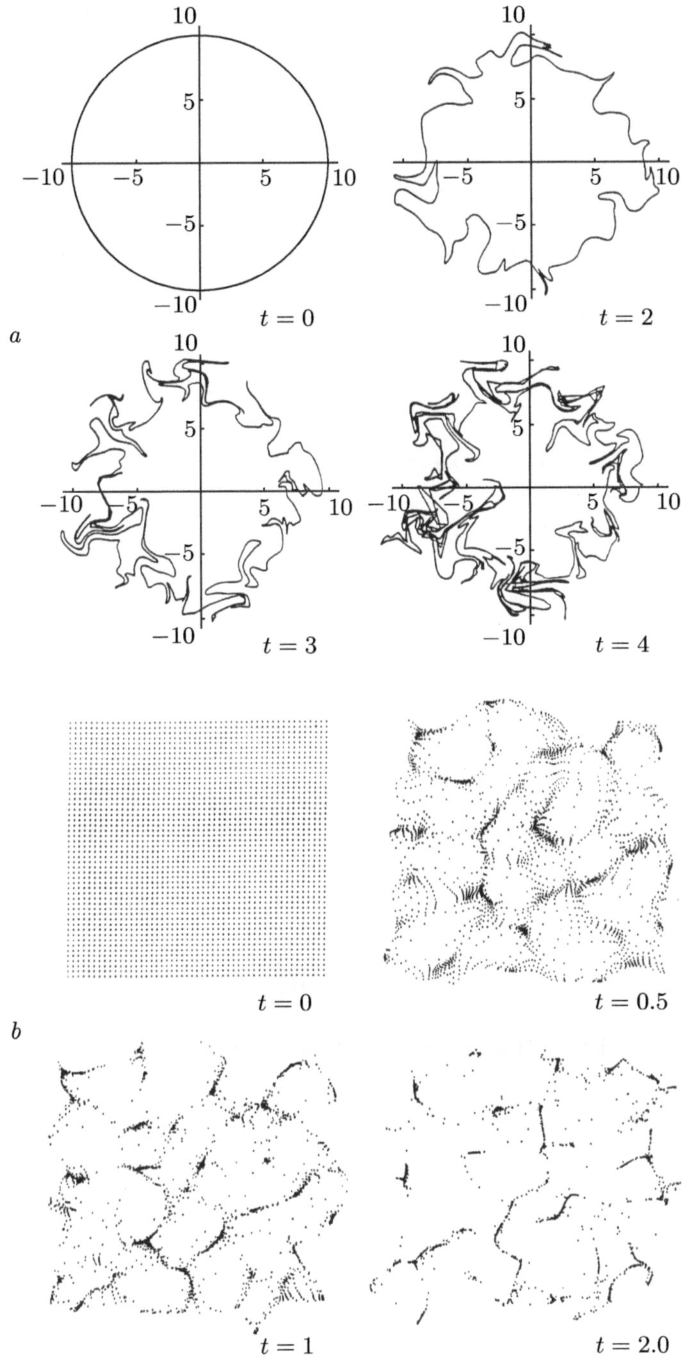

Fig. 1.1 Diffusion of a system of particles described by Eqs. (1.2) numerically simulated for (*a*) nondivergent and (*b*) potential random steady velocity field $\boldsymbol{u}(\boldsymbol{r})$

is the deterministic function of one variable. Note that this form of function $f(\boldsymbol{r})$ corresponds to the first term of the expansion in harmonic components and is commonly used in numerical simulations [76, 142].

In this case, Eq. (1.1) can be written in the form

$$\frac{d}{dt}\boldsymbol{r}(t) = \boldsymbol{v}(t)\sin(2\boldsymbol{k}\boldsymbol{r}), \quad \boldsymbol{r}(0) = \boldsymbol{r}_0.$$

In the context of this model, motions of a particle along vector $\boldsymbol{k}$ and in the plane perpendicular to vector $\boldsymbol{k}$ are independent and can be separated. If we direct the x-axis along vector $\boldsymbol{k}$, then the equations assume the form

$$\begin{aligned} \frac{d}{dt}x(t) &= v_x(t)\sin(2kx), \qquad x(0) = x_0, \\ \frac{d}{dt}\boldsymbol{R}(t) &= \boldsymbol{v}_{\boldsymbol{R}}(t)\sin(2kx), \quad \boldsymbol{R}(0) = \boldsymbol{R}_0. \end{aligned} \tag{1.5}$$

The solution of the first equation in (1.5) is

$$x(t) = \frac{1}{k}\arctan\left[e^{T(t)}\tan(kx_0)\right], \tag{1.6}$$

where

$$T(t) = 2k\int\limits_0^t d\tau v_x(\tau). \tag{1.7}$$

Taking into account the equality following from (1.6)

$$\begin{aligned} \sin(2kx) &= \frac{\sin(2kx_0)}{e^{-T(t)}\cos^2(kx_0) + e^{T(t)}\sin^2(kx_0)}, \\ \cos(2kx) &= \frac{1 - e^{2T(t)}\tan^2(kx_0)}{1 + e^{2T(t)}\tan^2(kx_0)}, \end{aligned} \tag{1.8}$$

we can rewrite the second equation in Eqs. (1.5) in the form

$$\frac{d}{dt}\boldsymbol{R}(t|\boldsymbol{r}_0) = \sin(2kx_0)\frac{\boldsymbol{v}_{\boldsymbol{R}}(t)}{e^{-T(t)}\cos^2(kx_0) + e^{T(t)}\sin^2(kx_0)}.$$

As a result, we have

$$\boldsymbol{R}(t|\boldsymbol{r}_0) = \boldsymbol{R}_0 + \sin(2kx_0)\int\limits_0^t d\tau\frac{\boldsymbol{v}_{\boldsymbol{R}}(\tau)}{e^{-T(\tau)}\cos^2(kx_0) + e^{T(\tau)}\sin^2(kx_0)}. \tag{1.9}$$

Consequently, if the initial particle position x_0 is such that

$$kx_0 = n\frac{\pi}{2}, \tag{1.10}$$

where $n = 0, \pm 1, \cdots$, then the particle will be the fixed particle and $\boldsymbol{r}(t) \equiv \boldsymbol{r}_0$.

Equalities (1.10) define planes in the general case and points in the one-dimensional case. They correspond to zeros of the field of velocities. Stability of these points depends on the sign of function $\boldsymbol{v}(t)$, and this sign changes during the evolution process. It can be expected that particles will be concentrated around these points if $v_x(t) \neq 0$, which just corresponds to clustering of particles.

In the case of a nondivergent velocity field, $v_x(t) = 0$ and, consequently, $T(t) \equiv 0$; as a result, we have

$$x(t|x_0) \equiv x_0, \quad \boldsymbol{R}(t|\boldsymbol{r}_0) = \boldsymbol{R}_0 + \sin(2kx_0)\int_0^t d\tau \boldsymbol{v}_{\boldsymbol{R}}(\tau),$$

which means that no clustering occurs.

In numerical simulations, we will describe velocity $\boldsymbol{v}(t)$ by the model of the Gaussian random delta-correlated vector field with the parameters

$$\langle \boldsymbol{v}(t)\rangle = 0 \quad \langle v_i(t)v_j(t')\rangle = 2\sigma^2\delta_{ij}\tau_0\delta(t-t'), \tag{1.11}$$

where σ^2 is the variance of every component of the velocity and τ_0 is its temporal radius of correlation. Below, we will use the nondimensional variables

$$t \to k^2\sigma^2\tau_0 t, \quad x \to kx, \quad \langle v_i(t)v_j(t')\rangle \to 2\delta_{ij}\delta(t-t'). \tag{1.12}$$

Figure 1.2*a* shows a fragment of the realization of random process $T(t)$ obtained by numerical integration of Eq. (1.7) for a realization of random process $v_x(t)$; we used this fragment for simulating the temporal evolution of coordinates of four particles $x(t)$, $x \in (0, \pi/2)$ initially located at coordinates $x_0(i) = \frac{\pi}{2}\frac{i}{5}$ $(i = 1, 2, 3, 4)$ (see Fig. 1.2*b*). Figure 1.2*b* shows that particles form a cluster in the vicinity of point $x = 0$ at the dimensionless time $t \approx 4$. Further, at time $t \approx 16$ the initial cluster disappears and new one appears in the vicinity of point $x = \pi/2$. At moment $t \approx 40$, the cluster appears again in the vicinity of point $x = 0$, and so on. In this process, particles in clusters remember their past history and significantly diverge during intermediate temporal segments (see Fig. 1.2*c*).

Thus, we see in this example that the cluster does not move from one region to another; instead, it first collapses and then a new cluster appears. Moreover, the lifetime of clusters significantly exceeds the duration of intermediate segments. It seems that this feature is characteristic of the particular model of the velocity field and follows from steadiness of points (1.10).

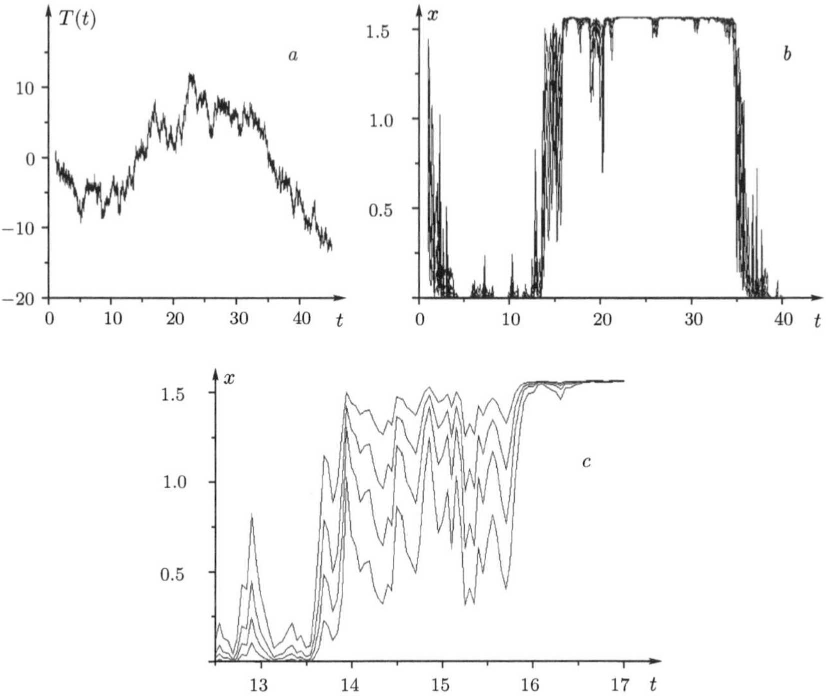

Fig. 1.2 (*a*) Segment of a realization of random process $T(t)$ obtained by numerically integrating Eq. (1.7) for a realization of random process $v_x(t)$; (*b*), (*c*) x-coordinates simulated with this segment for four particles versus time.

As regards the particle diffusion along the y-direction, no cluster occurs here.

Note that such clustering in a system of particles was found, to all appearance for the first time, in papers [104–107] as a result of simulating the so-called *Eole experiment* with the use of the simplest equations of atmospheric dynamics.

In this global experiment, 500 constant-density balloons were launched in Argentina in 1970-1971; these balloons traveled at a height of about 12 *km* and spread along the whole of the southern hemisphere.

Figure 1.3 shows the balloon distribution over the southern hemisphere for day 105 from the beginning of this process simulation [106]; this distribution clearly shows that balloons are concentrated in groups, which just corresponds to clustering. Results of statistical processing of balloon arrangement can be found, for example, in papers [16, 110].

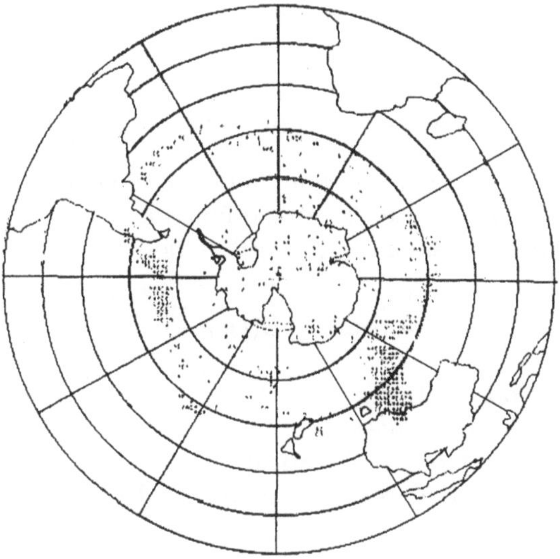

Fig. 1.3 Balloon distribution in the atmosphere for day 105 from the beginning of process simulation.

1.1.2 Particles under Random Forces

The system of equations (1.1) describes also the behavior of a particle under the field of random external forces $\boldsymbol{f}(\boldsymbol{r}, t)$. In the simplest case, the behavior of a particle in the presence of linear friction is described by the differential equation of the second order (Newton equation)

$$\frac{d^2}{dt^2}\boldsymbol{r}(t) = -\lambda \frac{d}{dt}\boldsymbol{r}(t) + \boldsymbol{f}(\boldsymbol{r}, t), \quad \boldsymbol{r}(0) = \boldsymbol{r}_0, \quad \frac{d}{dt}\boldsymbol{r}(0) = \boldsymbol{v}_0, \tag{1.13}$$

or the system of differential equations of the first order

$$\frac{d}{dt}\boldsymbol{r}(t) = \boldsymbol{v}(t), \quad \frac{d}{dt}\boldsymbol{v}(t) = -\lambda \boldsymbol{v}(t) + \boldsymbol{f}(\boldsymbol{r}, t),$$
$$\boldsymbol{r}(0) = \boldsymbol{r}_0, \quad \boldsymbol{v}(0) = \boldsymbol{v}_0. \tag{1.14}$$

Results of numerical simulations of stochastic system (1.14) can be found in Refs. [102, 136].

The behavior of a particle under the deterministic potential field in the presence of linear friction and random forces is described by the system of equations

$$\frac{d}{dt}\boldsymbol{r}(t) = \boldsymbol{v}(t), \quad \frac{d}{dt}\boldsymbol{v}(t) = -\lambda\boldsymbol{v}(t) - \frac{\partial U(\boldsymbol{r},t)}{\partial \boldsymbol{r}} + \boldsymbol{f}(\boldsymbol{r},t), \tag{1.15}$$

$$\boldsymbol{r}(0) = \boldsymbol{r}_0, \quad \boldsymbol{v}(0) = \boldsymbol{v}_0,$$

which is the simplest example of *Hamiltonian systems* with linear friction.If friction and external forces are absent and function U is independent of time, $U(\boldsymbol{r},t) = U(\boldsymbol{r})$, the system has an integral of motion

$$\frac{d}{dt}E(t) = \text{const}, \quad E(t) = \frac{\boldsymbol{v}^2}{2} + U(\boldsymbol{r})$$

expressing energy conservation.

In statistical problems, equations of type (1.14), (1.15) are widely used to describe the *Brownian motion* of particles.

In the general case, Hamiltonian systems are described by the system of equations

$$\frac{d}{dt}\boldsymbol{r}(t) = \frac{\partial H(\boldsymbol{r},\boldsymbol{v},t)}{\partial \boldsymbol{v}}, \qquad \frac{d}{dt}\boldsymbol{v}(t) = -\frac{\partial H(\boldsymbol{r},\boldsymbol{v},t)}{\partial \boldsymbol{r}}, \tag{1.16}$$

$$\boldsymbol{r}(0) = \boldsymbol{r}_0, \qquad \boldsymbol{v}(0) = \boldsymbol{v}_0,$$

where $H(\boldsymbol{r},\boldsymbol{v},t) = H(\boldsymbol{r}(t),\boldsymbol{v}(t),t)$ is the Hamiltonian function. In the case of conservative Hamiltonian systems, function $H(\boldsymbol{r},\boldsymbol{v},t)$ has no explicit dependence on time, $H(\boldsymbol{r},\boldsymbol{v},t) = H(\boldsymbol{r},\boldsymbol{v})$, and the system has the integral of motion

$$H(\boldsymbol{r},\boldsymbol{v}) = \text{const}.$$

1.1.3 The Hopping Phenomenon

Now, we dwell on another stochastic aspect related to dynamic equations of type (1.1); namely, we consider the *hopping phenomenon*caused by random fluctuations.

Consider the one-dimensional nonlinear equation

$$\frac{d}{dt}x(t) = x\left(1 - x^2\right) + f(t), \quad x(0) = x_0, \tag{1.17}$$

where $f(t)$ is the random function of time. In the absence of randomness ($f(t) \equiv 0$), the solution of Eq. (1.17) has two stable steady states $x = \pm 1$ and one instable state $x = 0$. Depending on the initial value, solution of Eq. (1.17) arrives at one of the stable states. However, in the presence of small random disturbances $f(t)$, dynamic system (1.17) will first approach the vicinity of one of the stable states and then, after the lapse of certain time, it will be transferred into the vicinity of another stable state.

Note that Eq. (1.17) corresponds to limit process $\lambda \to \infty$ in the equation

$$\frac{d^2}{dt^2}x(t) + \lambda\frac{d}{dt}x(t) - \lambda\left\{\frac{dU(x)}{dx} + f(t)\right\} = 0,$$

that is known as the *Duffing equation* and is the special case of the one-dimensional Hamiltonian system (1.15)

$$\frac{d}{dt}x(t) = v(t), \quad \frac{d}{dt}v(t) = -\lambda\left\{v(t) - \frac{dU(x)}{dx} - f(t)\right\} \tag{1.18}$$

with the potential function

$$U(x) = \frac{x^2}{2} - \frac{x^4}{4}.$$

In other words, Eq. (1.17) corresponds to great friction coefficients λ.

Statistical description of this problem will be considered in Sect.11.3.1, page 357. Additionally, we note that, in the context of statistical description, reduction of the Hamiltonian system (1.18) to the 'short-cut equation' is called the *Kramers problem* (see Sect. 11.2.2, page 347).

It is clear that the similar behavior can occur in more complicated situations.

Hydrodynamic-Type Nonlinear Systems

An important problem of studying large-scale processes in the atmosphere considered as a single physical system consists in revealing the mechanism of energy exchange between different 'degrees of freedom'.

The analysis of such nonlinear processes on the base of simple models described by a small number of parameters (degrees of freedom) is recently of great attention [97,98,101]. In this connection, A.M. Obukhov [115,116,119] (see also [14]) introduced the concept of *hydrodynamic-type systems* (*HTS*). These systems have a finite number of parameters $v_1, \cdots, v_n$, but the general features of the dynamic equations governing system motions coincide with those characteristic of the hydrodynamic equations of perfect incompressible fluid, including quadratic nonlinearity, energy conservation, and regularity (phase volume invariance during system motions). The general description of HTS is given in Sect. 11.3.1, page 357. Here, we dwell only on the dynamic description of simplest HTS systems.

The simplest system of such type (S_3) is equivalent to the *Euler equations* in the dynamics of solids; it describes the known problem on fluid motions in an ellipsoidal cavity [13, 14, 23]. Any finite-dimensional approximation of hydrodynamic equations also belongs to the class of HTS if it possesses the above features.

To model the cascade mechanism of energy transformation in a turbulent flow, Obukhov [117] suggested a multistage HTS. Each stage of this system consists of identical-scale triplets; at every next stage, the number of triplets is doubled and the scale is decreased in geometrical progression with ratio $Q \gtrsim 1$. As a result, this model describes interactions between the motions of different scales.

The first stage consists of a singe triplet whose unstable mode v_{01} is excited by an external force $f_0(t)$ applied to the system (Fig. 1.4*a*). The stable modes of this triplet $v_{1,1}$ and $v_{1,2}$ are the unstable modes of two triplets of the second stage; their stable modes $v_{2,1}$, $v_{2,2}$, $v_{2,3}$, and $v_{2,4}$ are, in turn, the unstable modes of four triplets of the third stage; and so on (Fig. 1.4*b*).

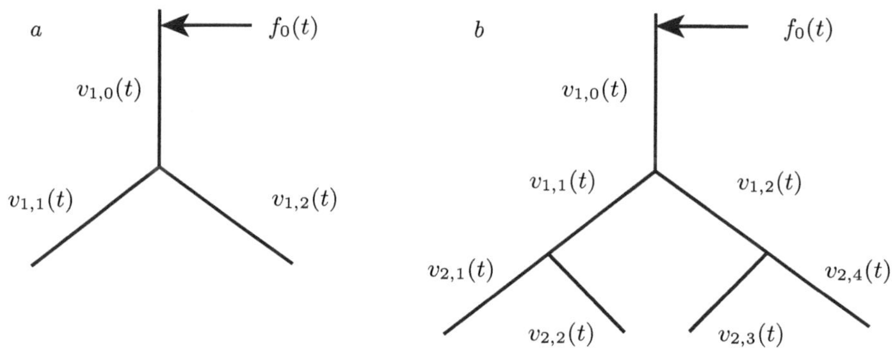

Fig. 1.4 Diagrams of (*a*) three- and (*b*) seven-mode HTS

It should be noted however that physical processes described in terms of macroscopic equations occur in actuality against the background of processes characterized by shorter scales (noises). Among these processes are, for example, the molecular noises (in the context of macroscopic hydrodynamics), microturbulence (against the large-scale motions), and the effect of truncated (small-scale) terms in the finite-dimensional approximation of hydrodynamic equations. The effect of these small-scale noises should be analyzed in statistical terms. Such a consideration can be performed in terms of macroscopic variables. With this goal, one must include external random forces with certain statistical characteristics in the corresponding macroscopic equations. The models considered here require additionally the inclusion of dissipative terms in the equations of motion to ensure energy outflow to small-scale modes.

Accordingly, the simplest hydrodynamic models that allow simulating actual processes are the HTS with random forces and linear friction.

An important problem appeared, for example, in the theory of climate consists in the determination of a possibility of significantly different circulation processes occurred under the same distribution of incoming heat, i.e., the problem on the existence of different regimes of motion of a given hydrodynamic system under the same 'external conditions'. A quite natural phenomenon that should be considered in this context is the 'hopping' phenomenon that consists in switching between such regimes of motion. Characteristic of these regimes is the fact that the duration of switching is small in comparison with the lifetime of the corresponding regimes.

It is expedient to study these problems using the above simple models. The corresponding systems of quadratically nonlinear ordinary differential equations may generally have several stable regimes, certain parameters describing external conditions being identical. The hopping events are caused by variations of these conditions in time and by the effect of random noises [24]. If the system has no stable regimes, its behavior can appear extremely difficult and require statistical description as it is the case in the *Lorentz model* [98]. In the case of availability of indifferent equilibrium states, the system may allow quasisteady-state regimes of motion [25]. Switching between such regimes can be governed by the dynamic structure of the system, and system behavior can appear 'stochastic' in this case, too.

In what follows, we study these hopping phenomena with the use of the simplest hydrodynamic models.

Dynamics of a Triplet (Gyroscope)

Consider first the case of a single stage, i.e., the triplet in the regime of forced motion (Fig. 1.4*a*). With this goal, we set force $f_0(t) = f_0 = \text{const}$ and assume the availability of dissipative forces acting on the stable modes of the triplet. In this case, equations of motion have the form

$$\begin{aligned}
\frac{d}{dt}v_{1,0}(t) &= \mu\left(v_{1,1}^2(t) - v_{1,2}^2(t)\right) + f_0, \\
\frac{d}{dt}v_{1,1}(t) &= -\mu v_{1,0}(t)v_{1,1}(t) - \lambda v_{1,1}(t), \\
\frac{d}{dt}v_{1,2}(t) &= \mu v_{1,0}(t)v_{1,2}(t) - \lambda v_{1,2}(t).
\end{aligned} \tag{1.19}$$

If $f_0 > 0$, component $v_{1,1}(t)$ vanishes with time, so that the motion of the triplet (1.19) is described in the dimensionless variables

$$x = \sqrt{\frac{\mu}{f_0}}v_{1,0} - \frac{\lambda}{\sqrt{\mu f_0}}, \quad y = \sqrt{\frac{\mu}{f_0}}v_{1,2}, \quad \tau = \sqrt{\mu f_0}t \tag{1.20}$$

by the two-mode system

$$\frac{d}{dt}x(t) = -y^2(t) + 1, \quad \frac{d}{dt}y(t) = x(t)y(t). \tag{1.21}$$

This system has the integral of motion $H_1 = x^2(t) + y^2(t) - 2\ln y(t)$. The change of variables $p(t) = x(t)$, $q(t) = \ln y(t)$ reduces the system to the Hamiltonian form

$$\frac{d}{dt}p(t) = -\frac{\partial \mathfrak{H}(p,q)}{\partial q}, \quad \frac{d}{dt}q(t) = \frac{\partial \mathfrak{H}(p,q)}{\partial p}$$

with the Hamiltonian $\mathfrak{H}(p,q) = \dfrac{p^2(t)}{2} + \dfrac{1}{2}e^{2q(t)} - q(t)$.

Thus, the behavior of a system with friction (1.19) under the action of a constant external force is described in terms of the Hamiltonian system.

Stationary points (0, 1) and (0, −1) of system (1.21) are the centers. If $H_1 - 1 \ll 1$, the period T_1 of motion along closed trajectories around every of these singular points is determined by the asymptotic formula (it is assumed that the sign of $y(t)$ remains intact during this motion)

$$T_1 \approx \sqrt{2}\pi \left[1 + \frac{H_1 - 1}{12}\right].$$

In the opposite limiting case $H_1 \gg 1$ (the trajectories are significantly distant from the mentioned centers), we obtain

$$T_1 \approx \frac{1}{\sqrt{H_1}}\left[2H_1 + \ln H_1\right].$$

Supplement now the dynamic system (1.19) with the linear friction acting on component $v_{01}(t)$:

$$\begin{aligned}
\frac{d}{dt}v_{1,0}(t) &= \mu\left(v_{1,1}^2(t) - v_{1,2}^2(t)\right) - \lambda v_{1,0}(t) + f_0, \\
\frac{d}{dt}v_{1,1}(t) &= -\mu v_{1,0}(t)v_{1,1}(t) - \lambda v_{1,1}(t), \\
\frac{d}{dt}v_{1,2}(t) &= \mu v_{1,0}(t)v_{1,2}(t) - \lambda v_{1,2}(t).
\end{aligned} \tag{1.22}$$

Introducing again the dimensionless variables

$$t \to t/\lambda, \quad v_{1,0}(t) \to \frac{\lambda}{\mu}v_0(t), \quad v_{1,2}(t) \to \frac{\lambda}{\mu}v_1(t), \quad v_{1,1}(t) \to \frac{\lambda}{\mu}v_2(t),$$

we arrive at the system of equations

$$\frac{d}{dt}v_0(t) = v_2^2(t) - v_1^2(t) - v_0(t) + R,$$
$$\frac{d}{dt}v_1(t) = v_0(t)v_1(t) - v_1(t), \quad \frac{d}{dt}v_2(t) = -v_0(t)v_2(t) - v_2(t), \tag{1.23}$$

where quantity $R = \dfrac{\mu f_0}{\lambda^2}$ is the analog of the Reynolds number.

Dynamic system (1.23) has steady-state solutions that depend now on parameter R, and $R = R_{\rm cr} = 1$ is the critical value.

For $R < 1$, the system has the stable steady-state solution

$$v_1 = v_2 = 0, \quad v_0 = R.$$

For $R > 1$, this solution becomes unstable with respect to small disturbances of parameters, and the steady-state regimes

$$v_0 = 1, \quad v_2 = 0, \quad v_1 = \pm\sqrt{R-1} \tag{1.24}$$

become available. Here, we have an element of randomness because component v_1 can be either positive or negative, depending on the amplitude of small disturbance.

Assume now that all components of the triplet are acted on by random forces. This means that system (1.23) is replaced with the system of equations

$$\frac{d}{dt}v_0(t) = v_2^2(t) - v_1^2(t) - v_0(t) + R + f_0(t),$$
$$\frac{d}{dt}v_1(t) = v_0(t)v_1(t) - v_1(t) + f_1(t), \tag{1.25}$$
$$\frac{d}{dt}v_2(t) = -v_0(t)v_2(t) - v_2(t) + f_2(t).$$

This system describes the motion of a triplet(gyroscope) with the linear isotropic friction, which is driven by the force acting on the instable mode and having both regular (R) and random ($\boldsymbol{f}(t)$) components. Such a situation occurs, for example, for a fluid moving in the ellipsoidal cavity [13, 14, 23].

For $R > 1$, dynamic system (1.25) under the action of random disturbances will first reach the vicinity of one of the stable states (1.24), and then, after the lapse of certain time, it will be hoppingly set into the vicinity of the other stable state. Figure 1.5 shows the results of simulations of this phenomenon for $R = 6$ and different realizations of random force $\boldsymbol{f}(t)$, whose components were simulated as the Gaussian random processes. (See Sect. 11.3.1, page 347 for the statistical description of this problem.)

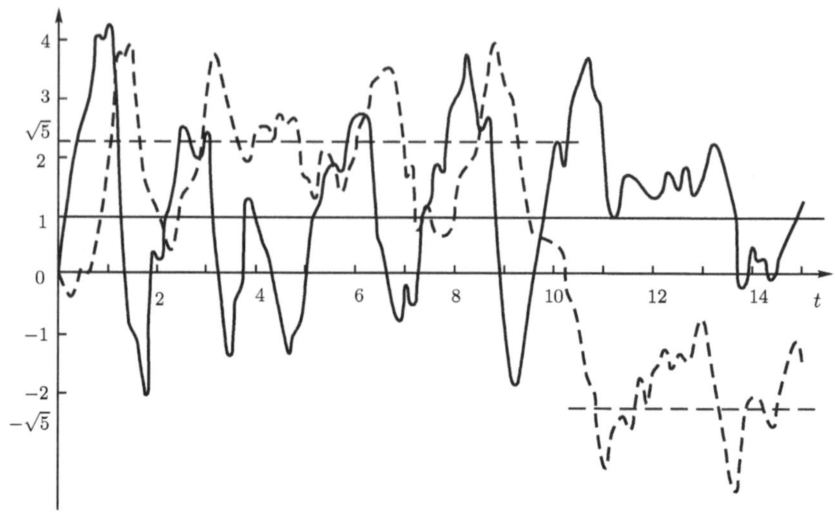

Fig. 1.5 Hopping phenomenon simulated from system (1.25) for $R = 6$ and variance $\sigma = 0.1$ (the solid and dashed lines show components $v_0(t)$ and $v_1(t)$, respectively).

Thus, within framework of the dynamics of the first stage, the hopping phenomenon can occur only due to the effect of external random forces acting on all modes.

Hopping between Quasisteady-State Regimes

The simplest two-stage system can be represented in the form

$$\frac{d}{dt}v_{1,0}(t) = v_{1,1}^2(t) - v_{1,2}^2(t) + 1,$$

$$\frac{d}{dt}v_{1,1}(t) = -v_{1,0}(t)v_{1,1}(t) + Q\left(v_{2,1}^2(t) - v_{2,2}^2(t)\right),$$

$$\frac{d}{dt}v_{1,2}(t) = -v_{1,0}(t)v_{1,2}(t) + Q\left(v_{2,3}^2(t) - v_{2,4}^2(t)\right), \tag{1.26}$$

$$\frac{d}{dt}v_{2,1}(t) = -Qv_{1,1}(t)v_{2,1}(t), \quad \frac{d}{dt}v_{2,2}(t) = Qv_{1,1}(t)v_{2,2}(t),$$

$$\frac{d}{dt}v_{2,3}(t) = -Qv_{1,2}(t)v_{2,3}(t), \quad \frac{d}{dt}v_{2,4}(t) = Qv_{1,2}(t)v_{2,4}(t),$$

where we used the dimensionless variables similar to (1.20). Only the components $v_{1,0}(t)$, $v_{1,2}(t)$, $v_{2,3}(t)$ and $v_{2,4}(t)$ survive for $f_0 > 0$ (see Fig. 1.6). These components satisfy the system of equations

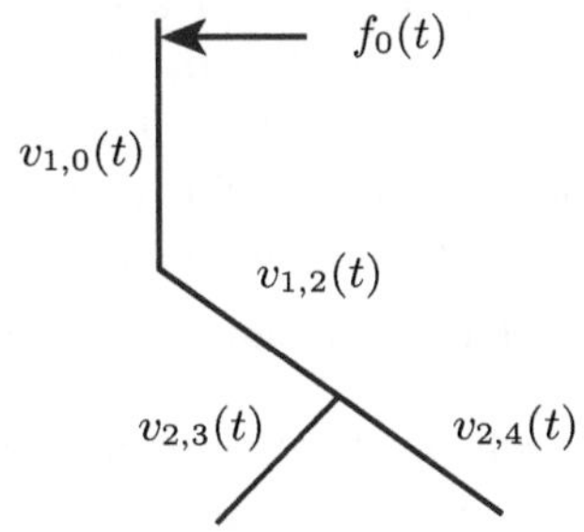

Fig. 1.6 Diagram of the excited seven-mode HTS

$$\frac{d}{dt}v_{1,0}(t) = -v_{1,2}^2(t) + 1,$$

$$\frac{d}{dt}v_{1,2}(t) = -v_{1,0}(t)v_{1,2}(t) + Q\left(v_{2,3}^2(t) - v_{2,4}^2(t)\right),$$

$$\frac{d}{dt}v_{2,3}(t) = -Qv_{1,2}(t)v_{2,3}(t), \quad \frac{d}{dt}v_{2,4}(t) = Qv_{1,2}(t)v_{2,4}(t)$$

that has the integral of motion

$$v_{2,3}(t)v_{2,4}(t) = I = \text{ const.}$$

Introducing notation

$$x(t) = v_{1,0}(t), \quad y(t) = v_{1,2}(t), \quad z(t) = \frac{v_{2,4}(t)}{v_{2,3}(t)},$$

we arrive at the system of three equations

$$\frac{d}{dt}x(t) = -y^2(t) + 1,$$

$$\frac{d}{dt}y(t) = x(t)y(t) + QI\left(\frac{1}{z(t)} - z(t)\right), \tag{1.27}$$

$$\frac{d}{dt}z(t) = y(t)z(t)$$

that describes the behavior of the seven-mode model (1.26).

Inclusion of the second stage significantly changes the dynamics of the first stage. Indeed, the initial values of components $v_{2,3}$ and $v_{2,4}$ being arbitrarily small, they nevertheless determine certain value of the integral of motion I, and, by virtue of Eqs. (1.27), variable y will repeatedly change the sign.

Consider the case of small values of constant I in more detail. Figure 1.7 shows the numerical solution of Eqs. (1.27) with the initial conditions $x = 0.05$, $y = 1$, and $z = 1$ at $Q = \sqrt{8}$ and $I = 10^{-20}$. As may be seen, two types of motion are characteristic of system (1.27) for small values of constant I; namely, 'fast' motions occur in a small vicinity of either closed trajectory of the Hamiltonian system (1.21) near the plane $z = 1$, and relatively rare hopping events occur due to the changes of sign of variable y at $z \sim I$ or $z \sim -\frac{1}{I}$. Every such hopping event significantly changes the parameters of fast motion trajectories of system (1.27), so that the motion of the system acquires the features characteristic of the dynamic systems with strange attractors (see, e.g., [98]).

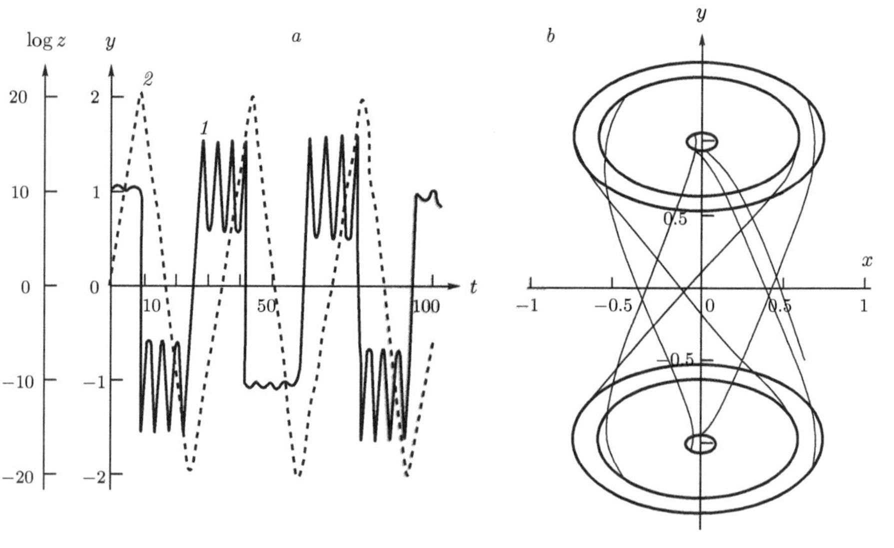

Fig. 1.7 (*a*) Time-dependent behavior of components of system (1.27) (curve *1* for y and curve *2* for $\log z$) and (*b*) projection of the phase trajectory of system (1.27) on plane (x, y)

To describe 'slow' motions, we introduce variables X and Y by the formulas

$$x(t) = X(t)x_1(t), \quad y(t) = Y(t)y_1(t),$$

where $(x_1(t),\, y_1(t))$ is a solution to Eqs. (1.21). According to Eqs. (1.27), $X(t)$, $Y(t)$, and $z(t)$ satisfy the equations

$$\frac{d}{dt}X(t) = \frac{1}{x_1(t)}\left[\left(X^2(t) - Y^2(t)\right) y_1^2(t) + 1 - X(t)\right],$$

$$\frac{d}{dt}Y(t) = (X(t) - 1)\, Y(t) x_1(t) + \frac{QI}{y_1(t)}\left(\frac{1}{z(t)} - z(t)\right),$$

$$\frac{d}{dt}z(t) = 2QY(t) y_1(t) z(t).$$

Averaging this equations in slow varying quantities over the period[1] $T_1 = \sqrt{2}\pi$, we obtain

$$\frac{d}{dt}X(t) = 0, \quad \frac{d}{dt}Y(t) = QI\left(\frac{1}{Z(t)} - Z(t)\right), \quad \frac{d}{dt}Z(t) = 2QY(t)Z(t), \tag{1.28}$$

where $Z = \tilde{z}(t)$. The system (1.28) has the integral of motion

$$H_2 = Y^2(t) + I\left(\frac{1}{Z(t)} + Z(t)\right), \tag{1.29}$$

so that we can use the corresponding variables to rewrite it in the Hamiltonian form, as it was done for the system (1.21). The motion of system (1.28) along closed trajectories around stationary point (0, 1) is characterized by the half-period T_2 (the time between hopping events) given by the formula

$$T_2 = \frac{2}{Q}\int_{Z_1}^{Z_2} \frac{dZ}{Z\sqrt{H_2 - I\left(\frac{1}{Z} + Z\right)}} = \frac{1}{Q\sqrt{I\left(r + \sqrt{r^2 - 1}\right)}} K\left(\sqrt{\frac{2\sqrt{r^2 - 1}}{r + \sqrt{r^2 - 1}}}\right),$$

where Z_1 and Z_2 are the roots of Eq. (1.29) at $Y = 0$, $r = 2H_2/I$, and $K(z)$ is the complete elliptic integral of the first kind. For small I, we have

$$T_2 \approx \frac{1}{Q\sqrt{H_2}} \ln\left(\frac{4H_2}{I}\right). \tag{1.30}$$

Figure 1.8 shows the numerical solution of Eqs. (1.28) with the initial conditions $Y = 1$, $Z = 1$ (they correspond to the initial conditions used earlier for solving Eqs. (1.27)), constants Q and I also being coincident with those

[1] This expression for T_1 appears to be quite adequate for the solution shown in Fig. 1.7. On closed trajectories of system (1.21) near which this solution passes, the values of Hamiltonian do not exceed 2. Because these trajectories are located in small vicinities of critical points (0, 1) and (0, −1) of this system, we performed averaging assuming that $\widetilde{y}_1(t) = 1$ and $\widetilde{1/y_1(t)} = 1$.

used for solving Eqs. (1.27). The comparison of curves in Figs. 1.7 and 1.8 shows that system (1.28) satisfactorily describes hopping events in system (1.27). The values of half-period T_2 determined by Eq. (1.30) and obtained from the numerical integration of Eqs. (1.27) are 33.54 and 33.51, respectively. We note that system (1.27) has an additional characteristic time $T_3 \sim 1/Q$, whose meaning is the duration of the hopping event.

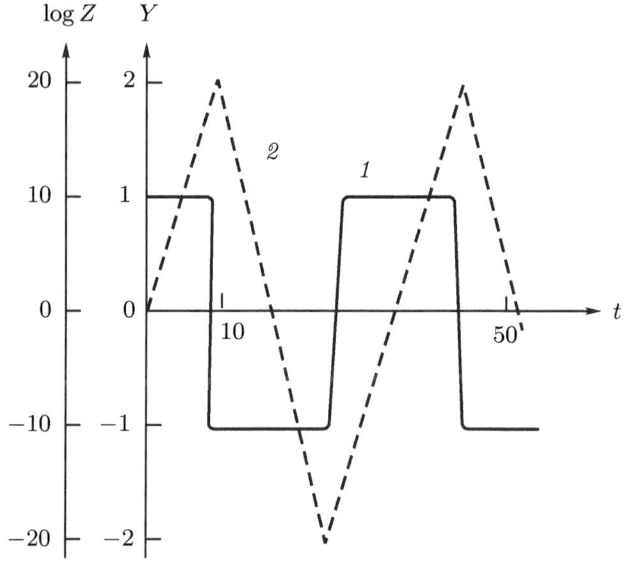

Fig. 1.8 Time-dependent behavior of components of system (1.28) (curve *1* for Y and curve *2* for $\log Z$)

1.1.4 Systems with Blow-Up Singularities

The simplest stochastic system showing singular behavior in time is described by the following equation commonly used in the statistical theory of waves

$$\frac{d}{dt}x(t) = -\lambda x^2(t) + f(t), \quad x(0) = x_0, \quad \lambda > 0, \tag{1.31}$$

where $f(t)$ is the random function of time.

In the absence of randomness ($f(t) = 0$), the solution to Eq. (1.31) has the form

$$x(t) = \frac{1}{\lambda\,(t - t_0)}, \quad t_0 = -\frac{1}{\lambda x_0}.$$

For $x_0 > 0$, we have $t_0 < 0$, and solution $x(t)$ monotonically tends to zero with increasing time. On the contrary, for $x_0 < 0$, solution $x(t)$ reaches $-\infty$ within a finite time $t_0 = -1/\lambda x_0$, which means that the solution becomes *singular* and shows the *blow-up behavior*. In this case, random force $f(t)$ has insignificant effect on the behavior of the system. The effect becomes significant only for positive parameter x_0.

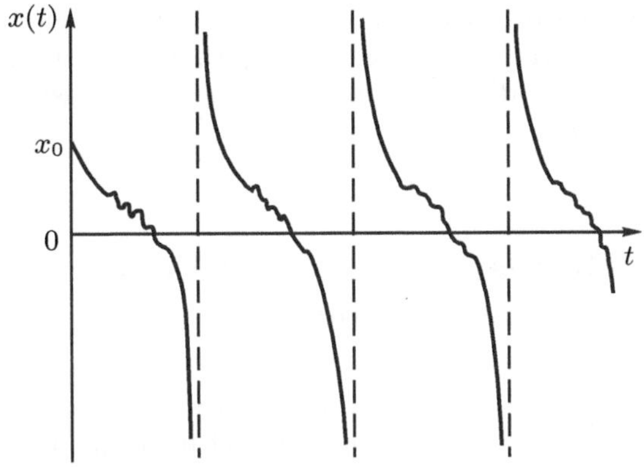

Fig. 1.9 Typical realization of the solution to Eq. (1.31)

Here, the solution, slightly fluctuating, decreases with time as long as it remains positive. On reaching sufficiently small value $x(t)$, the impact of force $f(t)$ can cause the solution to hop into the region of negative values of x, where it reaches the value of $-\infty$ within a certain finite time.

Thus, in the stochastic case, the solution to problem (1.31) shows the blow-up behavior for arbitrary values of parameter x_0 and always reaches $-\infty$ within a finite time t_0. Figure 1.9 schematically shows the temporal realization of the solution $x(t)$ to problem(1.31) for $t > t_0$; its behavior resembles a *quasi-periodic* structure.

1.1.5 Oscillator with Randomly Varying Frequency (Stochastic Parametric Resonance)

In the above stochastic examples, we considered the effect of additive random impacts (forces) on the behavior of systems. The simplest nontrivial system with multiplicative (parametric) impact can be illustrated using the *stochastic parametric resonance* as an example. Such a system is described by the second-order equation

$$\frac{d^2}{dt^2}x(t) + \omega_0^2[1 + z(t)]x(t) = 0,$$
$$x(0) = x_0, \quad \frac{d}{dt}x(0) = v_0, \tag{1.32}$$

where $z(t)$ is the random function of time. This equation is characteristic of almost all fields of physics. It is physically obvious that dynamic system (1.32) is capable of parametric excitation, because random process $z(t)$ has harmonic components of all frequencies, including frequencies $2\omega_0/n\,(n = 1, 2, \cdots)$ that exactly correspond to the frequencies of parametric resonance in the system with periodic function $z(t)$, as it is the case, for example, in the *Mathieu equation.*

1.2 Linear Ordinary Differential Equations: Boundary-Value Problems

In the previous section, we considered several dynamic systems described by a system of ordinary differential equations with given initial values. Now, we consider the simplest linear boundary-value problem, namely, the steady one-dimensional wave problem.

1.2.1 Plane Waves in Layered Media: A Wave Incident on a Medium Layer

Let the layer of inhomogeneous medium occupies the part of space $L_0 < x < L$ and let the unit-amplitude plane wave $u_0(x) = e^{-ik(x-L)}$ is incident on this layer from the region $x > L$ (Fig. 1.10a).

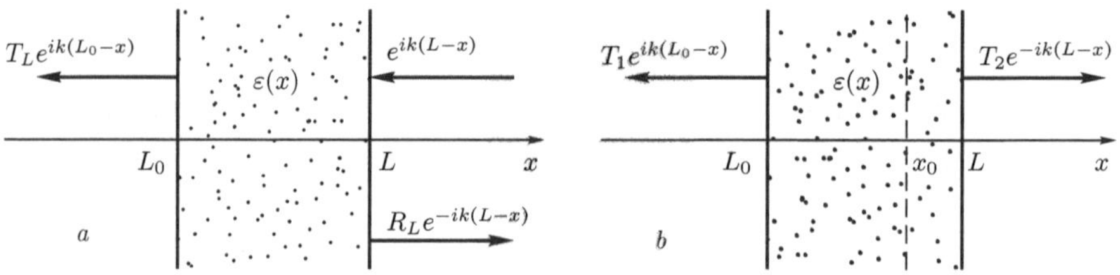

Fig. 1.10 (a) Plane wave incident on the medium layer and (b) source inside the medium layer

The wavefield satisfies the *Helmholtz equation*,

$$\frac{d^2}{dx^2}u(x) + k^2(x)u(x) = 0, \tag{1.33}$$

where

$$k^2(x) = k^2[1 + \varepsilon(x)]$$

and function $\varepsilon(x)$ describes medium inhomogeneities. We assume that $\varepsilon(x) = 0$, i.e., $k(x) = k$ outside the layer; inside the layer, we set $\varepsilon(x) = \varepsilon_1(x) + i\gamma$, where $\varepsilon_1(x)$ is the real part responsible for wave scattering in the medium and the imaginary part $\gamma \ll 1$ describes the absorption of the wave in the medium.

In region $x > L$, the wavefield has the structure

$$u(x) = e^{-ik(x-L)} + R_L e^{ik(x-L)},$$

where R_L is the complex reflection coefficient. In region $x < L_0$, the structure of the wavefield is

$$u(x) = T_L e^{ik(L_0 - x)},$$

where T_L is the complex transmission coefficient. Boundary conditions for Eq. (1.33) are the continuity conditions for the field and the field derivative at layer boundaries; they can be written as follows

$$u(L) + \frac{i}{k}\frac{du(x)}{dx}\bigg|_{x=L} = 2, \qquad u(L_0) - \frac{i}{k}\frac{du(x)}{dx}\bigg|_{x=L_0} = 0. \tag{1.34}$$

Thus, the wavefield in the layer of an inhomogeneous medium is described by the boundary-value problem (1.33), (1.34). Dynamic equation (1.33) coincides in form with Eq. (1.32). Note that the problem under consideration assumes that function $\varepsilon(x)$ is discontinuous at layer boundaries. We will call the boundary-value problem (1.33), (1.34) the unmatched boundary-value problem. In such problems, wave scattering is caused not only by medium inhomogeneities, but also by discontinuities of function $\varepsilon(x)$ at layer boundaries.

If medium parameters (function $\varepsilon_1(x)$) are specified in the statistical form, then solving the stochastic problem (1.33), (1.34) consists in obtaining statistical characteristics of the reflection and transmission coefficients, which are related to the wavefield values at layer boundaries by the relationships

$$R_L = u(L) - 1, \quad T_L = u(L_0),$$

and the wavefield intensity

$$I(x) = |u(x)|^2$$

inside the inhomogeneous medium. Determination of these characteristics constitutes the subject of the *statistical theory of radiative transfer*.

Note that, for $x < L$, from Eq. (1.33) follows the equality

$$k\gamma I(x) = \frac{d}{dx} S(x),$$

where energy-flux density $S(x)$ is determined by the relationship

$$S(x) = \frac{i}{2k}\left[u(x)\frac{d}{dx}u^*(x) - u^*(x)\frac{d}{dx}u(x)\right].$$

By virtue of boundary conditions, we have $S(L) = 1 - |R_L|^2$ and $S(L_0) = |T_L|^2$.

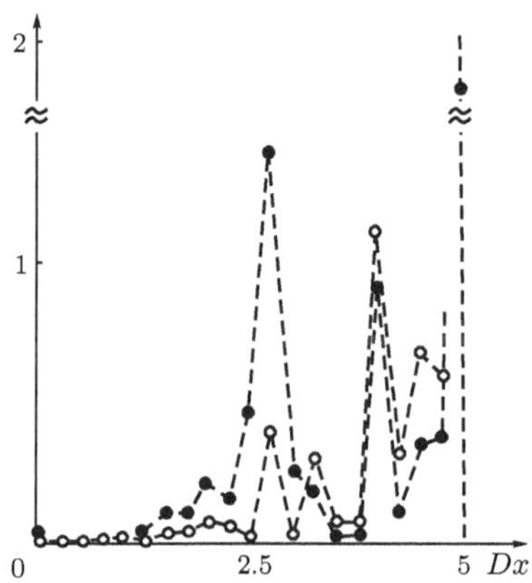

Fig. 1.11 Dynamic localization phenomenon simulated for two realizations of medium inhomogeneities

For non-absorptive media ($\gamma = 0$), conservation of energy-flux density is expressed by the equality

$$|R_L|^2 + |T_L|^2 = 1. \tag{1.35}$$

Consider some features characteristic of solutions to the stochastic boundary-value problem (1.33), (1.34). On the assumption that medium inhomogeneities are absent ($\varepsilon_1(x) = 0$) and absorption γ is sufficiently small, the intensity of the wavefield in the medium slowly decays with distance according to the exponential law

$$I(x) = |u(x)|^2 = e^{-k\gamma(L-x)}. \tag{1.36}$$

Figure 1.11 shows two realizations of the intensity of a wave in a sufficiently thick layer of medium. These realizations were simulated for two realizations of medium inhomogeneities [137]. The difference between them consists in the fact that the corresponding functions $\varepsilon_1(x)$ have different signs in the middle of the layer at a distance of the wavelength. This offers a possibility of estimating the effect of a small medium mismatch on the solution of the boundary problem. Omitting the detailed description of problem parameters, we mention only that this figure clearly shows the prominent tendency of a sharp exponential decay (accompanied by significant spikes toward both higher and nearly zero-valued intensity values), which is caused by multiple reflections of the wave in the chaotically inhomogeneous random medium (the phenomenon of *dynamic localization*). Recall that absorption is small ($\gamma \ll 1$), so that it cannot significantly affect the dynamic localization.

It is well known that the introduction of new function

$$\psi(x) = \frac{i}{k}\frac{d}{dx}\ln u(x)$$

reduces the second-order equation (1.33) to two first-order equations, and this function satisfies the closed equation following from Eq. (1.33):

$$\frac{d}{dx}\psi(x) = ik\left[\psi^2(x) - 1 - \varepsilon(x)\right], \quad \psi(L_0) = 1. \tag{1.37}$$

From the condition at boundary $x = L$ follows that

$$u(L) = \frac{2}{1+\psi(L)}$$

and, consequently, the reflection coefficient is determined from the solution to Eq. (1.37) by the formula

$$R_L = \frac{1-\psi(L)}{1+\psi(L)}.$$

Introducing the function

$$R(x) = \frac{1-\psi(x)}{1+\psi(x)}, \quad \psi(x) = \frac{1-R(x)}{1+R(x)},$$

we can rewrite Eq. (1.37) in the form of the equation

$$\frac{d}{dx}R(x) = 2ikR(x) + \frac{i}{2k}\varepsilon(x)\left(1+R(x)\right)^2, \quad R(L_0) = 0 \tag{1.38}$$

whose solution at $x = L$ coincides with the reflection coefficient, i.e.,

$$R_L = R(L).$$

In terms of function $R(x)$, the wavefield $u(x)$ inside the medium is now expressed by the equality

$$u(x) = [1 + R(L)] \exp\left[ik \int\limits_{x}^{L} d\xi \frac{1 - R(\xi)}{1 + R(\xi)}\right]. \tag{1.39}$$

Figure 1.12a shows the traditional procedure of solving the problem. One solves Eq. (1.38) first and then reconstructs the wavefield by the formula (1.39). This is the well known approach called the sweep method. However, it is inappropriate for analyzing statistical problems.

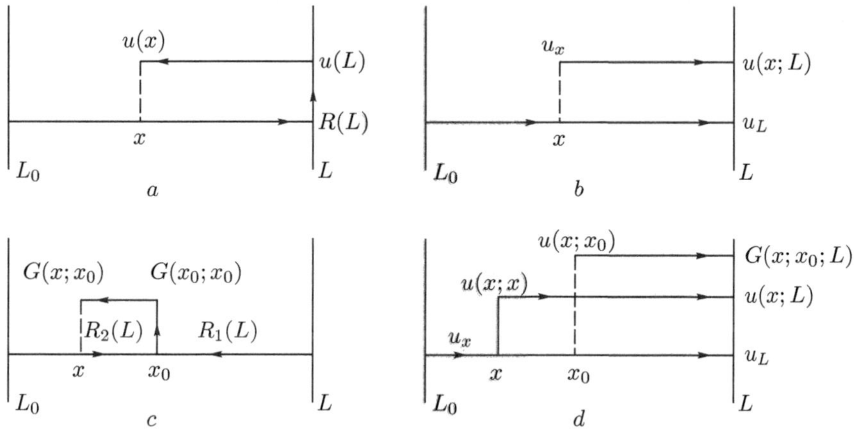

Fig. 1.12 Schematic of solving boundary problem (1.33), (1.34) by (a) the sweep method and (b) the imbedding method and boundary-value problem (1.49) by (c) the sweep method and (d) the imbedding method

Alternatively, the wavefield inside the medium can be represented in the form

$$u(x) = u_1(x) + u_2(x),$$

$$\frac{d}{dx}u(x) = -ik[u_1(x) - u_2(x)],$$

where $u_1(x)$ and $u_2(x)$ are the complex *contradirectional modes*. Because these modes are related to the wavefield by the expressions

$$\begin{aligned} u_1(x) &= \frac{1}{2}\left[1 + \frac{i}{k}\frac{d}{dx}\right]u(x), \quad u_1(L) = 1, \\ u_2(x) &= \frac{1}{2}\left[1 - \frac{i}{k}\frac{d}{dx}\right]u(x), \quad u_2(L_0) = 0, \end{aligned} \tag{1.40}$$

we can rewrite the boundary-value problem (1.33), (1.34) in the form

$$\left(\frac{d}{dx}+ik\right)u_1(x)=-\frac{ik}{2}\varepsilon(x)\left[u_1(x)+u_2(x)\right],\quad u_1(L)=1,$$
$$\left(\frac{d}{dx}-ik\right)u_2(x)=-\frac{ik}{2}\varepsilon(x)\left[u_1(x)+u_2(x)\right],\quad u_2(L_0)=0. \tag{1.41}$$

Note that function $R(x)$ introduced earlier is expressed in terms of modes $u_1(x)$ and $u_2(x)$ simply as the ratio

$$R(x)=\frac{u_2(x)}{u_1(x)}.$$

The *imbedding method* offers a possibility of reformulating the boundary-value problem (1.33), (1.34) to the dynamic problem with the initial values for parameter L (this parameter corresponds to the geometrical position of the layer right-hand boundary) by considering the solution to the boundary-value problem as a function of parameter L [49,50,52]. On such reformulation, the reflection coefficient R_L satisfies the *Riccati equation*

$$\frac{d}{dL}R_L=2ikR_L+\frac{ik}{2}\varepsilon(L)\left(1+R_L\right)^2,\quad R_{L_0}=0 \tag{1.42}$$

that coincides, naturally, with Eq. (1.41), and the wavefield in the medium layer $u(x)\equiv u(x;L)$ satisfies the linear equation

$$\frac{\partial}{\partial L}u(x;L)=iku(x;L)+\frac{ik}{2}\varepsilon(L)\left(1+R_L\right)u(x;L),$$
$$u(x;x)=1+R_x, \tag{1.43}$$

which can be derived, for example, by differentiating Eq. (1.39) with respect to parameter L. Figure 1.12*b* shows the procedure of solving the problem in this formulation. Comparing this procedure with that of the sweep method (Fig. 1.12*a*), we see that solving procedure has changed the direction, and namely this fact will offer a possibility of constructing the statistical description of the solution to the problem in the stochastic formulation.

The equation for the squared modulus of the reflection coefficient $W_L=|R_L|^2$ follows from Eq. (1.42):

$$\frac{d}{dL}W_L=-2k\gamma W_L-\frac{ik}{2}\varepsilon_1(L)\left(R_L-R_L^*\right)\left(1-W_L\right),\quad W_{L_0}=0. \tag{1.44}$$

Note that condition $W_{L_0}=1$ will be the initial value to Eq. (1.44) in the case of totally reflecting boundary at L_0. In this case, the wave incident on the layer of a non-absorptive medium ($\gamma=0$) is totally reflected from the layer, i.e., $W_L=1$, so that the reflection coefficient can be written in the form $R_L=e^{i\phi_L}$. For the phase of the reflection coefficient, we have the dynamic equation following from Eq. (1.42)

$$\frac{d}{dL}\phi_L = 2k + k\varepsilon_1(L)\left(1+\cos\phi_L\right). \tag{1.45}$$

It governs phase variations in the whole range of values $(-\infty, +\infty)$. At the same time, the equation for the wavefield (1.43) depends only on trigonometric functions of the phase of reflection coefficient. For this reason, it would be desirable to deal with the phase varying in interval $(-\pi, \pi)$. We can do this by introducing new function $z_L = \tan(\phi_L/2)$. This function satisfies the dynamic equation of type (1.31)

$$\frac{d}{dL}z_L = k\left(1+z_L^2\right) + k\varepsilon_1(L),$$

and its solution shows the singular behavior.

In the general case of arbitrarily reflecting boundary L_0, the steady-state (independent of L) solution $W_L = 1$ corresponding to the total reflection of incident wave formally exists for a half-space ($L_0 \to -\infty$) filled with non-absorptive random medium, too. This solution is actually realized in the statistical problem with probability equal to one [49, 50, 52].

It is obvious that the division of the field into contradirectional modes (1.40) is of arbitrary nature; this is nothing more than the mathematical technique that reduces the second-order equation (1.33) to two first-order equations with simplest boundary conditions.

If, in contrast to the above problem, we assume that function $k(x)$ is continuous at boundary $x = L$, i.e., if we assume that the wave number in the free half-space $x > L$ is equal to $k(L)$, then boundary conditions (1.34) of problem (1.33) will be replaced with the conditions

$$u(L) + \frac{i}{k(L)}\frac{du(x)}{dx}\bigg|_{x=L} = 2, \qquad u(L_0) - \frac{i}{k(L_0)}\frac{du(x)}{dx}\bigg|_{x=L_0} = 0. \tag{1.46}$$

We will call the boundary-value problem (1.33), (1.46) the *matched boundary-value problem*. In this case, it is convenient to represent the wavefield in the form

$$u(x) = u_1(x) + u_2(x),$$

$$\frac{du(x)}{dx} = -ik(x)\left[u_1(x) - u_2(x)\right],$$

where the complex *contradirectional modes* $u_1(x)$ and $u_2(x)$ are now related to the wavefield by the expressions

$$u_1(x) = \frac{1}{2}\left[1 + \frac{i}{k(x)}\frac{d}{dx}\right]u(x), \quad u_1(L) = 1,$$

$$u_2(x) = \frac{1}{2}\left[1 - \frac{i}{k(x)}\frac{d}{dx}\right]u(x), \quad u_2(L_0) = 0$$

and satisfy the boundary-value problem

$$\left(\frac{d}{dx} + ik(x)\right)u_1(x) = -\frac{k'(x)}{k(x)}\left[u_1(x) - u_2(x)\right], \quad u_1(L) = 1,$$

$$\left(\frac{d}{dx} - ik(x)\right)u_2(x) = \frac{k'(x)}{k(x)}\left[u_1(x) - u_2(x)\right], \quad u_2(L_0) = 0,$$

where $k'(x) = \dfrac{dk(x)}{dx}$. Function $R(x) = u_2(x)/u_1(x)$ is now described by the Riccati equation

$$\frac{d}{dx}R(x) = 2ikR(x) + \frac{k'(x)}{2k}\left[1 - R^2(x)\right], \quad R(L_0) = 0, \tag{1.47}$$

and the reflection coefficient is determined in terms of the solution to Eq. (1.47) from the relationship

$$R_L = R(L).$$

In the case of sufficiently small function $\varepsilon(x)$, we can rewrite Eq. (1.47) in the form

$$\frac{dR(x)}{dx} = 2ikR(x) + \frac{1}{4}\varepsilon'(x)\left[1 - R^2(x)\right],$$

where the derivative of function $\varepsilon(x)$ appears as distinct from Eq. (1.41).

Note that, for the matched boundary-value problem (1.33), (1.46), the equations of the imbedding method have the form

$$\begin{aligned}
&\frac{d}{dL}R_L = 2ikR_L + \frac{1}{4}\varepsilon'(L)\left(1 - R_L^2\right), \quad R_{L_0} = 0,\\
&\frac{\partial}{\partial L}u(x,L) = 2iku(x,L) + \frac{1}{4}\varepsilon'(L)\left(1 - R_L\right)u(x,L),\\
&u(x,x) = 1 + R_x.
\end{aligned} \tag{1.48}$$

1.2.2 *Plane Waves in Layered Media: Source Inside the Medium*

The field of a point source located in the layer of random medium is described by the similar boundary-value problem for Green's function of the Helmholtz equation:

$$\frac{d^2}{dx^2}G(x;x_0)+k^2[1+\varepsilon(x)]G(x;x_0)=2ik\delta(x-x_0),$$

$$G(L;x_0)+\frac{i}{k}\frac{dG(x;x_0)}{dx}\bigg|_{x=L}=0,\quad G(L_0;x_0)-\frac{i}{k}\frac{dG(x;x_0)}{dx}\bigg|_{x=L_0}=0. \tag{1.49}$$

Outside the layer, the solution has here the form of outgoing waves (Fig. 1.10*b*)

$$G(x;x_0)=T_1e^{ik(x-L)}\quad (x\geq L),\qquad G(x;x_0)=T_2e^{-ik(x-L_0)}\quad (x\leq L_0).$$

Note that, for the source located at layer boundary $x_0=L$, this problem coincides with the boundary-value problem (1.33), (1.34) on the wave incident on the layer, which yields

$$G(x;L)=u(x;L).$$

The solution to the boundary-value problem (1.49) has the structure

$$G(x;x_0)=G(x_0;x_0)\begin{cases}\exp\left[ik\int\limits_{x}^{x_0}\psi_1(\xi)\,d\xi\right], & x_0\geq x,\\[2ex] \exp\left[ik\int\limits_{x_0}^{x}\psi_2(\xi)\,d\xi\right] & x_0\leq x,\end{cases} \tag{1.50}$$

where the field at the source location, by virtue of the derivative gap condition

$$\frac{dG(x;x_0)}{dx}\bigg|_{x=x_0+0}-\frac{dG(x;x_0)}{dx}\bigg|_{x=x_0-0}=2ik,$$

is determined by the formula

$$G(x_0;x_0)=\frac{2}{\psi_1(x_0)+\psi_2(x_0)}$$

and functions $\psi_i(x)$ satisfy the Riccati equations

$$\begin{aligned}\frac{d}{dx}\psi_1&=ik\left[\psi_1^2-1-\varepsilon(x)\right],\quad \psi_1(L_0)=1,\\ \frac{d}{dx}\psi_2&=-ik\left[\psi_2^2-1-\varepsilon(x)\right],\quad \psi_2(L)=1.\end{aligned} \tag{1.51}$$

Figure 1.12*c* shows the procedure of solving this problem by the sweep method. One solves two equations (1.51) first and then reconstructs the wavefield using Eq. (1.50).

Introduce new functions $R_i(x)$ related to functions $\psi_i(x)$ by the formula

$$\psi_i(x) = \frac{1 - R_i(x)}{1 + R_i(x)}, \qquad i = 1, 2.$$

With these functions, the wavefield in region $x < x_0$ can be written in the form

$$G(x; x_0) = \frac{[1 + R_1(x_0)][1 + R_2(x_0)]}{1 - R_1(x_0) R_2(x_0)} \exp\left[ik \int_x^{x_0} d\xi \frac{1 - R_1(\xi)}{1 + R_1(\xi)}\right], \tag{1.52}$$

where function $R_1(x)$ satisfies the Riccati equation (1.38).

For $x_0 = L$, expression (1.52) becomes

$$G(x; L) = u(x; L) = [1 + R_1(L)] \exp\left[ik \int_x^{L} d\xi \frac{1 - R_1(\xi)}{1 + R_1(\xi)}\right], \tag{1.53}$$

so that parameter $R_1(L) = R_L$ is the reflection coefficient of the plane wave incident on the layer from region $x > L$. In a similar way, quantity $R_2(x_0)$ is the reflection coefficient of the wave incident on the medium layer (x_0, L) from the homogeneous half-space $x < x_0$ (i.e., from region with $\varepsilon = 0$).

Using Eq. (1.53), we can rewrite Eq. (1.52) in the form

$$G(x; x_0) = \frac{1 + R_2(x_0)}{1 - R_1(x_0) R_2(x_0)} u(x; x_0), \qquad x \le x_0,$$

where $u(x; x_0)$ is the wavefield inside the inhomogeneous layer (L_0, x_0) in the case of the incident wave coming from the free half-space $x > x_0$.

Thus, for $x < x_0$, the field of the point source is proportional to the wavefield generated by the plane wave incident on layer (L_0, x_0) from the free half-space $x > x_0$. The layer segment (x_0, L) affects only parameter $R_2(x_0)$.

Note that, considering the wavefield as a function of parameter L (i.e., setting $G(x; x_0) \equiv G(x; x_0; L)$), we can use the imbedding method to obtain the following system of equations with initial values:

$$\begin{gathered}
\frac{\partial}{\partial L}G(x;x_0;L) = i\frac{k}{2}\varepsilon(L)u(x_0;L)u(x;L), \\
G(x;x_0;L)|_{L=\max(x,x_0)} = \begin{cases} u\,(x;x_0)\,, \; x \geq x_0 \\ u\,(x_0;x)\,, \; x \leq x_0 \end{cases}, \\
\frac{\partial}{\partial L}u(x;L) = ik\left\{1+\varepsilon(L)u(L;L)\right\}u(x;L), \\
u(x;L)|_{L=x} = u(x;x), \\
\frac{d}{dL}u(L;L) = 2ik[u(L;L)-1] + i\frac{k}{2}\varepsilon(L)u^2(L;L), \\
u(L_0;L_0) = 1.
\end{gathered} \tag{1.54}$$

Here, two last equations describe the wavefield appearing in the problem on the wave incident on the medium layer. Figure 1.12*d* shows the procedure of solving this problem.

1.2.3 Plane Waves in Layered Media: Two-Layer Model

Investigators often faces with multidimensional situations in which one wave mode can originate other wave mode due to dependence of problem parameters on spatial coordinates. Sometimes, such problems allow a parametrization by selecting certain direction and dividing the medium in this direction into the layers characterized by discrete values of certain parameters, whereas other parameters may vary continuously in these layers. As an example, we mention the large-scale and low-frequency motions in Earth's atmosphere and ocean, such as the Rossby waves. These waves can be described within the framework of the quasi-geostrophic model that describes the atmosphere and ocean as thin multilayer films characterized in the vertical direction by thicknesses and densities of layers [120]. At the same time, other parameters

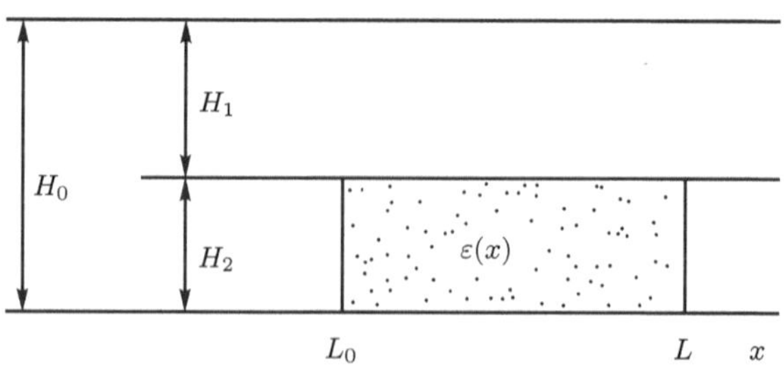

Fig. 1.13 Two-layer model of medium

vary continuously in these layers. It is quite possible that the reason of the local property of the Rossby waves consists in the spatial variation of bottom topography inhomogeneities in the horizontal plane. The simplest one-layer model is equivalent to the one-dimensional Helmholtz equation and describes barotropic motions of the medium; the two-layer model (Fig. 1.13) includes additionally the baroclinic effects [28, 54, 75].

In the context of two-layer media, the simplest model describing the propagation of interacting waves is the system of equations [27]

$$\begin{gathered}\frac{d^2}{dx^2}\psi_1 + k^2\psi_1 - \alpha_1 F\left(\psi_1 - \psi_2\right) = 0,\\ \frac{d^2}{dx^2}\psi_2 + k^2\left[1 + \varepsilon(x)\right]\psi_2 + \alpha_2 F\left(\psi_1 - \psi_2\right) = 0,\end{gathered} \tag{1.55}$$

where parameters $\alpha_1 = 1/H_1$, $\alpha_2 = 1/H_2$ (H_1 and H_2 are the thicknesses of the top and bottom layers), parameter F characterizes wave interaction, and function $\varepsilon(x)$ describes medium inhomogeneities in the bottom layer. Boundary conditions for system (1.55) are the radiation conditions at infinity.

Note that parameter F characterizing the medium parametrization in the vertical direction appears in system (1.55) as some sort of the horizontal scale responsible for generation of an additional wave. System (1.55) describes wave interaction (and, in particular, dependence of parameters α_i on layer thicknesses) in conformity with problems of geophysical hydrodynamics. For other problem types, the form of these relationships can change, which only slightly concerns the essence of the problem. The only essential point is the fact that wave interaction is the linear interaction.

Transition to the one-layer model is performed by setting $F = 0$, $\psi_1 = 0$ which transforms the corresponding wave equation to the Helmholtz equation (1.33). Limit process $H_1 \to 0$ also results in the transition to the one-layer model; in this case, $\psi_1 = \psi_2$. However, one can bear in mind that limit processes $L_0 \to -\infty$ (transition to the half-space) and $H_i \to 0$ do not commute in statistical problems. In this case, layer thicknesses H_i must be finite though arbitrarily small.

1.3 First-Order Partial Differential Equations

Consider now several dynamic systems (dynamic fields) described by partial differential equations.

1.3.1 Linear First-Order Partial Differential Equations

Diffusion of Density Field under Random Velocity Field

In the context of linear first-order partial differential equations, the simplest problems concern the equation of continuity for the density (concentration) of a conservative tracer and the transfer of a nonconservative passive tracer by a random velocity field $\boldsymbol{U}(\boldsymbol{r},t)$:

$$\left(\frac{\partial}{\partial t}+\frac{\partial}{\partial \boldsymbol{r}}\boldsymbol{U}(\boldsymbol{r},t)\right)\rho(\boldsymbol{r},t)=0,\quad \rho(\boldsymbol{r},0)=\rho_0(\boldsymbol{r}), \tag{1.56}$$

$$\left(\frac{\partial}{\partial t}+\boldsymbol{U}(\boldsymbol{r},t)\frac{\partial}{\partial \boldsymbol{r}}\right)q(\boldsymbol{r},t)=0,\quad q(\boldsymbol{r},0)=q_0(\boldsymbol{r}). \tag{1.57}$$

The conservative tracer is a tracer whose total mass remains intact

$$M_0=\int d\boldsymbol{r}\rho(\boldsymbol{r},t)=\int d\boldsymbol{r}\rho_0(\boldsymbol{r}) \tag{1.58}$$

We can use the method of characteristics to solve the linear first-order partial differential equations (1.56), (1.57). Introducing *characteristic curves* (particles)

$$\frac{d}{dt}\boldsymbol{r}(t)=\boldsymbol{U}(\boldsymbol{r},t),\qquad \boldsymbol{r}(0)=\boldsymbol{r}_0, \tag{1.59}$$

we can write these equations in the form

$$\begin{aligned}\frac{d}{dt}\rho(t)&=-\frac{\partial \boldsymbol{U}(\boldsymbol{r},t)}{\partial \boldsymbol{r}}\rho(t),\quad \rho(0)=\rho_0(\boldsymbol{r}_0),\\ \frac{d}{dt}q(t)&=0,\quad q(0)=q_0(\boldsymbol{r}_0).\end{aligned} \tag{1.60}$$

This formulation of the problem corresponds to the *Lagrangian description*, while the initial dynamic equations (1.56), (1.57) correspond to the *Eulerian description*.

Here, we introduced the characteristic vector parameter $\boldsymbol{r}_0$ in the system of equations (1.59), (1.60). With this parameter, Eq. (1.59) coincides with Eq. (1.1) that describes particle dynamics in random velocity field.

The solution of the system of equations (1.59), (1.60) depends on the initial value $\boldsymbol{r}_0$,

$$\boldsymbol{r}(t)=\boldsymbol{r}(t|\boldsymbol{r}_0),\quad \rho(t)=\rho(t|\boldsymbol{r}_0), \tag{1.61}$$

which we will isolate by the vertical bar symbol.

The first equality in Eq. (1.61) can be considered as an algebraic equation in characteristic parameter; the solution of this equation

$$\boldsymbol{r}_0 = \boldsymbol{r}_0(\boldsymbol{r}, t)$$

exists because *divergence* $j(t|\boldsymbol{r}_0) = \det \|\partial r_i(t|\boldsymbol{r}_0)/\partial r_{0k}\|$ is different from zero. Consequently, we can write the solution of the initial equation (1.56) in the form

$$\rho(\boldsymbol{r}, t) = \rho(t|\boldsymbol{r}_0(\boldsymbol{r}, t)) = \int d\boldsymbol{r}_0 \rho(t|\boldsymbol{r}_0) j(t|\boldsymbol{r}_0) \delta\left(\boldsymbol{r}(t|\boldsymbol{r}_0) - \boldsymbol{r}\right).$$

From Eq. (1.59) follows the equation for divergence $j(t|\boldsymbol{r}_0)$

$$\frac{d}{dt} j(t|\boldsymbol{r}_0) = \frac{\partial \boldsymbol{U}(\boldsymbol{r}, t)}{\partial \boldsymbol{r}} j(t|\boldsymbol{r}_0), \quad j(0) = 1. \tag{1.62}$$

Correlating it with Eq. (1.60), we see that

$$\rho(t|\boldsymbol{r}_0) = \frac{\rho_0(\boldsymbol{r}_0)}{j(t|\boldsymbol{r}_0)} \tag{1.63}$$

and, consequently, the density field can be rewritten in the form of the equality

$$\rho(\boldsymbol{r}, t) = \int d\boldsymbol{r}_0 \rho(t|\boldsymbol{r}_0) j(t|\boldsymbol{r}_0) \delta\left(\boldsymbol{r}(t|\boldsymbol{r}_0) - \boldsymbol{r}\right) = \int d\boldsymbol{r}_0 \rho_0(\boldsymbol{r}_0) \delta\left(\boldsymbol{r}(t|\boldsymbol{r}_0) - \boldsymbol{r}\right) \tag{1.64}$$

that states the relationship between the Lagrangian and Eulerian characteristics. For the position of the Lagrangian particle, the delta-function appeared in the right-hand side of this equality is the *indicator function*(see the next chapter). Consequently, averaging this equality over an ensemble of realizations of random velocity field, we obtain the well-known relationship between the average density in the Eulerian description and the one-time probability density $P(t, \boldsymbol{r}|\boldsymbol{r}_0) = \langle \delta\left(\boldsymbol{r}(t|\boldsymbol{r}_0) - \boldsymbol{r}\right) \rangle$ of the Lagrangian particle (see, e.g., [112]):

$$\langle \rho(\boldsymbol{r}, t) \rangle = \int d\boldsymbol{r}_0 \rho_0(\boldsymbol{r}_0) P\left(t, \boldsymbol{r}|\boldsymbol{r}_0\right). \tag{1.65}$$

For a nondivergent velocity field ($\operatorname{div} \boldsymbol{U}(\boldsymbol{r}, t) = 0$), both particle divergence and particle density are conserved, i.e.,

$$j(t|\boldsymbol{r}_0) = 1, \quad \rho(t|\boldsymbol{r}_0) = \rho_0(\boldsymbol{r}_0), \quad q(t|\boldsymbol{r}_0) = q_0(\boldsymbol{r}_0).$$

Consider now stochastic features of the solutions to problem (1.56). A convenient way of analyzing random field dynamics consists in using *topographic* concepts. Indeed, in the case of the nondivergent velocity field, temporal evolution of the contour of constant concentration $\rho =$ const coincides with the dynamics of particles in this velocity field and, consequently, coincides with the dynamics shown in Fig. 1.1*a*, page 5. In this case, the area within the contour remains constant and, as it is seen from Fig. 1.1*a*, the pattern becomes highly indented, which is manifested in gradient sharpening and the appearance of contour dynamics for progressively shorter scales. In the other limiting case (the divergent velocity field), the area within the contour tends to zero, and the field of density condenses forming clusters. One can find examples simulated for this case in papers [76,142]. These features of particle dynamics disappear on averaging over an ensemble of realizations.

Cluster formation in the Eulerian description can be traced using the random velocity field of form (1.3), (1.4), page 4. If $v_x(t) \neq 0$, then concentration field in Lagrangian description $\rho(t|\boldsymbol{r}_0)$ in the particular case of uniform (independent of $\boldsymbol{r}$) initial distribution $\rho_0(\boldsymbol{r}) = \rho_0$ can be described by the following equation

$$\frac{d}{dt}\rho(t) = -2kv_x(t)\cos(2kx)\rho(t), \quad \rho(0) = \rho_0,$$

which can be rewritten, by virtue of Eq. (1.6), page 6, in the form

$$\frac{d}{dt}\rho(t|\boldsymbol{r}_0) = -2kv_x(t)\frac{1 - e^{2T(t)}\tan^2(kx_0)}{1 + e^{2T(t)}\tan^2(kx_0)}\rho(t|\boldsymbol{r}_0), \tag{1.66}$$

where function $T(t)$ is given by Eq. (1.7), page 6. Integrating Eq. (1.66), we obtain the Lagrangian representation of the velocity field in the framework of the model under consideration

$$\rho(t|x_0)/\rho_0 = \left[e^{-T(t)}\cos^2(kx_0) + e^{T(t)}\sin^2(kx_0)\right].$$

Eliminating characteristic parameter x_0 with the use of equalities

$$\begin{aligned} \sin^2(kx_0) &= \frac{e^{-T(t)}\sin^2(kx(t))}{e^{T(t)}\cos^2(kx(t)) + e^{-T(t)}\sin^2(kx(t))}, \\ \cos^2(kx_0) &= \frac{e^{T(t)}\cos^2(kx(t))}{e^{T(t)}\cos^2(kx(t)) + e^{-T(t)}\sin^2(kx(t))}, \end{aligned} \tag{1.67}$$

following from Eq. (1.6), page 6, we pass to the Eulerian description

$$\rho(\boldsymbol{r},t)/\rho_0 = \frac{1}{e^{T(t)}\cos^2(kx) + e^{-T(t)}\sin^2(kx)}. \tag{1.68}$$

Expression (1.68) shows that the density field is low everywhere excluding the neighborhoods of points $kx = n\frac{\pi}{2}$, where $\rho(x,t)/\rho_0 = e^{\pm T(t)}$ and is sufficiently high if random factor $T(t)$ has appropriate sign.

Thus, in the problem under consideration, the cluster structure of the density field in the Eulerian description is formed in the neighborhoods of points

$$kx = n\frac{\pi}{2} \quad (n = 0, \pm 1, \pm 2, \ldots).$$

Note that Eulerian density field (1.68) averaged over spatial variables is independent of random factor $T(t)$,

$$\overline{\rho(x,t)/\rho_0} = 1,$$

and the average square of the density mainly grows with time

$$\overline{(\rho(x,t)/\rho_0)^2} = \frac{1}{2}\left(e^{T(t)} + e^{-T(t)}\right),$$

Figure 1.14 shows the Eulerian concentration field $1 + \rho(\boldsymbol{r},t)/\rho_0$ and its space-time evolution calculated by Eq. (1.68) in the dimensionless space-time variables (the density field is added with a unity to avoid the difficulties of dealing with nearly zero-valued concentrations in the logarithmic scale). This figure shows successive patterns of concentration field rearrangement toward narrow neighborhoods of points $x \approx 0$ and $x \approx \pi/2$, i.e., the formation of clusters, in which relative density is as high as $10^3 - 10^4$, while relative density is practically zero in the whole other space. Note that the realization of the density field passes through the initial homogeneous state at the instants t such that $T(t) = 0$. As is seen from figures, the lifetimes of such clusters coincide on the order of magnitude with the time of cluster formation.

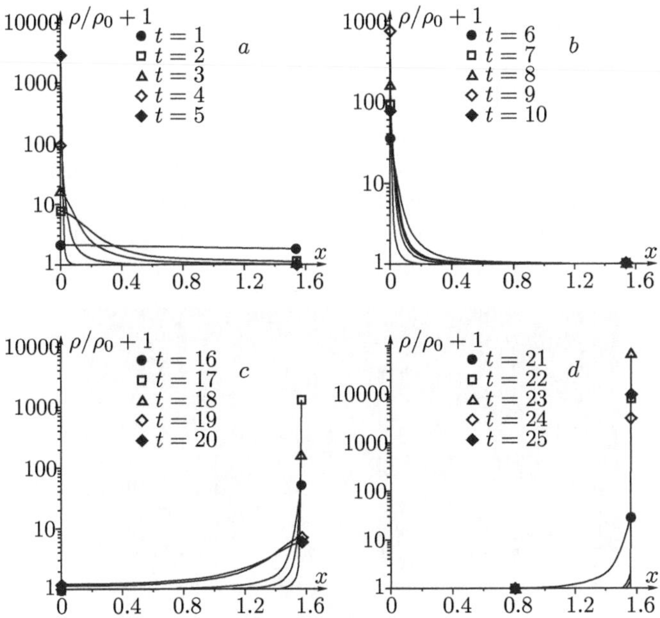

Fig. 1.14 Space-time evolution of the Eulerian density field given by Eq. (1.68)

This model provides an insight into the difference between the diffusion processes in divergent and nondivergent velocity fields. In nondivergent (incompressible) velocity fields, particles (and, consequently, density field) have no time for attracting to stable centers of attraction during the lifetime of these centers, and particles slightly fluctuate relative their initial location. On the contrary, in the divergent (compressible) velocity field, lifetime of stable centers of attraction is sufficient for particles to attract to them, because the speed of attraction increases exponentially, which is clearly seen from Eq. (1.68).

From the above description, it becomes obvious that dynamic equation (1.56) considered as a model equation describing actual physical phenomena can be used only on finite temporal intervals. A more complete analysis assumes the consideration of the field of tracer concentration gradient $\boldsymbol{p}(\boldsymbol{r},t) = \boldsymbol{\nabla}\rho(\boldsymbol{r},t)$ that satisfies the equation

$$\left(\frac{\partial}{\partial t} + \frac{\partial}{\partial \boldsymbol{r}}\boldsymbol{U}(\boldsymbol{r},t)\right) p_i(\boldsymbol{r},t) = -p_k(\boldsymbol{r},t)\frac{\partial U_k(\boldsymbol{r},t)}{\partial r_i} - \rho(\boldsymbol{r},t)\frac{\partial^2 \boldsymbol{U}(\boldsymbol{r},t)}{\partial r_i \partial \boldsymbol{r}},$$
$$\boldsymbol{p}(\boldsymbol{r},0) = \boldsymbol{p}_0(\boldsymbol{r}) = \boldsymbol{\nabla}\rho_0(\boldsymbol{r}). \tag{1.69}$$

In addition, one should also include the effect of the dynamic diffusion (with the dynamic diffusion coefficient μ) that smooths the mentioned sharpen of the gradient; this effect is described by the linear second-order partial differential equation

$$\left(\frac{\partial}{\partial t} + \frac{\partial}{\partial \boldsymbol{r}}\boldsymbol{U}(\boldsymbol{r},t)\right) \rho(\boldsymbol{r},t) = \mu\Delta\rho(\boldsymbol{r},t), \quad \rho(\boldsymbol{r},0) = \rho_0(\boldsymbol{r}). \tag{1.70}$$

In conclusion of this Section I give a pattern taken from Internet (Fig. 1.15) that shows the cluster structure of the Universe and is seemingly a direct consequence of clustering the cosmic matter in random velocity fields.

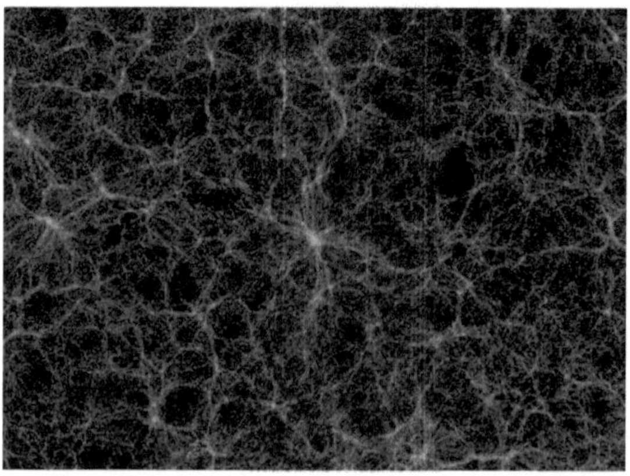

Fig. 1.15 Cluster structure of the Universe

Diffusion of Magnetic Field in a Random Velocity Field

The diffusion of such passive fields as the scalar density (particle concentration) field and the magnetic field is an important problem of the theory of turbulence in magnetohydrodynamics. Here, the basic stochastic equations are the continuity equation (1.56) for density field $\rho(\boldsymbol{r},t)$ (see previous section) and the induction equation for nondivergent magnetic field $\boldsymbol{H}(\boldsymbol{r},t)$ ($\operatorname{div}\boldsymbol{H}(\boldsymbol{r},t)=0$) [82]

$$\frac{\partial}{\partial t}\boldsymbol{H}(\boldsymbol{r},t)=\operatorname{curl}\left[\boldsymbol{U}(\boldsymbol{r},t)\times\boldsymbol{H}(\boldsymbol{r},t)\right],\quad \boldsymbol{H}(\boldsymbol{r},0)=\boldsymbol{H}_0(\boldsymbol{r}). \tag{1.71}$$

In Eq. (1.71), $\boldsymbol{U}(\boldsymbol{r},t)$ is the hydrodynamic velocity field and pseudovector $\boldsymbol{C}=\boldsymbol{A}\times\boldsymbol{B}$ is the vector product of vectors $\boldsymbol{A}$ and $\boldsymbol{B}$, $C_i=\varepsilon_{ijk}A_jB_k$, where ε_{ijk} is the pseudotensor such that $\varepsilon_{ijk}=0$ if indices i, j, and k are not all different and $\varepsilon_{ijk}=1$ or $\varepsilon_{ijk}=-1$, if indices i, j, and k are all different and form cyclic or anticyclic sequence (see, e.g., [7]). The operator

$$\operatorname{curl}\boldsymbol{C}(\boldsymbol{r},t)=[\boldsymbol{\nabla}\times\boldsymbol{C}(\boldsymbol{r},t)],\quad \operatorname{curl}\boldsymbol{C}(\boldsymbol{r},t)\big|_i=\varepsilon_{ijk}\frac{\partial}{\partial r_j}C_k(\boldsymbol{r},t)$$

is called the *vortex* of field $\boldsymbol{C}(\boldsymbol{r},t)$.

In magnetohydrodynamics, velocity field $\boldsymbol{U}(\boldsymbol{r},t)$ is generally described by the Navier–Stokes equation complemented with the density of extraneous electromagnetic forces

$$\boldsymbol{f}(\boldsymbol{r},t)=\frac{1}{4\pi}\left[\operatorname{curl}\boldsymbol{H}(\boldsymbol{r},t)\times\boldsymbol{H}(\boldsymbol{r},t)\right].$$

Nevertheless we, as earlier, will consider velocity field $\boldsymbol{U}(\boldsymbol{r},t)$ as random field whose statistical parameters are given.

Remark 1.1. ***Important Formulas of the Vector Analysis***

The product of two pseudotensors is a tensor, and, in the case of pseudotensors ε, we have the following equality

$$\varepsilon_{ilm}\varepsilon_{jpq}=\delta_{ij}\delta_{lp}\delta_{mq}+\delta_{ip}\delta_{lq}\delta_{mj}+\delta_{iq}\delta_{lj}\delta_{mp}-\delta_{ij}\delta_{lq}\delta_{mp}-\delta_{ip}\delta_{lj}\delta_{mq}-\delta_{iq}\delta_{lp}\delta_{mj}, \tag{1.72}$$

Setting $j=m$ (repeated indices assume summation), we obtain

$$\varepsilon_{ilm}\varepsilon_{mpq}=(d-2)(\delta_{ip}\delta_{lq}-\delta_{iq}\delta_{lp}), \tag{1.73}$$

where d is the dimension of space, so that the above convolution becomes zero in the two-dimensional case.

Thus, the double vector product is given by the formula

$$\left[\boldsymbol{C}\times[\boldsymbol{A}\times\boldsymbol{B}]\right]_i=\varepsilon_{ilm}\varepsilon_{mpq}C_lA_pB_q=C_qA_iB_q-C_pA_pB_i. \tag{1.74}$$

If fields $\boldsymbol{C}$, $\boldsymbol{A}$, and $\boldsymbol{B}$ are the conventional vector fields, Eq. (1.74) assumes the form

$$[\boldsymbol{C}\times[\boldsymbol{A}\times\boldsymbol{B}]]=(\boldsymbol{C}\cdot\boldsymbol{B})\,\boldsymbol{A}-(\boldsymbol{C}\cdot\boldsymbol{A})\,\boldsymbol{B}. \tag{1.75}$$

In the case of operator vector field $\boldsymbol{C}=\boldsymbol{\nabla}=\dfrac{\partial}{\partial\boldsymbol{r}}$, Eq. (1.74) results in the expression

$$\operatorname{curl}\left[\boldsymbol{A}(\boldsymbol{r})\times\boldsymbol{B}(\boldsymbol{r})\right]=\left(\frac{\partial}{\partial\boldsymbol{r}}\cdot\boldsymbol{B}(\boldsymbol{r})\right)\boldsymbol{A}(\boldsymbol{r})-\left(\frac{\partial}{\partial\boldsymbol{r}}\cdot\boldsymbol{A}(\boldsymbol{r})\right)\boldsymbol{B}(\boldsymbol{r}). \tag{1.76}$$

Note that, if vector field $\boldsymbol{A}$ is an operator in Eq. (1.74), $\boldsymbol{A}=\boldsymbol{\nabla}=\dfrac{\partial}{\partial\boldsymbol{r}}$, then we have

$$[\boldsymbol{C}(\boldsymbol{r})\times\operatorname{curl}\boldsymbol{B}(\boldsymbol{r})]=C_q(\boldsymbol{r})\frac{\partial}{\partial\boldsymbol{r}}B_q(\boldsymbol{r})-\left(\boldsymbol{C}(\boldsymbol{r})\cdot\frac{\partial}{\partial\boldsymbol{r}}\right)\boldsymbol{B}(\boldsymbol{r})$$

and, in particular,

$$[\boldsymbol{B}(\boldsymbol{r})\times\operatorname{curl}\boldsymbol{B}(\boldsymbol{r})]=\frac{1}{2}\frac{\partial}{\partial\boldsymbol{r}}\boldsymbol{B}^2(\boldsymbol{r})-\left(\boldsymbol{B}(\boldsymbol{r})\cdot\frac{\partial}{\partial\boldsymbol{r}}\right)\boldsymbol{B}(\boldsymbol{r}). \tag{1.77}$$

♦

Using Eq. (1.76), we can rewrite Eq. (1.71) in the form

$$\begin{gathered}\left(\frac{\partial}{\partial t}+\frac{\partial}{\partial\boldsymbol{r}}\boldsymbol{u}(\boldsymbol{r},t)\right)\boldsymbol{H}(\boldsymbol{r},t)=\left(\boldsymbol{H}(\boldsymbol{r},t)\cdot\frac{\partial}{\partial\boldsymbol{r}}\right)\boldsymbol{u}(\boldsymbol{r},t),\\ \boldsymbol{H}(\boldsymbol{r},0)=\boldsymbol{H}_0(\boldsymbol{r}).\end{gathered} \tag{1.78}$$

Dynamic system (1.78) is a conservative system, and magnetic field flux $\int d\boldsymbol{r}\,\boldsymbol{H}(\boldsymbol{r},t)$ remains constant during evolution.

We are interested in the evolution of magnetic field in space and time from given smooth initial distributions and, in particular, simply homogeneous ones, $\boldsymbol{H}_0(\boldsymbol{r})=\boldsymbol{H}_0$. Clearly, at the initial evolutionary stages of the process, the effects of dynamic diffusion are insignificant, and Eq. (1.78) describes namely this case. Further stages of evolution require consideration of the effects of dynamic diffusion; these effects are described by the equation

$$\left(\frac{\partial}{\partial t}+\frac{\partial}{\partial\boldsymbol{r}}\boldsymbol{u}(\boldsymbol{r},t)\right)\boldsymbol{H}(\boldsymbol{r},t)=\left(\boldsymbol{H}(\boldsymbol{r},t)\cdot\frac{\partial}{\partial\boldsymbol{r}}\right)\boldsymbol{u}(\boldsymbol{r},t)+\mu_{\boldsymbol{H}}\Delta\boldsymbol{H}(\boldsymbol{r},t), \tag{1.79}$$

where $\mu_{\boldsymbol{H}}$ is the dynamic diffusion coefficient for the magnetic field.

Note that, similarly to the case of tracer density, velocity filed model (1.3), (1.4), page 4 allows obtaining the magnetic field in an explicit form if dynamic diffusion can be neglected [67]. With the use of this model, the induction equation for homogeneous initial condition (1.78) assumes the form

$$\left(\frac{\partial}{\partial t}+v_x(t)\sin(2kx)\frac{\partial}{\partial x}\right)\boldsymbol{H}(x,t)=2k\cos(2kx)\left[\boldsymbol{v}(t)H_x(x,t)-v_x(t)\boldsymbol{H}(x,t)\right],$$
$$\boldsymbol{H}(x,0)=\boldsymbol{H}_0,$$

from which follows that the x-component of the magnetic field remains constant ($H_x(x,t)=H_{x0}$), and the existence of the H_{x0} component causes the appearance of an additional source of magnetic field in the transverse (y,z)-plane

$$\left(\frac{\partial}{\partial t}+v_x(t)\sin(2kx)\frac{\partial}{\partial x}\right)\boldsymbol{H}_\perp(\boldsymbol{r},t)=2k\cos(2kx)\left[\boldsymbol{v}_\perp(t)H_{x0}-v_x(t)\boldsymbol{H}_\perp(x,t)\right],$$
$$\boldsymbol{H}_\perp(x,0)=\boldsymbol{H}_{\perp 0}. \tag{1.80}$$

Equation (1.80) is a partial differential equation, and we can solve it using the method of characteristics (the Lagrangian description). The characteristics satisfy the equations

$$\frac{d}{dt}x(t|x_0)=v_x(t)\sin(2kx(t|x_0)),\quad x(0|x_0)=x_0,$$
$$\frac{d}{dt}\boldsymbol{H}_\perp(t|x_0)=2k\cos(2kx|x_0)\left[\boldsymbol{v}_\perp(t)H_{x0}-v_x(t)\boldsymbol{H}_\perp(t|x_0)\right],\quad \boldsymbol{H}_\perp(0|x_0)=\boldsymbol{H}_{\perp 0}, \tag{1.81}$$

where the vertical bar separates the dependence on characteristic parameter x_0.

The first equation in Eqs. (1.81) describes particle diffusion, and its solution has the form

$$x(t|x_0)=\frac{1}{k}\arctan\left[e^{T(t)}\tan(kx_0)\right],$$

where function $T(t)$ is given by Eq. (1.7), page 6. The solution of the equation in the magnetic field has the form

$$\boldsymbol{H}_\perp(t|x_0)=\left[e^{-T(t)}\cos^2(kx_0)+e^{T(t)}\sin^2(kx_0)\right]\boldsymbol{H}_{\perp 0}$$
$$+2k\left[e^{-T(t)}\cos^2(kx_0)+e^{T(t)}\sin^2(kx_0)\right]$$
$$\times\int\limits_0^t d\tau\frac{\left[e^{-T(\tau)}\cos^2(kx_0)-e^{T(\tau)}\sin^2(kx_0)\right]}{\left[e^{-T(\tau)}\cos^2(kx_0)+e^{T(\tau)}\sin^2(kx_0)\right]^2}v_x(\tau)\boldsymbol{v}_\perp(\tau)H_{x0}.$$

Eliminating now characteristic parameter x_0 with the use of Eqs. (1.67), page 36, we pass to the Eulerian description

$$\boldsymbol{H}_{\perp}(x,t) = \frac{\rho(x,t)}{\rho_0}\boldsymbol{H}_{\perp 0}$$
$$+ 2kH_{x0}\int\limits_0^t d\tau \frac{\left[e^{T(t)-T(\tau)}\cos^2(kx) - e^{-T(t)+T(\tau)}\sin^2(kx)\right]}{\left[e^{T(t)-T(\tau)}\cos^2(kx) + e^{-T(t)+T(\tau)}\sin^2(kx)\right]^2} v_x(\tau)\boldsymbol{v}_{\perp}(\tau), \tag{1.82}$$

where the density of passive tracer $\rho(x,t)$ is described by Eq. (1.68). Making now the change of integration variables $t-\tau=\lambda$ in Eq. (1.82), we can rewrite it as

$$\boldsymbol{H}_{\perp}(x,t) = \frac{\rho(x,t)}{\rho_0}\boldsymbol{H}_{\perp 0}$$
$$+2kH_{x0}\int\limits_0^t d\lambda \frac{\left[e^{T(t)-T(\tau)}\cos^2(kx) - e^{-T(t)+T(\tau)}\sin^2(kx)\right]}{\left[e^{T(t)-T(\tau)}\cos^2(kx) + e^{-T(t)+T(\tau)}\sin^2(kx)\right]^2} v_x(t-\lambda)\boldsymbol{v}_{\perp}(t-\lambda),$$

where

$$T(t) - T(\tau) = \int\limits_\tau^t d\xi v_x(\xi) = \int\limits_0^{t-\tau} d\eta v_x(t-\eta) = \int\limits_0^{\lambda} d\eta v_x(t-\eta).$$

Hence, dealing with the one-time statistical characteristics of magnetic field, we can replace $v_x(t-\lambda)$ with $v_x(\lambda)$ in view of stationarity of the velocity field (see section 4.2.1, 5.3 page 125) and rewrite Eq. (1.82) in a statistically equivalent form,

$$\boldsymbol{H}_{\perp}(x,t) = \frac{\rho(x,t)}{\rho_0}\boldsymbol{H}_{\perp 0} + 2kH_{x0}\int\limits_0^t d\tau \frac{\left[e^{T(\tau)}\cos^2(kx) - e^{-T(\tau)}\sin^2(kx)\right]}{\left[e^{T(\tau)}\cos^2(kx) + e^{-T(\tau)}\sin^2(kx)\right]^2} v_x(\tau)\boldsymbol{v}_{\perp}(\tau). \tag{1.83}$$

The first term describes magnetic field clustering like the density field clustering if $\boldsymbol{H}_{\perp 0} \neq 0$. The second term describes the generation of a magnetic field in the transverse (y,z)-plane due to the presence of an initial field H_{x0}. At $\boldsymbol{H}_{\perp 0} = 0$, this term, proportional to the square of the random velocity field, determines the situation. Like the density field, the structure of this field is also clustered.

Figures 1.16, 1.17 present the calculated space–time evolution of a realization of energy of the magnetic field generated in the transverse plane, $E(x,t) = \boldsymbol{H}_{\perp}^2(x,t)$, in dimensionless variables (see 1.12, page 7) at $\boldsymbol{H}_{\perp 0} = 0$ for the same realization of the random process $T(t)$ as that presented previously in Fig. 1.2 *a*.

First of all, we note that the total energy of the generated magnetic field concentrated in the segment $[0,\ \pi/2]$ increases rapidly with time (Fig. 1.16a).

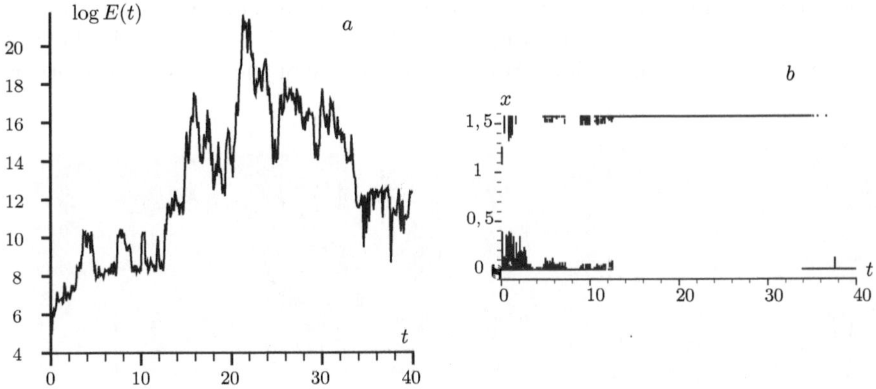

Fig. 1.16 (a) Temporal evolution of the total magnetic energy in segment $[0, \pi/2]$ and (b) cluster structure in the (x, t)-plane

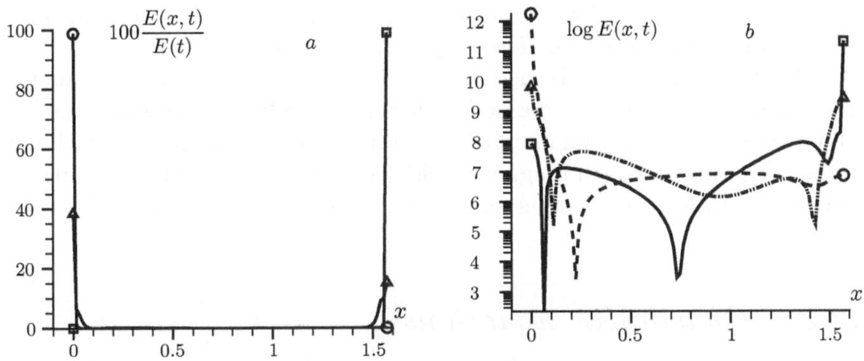

Fig. 1.17 Dynamics of disappearance of a cluster at point 0 and appearance of a cluster at point $\pi/2$. The circles, triangles, and squares mark the curves corresponding to time instants $t = 10.4$, 10.8, and 11.8, respectively.

A general space–time structure of the magnetic energy clustering is shown in Fig. 1.16b. This structure was calculated in the following way. The coordinates of points x_i are plotted along the x-axis and the time is plotted along the t-axis. The points are marked as white squares (unseen) if they contain less than 1 of the energy available in the entire layer at current t and as black squares if they contain more than 1 of the energy available in the entire layer at the time in question. There are a total of 40000 points (100 steps in x and 400 steps in time).

A more detailed pattern of the evolution of clusters with time is presented in Fig. 1.17. Figure 1.17*a* shows the percentage ratio of the generated magnetic energy field contained in a cluster to the total energy in the layer at the time under consideration; Fig. 1.17*b* shows the dynamics of the flow of magnetic energy perturbations from one boundary of the region to the other.

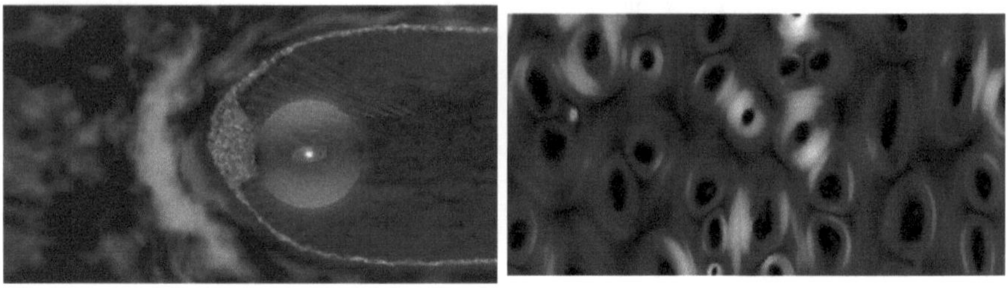

Fig. 1.18 Artist's interpretation depicting the new view of the heliosphere (left). Computer simulation of the magnetic reconnection in the heliosheats, which look like bubbles, or sausages (right).

I illustrate structure formation in magnetic field by the extract from an internet-page: *What does puzzle astrophysicists so strongly?*

Contrary to hypotheses formed for fifty years, at the boundary of planetary system observers encountered a boiling foam of locally magnetized areas each of hundreds of millions kilometers in extent, which form a non-stationary cellular structure in which magnetic field lines are permanently breaking and recombining to form new areas—*magnetic "bubbles"* (see fig. 1.18).[2]

1.3.2 Quasilinear Equations

Consider now the simplest quasilinear equation for scalar quantity $q(\boldsymbol{r},t)$, which we write in the form

$$\left(\frac{\partial}{\partial t} + \boldsymbol{U}(t,q)\frac{\partial}{\partial \boldsymbol{r}}\right) q(\boldsymbol{r},t) = Q\left(t,q\right), \qquad q(\boldsymbol{r},0) = q_0(\boldsymbol{r}), \tag{1.84}$$

where we assume for simplicity that functions $\boldsymbol{U}(t,q)$ and $Q(t,q)$ are explicitly independent of spatial variable $\boldsymbol{r}$.

[2] Nancy Atkinson, Voyagers Find Giant Jacuzzi-like Bubbles at Edge of Solar System, `http://www.universetoday.com/86446/voyagers-find-giant-jacuzzi-like-bubbles-at-edge-of-solar-system/`

Supplement Eq. (1.84) with the equation for the gradient $\boldsymbol{p}(\boldsymbol{r},t) = \boldsymbol{\nabla}q(\boldsymbol{r},t)$, which follows from Eq. (1.84), and the equation of continuity for conserved quantity $I(\boldsymbol{r},t)$:

$$\left(\frac{\partial}{\partial t} + \boldsymbol{U}(t,q)\frac{\partial}{\partial \boldsymbol{r}}\right)p(r,t) + \frac{\partial\left\{\boldsymbol{U}(t,q)\boldsymbol{p}(\boldsymbol{r},t)\right\}}{\partial q}p(r,t) = \frac{\partial Q(t,q)}{\partial q}p(r,t),$$
$$\frac{\partial}{\partial t}I(r,t) + \frac{\partial}{\partial \boldsymbol{r}}\left\{\boldsymbol{U}(t,q)I(\boldsymbol{r},t)\right\} = 0, \quad I(\boldsymbol{r},0) = I_0(\boldsymbol{r}). \tag{1.85}$$

From Eqs. (1.85) follows that

$$\int d\boldsymbol{r}I(\boldsymbol{r},t) = \int d\boldsymbol{r}I_0(\boldsymbol{r}). \tag{1.86}$$

In terms of characteristic curves determined from the system of ordinary differential equations, Eqs. (1.84) and (1.85) can be written in the form

$$\frac{d}{dt}\boldsymbol{r}(t) = \boldsymbol{U}\left(t,q\right), \quad \frac{d}{dt}q(t) = Q\left(t,q\right), \quad \boldsymbol{r}\left(0\right) = \boldsymbol{r}_0, \quad q\left(0\right) = q_0\left(\boldsymbol{r}_0\right),$$
$$\frac{d}{dt}\boldsymbol{p}(t) = -\frac{\partial\left\{\boldsymbol{U}\left(t,q\right)\boldsymbol{p}(t)\right\}}{\partial q}\boldsymbol{p}(t) + \frac{\partial Q(t,q)}{\partial q}\boldsymbol{p}(t), \quad \boldsymbol{p}\left(0\right) = \frac{\partial q_0(\boldsymbol{r}_0)}{\partial \boldsymbol{r}_0}, \tag{1.87}$$
$$\frac{d}{dt}I(t) = -\frac{\partial\left\{\boldsymbol{U}\left(t,q\right)\boldsymbol{p}(t)\right\}}{\partial q}I(t), \quad I\left(0\right) = I_0\left(\boldsymbol{r}_0\right).$$

Thus, the Lagrangian description considers the system (1.87) as the initial-value problem. In this description, the two first equations form the closed system that defines characteristic curves .

Expressing now characteristic parameter $\boldsymbol{r}_0$ in terms of t and $\boldsymbol{r}$, one can write the solution to Eqs. (1.84) and (1.85) in the Eulerian description as

$$q\left(\boldsymbol{r},t\right) = \int d\boldsymbol{r}_0 q\left(t|\boldsymbol{r}_0\right)j\left(t|\boldsymbol{r}_0\right)\delta\left(\boldsymbol{r}\left(t|\boldsymbol{r}_0\right) - \boldsymbol{r}\right),$$
$$I\left(\boldsymbol{r},t\right) = \int d\boldsymbol{r}_0 I\left(t|\boldsymbol{r}_0\right)j\left(t|\boldsymbol{r}_0\right)\delta\left(\boldsymbol{r}\left(t|\boldsymbol{r}_0\right) - \boldsymbol{r}\right). \tag{1.88}$$

The feature of the transition from the Lagrangian description (1.87) to the Eulerian description (1.88) consists in the general appearance of ambiguities, which yields discontinuous solutions. These ambiguities are related to the fact that the divergence $j\left(t|\boldsymbol{r}_0\right) = \det\left\|\frac{\partial}{\partial r_{0k}}r_i\left(t|\boldsymbol{r}_0\right)\right\|$ can now vanish at certain moments.

Quantities $I(t|\boldsymbol{r}_0)$ and $j(t|\boldsymbol{r}_0)$ are not independent. Indeed, integrating $I\left(\boldsymbol{r},t\right)$ in Eq. (1.88) over $\boldsymbol{r}$ and taking into account Eq. (1.86), we see that there exists the evolution integral

$$j\left(t|\boldsymbol{r}_0\right) = \frac{I_0\left(\boldsymbol{r}_0\right)}{I\left(t|\boldsymbol{r}_0\right)}, \tag{1.89}$$

from which follows that zero-valued divergence $j(t|\boldsymbol{r}_0)$ is accompanied by the infinite value of conservative quantity $I(t|\boldsymbol{r}_0)$.

It is obvious that all these results can be easily extended to the case in which functions $\boldsymbol{U}(\boldsymbol{r},t,q)$ and $Q(\boldsymbol{r},t,q)$ explicitly depend on spatial variable $\boldsymbol{r}$ and Eq. (1.84) itself is the vector equation. As a particular physical example, we consider the equation for the velocity field $\boldsymbol{V}(\boldsymbol{r},t)$ of low-inertia particles moving in the hydrodynamic flow whose velocity field is $\boldsymbol{u}(\boldsymbol{r},t)$ (see, e.g., [100])

$$\left(\frac{\partial}{\partial t} + \boldsymbol{V}(\boldsymbol{r},t)\frac{\partial}{\partial \boldsymbol{r}}\right)\boldsymbol{V}(\boldsymbol{r},t) = -\lambda\left[\boldsymbol{V}(\boldsymbol{r},t) - \boldsymbol{u}(\boldsymbol{r},t)\right]. \tag{1.90}$$

We will assume this equation the phenomenological equation.

In the general case, the solution to Eq. (1.90) can be nonunique, it can have discontinuities, etc. However, in the case of asymptotically small inertia property of particles (parameter $\lambda \to \infty$), which is of our concern here, the solution will be unique during reasonable temporal intervals. Note that, in the right-hand side of Eq. (1.90), term $\boldsymbol{F}(\boldsymbol{r},t) = \lambda\boldsymbol{V}(\boldsymbol{r},t)$ linear in the velocity field $\boldsymbol{V}(\boldsymbol{r},t)$ is, according to the known *Stokes formula*, the resistance force acting on a slowly moving particle. If we approximate the particle by the sphere of radius a, parameter λ will be $\lambda = 6\pi a\eta/m_{\mathrm{p}}$, where η is the coefficient of dynamic viscosity and m_{p} is the mass of the particle (see, e.g., [79,83]).

From Eq. (1.90) follows that velocity field $\boldsymbol{V}(\boldsymbol{r},t)$ is the divergent field even if hydrodynamic flow is nondivergent (div $\boldsymbol{u}(\boldsymbol{r},t) = 0$). As a consequence, particle number density $n(\boldsymbol{r},t)$ in nondivergent hydrodynamic flows, which satisfies the linear equation of continuity

$$\left(\frac{\partial}{\partial t} + \frac{\partial}{\partial \boldsymbol{r}}\boldsymbol{V}(\boldsymbol{r},t)\right)n(\boldsymbol{r},t) = 0, \quad n(\boldsymbol{r},0) = n_0(\boldsymbol{r}) \tag{1.91}$$

similar to Eq. (1.56), shows the cluster behavior.

For large parameters $\lambda \to \infty$ (inertialess particles), we have

$$\boldsymbol{V}(\boldsymbol{r},t) \approx \boldsymbol{u}(\boldsymbol{r},t), \tag{1.92}$$

which means that particle number density $n(\boldsymbol{r},t)$ shows no cluster behavior in nondivergent hydrodynamic flows.

The first-order partial differential equation (1.90) (Eulerian description) is equivalent to the system of ordinary differential characteristic equations (Lagrangian description)

$$\begin{aligned} \frac{d}{dt}\boldsymbol{r}(t) &= \boldsymbol{V}\left(\boldsymbol{r}(t),t\right), \quad \boldsymbol{r}(0) = \boldsymbol{r}_0, \\ \frac{d}{dt}\boldsymbol{V}(t) &= -\lambda\left[\boldsymbol{V}(t) - \boldsymbol{u}\left(\boldsymbol{r}(t),t\right)\right], \quad \boldsymbol{V}(0) = \boldsymbol{V}_0(\boldsymbol{r}_0), \end{aligned} \tag{1.93}$$

that describes the diffusion of a particle under the random external force and linear friction and coincide with Eq. (1.14), page 9.

Conventional statistical description usually assumes that fluctuations of the hydrodynamic velocity field are sufficiently small. For this reason, we can linearize the system of equations (1.93) and rewrite it in the form (for simplicity, we assume that the mean flow is absent and use the zero-valued initial conditions)

$$\begin{gathered}\frac{d}{dt}\boldsymbol{r}(t) = \boldsymbol{v}(t), \quad \frac{d}{dt}\boldsymbol{v}(t) = -\lambda\left[\boldsymbol{v}(t) - \boldsymbol{f}(t)\right], \\ \boldsymbol{r}(0) = \boldsymbol{0}, \quad \boldsymbol{v}(0) = \boldsymbol{0},\end{gathered} \tag{1.94}$$

the stochastic solution to which has the form

$$\boldsymbol{v}(t) = \lambda \int_0^t d\tau e^{-\lambda(t-\tau)} \boldsymbol{f}(\tau), \quad \boldsymbol{r}(t) = \int_0^t d\tau \left[1 - e^{-\lambda(t-\tau)}\right] \boldsymbol{f}(\tau).$$

Note that the closed linear equation of the first order in velocity $\boldsymbol{v}(t)$ is called the *Langevin equation.*

Now, we turn back to Eq. (1.90). Setting $\lambda = 0$, we arrive at the equation

$$\left(\frac{\partial}{\partial t} + \boldsymbol{V}(\boldsymbol{r},t)\frac{\partial}{\partial \boldsymbol{r}}\right)\boldsymbol{V}(\boldsymbol{r},t) = 0, \quad \boldsymbol{V}(\boldsymbol{r},0) = \boldsymbol{V}_0(\boldsymbol{r}) \tag{1.95}$$

called the *Riemann equation*. It describes free propagation of the *nonlinear Riemann wave*. The solution to this equation obviously satisfies the transcendental equation

$$\boldsymbol{V}(\boldsymbol{r},t) = \boldsymbol{V}_0\left(\boldsymbol{r} - t\boldsymbol{V}(\boldsymbol{r},t)\right).$$

Example. The One-Dimensional Riemann Equation

In the special case of the one-dimensional equation

$$\left(\frac{\partial}{\partial t} + q(x,t)\frac{\partial}{\partial x}\right)q(x,t) = 0, \quad q(x,0) = q_0(x), \tag{1.96}$$

the solution can be drawn in both implicit and explicit form. This equation coincides with Eq. (1.84) at $G(t,q) = 0$, $U(t,q) = q(x,t)$.

The method of characteristics applied to Eq. (1.96) gives

$$q(t|x_0) = q_0(x_0), \quad x(t|x_0) = x_0 + tq_0(x_0),$$

so that the solution of Eq. (1.96) can be written in the form of the transcendental equation

$$q(x,t) = q_0\left(x - tq(x,t)\right)$$

from which follows an expression for spatial derivative

$$p(x,t) = \frac{\partial}{\partial x} q(x,t) = \frac{q_0'(x_0)}{1 + t q_0'(x_0)}, \tag{1.97}$$

where

$$x_0 = x - tq(x,t) \quad \text{and} \quad q_0'(x_0) = \frac{d}{dx_0} q_0(x_0).$$

The function $p(x,t)$ by itself satisfies here the equation

$$\left(\frac{\partial}{\partial t} + q(x,t)\frac{\partial}{\partial x}\right) p(x,t) = -p^2(x,t), \quad p(x,0) = p_0(x) = q_0'(x). \tag{1.98}$$

For completeness, we give the equation of continuity for the density field $\rho(x,t)$

$$\left(\frac{\partial}{\partial t} + q(x,t)\frac{\partial}{\partial x}\right) \rho(x,t) = -p(x,t)\rho(x,t), \quad \rho(x,0) = \rho_0(x), \tag{1.99}$$

and its logarithm $\chi(x,t) = \ln \rho(x,t)$

$$\left(\frac{\partial}{\partial t} + q(x,t)\frac{\partial}{\partial x}\right) \chi(x,t) = -p(x,t), \quad \chi(x,0) = \chi_0(x), \tag{1.100}$$

which are related to the Riemann equation (1.96). The solution to Eq. (1.96) has the form

$$\rho(x,t) = \frac{\rho_0(x_0)}{1 + tp_0(x_0)} = \frac{\rho_0(x - tq(x,t))}{1 + tp_0(x - tq(x,t))}.$$

If $q_0'(x_0) < 0$, then derivative $\frac{\partial}{\partial x} q(x,t)$ and solution of Eq. (1.96) becomes discontinuous. For times prior to t_0, the solution is unique and representable in the form of a quadrature. To show this fact, we calculate the variational derivative (for variational derivative definitions and the corresponding operation rules, see Chapter 2)

$$\frac{\delta q(x,t)}{\delta q_0(x_0)} = \frac{1}{1 + t q_0'(x_0)} \delta(x - tq(x,t) - x_0).$$

Because $q(x,t) = q_0(x_0)$ and $x - tq_0(x_0) = x_0$, the argument of delta function vanishes at $x = F(x_0,t) = x_0 + tq_0(x_0)$. Consequently, we have

$$\frac{\delta q(x,t)}{\delta q_0(x_0)} = \delta(x - F(x_0,t)) = \frac{1}{2\pi} \int\limits_{-\infty}^{\infty} dk\, e^{ik(x-x_0) - iktq_0(x_0)}.$$

We can consider this equality as the functional equation in variable $q_0(x_0)$. Then, integrating this equation with the initial value

$$q\left(x,t\right)|_{q_0(x_0)=0} = 0,$$

in the functional space, we obtain the solution of the Riemann equation in the form of the quadrature (see e.g., [121])

$$q\left(x,t\right) = \frac{i}{2\pi t}\int\limits_{-\infty}^{\infty}\frac{dk}{k}\int\limits_{-\infty}^{\infty}d\xi e^{ik(x-x_0)}\left[e^{-iktq_0(\xi)}-1\right].$$

The mentioned ambiguity can be eliminated by considering the *Burgers equation*

$$\frac{\partial}{\partial t}q(x,t) + q(x,t)\frac{\partial}{\partial x}q(x,t) = \mu\frac{\partial^2}{\partial x^2}q(x,t), \quad q(x,0) = q_0(x)$$

(it includes the molecular viscosity and also can be solved in quadratures) followed by the limit process $\mu \to 0$.

♦

1.3.3 Boundary-Value Problems for Nonlinear Ordinary Differential Equations

Note that, using the imbedding method (see Chapter 2), the boundary-value problems for nonlinear ordinary differential equations also can be reduced to quasilinear equations. This is the case, for example, for the nonlinear vector boundary-value problem

$$\frac{d}{dt}\boldsymbol{x}(t) = \boldsymbol{U}\left(t,\boldsymbol{x}(t)\right),$$

defined on segment $t \in [0,T]$ with boundary conditions

$$G\boldsymbol{x}(0) + H\boldsymbol{x}(T) = \boldsymbol{v},$$

where G and H are constant matrices. Consider the solution of this problem as a function of parameters T and $\boldsymbol{v}$, i.e., $\boldsymbol{x}(t) = \boldsymbol{x}(t;T,\boldsymbol{v})$. Then, function $\boldsymbol{R}(T,\boldsymbol{v}) = \boldsymbol{x}(T;T,\boldsymbol{v})$ as a function of parameters T and $\boldsymbol{v}$ is described by the quasilinear vector equation [26, 50]

$$\left(\frac{\partial}{\partial T} + \left[H\boldsymbol{U}\left(T,\boldsymbol{R}\left(t,\boldsymbol{v}\right)\right)\right]\frac{\partial}{\partial \boldsymbol{v}}\right)\boldsymbol{R}\left(T,\boldsymbol{v}\right) = \boldsymbol{U}\left(T,\boldsymbol{R}\left(t,\boldsymbol{v}\right)\right)$$

with boundary condition for $T \to 0$

$$\boldsymbol{R}\left(T,\boldsymbol{v}\right)|_{T=0} = \left(G+H\right)^{-1}\boldsymbol{v},$$

and function $\boldsymbol{x}(t;T,\boldsymbol{v})$ itself satisfies the linear equation

$$\frac{\partial x_i(t;T,\boldsymbol{v})}{\partial T} = -H_{kl}U_l\left(T,\boldsymbol{R}(T,\boldsymbol{v})\right)\frac{\partial x_i(t;T,\boldsymbol{v})}{\partial v_k}$$

with boundary condition

$$\boldsymbol{x}(t;T,\boldsymbol{v})|_{T=t} = \boldsymbol{R}(t,\boldsymbol{v}).$$

1.3.4 Nonlinear First-Order Partial Differential Equations

In the general case, a nonlinear scalar first-order partial differential equation can be written in the form

$$\frac{\partial}{\partial t}q(\boldsymbol{r},t) + H\left(\boldsymbol{r},t,q,\boldsymbol{p}\right) = 0, \quad q\left(\boldsymbol{r},0\right) = q_0\left(\boldsymbol{r}\right), \tag{1.101}$$

where $\boldsymbol{p}(\boldsymbol{r},t) = \boldsymbol{\nabla}q(\boldsymbol{r},t)$.

In terms of the Lagrangian description, this equation can be rewritten in the form of the system of characteristic equations (see, e.g., [135]):

$$\begin{aligned}
\frac{d}{dt}\boldsymbol{r}(t|\boldsymbol{r}_0) &= \frac{\partial}{\partial \boldsymbol{p}}H\left(\boldsymbol{r},t,q,\boldsymbol{p}\right), \quad \boldsymbol{r}(0|\boldsymbol{r}_0) = \boldsymbol{r}_0;\\
\frac{d}{dt}\boldsymbol{p}(t|\boldsymbol{r}_0) &= -\left(\frac{\partial}{\partial \boldsymbol{r}} + \boldsymbol{p}\frac{\partial}{\partial q}\right)H\left(\boldsymbol{r},t,q,\boldsymbol{p}\right), \quad \boldsymbol{p}(0|\boldsymbol{r}_0) = \boldsymbol{p}_0(\boldsymbol{r}_0);\\
\frac{d}{dt}q(t|\boldsymbol{r}_0) &= \left(\boldsymbol{p}\frac{\partial}{\partial \boldsymbol{p}} - 1\right)H\left(\boldsymbol{r},t,q,\boldsymbol{p}\right), \quad q(0|\boldsymbol{r}_0) = \boldsymbol{q}_0(\boldsymbol{r}_0).
\end{aligned} \tag{1.102}$$

Now, we supplement Eq. (1.101) with the equation for the conservative quantity $I(\boldsymbol{r},t)$

$$\frac{\partial}{\partial t}I(\boldsymbol{r},t) + \frac{\partial}{\partial \boldsymbol{r}}\left\{\frac{\partial H\left(\boldsymbol{r},t,q,\boldsymbol{p}\right)}{\partial \boldsymbol{p}}I(\boldsymbol{r},t)\right\} = 0, \quad I\left(\boldsymbol{r},0\right) = I_0\left(\boldsymbol{r}\right). \tag{1.103}$$

From Eq. (1.103) follows that

$$\int d\boldsymbol{r}I(\boldsymbol{r},t) = \int d\boldsymbol{r}I_0(\boldsymbol{r}). \tag{1.104}$$

In the Lagrangian description, the corresponding quantity satisfies the equation

$$\frac{d}{dt}I(t|\boldsymbol{r}_0) = -\frac{\partial^2 H\left(\boldsymbol{r},t,q,\boldsymbol{p}\right)}{\partial \boldsymbol{r}\partial \boldsymbol{p}}I(\boldsymbol{r},t), \quad I\left(0|\boldsymbol{r}_0\right) = I_0\left(\boldsymbol{r}_0\right),$$

so that the solution to Eq. (1.103) has the form

$$I(\boldsymbol{r},t) = I(t|\boldsymbol{r}_0(t,\boldsymbol{r})) = \int d\boldsymbol{r}_0 I(t|\boldsymbol{r}_0)\, j(t|\boldsymbol{r}_0)\, \delta(\boldsymbol{r}(t|\boldsymbol{r}_0) - \boldsymbol{r}), \qquad (1.105)$$

where $j(t|\boldsymbol{r}_0) = \det \|\partial r_i(t|\boldsymbol{r}_0)/\partial r_{0j}\|$ is the divergence (Jacobian).

Quantities $I(t|\boldsymbol{r}_0)$ and $j(t|\boldsymbol{r}_0)$ are related to each other. Indeed, substituting Eq. (1.105) for $I(\boldsymbol{r},t)$ in Eq. (1.104), we see that there exists the evolution integral

$$j(t|\boldsymbol{r}_0) = \frac{I_0(\boldsymbol{r}_0)}{I(t|\boldsymbol{r}_0)},$$

and Eq. (1.105) assumes the form

$$I(\boldsymbol{r},t) = \int d\boldsymbol{r}_0 I_0(\boldsymbol{r}_0)\, \delta(\boldsymbol{r}(t|\boldsymbol{r}_0) - \boldsymbol{r}).$$

Example. Hamilton–Jacobi Equation

In the case of function $H(\boldsymbol{r},t,q,\boldsymbol{p})$ specified as

$$H(\boldsymbol{r},t,q,\boldsymbol{p}) = \frac{1}{2}\boldsymbol{p}^2(\boldsymbol{r},t) + U(\boldsymbol{r},t),$$

Eqs. (1.102) correspond to the Hamilton equations

$$\frac{d}{dt}\boldsymbol{r}(t) = \boldsymbol{p}(t), \quad \frac{d}{dt}\boldsymbol{p}(t) = -\frac{\partial}{\partial \boldsymbol{r}}U(\boldsymbol{r},t), \quad \frac{d}{dt}q(t) = -U(\boldsymbol{r},t),$$

whereas Eq. (1.101) becomes the *Hamilton–Jacobi equation*

$$\frac{\partial}{\partial t}q(\boldsymbol{r},t) + U(\boldsymbol{r},t) + \boldsymbol{p}^2(\boldsymbol{r},t) = 0, \quad q(\boldsymbol{r},0) = q_0(\boldsymbol{r})$$

and function $\boldsymbol{p}(\boldsymbol{r},t) = \boldsymbol{\nabla} q(\boldsymbol{r},t)$ satisfies the quasilinear equation

$$\left(\frac{\partial}{\partial t} + \boldsymbol{p}(\boldsymbol{r},t)\frac{\partial}{\partial \boldsymbol{r}}\right)\boldsymbol{p}(\boldsymbol{r},t) + \frac{\partial}{\partial \boldsymbol{r}}U(\boldsymbol{r},t) = 0, \quad \boldsymbol{p}(\boldsymbol{r},0) = \boldsymbol{\nabla} q_0(\boldsymbol{r}). \qquad \blacklozenge$$

1.4 Partial Differential Equations of Higher Orders

1.4.1 *Fundamental Solutions of Wave Problems in Free Space and Layered Media*

Here we discuss several properties of fundamental solutions (Green's functions) of wave equations in free space and layered media following monograph [50] and papers [31, 52].

Free Space

First of all, we consider Green's function of the one-dimensional Helmholtz equation

$$\frac{d^2}{dx^2}g(x;x_0) + k^2 g(x;x_0) = \delta(x - x_0). \tag{1.106}$$

The solution of Eq. (1.106) satisfying radiation condition for $x \to \pm\infty$ has the form

$$g(x;x_0) = g(x - x_0) = \frac{1}{2ik}e^{ik|x-x_0|}. \tag{1.107}$$

The modulus $|x - x_0|$ appeared in the right-hand side of Eq. (1.107) by virtue of the fact that Eq. (1.106) is the equation of the second order in variable x. However, if we fix mutual order of observation points and source, then Green's function will satisfy the equality (for definiteness, we assume that $x_0 > x$)

$$\frac{\partial}{\partial x_0}g(x - x_0) = ikg(x - x_0)$$

that, being supplemented with the initial condition

$$g(x - x_0)|_{x_0=x} = g(0) = \frac{1}{2ik},$$

can be considered the first-order differential equation.

Thus, the order of the equation for Green's function decreases if source and observation points obey certain order. This property is generic of wave problems (*factorization property of wave equations*) and follows from the fact that the wave radiated in direction $x < x_0$ (or $x > x_0$) travels in free space without changing the direction.

In the general case, Green's function satisfies the second-order operator equation

$$\left\{\frac{\partial^2}{\partial x^2} + \widehat{M}^2(\boldsymbol{\eta})\right\} g(x - x_0, \boldsymbol{\eta} - \boldsymbol{\eta}_0) = \delta(x - x_0)g(\boldsymbol{\eta} - \boldsymbol{\eta}_0), \tag{1.108}$$

where operator $\widehat{M}(\boldsymbol{\eta})$ acts on the temporal and other spatial variables denoted by $\boldsymbol{\eta}$. For example, operator $\widehat{M}^2(\boldsymbol{\eta})$ in Eq. (1.106) is the number $\widehat{M}^2(\boldsymbol{\eta}) = k^2$.

Structurally, Green's function is similar to Eq. (1.107),

$$g(x - x_0, \boldsymbol{\eta} - \boldsymbol{\eta}_0) = e^{i|x-x_0|\widehat{M}(\boldsymbol{\eta})} g(0, \boldsymbol{\eta} - \boldsymbol{\eta}_0) = e^{i|x-x_0|\widehat{M}(-\boldsymbol{\eta}_0)} g(0, \boldsymbol{\eta} - \boldsymbol{\eta}_0). \tag{1.109}$$

As a consequence, it can be described for $x < x_0$ by the operator equation of the first order in variable x (or x_0)

$$\frac{\partial}{\partial x_0} g(x - x_0, \boldsymbol{\eta} - \boldsymbol{\eta}_0) = -\frac{\partial}{\partial x} g(x - x_0, \boldsymbol{\eta} - \boldsymbol{\eta}_0)$$
$$= i\widehat{M}(\boldsymbol{\eta}) g(x - x_0, \boldsymbol{\eta} - \boldsymbol{\eta}_0) = i\widehat{M}(-\boldsymbol{\eta}_0) g(x - x_0, \boldsymbol{\eta} - \boldsymbol{\eta}_0)$$

with the initial condition

$$g(x - x_0, \boldsymbol{\eta} - \boldsymbol{\eta}_0)|_{x_0 = x} = g(0, \boldsymbol{\eta} - \boldsymbol{\eta}_0) \equiv g(\boldsymbol{\eta} - \boldsymbol{\eta}_0).$$

For $x > x_0$, the equations are similar.

The solution of Eq. (1.108) is continuous in x, but its derivative with respect to x is discontinuous at the point of source location $x = x_0$

$$\left.\frac{\partial}{\partial x} g(x - x_0, \boldsymbol{\eta} - \boldsymbol{\eta}_0)\right|_{x = x_0 + 0} - \left.\frac{\partial}{\partial x} g(x - x_0, \boldsymbol{\eta} - \boldsymbol{\eta}_0)\right|_{x = x_0 - 0} = \delta(\boldsymbol{\eta} - \boldsymbol{\eta}_0). \tag{1.110}$$

Substituting Eq. (1.109) in Eq. (1.110), we obtain the expression

$$2i\widehat{M}(\boldsymbol{\eta}) g(0, \boldsymbol{\eta} - \boldsymbol{\eta}_0) = \delta(\boldsymbol{\eta} - \boldsymbol{\eta}_0). \tag{1.111}$$

In the general case, operator $\widehat{M}(\boldsymbol{\eta})$ can be considered as an integral operator. Indeed, action of operator $\widehat{M}(\boldsymbol{\eta})$ on arbitrary function $f(\boldsymbol{\eta})$ is representable in the form

$$\widehat{M}(\boldsymbol{\eta}) f(\boldsymbol{\eta}) = \int\limits_{-\infty}^{\infty} d\boldsymbol{\xi}\, \widehat{M}(\boldsymbol{\eta}) \delta(\boldsymbol{\eta} - \boldsymbol{\xi}) f(\boldsymbol{\xi}) = \int\limits_{-\infty}^{\infty} d\boldsymbol{\xi}\, M(\boldsymbol{\eta} - \boldsymbol{\xi}) f(\boldsymbol{\xi}),$$

where the kernel of the integral operator is defined by the equality

$$M(\boldsymbol{\eta} - \boldsymbol{\xi}) = \widehat{M}(\boldsymbol{\eta}) \delta(\boldsymbol{\eta} - \boldsymbol{\xi}). \tag{1.112}$$

The inverse operator $\widehat{M}^{-1}(\boldsymbol{\eta})$ also can be introduced by the corresponding choice of kernel $M^{-1}(\boldsymbol{\eta} - \boldsymbol{\xi})$.

Applying operator $\widehat{M}(\boldsymbol{\eta})$ to Eq. (1.111), we obtain, according to Eq. (1.112), the kernel of the integral operator in the form

$$M(\boldsymbol{\eta}-\boldsymbol{\eta}_0)=2i\widehat{M}^2(\boldsymbol{\eta})g(0,\boldsymbol{\eta}-\boldsymbol{\eta}_0). \tag{1.113}$$

The kernel of the inverse integral operator

$$M^{-1}(\boldsymbol{\eta}-\boldsymbol{\eta}_0)=\widehat{M}^{-1}(\boldsymbol{\eta})\delta(\boldsymbol{\eta}-\boldsymbol{\eta}_0)=2ig(0,\boldsymbol{\eta}-\boldsymbol{\eta}_0). \tag{1.114}$$

is obtained by applying the inverse operator $\widehat{M}^{-1}(\boldsymbol{\eta})$ to Eq. (1.111).

Thus, kernels of integral operators $\widehat{M}(\boldsymbol{\eta})$ and $\widehat{M}^{-1}(\boldsymbol{\eta})$ are expressed in terms of the wave equation fundamental solution.

Consider now several specific wave problems.

1. We represent the Helmholtz equation in the form

$$\left(\frac{\partial^2}{\partial x^2}+\Delta_{\boldsymbol{R}}+k^2\right)g(x-x_0,\boldsymbol{R}-\boldsymbol{R}_0)=\delta(x-x_0)\delta(\boldsymbol{R}-\boldsymbol{R}_0), \tag{1.115}$$

where vector $\boldsymbol{R}$ denotes the coordinates in the plane perpendicular to the x-axis.

The solution of Eq. (1.115) satisfying radiation conditions at infinity has the form

$$g(\boldsymbol{r}-\boldsymbol{r}_0)=-\frac{1}{4\pi|\boldsymbol{r}-\boldsymbol{r}_0|}e^{ik|\boldsymbol{r}-\boldsymbol{r}_0|},\quad \boldsymbol{r}=\{x,\boldsymbol{R}\}.$$

Function $g(\boldsymbol{r})$ can be represented in the integral form

$$g(x,\boldsymbol{R})=\frac{1}{8i\pi^2}\int\frac{d\boldsymbol{q}}{\sqrt{k^2-\boldsymbol{q}^2}}\exp\left\{i\sqrt{k^2-\boldsymbol{q}^2}|x|+i\boldsymbol{q}\boldsymbol{R}\right\},$$

from which follows that operator $\widehat{M}(\boldsymbol{R})$ has in this case the form

$$\widehat{M}(\boldsymbol{R})=\sqrt{k^2+\Delta_{\boldsymbol{R}}},\quad \widehat{M}(\boldsymbol{R}_0)=\sqrt{k^2+\Delta_{\boldsymbol{R}_0}},$$

and the corresponding kernels of the integral operators are given, according to Eq. (1.113) and (1.114), by the expressions

$$\begin{aligned}M(\boldsymbol{R})&=2i\left(k^2+\Delta_{\boldsymbol{R}}\right)g(\boldsymbol{R})=-\frac{i}{2\pi R^2}\left(\frac{1}{R}-ik\right)e^{ikR},\\ M^{-1}(\boldsymbol{R})&=-\frac{i}{2\pi R}e^{ikR}.\end{aligned} \tag{1.116}$$

In the two-dimensional case, we have

$$g(\boldsymbol{r}-\boldsymbol{r}_0)=-\frac{i}{4}H_0^{(1)}\left(k|\boldsymbol{r}-\boldsymbol{r}_0|\right)\quad(\boldsymbol{r}=\{x,y\}),$$

where $H_0^{(1)}(k|\boldsymbol{r}|)$ is the Hankel function. As a consequence, kernels of the corresponding integral operators

$$\widehat{M}(y) = \sqrt{k^2 + \frac{\partial^2}{\partial y^2}}, \quad \widehat{M}^{-1}(y) = \frac{1}{\sqrt{k^2 + \dfrac{\partial^2}{\partial y^2}}}$$

are given by the expressions

$$M(y) = \frac{k}{2|y|} H_1^{(1)}(k|y|), \quad M^{-1}(y) = \frac{1}{2} H_0^{(1)}(k|y|). \tag{1.117}$$

As we mentioned earlier, in the one-dimensional case, operators $\widehat{M}$ and $\widehat{M}^{-1}$ are simply the numbers.

2. We represent the nonstationary wave equation in the form of Eq. (1.108)

$$\left(\frac{\partial^2}{\partial x^2} + \Delta_{\boldsymbol{R}} - \frac{1}{c^2}\frac{\partial^2}{\partial t^2}\right) g(x - x_0, \boldsymbol{R} - \boldsymbol{R}_0, t - t_0) = \delta(x - x_0)\delta(\boldsymbol{R} - \boldsymbol{R}_0)\delta(t - t_0). \tag{1.118}$$

In this case, operator $\widehat{M}^2(\boldsymbol{\eta})$ is the differential operator

$$\widehat{M}^2(\boldsymbol{R}, t) = \Delta_{\boldsymbol{R}} - \frac{1}{c^2}\frac{\partial^2}{\partial t^2}.$$

In the three-dimensional case, the solution of Eq. (1.118) satisfying radiation conditions (the retarded solution) has the form

$$g(x, \boldsymbol{R}, t) = -\frac{c}{2\pi}\theta(t)\delta\left(c^2t^2 - x^2 - \boldsymbol{R}^2\right),$$

where $\theta(t)$ is the Heaviside step function . As a consequence, kernels of the corresponding integral operators are given by the formulas

$$M(\boldsymbol{R}, t) = \frac{i}{\pi ct}\theta(t)\frac{\partial}{\partial t}\delta\left(c^2t^2 - \boldsymbol{R}^2\right), \quad M^{-1}(\boldsymbol{R}, t) = -\frac{ic}{\pi}\theta(t)\delta\left(c^2t^2 - \boldsymbol{R}^2\right). \tag{1.119}$$

In the two-dimensional case,

$$g(x, y, t) = -\frac{c}{2\pi}\frac{\theta\left(ct - \sqrt{x^2 + y^2}\right)}{\sqrt{c^2t^2 - x^2 - y^2}} = -\frac{c}{2\pi}\frac{\theta\left(c^2t^2 - x^2 - y^2\right)}{\sqrt{c^2t^2 - x^2 - y^2}}$$

and, consequently,

$$M(y, t) = \frac{i}{\pi ct}\theta(t)\frac{\partial}{\partial t}\frac{\theta\left(c^2t^2 - y^2\right)}{\sqrt{c^2t^2 - y^2}}, \quad M^{-1}(y, t) = -\frac{ic}{\pi}\frac{\theta(ct - y)}{\sqrt{c^2t^2 - y^2}}. \tag{1.120}$$

In the one-dimensional case,

$$g(x, t) = -\frac{c}{2}\theta(ct - |x|)$$

and, consequently,

$$M(t) = \frac{i}{c}\delta'(t), \quad M^{-1}(t) = -ic\theta(t). \tag{1.121}$$

Here, we considered certain properties of the fundamental solutions (Green's functions) of wave equations describing the field of the point source in the unbounded free space. Note that the similar analysis for problems on the point source field in a finite layer of free or layered space differs from the above analysis only in insignificant details.

Layered Space

For a layered medium in which $\varepsilon(x, y, z) = \varepsilon(z)$, wave equations can be factorized because waves spread in plane (x, y) and do not scatter in the backward direction.

We denote $G^{(1)}(z; z_0)$ the point source field in the one-dimensional space. This function satisfies the equation

$$\left[\frac{d^2}{dz^2} + k^2(z)\right] G^{(1)}(z; z_0) = \delta(z - z_0),$$

whose solution can be represented in the operator form,

$$G^{(1)}(z; z_0) = \widehat{L}^{-2}(z)\delta(z - z_0),$$

where

$$\widehat{L}^2(z) = \frac{d^2}{dz^2} + k^2(z).$$

In the two-dimensional space, the wave field of the point source is described by Green's function $G^{(2)}(x, z; z_0)$ satisfying the equation

$$\left[\frac{\partial^2}{\partial x^2} + \widehat{L}^2(z)\right] G^{(2)}(x, z; z_0) = \delta(x)\delta(z - z_0).$$

The solution of this equation has the form

$$G^{(2)}(x, z; z_0) = e^{i|x|\widehat{L}(z)} G^{(2)}(0, z; z_0), \tag{1.122}$$

where function $G^{(2)}(0, z; z_0)$ describes the wave field on axis $x = 0$. The discontinuity of derivative $\dfrac{\partial}{\partial x} G^{(2)}(x, z; z_0)$ at $x = 0$ is given by the expression

$$\left.\frac{\partial}{\partial x} G^{(2)}(x, z; z_0)\right|_{x=+0} - \left.\frac{\partial}{\partial x} G^{(2)}(x, z; z_0)\right|_{x=-0} = \delta(z - z_0).$$

Being combined with Eq. (1.122), this discontinuity yields the equality

$$2i\widehat{L}(z)G^{(2)}(0,z;z_0) = \delta(z - z_0), \tag{1.123}$$

from which follows that

$$G^{(2)}(0,z;z_0) = \frac{1}{2i}\widehat{L}^{-1}(z)\delta(z - z_0). \tag{1.124}$$

Applying now operator $\widehat{L}^2(z)$ to Eq. (1.123), we obtain the equality

$$\widehat{L}^2(z)G^{(2)}(0,z;z_0) = \frac{1}{2i}\widehat{L}(z)\delta(z - z_0). \tag{1.125}$$

We can consider operators $\widehat{L}(z)$ and $\widehat{L}^{-1}(z)$ as the integral operators; in this case, Eq. (1.125), (1.124) define the kernels of these operators. With this fact in mind, we see that Eq. (1.124) is the nonlinear integral equation in function $G^{(2)}(0,z;z_0)$ describing the wave field on axis $x = 0$,

$$\int\limits_{-\infty}^{\infty} d\xi G^{(2)}(0,z;\xi)G^{(2)}(0,\xi;z_0) = -\frac{1}{4}G^{(1)}(z;z_0),$$

where $G^{(1)}(z;z_0)$ is Green's function of the one-dimensional problem.

In the three-dimensional case, Green's function of layered medium satisfies the equation

$$\left[\frac{\partial^2}{\partial x^2}+\frac{\partial^2}{\partial y^2}+\widehat{L}^2(z)\right] G^{(3)}(x,y,z;z_0) = \delta(x)\delta(y)\delta(z - z_0).$$

We represent the solution to this equation in the form

$$G^{(3)}(x,y,z;z_0) = e^{i|x|\widehat{L}(y,z)}G^{(3)}(0,y,z;z_0), \tag{1.126}$$

where $\widehat{L}(y,z) = \dfrac{\partial^2}{\partial y^2} + \widehat{L}^2(z)$ and function $G^{(3)}(0,y,z;z_0)$ describes the wave field in plane (y,z). The condition of discontinuity of derivative $\dfrac{\partial}{\partial x}G^{(3)}(x,y,z;z_0)$ in plane $x = 0$ yields the operator equality

$$G^{(3)}(0,y,z;z_0) = \frac{1}{2i}\widehat{L}^{-1}(y,z)\delta(y)\delta(z - z_0),$$

which can be rewritten in terms of the Hankel function of the first kind

$$G^{(3)}(0,y,z;z_0) = -\frac{i}{4}H_0^{(1)}\left[y\widehat{L}(z)\right]\delta(z - z_0).$$

Using the Hankel function integral representation

$$H_0^{(1)}(\beta\mu) = \frac{1}{i\pi}\int_0^\infty \frac{dx}{x}\exp\left\{i\frac{\mu}{2}\left(x+\frac{\beta^2}{x}\right)\right\},$$

we obtain that function $G^{(3)}(0, y, z; z_0)$ is related to the solution of the parabolic equation

$$\frac{\partial}{\partial t}u(t, z; z_0) = \frac{i}{2k}\widehat{L}^2(z)u(t, z; z_0), \quad u(0, z; z_0) = \delta(z - z_0)$$

with respect to auxiliary parameter t by the quadrature

$$G^{(3)}(0, y, z; z_0) = -\frac{1}{4\pi}\int_0^\infty \frac{dt}{t}\exp\left[i\frac{k}{2t}y^2\right]u(t, z; z_0),$$

or by the expression

$$G^{(3)}(0, y, z; z_0) = -\frac{1}{4\pi}\int_0^\infty \frac{dt}{t}\exp\left\{i\frac{k}{2t}\left(y^2 + t^2\right)\right\}\psi(t, z; z_0),$$

where function $\psi(t, z; z_0)$ is the solution to the parabolic equation

$$\frac{\partial}{\partial t}\psi(t, z; z_0) = \frac{i}{2k}\left[\frac{\partial^2}{\partial z^2} + k^2(z) - k^2\right]\psi(t, z; z_0),$$
$$\psi(0, z; z_0) = \delta(z - z_0). \tag{1.127}$$

In view of arbitrary direction of the x-axis, we obtain that, for $y > 0$, function

$$G^{(3)}(x, y, z; z_0) = G^{(3)}(\rho, z; z_0),$$

where $\rho^2 = x^2 + y^2$, defines Green's function in the whole of the space,

$$G^{(3)}(x, y, z; z_0) = -\frac{1}{4\pi}\int_0^\infty \frac{dt}{t}\exp\left\{i\frac{k}{2t}\left(x^2 + y^2 + t^2\right)\right\}\psi(t, z; z_0). \tag{1.128}$$

Integrating Eq. (1.128) first over y and x, we obtain the corresponding integral representations of two- and one-dimensional Green's functions

$$G^{(2)}(x, z; z_0) = \frac{1}{2ik}\left(\frac{k}{2\pi i}\right)^{1/2}\int_0^\infty \frac{dt}{\sqrt{t}}\exp\left\{i\frac{k}{2t}\left(x^2 + t^2\right)\right\}\psi(t, z; z_0), \tag{1.129}$$

$$G^{(1)}(z;z_0) = \frac{1}{2ik}\int_0^\infty dt \exp\left\{i\frac{kt}{2}\right\}\psi(t,z;z_0). \tag{1.130}$$

1.4.2 Stationary Problems for Maxwell's Equations

In the steady inhomogeneous medium, propagation of a monochromatic electromagnetic wave of frequency ω is described by *Maxwell's equations* (see, e.g., [130])

$$\operatorname{curl}\boldsymbol{E}(\boldsymbol{r}) = ik\boldsymbol{H}(\boldsymbol{r}),\ \operatorname{curl}\boldsymbol{H}(\boldsymbol{r}) = -ik\varepsilon(\boldsymbol{r})\boldsymbol{E}(\boldsymbol{r}),\ \operatorname{div}\{\varepsilon(\boldsymbol{r})\boldsymbol{E}(\boldsymbol{r})\} = 0, \tag{1.131}$$

where $\boldsymbol{E}(\boldsymbol{r})$ and $\boldsymbol{H}(\boldsymbol{r})$ are the electric and nondivergent magnetic strengths, $\varepsilon(\boldsymbol{r})$ is the dielectric permittivity of the medium, and $k = \omega/c = 2\pi/\lambda$ is the wave number (λ is the wavelength and c is the velocity of wave propagation). Here, we assumed that magnetic permeability $\mu = 1$, medium conductivity $\sigma = 0$, and specified temporal factor $e^{-i\omega t}$ for all fields.

Equations (1.131) can be rewritten in the form of the equation closed in the electric field $\boldsymbol{E}(\boldsymbol{r})$

$$\left[\Delta + k^2\varepsilon(\boldsymbol{r})\right]\boldsymbol{E}(\boldsymbol{r}) = -\boldsymbol{\nabla}\left(\boldsymbol{E}(\boldsymbol{r})\boldsymbol{\nabla}\ln\varepsilon(\boldsymbol{r})\right). \tag{1.132}$$

In this case, the magnetic field $\boldsymbol{H}(\boldsymbol{r})$ is calculated by the equality

$$\boldsymbol{H}(\boldsymbol{r}) = \frac{1}{ik}\operatorname{curl}\boldsymbol{E}(\boldsymbol{r}). \tag{1.133}$$

We restrict ourselves with electromagnetic wave propagation in media with weakly fluctuating dielectric permittivity. We set

$$\varepsilon(\boldsymbol{r}) = 1 + \varepsilon_1(\boldsymbol{r}),$$

where $\varepsilon_1(\boldsymbol{r})$ stands for small fluctuations of dielectric permittivity ($\langle\varepsilon_1(\boldsymbol{r})\rangle = 0$). Smallness of fluctuations $\varepsilon_1(\boldsymbol{r})$ assumes that

$$\langle|\varepsilon_1(\boldsymbol{r})|\rangle \ll 1\,.$$

With this assumption, Eq. (1.132) can be rewritten in the simplified form

$$\left[\Delta + k^2\right]\boldsymbol{E}(\boldsymbol{r}) = -k^2\varepsilon_1(\boldsymbol{r})\boldsymbol{E}(\boldsymbol{r}) - \boldsymbol{\nabla}\left(\boldsymbol{E}(\boldsymbol{r})\boldsymbol{\nabla}\varepsilon_1(\boldsymbol{r})\right). \tag{1.134}$$

Using the theory of perturbations, Tatarskii [128] and Kravtsov [77] estimated light wave depolarization at propagation paths of about 1 km in the conditions of the actual atmosphere and showed that depolarization is very small. In these conditions, we can neglect the last term in the right-hand side of Eq. (1.134). As a result, the problem reduces in fact to the scalar *Helmholtz equation*

$$\left[\Delta + k^2\right] U(\boldsymbol{r}) = -k^2 \varepsilon_1(\boldsymbol{r}) U(\boldsymbol{r}). \tag{1.135}$$

For Eq. (1.135) be meaningful, one must formulate boundary conditions and specify the source of radiation.

1.4.3 The Helmholtz Equation (Boundary-Value Problem) and the Parabolic Equation of Quasi-Optics (Waves in Randomly Inhomogeneous Media)

Let the layer of inhomogeneous medium occupies spatial segment $L_0 < x < L$ and let the point source is located at point $(x_0, \boldsymbol{R}_0)$, where $\boldsymbol{R}_0$ stands for the coordinates in the plane perpendicular to the x-axis. In this case, the field inside the layer $G(x, \boldsymbol{R}; x_0, \boldsymbol{R}_0)$ satisfies the equation for Green's function

$$\left\{\frac{\partial^2}{\partial x^2} + \Delta_{\boldsymbol{R}} + k^2\left[1 + \varepsilon(x, \boldsymbol{R})\right]\right\} G(x, \boldsymbol{R}; x_0, \boldsymbol{R}_0) = \delta(x - x_0)\,\delta(\boldsymbol{R} - \boldsymbol{R}_0), \tag{1.136}$$

where k is the wave number, $\Delta_{\boldsymbol{R}} = \partial^2/\partial \boldsymbol{R}^2$, and $\varepsilon_1(\boldsymbol{r}) = \varepsilon(x, \boldsymbol{R})$ is the deviation of the refractive index (or dielectric permittivity) from unity.

Let $\varepsilon(x, \boldsymbol{R}) = 0$ outside the layer. Then, the wavefield outside the layer satisfies the Helmholtz equation

$$\left\{\frac{\partial^2}{\partial x^2} + \Delta_{\boldsymbol{R}} + k^2\right\} G(x, \boldsymbol{R}; x_0, \boldsymbol{R}_0) = 0,$$

and continuity conditions for functions G and $\partial G/\partial x$ at layer boundaries. Furthermore, the solution to Eq. (1.136) must satisfy the radiation conditions for $x \to \pm\infty$.

The wavefield outside the layer can obviously be represented in the form

$$G(x, \boldsymbol{R}; x_0, \boldsymbol{R}_0) = \begin{cases} \int d\boldsymbol{q}\, T_1(\boldsymbol{q}) \exp\left[-i\sqrt{k^2 - q^2}\,(x - L_0) + i\boldsymbol{q}\boldsymbol{R}\right], & x \le L_0; \\ \int d\boldsymbol{q}\, T_2(\boldsymbol{q}) \exp\left[i\sqrt{k^2 - q^2}\,(x - L) + i\boldsymbol{q}\boldsymbol{R}\right], & x \ge L. \end{cases}$$

Consequently, the boundary condition for Eq. (1.136) at $x = L_0$ can be written as

$$\left(\frac{\partial}{\partial x} + i\sqrt{k^2 + \Delta_{\boldsymbol{R}}}\right) G(x, \boldsymbol{R}; x_0, \boldsymbol{R}_0)\Big|_{x=L_0} = 0. \tag{1.137}$$

Similarly, the boundary condition at $x = L$ has the form

$$\left(\frac{\partial}{\partial x} - i\sqrt{k^2 + \Delta_{\boldsymbol{R}}}\right) G\left(x, \boldsymbol{R}; x_0, \boldsymbol{R}_0\right)\Big|_{x=L} = 0. \qquad (1.138)$$

In the case of space infinite in coordinates $\boldsymbol{R}$, operator $\sqrt{k^2 + \Delta_{\boldsymbol{R}}}$ appeared in Eqs. (1.137), (1.138) can be defined in terms of the *Fourier transform*. Alternatively, this operator can be also treated as the linear integral operator whose kernel is expressed in terms of Green's function for free space (see sect. 1.4.1).

Thus, the field of the point source in inhomogeneous medium is described by the boundary-value problem (1.136)–(1.138). This problem is equivalent to the integral equation

$$G\left(x, \boldsymbol{R}; x_0, \boldsymbol{R}_0\right) = g\left(x - x_0, \boldsymbol{R} - \boldsymbol{R}_0\right)$$
$$+ \int\limits_{L_0}^{L} dx' \int d\boldsymbol{R}' g\left(x - x', \boldsymbol{R} - \boldsymbol{R}'\right) \varepsilon\left(x', \boldsymbol{R}'\right) G\left(x', \boldsymbol{R}'; x_0, \boldsymbol{R}_0\right), \quad (1.139)$$

where $g(x, \boldsymbol{R})$ is Green's function in the free space. In the three-dimensional case, we have

$$g\left(x, \boldsymbol{R}\right) = -\frac{1}{4\pi r} e^{ikr}, \quad r = \sqrt{x^2 + \boldsymbol{R}^2}.$$

The integral representation of this Green's function is as follows

$$g\left(x, \boldsymbol{R}\right) = \int d\boldsymbol{q}\, g\left(\boldsymbol{q}\right) \exp\left\{i\sqrt{k^2 - q^2}\left|x\right| + i\boldsymbol{q}\boldsymbol{R}\right\},$$
$$g\left(\boldsymbol{q}\right) = \frac{1}{8i\pi^2\sqrt{k^2 - q^2}}. \qquad (1.140)$$

It can be shown that operator $\sqrt{k^2 + \Delta_{\boldsymbol{R}}}$ applied to arbitrary function $F(\boldsymbol{R})$ acts as the integral operator

$$\sqrt{k^2 + \Delta_{\boldsymbol{R}}} F\left(\boldsymbol{R}\right) = \int d\boldsymbol{R}' K\left(\boldsymbol{R} - \boldsymbol{R}'\right) F\left(\boldsymbol{R}'\right) \qquad (1.141)$$

whose kernel is

$$K\left(\boldsymbol{R} - \boldsymbol{R}'\right) = \sqrt{k^2 + \Delta_{\boldsymbol{R}}}\,\delta\left(\boldsymbol{R} - \boldsymbol{R}'\right) = 2i\left(k^2 + \Delta_{\boldsymbol{R}}\right) g\left(0, \boldsymbol{R} - \boldsymbol{R}'\right). \qquad (1.142)$$

The corresponding kernel of the inverse operator is

$$L\left(\boldsymbol{R} - \boldsymbol{R}'\right) = \left(k^2 + \Delta_{\boldsymbol{R}}\right)^{-1/2} \delta\left(\boldsymbol{R} - \boldsymbol{R}'\right) = 2ig\left(0, \boldsymbol{R} - \boldsymbol{R}'\right). \qquad (1.143)$$

If the point source resides at the layer boundary $x_0 = L$, then the wavefield inside the layer $L_0 < x < L$ satisfies the equation

$$\left\{\frac{\partial^2}{\partial x^2}+\Delta_{\boldsymbol{R}}+k^2\left[1+\varepsilon\left(x,\boldsymbol{R}\right)\right]\right\}G\left(x,\boldsymbol{R};L,\boldsymbol{R}_0\right)=0 \tag{1.144}$$

with the boundary conditions following from conditions (1.137), (1.138)

$$\begin{aligned}\left(\frac{\partial}{\partial x}+i\sqrt{k^2+\Delta_{\boldsymbol{R}}}\right)G\left(x,\boldsymbol{R};L,\boldsymbol{R}_0\right)\Big|_{x=L_0}&=0,\\ \left(\frac{\partial}{\partial x}-i\sqrt{k^2+\Delta_{\boldsymbol{R}}}\right)G\left(x,\boldsymbol{R};L,\boldsymbol{R}_0\right)\Big|_{x=L}&=-\delta(\boldsymbol{R}-\boldsymbol{R}_0).\end{aligned} \tag{1.145}$$

Boundary-value problem (1.144), (1.145) can be reduced to an equivalent integral equation

$$\begin{aligned}G\left(x,\boldsymbol{R};L,\boldsymbol{R}_0\right)&=g\left(x-L,\boldsymbol{R}-\boldsymbol{R}_0\right)\\ &+\int\limits_{L_0}^{L}dx'\int d\boldsymbol{R}'g\left(x-x',\boldsymbol{R}-\boldsymbol{R}'\right)\varepsilon\left(x',\boldsymbol{R}'\right)G\left(x',\boldsymbol{R}';L,\boldsymbol{R}_0\right)\end{aligned} \tag{1.146}$$

coinciding with Eq. (1.139) for $x_0=L$.

If the wave $u_0(x,\boldsymbol{R})$ is incident on the layer from region $x>L$ (in the negative direction of the x-axis), then the wavefield $U(x,\boldsymbol{R})$ inside the layer satisfies the Helmholtz equation

$$\left\{\frac{\partial^2}{\partial x^2}+\Delta_{\boldsymbol{R}}+k^2\left[1+\varepsilon\left(x,\boldsymbol{R}\right)\right]\right\}U\left(x,\boldsymbol{R}\right)=0, \tag{1.147}$$

with the boundary conditions

$$\begin{aligned}\left(\frac{\partial}{\partial x}+i\sqrt{k^2+\Delta_{\boldsymbol{R}}}\right)U\left(x,\boldsymbol{R}\right)\Big|_{x=L_0}&=0,\\ \left(\frac{\partial}{\partial x}-i\sqrt{k^2+\Delta_{\boldsymbol{R}}}\right)U\left(x,\boldsymbol{R}\right)\Big|_{x=L}&=-2i\sqrt{k^2+\Delta_{\boldsymbol{R}}}u_0\left(L,\boldsymbol{R}\right).\end{aligned} \tag{1.148}$$

Similarly to the one-dimensional case, we can represent field $U(x,\boldsymbol{R})$ in the form

$$\begin{aligned}U\left(x,\boldsymbol{R}\right)&=u_1\left(x,\boldsymbol{R}\right)+u_2\left(x,\boldsymbol{R}\right),\\ \frac{\partial}{\partial x}U\left(x,\boldsymbol{R}\right)&=-ik\sqrt{k^2+\Delta_{\boldsymbol{R}}}\left\{u_1\left(x,\boldsymbol{R}\right)+u_2\left(x,\boldsymbol{R}\right)\right\},\end{aligned} \tag{1.149}$$

where we replaced function $U(x,\boldsymbol{R})$ with the sum of two functions $u_1(x,\boldsymbol{R})$ and $u_2(x,\boldsymbol{R})$ corresponding to the waves propagating in the negative and positive directions of the x-axis, respectively. These functions are related to field $U(x,\boldsymbol{R})$ through the expressions

$$u_1(x,\boldsymbol{R}) = \frac{i}{2\sqrt{k^2+\Delta_{\boldsymbol{R}}}}\left(\frac{\partial}{\partial x} - i\sqrt{k^2+\Delta_{\boldsymbol{R}}}\right)U(x,\boldsymbol{R}),$$
$$u_2(x,\boldsymbol{R}) = -\frac{i}{2\sqrt{k^2+\Delta_{\boldsymbol{R}}}}\left(\frac{\partial}{\partial x} + i\sqrt{k^2+\Delta_{\boldsymbol{R}}}\right)U(x,\boldsymbol{R}) \tag{1.150}$$

following from Eq. (1.149).

Differentiating Eq. (1.150) with respect to x and using Eq. (1.147), we obtain the system of equations in functions $u_1(x,\boldsymbol{R})$ and $u_2(x,\boldsymbol{R})$; the corresponding boundary conditions are derived from Eq. (1.148)

$$\left(\frac{\partial}{\partial x} + i\sqrt{k^2+\Delta_{\boldsymbol{R}}}\right)u_1(x,\boldsymbol{R}) = -\frac{ik^2}{2\sqrt{k^2+\Delta_{\boldsymbol{R}}}}\left\{\varepsilon(x,\boldsymbol{R})U(x,\boldsymbol{R})\right\},$$
$$\left(\frac{\partial}{\partial x} - i\sqrt{k^2+\Delta_{\boldsymbol{R}}}\right)u_2(x,\boldsymbol{R}) = \frac{ik^2}{2\sqrt{k^2+\Delta_{\boldsymbol{R}}}}\left\{\varepsilon(x,\boldsymbol{R})U(x,\boldsymbol{R})\right\},$$
$$u_1(L,\boldsymbol{R}) = u_0(L,\boldsymbol{R}), \qquad u_2(L_0,\boldsymbol{R}) = 0. \tag{1.151}$$

Function $u_2(x,\boldsymbol{R})$ describes the wave propagating in the direction inverse to the direction of the incident wave, i.e., it describes the *backscattered field*.

Neglecting the backscattering effects, i.e., setting $u_2(x,\boldsymbol{R}) = 0$, we obtain the *generalized parabolic equation*

$$\left(\frac{\partial}{\partial x} + i\sqrt{k^2+\Delta_{\boldsymbol{R}}}\right)U(x,\boldsymbol{R}) = -\frac{ik^2}{2\sqrt{k^2+\Delta_{\boldsymbol{R}}}}\left\{\varepsilon(x,\boldsymbol{R})U(x,\boldsymbol{R})\right\},$$
$$U(L,\boldsymbol{R}) = u_0(L,\boldsymbol{R}). \tag{1.152}$$

valid for waves scattered by arbitrary angles (less than $\pi/2$). In the case of small-angle scattering ($\Delta_{\boldsymbol{R}} \ll k^2$), we represent field $U(x,\boldsymbol{R})$ in the form

$$U(x,\boldsymbol{R}) = e^{-ik(x-L)}u(x,\boldsymbol{R}).$$

If we assume that the wave is incident on the inhomogeneous medium from half-space $x < 0$ (i.e., if we replace $L - x$ with x), then Eq. (1.152) reduces to the *parabolic equation of quasi-optics*,

$$\frac{\partial}{\partial x}u(x,\boldsymbol{R}) = \frac{i}{2k}\Delta_{\boldsymbol{R}}u(x,\boldsymbol{R}) + \frac{ik}{2}\varepsilon(x,\boldsymbol{R})u(x,\boldsymbol{R}), \qquad u(0,\boldsymbol{R}) = u_0(\boldsymbol{R}), \tag{1.153}$$

which concerns the wave propagation in media with large-scale three-dimensional inhomogeneities responsible for small-angle scattering. It was successfully used in many problems on wave propagation in Earth's atmosphere and ocean.

There is the waste literature on derivation and basing of both parabolic and generalized parabolic equations. Appendix of Vol. 2 gives such a derivation in terms of the imbedding method.

Introducing the amplitude-phase representation of the wavefield in Eq. (1.153) by the formula

$$u(x,\boldsymbol{R}) = A(x,\boldsymbol{R})e^{iS(x,\boldsymbol{R})},$$

we can write the equation for the wavefield intensity $I(x,\boldsymbol{R}) = |u(x,\boldsymbol{R})|^2$ in the form

$$\frac{\partial}{\partial x}I(x,\boldsymbol{R}) + \frac{1}{k}\boldsymbol{\nabla}_{\boldsymbol{R}}\left\{\boldsymbol{\nabla}_{\boldsymbol{R}}S(x,\boldsymbol{R})I(x,\boldsymbol{R})\right\} = 0, \qquad I(0,\boldsymbol{R}) = I_0(\boldsymbol{R}). \quad (1.154)$$

From this equation follows that the power of a wave in plane $x = \text{const}$ is conserved in the general case of arbitrary incident wave beam:

$$E_0 = \int I(x,\boldsymbol{R})d\boldsymbol{R} = \int I_0(\boldsymbol{R})d\boldsymbol{R}.$$

Equation (1.154) coincides in form with Eq. (1.56), page 34. Consequently, we can treat it as the equation of transfer of conservative tracer in the potential velocity field. However, this tracer can be considered the passive tracer only in the *geometrical optics approximation*, in which case the phase of the wave, the transverse gradient of the phase

$$\boldsymbol{p}(x,\boldsymbol{R}) = \frac{1}{k}\boldsymbol{\nabla}_{\boldsymbol{R}}S(x,\boldsymbol{R}),$$

and the matrix of the phase second derivatives

$$u_{ij}(x,\boldsymbol{R}) = \frac{1}{k}\frac{\partial^2}{\partial R_i \partial R_j}S(x,\boldsymbol{R})$$

characterizing the *curvature of the phase front* $S(x,\boldsymbol{R}) = \text{const}$ satisfy the closed system of equations [48,49]

$$\begin{gathered}
\frac{\partial}{\partial x}S(x,R) + \frac{k}{2}p^2(x,R) = \frac{k}{2}\varepsilon(x,R), \\
\left(\frac{\partial}{\partial x} + \boldsymbol{p}(x,\boldsymbol{R})\boldsymbol{\nabla}_{\boldsymbol{R}}\right)p(x,R) = \frac{1}{2}\nabla_{\boldsymbol{R}}\varepsilon(x,R), \\
\left(\frac{\partial}{\partial x} + \boldsymbol{p}(x,\boldsymbol{R})\boldsymbol{\nabla}_{\boldsymbol{R}}\right)u_{ij}(x,\boldsymbol{R}) + u_{ik}(x,\boldsymbol{R})u_{kj}(x,\boldsymbol{R}) = \frac{1}{2}\frac{\partial^2}{\partial R_i \partial R_j}\varepsilon(x,\boldsymbol{R}).
\end{gathered} \quad (1.155)$$

In the general case, i.e., with the inclusion of diffraction effects, this tracer becomes the active tracer.

According to the material of the previous section, realizations of intensity must show cluster behavior, which manifests itself in the appearance of *caustic* structures. An example demonstrating the appearance of wavefield caustic structures is given in Fig. 1.19, which is a fragment of the photo on the back of the cover — the flyleaf — of book [122] that shows the transverse section of the laser beam propagating in turbulent atmosphere (see also papers [18, 19, 30] and Chapter 20, page 272, Vol. 2 for the results of laboratory investigations and simulations).

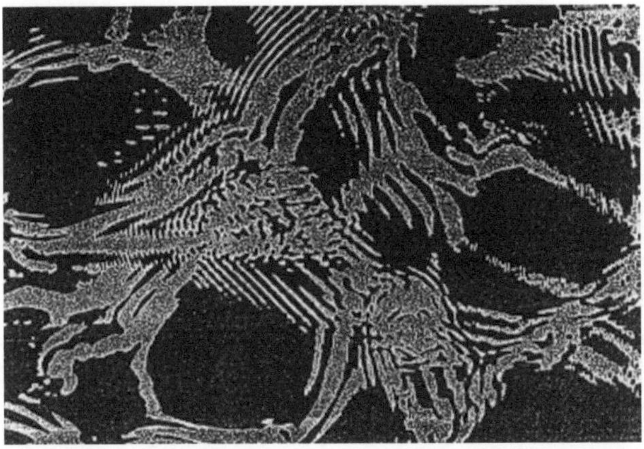

Fig. 1.19 Transverse section of a laser beam in turbulent medium

We mention that Eq. (1.153) with parameter x considered as time t coincides in form with the *Schrödinger equation* with a random potential. In a similar way, the *nonlinear parabolic equation* describing self-action of a harmonic wave field in multidimensional random media,

$$\frac{\partial}{\partial x}u(x,\boldsymbol{R}) = \frac{i}{2k}\Delta_{\boldsymbol{R}}u(x,\boldsymbol{R}) + \frac{ik}{2}\varepsilon(x,\boldsymbol{R};I(x,\boldsymbol{R}))u(x,\boldsymbol{R}), \quad u(0,\boldsymbol{R}) = u_0(\boldsymbol{R}) \tag{1.156}$$

coincides in form with the *nonlinear Schrödinger equation*. Consequently, clustering of wave field energy must occur in this case too, because Eq. (1.154) is formally independent of the shape of function $\varepsilon(x,\boldsymbol{R};I(x,\boldsymbol{R}))$. In particular, random initial condition $u_0(\boldsymbol{R})$ yields the caustic structure formation even if $\varepsilon(x,\boldsymbol{R}) = 0$. In this case, Eqs. (1.153) and (1.156) are simplified and assume the form

$$\frac{\partial}{\partial x}u(x,\boldsymbol{R}) = \frac{i}{2k}\Delta_{\boldsymbol{R}}u(x,\boldsymbol{R}), \quad u(0,\boldsymbol{R}) = u_0(\boldsymbol{R}) \tag{1.157}$$

that allows an analytical solution,

$$u(x,\boldsymbol{R}) = \exp\left\{\frac{ix}{2k}\Delta_{\boldsymbol{R}}\right\}u_0(\boldsymbol{R}) = \frac{k}{2\pi ix}\int d\boldsymbol{R}'\exp\left\{\frac{ik}{2x}(\boldsymbol{R}-\boldsymbol{R}')^2\right\}u_0(\boldsymbol{R}'). \tag{1.158}$$

For the plane incident wave, the initial condition to Eqs. (1.157) has the form $|u_0(\boldsymbol{R})| = 1$, i.e., $u_0(\boldsymbol{R}) = e^{iS_0(\boldsymbol{R})}$, where $S_0(\boldsymbol{R})$ is the field of the random initial phase. Here, the spatial fluctuations in the initial distribution of the phase are transformed into the caustic structure of the wave field intensity (*random phase screen*), which is well known and is regularly observed both in water pools and shallow waters (see fig. 1.20).

Fig. 1.20 Caustics in a pool and in a sea

For more detailed statistical description of this phenomenon, see Chapter 20, page 271 in Volume 2.

Interest in equation (1.153), (1.156) and (1.157) is motivated by the problem of so-called *rogue waves* (see e.g. [39]). These waves, undoubtedly, arise because of water mass clustering.

Replacing the imaginary-valued time *it* with t in the Schrödinger equation with a random potential $z(\boldsymbol{r},t)$, we arrive at the real diffusion equation

$$\frac{d}{dt}f(\boldsymbol{r},t) = z(\boldsymbol{r},t)f(\boldsymbol{r},t) + \mu_f \Delta f(\boldsymbol{r},t), \quad f(\boldsymbol{r},0) = f_0(\boldsymbol{r}), \tag{1.159}$$

where μ_f is the dynamic diffusivity coefficient for the field $f(\boldsymbol{r},t)$. It is noteworthy that Eq. (1.159), containing the terms responsible for random reproduction and diffusion, is also relevant for problems occurring in biology and the kinetics of chemical and nuclear reactions (see, e.g., [108]).

Note that at the initial stage of diffusion, the solution of problem (1.159) is given by function

$$f(\boldsymbol{r},t) = f_0(\boldsymbol{r})e^{w(\boldsymbol{r},t)}, \tag{1.160}$$

where $w(\boldsymbol{r},t) = \int\limits_0^t d\tau\ z(\boldsymbol{r},\tau)$.

Consider now the geometrical optics approximation (1.155) for parabolic equation (1.153). In this approximation, the equation for the phase of the wave is the Hamilton–Jacobi equation and the equation for the transverse gradient of the phase (1.155) is the closed quasilinear first-order partial differential equation, and we can solve it by the method of characteristics (see, e.g., [135]). Equations for the *characteristic curves* (*rays*) have the form

$$\frac{d}{dx}\boldsymbol{R}(x) = \boldsymbol{p}(x), \quad \frac{d}{dx}\boldsymbol{p}(x) = \frac{1}{2}\boldsymbol{\nabla}_{\boldsymbol{R}}\varepsilon(x, \boldsymbol{R}), \tag{1.161}$$

and the wavefield intensity and matrix of the phase second derivatives along the characteristic curves will satisfy the equations

$$\begin{aligned} \frac{d}{dx}I(x) &= -I(x)u_{ii}(x), \\ \frac{d}{dx}u_{ij}(x) + u_{ik}(x)u_{kj}(x) &= \frac{1}{2}\frac{\partial^2}{\partial R_i \partial R_j}\varepsilon(x, \boldsymbol{R}). \end{aligned} \tag{1.162}$$

Equations (1.161) coincide in appearance with the equations for a particle under random external forces in the absence of friction (1.14) and form the system of the Hamilton equations .

In the two-dimensional case ($R = y$), Eqs. (1.161), (1.162) become significantly simpler and assume the form

$$\begin{aligned} \frac{d}{dx}y(x) = p(x), \quad & \frac{d}{dx}p(x) = \frac{1}{2}\frac{\partial}{\partial y}\varepsilon(x, y), \\ \frac{d}{dx}I(x) = -I(x)u(x), \quad & \frac{d}{dx}u(x) + u^2(x) = \frac{1}{2}\frac{\partial^2}{\partial y^2}\varepsilon(x, y). \end{aligned} \tag{1.163}$$

The last equation for $u(x)$ in Eqs. (1.163) is similar to Eq. (1.31) whose solution shows the singular behavior. The only difference between these equations consists in the random term that has now more complicated structure. Nevertheless, it is quite clear that solutions to stochastic problem (1.163) will show the blow-up behavior; namely, function $u(x)$ will reach minus infinity and intensity will reach plus infinity at a finite distance. Such a behavior of a wavefield in randomly inhomogeneous media corresponds to *random focusing*, i.e., to the formation of caustics, which means the appearance of points of multivaluedness (and discontinuity) in the solutions to quasilinear equation (1.155) for the transverse gradient of the wavefield phase.

1.4.4 The Navier–Stokes Equation: Random Forces in Hydrodynamic Theory of Turbulence

Consider now the turbulent motion model that assumes the presence of external forces $\boldsymbol{f}(\boldsymbol{r}, t)$ acting on the fluid. Such a model is evidently only imaginary, because no actual analogues exist for these forces. However, assuming that forces $\boldsymbol{f}(\boldsymbol{r}, t)$ ensure an appreciable energy income only to large-scale velocity components, we can expect that, within the concepts of the theory of local isotropic turbulence, the imaginary nature of field $\boldsymbol{f}(\boldsymbol{r},t)$ will only slightly affect statistical properties of small-scale turbulent components [112]. Consequently, this model is quite appropriate for describing small-scale properties of turbulence.

Motion of an incompressible fluid under external forces is governed by the *Navier–Stokes equation*

$$\left(\frac{\partial}{\partial t}+\boldsymbol{u}(\boldsymbol{r},t)\frac{\partial}{\partial \boldsymbol{r}}\right)\boldsymbol{u}(\boldsymbol{r},t)=-\frac{1}{\rho_0}\frac{\partial}{\partial \boldsymbol{r}}p(\boldsymbol{r},t)+\nu\Delta\boldsymbol{u}(\boldsymbol{r},t)+\boldsymbol{f}(\boldsymbol{r},t),$$
$$\frac{\partial}{\partial \boldsymbol{r}}\boldsymbol{u}(\boldsymbol{r},t)=0,\quad \frac{\partial}{\partial \boldsymbol{r}}\boldsymbol{f}(\boldsymbol{r},t)=0. \tag{1.164}$$

Here, ρ_0 is the density of the fluid, ν is the kinematic viscosity, and pressure field $p(\boldsymbol{x},t)$ is expressed in terms of the velocity field at the same instant by the relationship

$$p(\boldsymbol{r},t)=-\rho_0\int\Delta^{-1}\left(\boldsymbol{r},\boldsymbol{r}'\right)\frac{\partial^2\left(u_i(\boldsymbol{r}',t)u_j(\boldsymbol{r}',t)\right)}{\partial r_i'\partial r_j'}d\boldsymbol{r}', \tag{1.165}$$

where $\Delta^{-1}(\boldsymbol{r},\boldsymbol{r}')$ is the integral operator inverse to the Laplace operator (repeated indices assume summation).

Note that the linearized equation (1.164)

$$\frac{\partial}{\partial t}\boldsymbol{u}(\boldsymbol{r},t)=\nu\Delta\boldsymbol{u}(\boldsymbol{r},t)+\boldsymbol{f}(\boldsymbol{r},t),\quad \boldsymbol{u}(\boldsymbol{r},0)=0 \tag{1.166}$$

describes the problem on turbulence degeneration for $t\to\infty$. The solution to this problem can be drawn in an explicit form:

$$\begin{aligned}\boldsymbol{u}(\boldsymbol{r},t)&=\int\limits_0^t d\tau\; e^{\nu(t-\tau)\Delta}\boldsymbol{f}(\boldsymbol{r},\tau)=\int\limits_0^t d\tau\; e^{\nu\tau\Delta}\int d\boldsymbol{r}'\;\delta(\boldsymbol{r}-\boldsymbol{r}')\boldsymbol{f}(\boldsymbol{r}',t-\tau)\\
&=\frac{1}{(2\pi)^3}\int\limits_0^t d\tau\; e^{\nu\tau\Delta}\int d\boldsymbol{r}'\int d\boldsymbol{q}\; e^{i\boldsymbol{q}(\boldsymbol{r}-\boldsymbol{r}')}\boldsymbol{f}(\boldsymbol{r}',t-\tau)\\
&=\frac{1}{(2\pi)^3}\int\limits_0^t d\tau\int d\boldsymbol{r}'\int d\boldsymbol{q}\; e^{-\nu\tau\boldsymbol{q}^2}e^{i\boldsymbol{q}\boldsymbol{r}'}\boldsymbol{f}(\boldsymbol{r}-\boldsymbol{r}',t-\tau)\\
&=\int d\boldsymbol{r}'\int\limits_0^t\frac{d\tau}{(4\pi\nu\tau)^{3/2}}\;\exp\left\{-\frac{\boldsymbol{r}'^2}{4\nu\tau}\right\}\boldsymbol{f}(\boldsymbol{r}-\boldsymbol{r}',t-\tau).\end{aligned} \tag{1.167}$$

The linear equation (1.166) is an extension of the Langevin equation in velocity (1.94), page 47 to random fields.

Neglecting the effect of viscosity and external random forces in Eq. (1.164), we arrive at the equation

$$\frac{\partial}{\partial t}\boldsymbol{u}(\boldsymbol{r},t)+(\boldsymbol{u}(\boldsymbol{r},t)\cdot\boldsymbol{\nabla})\,\boldsymbol{u}(\boldsymbol{r},t)=-\frac{1}{\rho}\boldsymbol{\nabla}p(\boldsymbol{r},t) \tag{1.168}$$

that describes dynamics of *perfect fluid* and is called the *Euler equation.* Using Eq. (1.77), page 40, this equation can be rewritten in terms of the velocity field vortex $\boldsymbol{\omega}=\operatorname{curl}\boldsymbol{u}(\boldsymbol{r},t)$ as

$$\frac{\partial}{\partial t}\boldsymbol{\omega}(\boldsymbol{r},t)=\operatorname{curl}\left[\boldsymbol{u}(\boldsymbol{r},t)\times\boldsymbol{\omega}(\boldsymbol{r},t)\right],\quad \boldsymbol{\omega}(\boldsymbol{r},0)=\boldsymbol{\omega}_0(\boldsymbol{r}), \tag{1.169}$$

or

$$\left(\frac{\partial}{\partial t}+\frac{\partial}{\partial \boldsymbol{r}}\boldsymbol{u}(\boldsymbol{r},t)\right)\boldsymbol{\omega}(\boldsymbol{r},t)=\left(\boldsymbol{\omega}(\boldsymbol{r},t)\cdot\frac{\partial}{\partial \boldsymbol{r}}\right)\boldsymbol{\omega}(\boldsymbol{r},t),\quad \boldsymbol{\omega}(\boldsymbol{r},0)=\boldsymbol{\omega}_0(\boldsymbol{r}). \tag{1.170}$$

These equations coincide with Eqs. (1.71) and (1.78), page 40 appeared in the problem on the diffusion of magnetic field.

However, this coincidence is only formal, because different boundary conditions to these problems results in drastically different behaviors of fields $\boldsymbol{\omega}(\boldsymbol{r},t)$ and $\boldsymbol{H}(\boldsymbol{r},t)$ as functions of the velocity field $\boldsymbol{u}(\boldsymbol{r},t)$.

If we substitute Eq. (1.165) in Eq. (1.164) to exclude the pressure field, then we obtain in the three-dimensional case that the Fourier transform of the velocity field with respect to spatial coordinates

$$\widehat{u}_i(\boldsymbol{k},t)=\int d\boldsymbol{r}u_i(\boldsymbol{r},t)e^{-i\boldsymbol{k}\boldsymbol{r}},\quad u_i(\boldsymbol{r},t)=\frac{1}{(2\pi)^3}\int d\boldsymbol{k}\widehat{u}_i(\boldsymbol{k},t)e^{i\boldsymbol{k}\boldsymbol{r}},$$

$(\widehat{u}_i^*(\boldsymbol{k},t)=\widehat{u}_i(-\boldsymbol{k},t))$ satisfies the nonlinear integro-differential equation

$$\frac{\partial}{\partial t}\widehat{u}_i\left(\boldsymbol{k},t\right)+\frac{i}{2}\int d\boldsymbol{k}_1\int d\boldsymbol{k}_2\Lambda_i^{\alpha\beta}\left(\boldsymbol{k}_1,\boldsymbol{k}_2,\boldsymbol{k}\right)\widehat{u}_\alpha\left(\boldsymbol{k}_1,t\right)\widehat{u}_\beta\left(\boldsymbol{k}_2,t\right)$$
$$-\nu k^2\widehat{u}_i\left(\boldsymbol{k},t\right)=\widehat{f}_i\left(\boldsymbol{k},t\right), \tag{1.171}$$

where

$$\Lambda_i^{\alpha\beta}\left(\boldsymbol{k}_1,\boldsymbol{k}_2,\boldsymbol{k}\right)=\frac{1}{(2\pi)^3}\left\{k_\alpha\Delta_{i\beta}\left(\boldsymbol{k}\right)+k_\beta\Delta_{i\alpha}\left(\boldsymbol{k}\right)\right\}\delta(\boldsymbol{k}_1+\boldsymbol{k}_2-\boldsymbol{k}),$$

$$\Delta_{ij}\left(\boldsymbol{k}\right)=\delta_{ij}-\frac{k_ik_j}{\boldsymbol{k}^2}\quad(i,\,\alpha,\,\beta\,=\,1,\,2,\,3),$$

and $\widehat{\boldsymbol{f}}(\boldsymbol{k},t)$ is the spatial Fourier harmonic of external forces,

$$\widehat{\boldsymbol{f}}(\boldsymbol{k},t)=\int d\boldsymbol{r}\boldsymbol{f}(\boldsymbol{r},t)e^{-i\boldsymbol{k}\boldsymbol{r}},\quad \boldsymbol{f}(\boldsymbol{r},t)=\frac{1}{(2\pi)^3}\int d\boldsymbol{k}\widehat{\boldsymbol{f}}(\boldsymbol{k},t)e^{i\boldsymbol{k}\boldsymbol{r}}.$$

A specific feature of the three-dimensional hydrodynamic motions consists in the fact that the absence of external forces and viscosity-driven effects is sufficient for energy conservation.

It appears convenient to describe the stationary turbulence in terms of the space–time Fourier harmonics of the velocity field

$$\widehat{u}_i(\boldsymbol{K})=\int d\boldsymbol{x}\int_{-\infty}^{\infty} dt u_i(\boldsymbol{x},t)e^{-i(\boldsymbol{k}\boldsymbol{x}+\omega t)},$$

$$u_i(\boldsymbol{x},t)=\frac{1}{(2\pi)^4}\int d\boldsymbol{k}\int_{-\infty}^{\infty} d\omega\widehat{u}_i(\boldsymbol{K})e^{i(\boldsymbol{k}\boldsymbol{x}+\omega t)},$$

where $\boldsymbol{K}$ is the four-dimensional wave vector $\{\boldsymbol{k},\omega\}$ and field

$$\widehat{u}_i^*(\boldsymbol{K})=\widehat{u}_i(-\boldsymbol{K})$$

because field $u_i(\boldsymbol{r},t)$ is real. In this case, we obtain the equation for component $\widehat{u}_i(\boldsymbol{K})$ by accomplishing the Fourier transformation of Eq. (1.171) with respect to time:

$$(i\omega+\nu\boldsymbol{k}^2)\widehat{u}_i(\boldsymbol{K})+\frac{i}{2}\int d^4\boldsymbol{K}_1\int d^4\boldsymbol{K}_2\Lambda_i^{\alpha\beta}(\boldsymbol{K}_1,\boldsymbol{K}_2,\boldsymbol{K})\widehat{u}_\alpha(\boldsymbol{K}_1)\widehat{u}_\beta(\boldsymbol{K})=\widehat{f}_i(\boldsymbol{K}),\tag{1.172}$$

where

$$\Lambda_i^{\alpha\beta}(\boldsymbol{K}_1,\boldsymbol{K}_2,\boldsymbol{K})=\frac{1}{2\pi}\Lambda_i^{\alpha\beta}(\boldsymbol{k}_1,\boldsymbol{k}_2,\boldsymbol{k})\,\delta(\omega_1+\omega_2-\omega),$$

and $\widehat{f}_i(\boldsymbol{K})$ are the space–time Fourier harmonics of external forces. The obtained Eq. (1.172) is the integral (and not integro-differential) nonlinear equation.

Waves on Sea Surface

Consider now the behavior of water boundary at the sea surface. In this case, equations of hydrodynamics must be supplemented with the kinematic boundary condition on the free sea surface $z=\xi(\boldsymbol{R},t)$, where $\boldsymbol{R}$ denotes the coordinates in the plane perpendicular to the vertical axis z (Fig. 1.21), which has the form

$$\frac{d}{dt}\xi(\boldsymbol{R},t)=w_z(\boldsymbol{R},z;t)\Big|_{z=\xi(\boldsymbol{R},t)}.\tag{1.173}$$

Here, $\frac{d}{dt}\xi(\boldsymbol{R},t)$ is the total derivative of the displacement of water surface and $w_z(\boldsymbol{R},z;t)$ is the vertical component of the hydrodynamic velocity of medium.

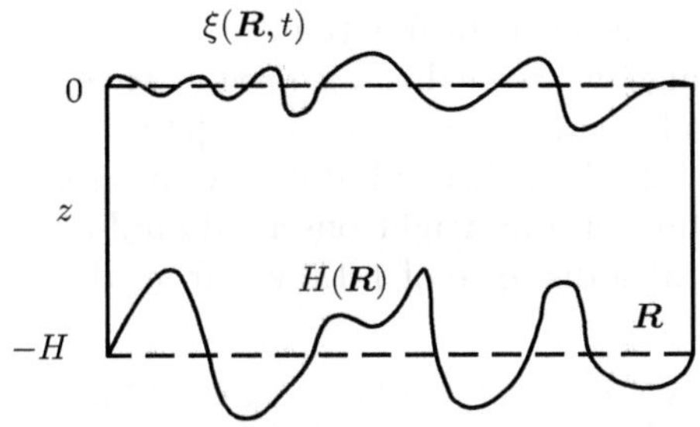

Fig. 1.21 Disturbance of water surface

If we denote the horizontal component of the velocity field of the medium as $\boldsymbol{u}(\boldsymbol{R},z,t)$, then we can consider condition (1.173) as a closed stochastic quasilinear eqiation in the kinematic approximation, i.e., at given statistical characteristics of the velocity field $\{\boldsymbol{u}(\boldsymbol{R},z,t),\ w(\boldsymbol{R},z;t)\}$,

$$\frac{\partial\xi(\boldsymbol{R},t)}{\partial t}+u_i(\boldsymbol{R},\xi(\boldsymbol{R},t),t)\frac{\partial\xi(\boldsymbol{R},t)}{\partial R_i}=w_z(\boldsymbol{R},\xi(\boldsymbol{R},t);t). \tag{1.174}$$

At the same time, Eq. (1.174) describes generation of waves on the sea surface which are driven by the vertical component of the hydrodynamic velocity field. Differentiating Eq. (1.173) with respect to $\boldsymbol{R}$, we obtain an equation in the gradient of surface displacement $p_k(\boldsymbol{R},t)=\frac{\partial\xi(\boldsymbol{R},t)}{\partial R_k}$, which is a characteristic of surface slopes,

$$\begin{aligned}&\frac{\partial p_k(\boldsymbol{R},t)}{\partial t}+\left[\left.\frac{\partial u_i(\boldsymbol{R},z;t)}{\partial R_k}\right|_{z=\xi(\boldsymbol{R},t)}+\frac{\partial u_i(\boldsymbol{R},\xi(\boldsymbol{R},t);t)}{\partial z}p_k(\boldsymbol{R},t)\right]p_i(\boldsymbol{R},t)\\&+u_i(\boldsymbol{R},\xi(\boldsymbol{R},t),t)\frac{\partial p_i(\boldsymbol{R},t)}{\partial R_k}=\left.\frac{\partial w_z(\boldsymbol{R},z;t)}{\partial R_k}\right|_{z=\xi(\boldsymbol{R},t)}+\frac{\partial w_z(\boldsymbol{R},\xi(\boldsymbol{R},t);t)}{\partial z}p_k(\boldsymbol{R},t).\end{aligned} \tag{1.175}$$

Note that the problem considered here has the second boundary condition associated with inhomogeneity of bottom topography (see Fig. 1.21). In the scope of the kinematic approximation, this boundary condition appears in the functional form; namely, the variational derivatives of the solution $\xi(\boldsymbol{R},t)$ and $\boldsymbol{p}(\boldsymbol{R},t)$ assume the form

$$
\begin{aligned}
\frac{\delta \xi(\boldsymbol{R},t)}{\delta \boldsymbol{u}(\boldsymbol{R}',z',t')} &\sim \theta\left(z'-H(\boldsymbol{R})\right)\theta\left(t-t'\right), \\
\frac{\delta \boldsymbol{p}(\boldsymbol{R},t)}{\delta \boldsymbol{u}(\boldsymbol{R}',z',t')} &\sim \theta\left(z'-H(\boldsymbol{R})\right)\theta\left(t-t'\right),
\end{aligned}
\tag{1.176}
$$

where $\theta(z)$ is the Heaviside theta function.

Note that some huge waves—called also rogue waves—are occasionally observed on the sea and ocean surface (see, e.g., [39]).

Figures 1.22 and 1.23 show three photos of an unusual narrow and lengthy immobile wave of about 4-5-m height observed on June 11, 2006 near Kamchatka Pacific coast at a distance of 1-1.5 km from the shoreline.

Fig. 1.22 Rogue wave. (*a*) Side view.

Fig. 1.23 Rogue wave. Front view of (*b*) *the head of the wave and* (*c*) *the middle of the wave.*

The author of these photos, M.M. Sokolovsky described this phenomenon in the following words: "It was certainly a strange wave, because it appeared and then disappeared a few times. No waves were observed around this wave, peace and quite."

What was this structure formation? Was it a rogue wave? No answer yet...

Plane Motion under the Action of a Periodic Force

Consider now the two-dimensional motion of an incompressible viscous fluid $\boldsymbol{U}(\boldsymbol{r},t) = \{u(\boldsymbol{r},t),\, v(\boldsymbol{r},t)\}$ in plane $\boldsymbol{r} = \{x,\, y\}$ under the action of a spatially periodic force directed along the x-axis,

$$f_x(\boldsymbol{r},t) = \gamma \sin py \ (\gamma > 0).$$

Such a flow is usually called the *Kolmogorov flow* (*stream*). The corresponding motion is described by the system of equations

$$\begin{gathered}
\frac{\partial u}{\partial t} + \frac{\partial u^2}{\partial x} + \frac{\partial uv}{\partial y} = -\frac{1}{\rho}\frac{\partial P}{\partial x} + \nu \Delta u + \gamma \sin py, \\
\frac{\partial v}{\partial t} + \frac{\partial uv}{\partial x} + \frac{\partial v^2}{\partial y} = -\frac{1}{\rho}\frac{\partial P}{\partial y} + \nu \Delta v, \\
\frac{\partial u}{\partial x} + \frac{\partial v}{\partial y} = 0,
\end{gathered} \tag{1.177}$$

where $P(\boldsymbol{r},t)$ is the pressure, ρ is the density, and ν is the kinematic viscosity. See papers [118, 119] for the laboratory modeling of the Kolmogorov flow.

The system of the Navier–Stokes and continuity equations (1.177) has the steady-state solution that corresponds to the laminar flow at constant pressure along the x-axis and has the following form

$$u_{\text{s-s}}(\boldsymbol{r}) = \frac{\gamma}{\nu p^2}\sin py, \quad v_{\text{s-s}}(\boldsymbol{r}) = 0, \quad P_{\text{s-s}}(\boldsymbol{r}) = \text{const}. \tag{1.178}$$

Introducing scales of length p^{-1}, velocity $p^{-2}\upsilon^{-1}\gamma$, and time $p\nu\gamma^{-1}$ and using dimensionless variables, we reduce system (1.177) to the form

$$\begin{gathered}
\frac{\partial u}{\partial t} + \frac{\partial u^2}{\partial x} + \frac{\partial uv}{\partial y} = -\frac{\partial P}{\partial x} + \frac{1}{R}\Delta u + \frac{1}{R}\sin y, \\
\frac{\partial v}{\partial t} + \frac{\partial uv}{\partial x} + \frac{\partial v^2}{\partial y} = -\frac{\partial P}{\partial y} + \frac{1}{R}\Delta v, \\
\frac{\partial u}{\partial x} + \frac{\partial v}{\partial y} = 0,
\end{gathered} \tag{1.179}$$

where $R = \dfrac{\gamma}{\nu^2 p^3}$ is the Reynolds number. In these variables, the steady-state solution has the form

$$u_{\text{s-s}}(\boldsymbol{r}) = \sin y, \quad v_{\text{s-s}}(\boldsymbol{r}) = 0, \quad P_{\text{s-s}}(\boldsymbol{r}) = \text{const}.$$

Introducing flow function $\psi(\boldsymbol{r},t)$ by the relationship

$$u(\boldsymbol{r},t) = \frac{\partial}{\partial y}\psi(\boldsymbol{r},t), \quad v(\boldsymbol{r},t) = -\frac{\partial}{\partial x}\psi(\boldsymbol{r},t),$$

we obtain that it satisfies the equation

$$\left(\frac{\partial}{\partial t} - \frac{\Delta}{R}\right)\Delta\psi - \frac{\partial\psi}{\partial x}\frac{\partial\Delta\psi}{\partial y} + \frac{\partial\psi}{\partial y}\frac{\partial\Delta\psi}{\partial x} = \frac{1}{R}\cos y \tag{1.180}$$

and

$$\psi_{\text{s-s}}(\boldsymbol{r}) = -\cos y.$$

It was shown [103,138] that, in the linear problem formulation, the steady-state solution (1.178) corresponding to the laminar flow is unstable with respect to small disturbances for certain values of parameter R. These disturbances rapidly increase in time getting the energy from the flow (1.178); this causes the Reynolds stresses described by the nonlinear terms in Eq. (1.180) to increase, which results in decreasing the amplitude of laminar flow until certain new steady-state flow (called usually the *secondary flow*) is formed.

Represent the hydrodynamic fields in the form

$$u(\boldsymbol{r},t) = U(y,t) + \widetilde{u}(\boldsymbol{r},t), \quad v(\boldsymbol{r},t) = \widetilde{v}(\boldsymbol{r},t),$$
$$P(\boldsymbol{r},t) = P_0 + \widetilde{P}(\boldsymbol{r},t), \quad \psi(\boldsymbol{r},t) = \Psi(y,t) + \widetilde{\psi}(\boldsymbol{r},t).$$

Here, $U(y,t)$ is the new profile of the steady-state flow to be determined together with the Reynolds stresses and the tilde denotes the corresponding ultimate disturbances. Abiding by the cited works, we will consider disturbances harmonic in variable x with wavelength $2\pi/\alpha$ ($\alpha > 0$). The new flow profile $U(y,t)$ is the result of averaging with respect to x over distances of about a wavelength.

One can easily see that, for $\alpha \geq 1$, the laminar flow (1.178) is unique and stable for all R [138] and instability can appear only for disturbances with $\alpha < 1$.

According to the linear theory of stability, we will consider first the nonlinear interaction only between the first harmonic of disturbances and the mean flow and neglect the generation of higher harmonics and their mutual interactions and interaction with the mean flow.

We represent all disturbances in the form

$$\widetilde{\varphi}(\boldsymbol{r},t) = \varphi^{(1)}(y,t)e^{i\alpha x} + \varphi^{(-1)}(y,t)e^{-i\alpha x},$$
$$\left(\widetilde{\varphi}(\boldsymbol{r},t) = \widetilde{u}(\boldsymbol{r},t),\, \widetilde{v}(\boldsymbol{r},t),\, \widetilde{P}(\boldsymbol{r},t),\, \widetilde{\psi}(\boldsymbol{r},t)\right),$$

where quantity $\varphi^{(-1)}(y,t)$ is the complex conjugated of $\varphi^{(1)}(y,t)$. Then, using this representation in system (1.179) and eliminating quantities $\widetilde{P}(\boldsymbol{r},t)$ and $\widetilde{u}(\boldsymbol{r},t)$, we obtain the system of equations in mean flow $U(y,t)$ and disturbances $v^{(1)}(y,t)$ [14, 43]:

$$\frac{\partial}{\partial t}U + \frac{i}{\alpha}\left(v^{(-1)}\frac{\partial^2 v^{(1)}}{\partial y^2} - v^{(1)}\frac{\partial^2 v^{(-1)}}{\partial y^2}\right) = \frac{1}{R}\frac{\partial^2 U}{\partial y^2} + \frac{1}{R}\sin y,$$
$$\left(\frac{\partial}{\partial t} - \frac{\Delta}{R}\right)\Delta v^{(1)} + i\alpha\left[U\Delta v^{(1)} - v^{(1)}\frac{\partial^2 U}{\partial y^2}\right] = 0. \tag{1.181}$$

The second equation in system (1.181) is the known *Orr–Sommerfeld equation*. A similar system can be derived for the flow function.

To examine the stability of the laminar regime (1.178), we set

$$U(y) = \sin y$$

in the second equation of system (1.181) to obtain

$$\left(\frac{\partial}{\partial t} - \frac{\Delta}{R}\right)\Delta v^{(1)}(y,t) + i\alpha\sin y[1+\Delta]v^{(1)}(y,t) = 0. \tag{1.182}$$

Representing disturbances $v^{(1)}(y,t)$ in the form

$$v^{(1)}(y,t) = \sum_{n=-\infty}^{\infty} v_n^{(1)} e^{\sigma t + iny} \tag{1.183}$$

and substituting this representation in Eq. (1.182), we arrive at the recurrent system in quantities $v_n^{(1)}$

$$\frac{2}{\alpha}\left(\alpha^2+n^2\right)\left[\sigma + \frac{\alpha^2+n^2}{R}\right]v_n^{(1)} + v_{n-1}^{(1)}\left[\alpha^2 - 1 + (n-1)^2\right]$$
$$- v_{n+1}^{(1)}\left[\alpha^2 - 1 + (n+1)^2\right] = 0, \qquad n = -\infty, \cdots, +\infty. \tag{1.184}$$

The analysis of system (1.184) showed [103,138] that, under certain restrictions on wave number α and Reynolds number R, positive values of parameter σ can exist, i.e., the solutions are unstable. The corresponding dispersion equation in σ has the form of an infinite continued fraction, and the critical Reynolds number is $R_{\text{cr}} = \sqrt{2}$ for $\alpha \to 0$. In other words, long-wave disturbances along the applied force appear to be most unstable. For this reason, we can consider parameter α as a small parameter of the problem at hand and integrate the Orr–Sommerfeld equation asymptotically. We will not dwell on details of this solution. Note only that the components of eigenvector $\left\{v_n^{(1)}\right\}$ of problem (1.184) have different orders of magnitude in parameter α. For example, all components of vector $\left\{v_n^{(1)}\right\}$ with $n = \pm 2, \pm 3, \cdots$ will have an order of α^4 at least. As a result, we can confine ourselves to the most significant harmonics with $n = 0, \pm 1$, which is, in essence, equivalent to the *Galerkin method* with the trigonometric coordinate functions. In this case,

$$U(y,t) = U(t)\sin y,$$

and the equation in $v^{(1)}$ assumes the form

$$\left(\frac{\partial}{\partial t} - \frac{\Delta}{R}\right)\Delta v^{(1)}(y,t) + i\alpha U(t)\sin y[1+\Delta]v^{(1)}(y,t) = 0.$$

Substituting the expansion

$$v^{(1)}(y,t) = \sum_{n=-1}^{1} v_n^{(1)}(t)e^{iny}$$

in Eqs. (1.181), we obtain that functions

$$U(t), \quad z_0(t) = v_0^{(1)}(t), \quad z_+(t) = v_1^{(1)}(t) + v_{-1}^{(1)}(t),$$

$$z_-(t) = \frac{v_1^{(1)}(t) - v_{-1}^{(1)}(t)}{2}$$

satisfy the system of equations [14, 43]

$$\left(\frac{d}{dt} + \frac{1}{R}\right)U(t) = \frac{1}{R} - \frac{4}{\alpha}z_0(t)z_-(t),$$

$$\left(\frac{d}{dt} + \frac{\alpha^2}{R}\right)z_0(t) = \alpha U(t)z_-(t), \tag{1.185}$$

$$\left(\frac{d}{dt} + \frac{1}{R}\right)z_-(t) = \frac{\alpha}{2}U(t)z_0(t), \quad \left(\frac{d}{dt} + \frac{1}{R}\right)z_+(t) = 0.$$

Equation in quantity $z_+(t)$ is independent of other equations; as a consequence, the corresponding disturbances can only decay with time. Three remaining equations form the simplest three-component hydrodynamic-type system (see Sect. 1.1.3, page 13). As was mentioned earlier, this system is equivalent to the dynamic system describing the motion of a gyroscope with anisotropic friction under the action of an external moment of force relative to the unstable axis. An analysis of system (1.185) shows that, for $R < R_{\text{cr}} = \sqrt{2}$, it yields the laminar regime with $U = 1, z_i = 0$. For $R > \sqrt{2}$, this regime becomes unstable, and new regime (the secondary flow) is formed that corresponds to the mean flow profile and steady-state Reynolds stresses

$$U = \frac{\sqrt{2}}{R}, \quad \frac{4}{\alpha}z_0z_- = \frac{R-\sqrt{2}}{R^2}, \quad z_+ = 0,$$

$$\left[v_0^{(1)}\right]^2 = \frac{R-\sqrt{2}}{2\sqrt{2}R^2}, \quad v_1^{(1)} = \frac{\alpha}{\sqrt{2}}v_0^{(1)}, \quad \alpha \ll 1, \quad R \geq \sqrt{2}.$$

Turning back to the dimensional quantities, we obtain

$$U(y) = \sqrt{2}\upsilon p \sin py, \quad \langle \widetilde{u}\widetilde{v} \rangle = -\frac{\gamma}{p}\frac{R-\sqrt{2}}{R}\cos py. \tag{1.186}$$

Note that the amplitude of the steady-state mean flow is independent of the amplitude of exciting force. Moreover, quantity $v_0^{(1)}$ can be both positive and negative, depending on the signs of the amplitudes of small initial disturbances.

Flow function of the steady-state flow has the form

$$\psi_1(x,y) = -\frac{\sqrt{2}}{R}\cos y - \frac{2}{\alpha}v_0^{(1)}\left[\sqrt{2}\alpha \sin y \cos \alpha x + \sin \alpha x\right].$$

Figure 1.24 shows the current lines

$$\alpha \cos y + \sqrt{2}\alpha \sin y \cos \alpha x + \sin \alpha x = C$$

of flow (1.186) at $R = 2R_{\rm cr} = 2\sqrt{2}$ ($v_0^{(1)} > 0$). In addition, Fig. 1.24 shows schematically the profile of the mean flow. As distinct from the laminar solution, systems of spatially periodic vortices appear here, and the tilt of longer axes of these vortices is determined by the sign of the derivative of the mean flow profile with respect to y.

Flow (1.186) was derived under the assumption that the nonlinear interactions between different harmonics of the disturbance are insignificant in comparison with their interactions with the mean flow. This assumption will hold if flow (1.186) is, in turn, stable with respect to small disturbances.

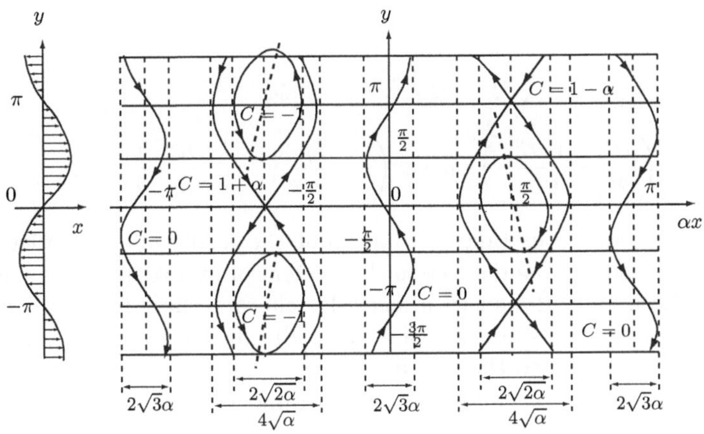

Fig. 1.24 Mean velocity and current lines of the secondary flow at $R = 2R_{\rm cr} = 2\sqrt{2}$ ($\widetilde{v}_0 > 0$)

The corresponding stability analysis can be carried out by the standard procedure, i.e., by linearizing the equation for flow function (1.180) relative to flow (1.186) [43]. The analysis shows that flow (1.186) is stable if we restrict ourselves to the harmonics of the types same as in solution (1.186).

However, the solution appears to be unstable with respect to small-scale disturbances. In this case, the nonlinear interaction of infinitely small disturbances governs the motion along with the interaction of the disturbances with the mean flow. Moreover, we cannot here content ourselves with a finite number of harmonics in x-coordinate and need to consider the infinite series. As regards the harmonics in y-coordinate, we, as earlier, can limit the consideration to the harmonics with $n = 0, \pm 1$.

Represent the flow function in the form

$$\psi(x,y,t) = \psi_{-1}(x,t)e^{-iy} + \psi_0(x,t) + \psi_1(x,t)e^{iy}$$
$$(\psi_1^*(x,t) = \psi_{-1}(x,t)), \tag{1.187}$$

where $\psi_i(x,t)$ are the periodic functions in x-coordinate with a period $2\pi/\alpha$. Substituting Eq. (1.187) in Eq. (1.180), neglecting the terms of order α^3 in interactions of harmonics and the terms of order α^2 in the dissipative terms of harmonics $\psi_{\pm 1}$, and introducing new functions

$$\psi_+(x,t) = \frac{\psi_{-1} + \psi_1}{2}, \quad \psi_-(x,t) = \frac{\psi_{-1} - \psi_1}{2i},$$

we obtain the system of equations

$$\left(\frac{\partial}{\partial t} + \frac{1}{R}\right)\psi_+ - \psi_- \frac{\partial \psi_0}{\partial x} = -\frac{1}{2R},$$

$$\left(\frac{\partial}{\partial t} + \frac{1}{R}\right)\psi_- + \psi_+ \frac{\partial \psi_0}{\partial x} = 0,$$

$$\left(\frac{\partial}{\partial t} - \frac{1}{R}\frac{\partial^2}{\partial x^2}\right)\psi_0 + 2\left(\psi_- \frac{\partial \psi_+}{\partial x} - \psi_+ \frac{\partial \psi_-}{\partial x}\right) = 0.$$

This system of equations takes into account the whole infinite series of harmonics in x-coordinate and extends the system of gyroscopic-type equations (1.185) to the infinite-dimensional case. Its characteristic feature consists in the absence of steady-state solutions periodic in x-coordinate (except the solution corresponding to the laminar flow).

Figure 1.25 shows an example of the Kolmogorov flow obtained experimentally (see, e.g., [13]).

Fig. 1.25 An example of the Kolmogorov flow obtained experimentally

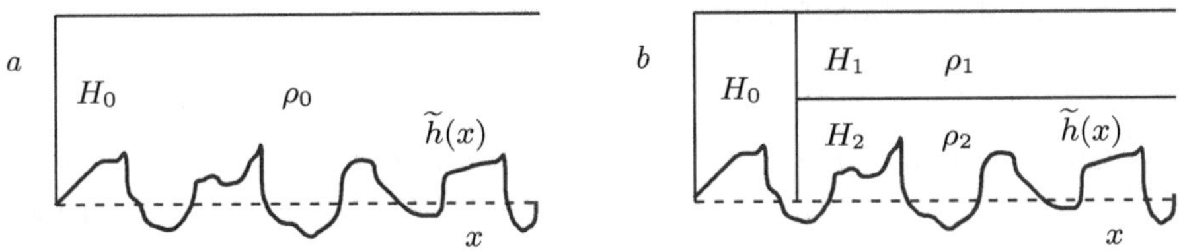

Fig. 1.26 Diagrammatic views of (*a*) one-layer and (*b*) two-layer models of hydrodynamic flows

1.4.5 Equations of Geophysical Hydrodynamics

Consider now the description of hydrodynamic flows on the rotating Earth in the so-called *quasi-geostrophic approximation* [120]. In the simplest case of the one-layer model, the incompressible fluid flow in the two-dimensional plane $\boldsymbol{R} = (x, y)$ is described by the *stream function* that satisfies the equation

$$\begin{gathered}\frac{\partial}{\partial t}\Delta\psi(\boldsymbol{R},t) + \beta_0\frac{\partial}{\partial x}\psi(\boldsymbol{R},t) = J\left\{\Delta\psi(\boldsymbol{R},t) + h(\boldsymbol{R});\psi(\boldsymbol{R},t)\right\},\\ \psi(\boldsymbol{R},0) = \psi_0(\boldsymbol{R}),\end{gathered} \tag{1.188}$$

where parameter β_0 is the derivative of the local *Coriolis parameter* f_0 with respect to latitude, $J\{\psi, \varphi\}$ is the Jacobian of two functions $\psi(\boldsymbol{R}, t)$ and $\varphi(\boldsymbol{R}, t)$

$$J\left\{\psi(\boldsymbol{R},t);\varphi(\boldsymbol{R},t)\right\} = \frac{\partial\psi(\boldsymbol{R},t)}{\partial x}\frac{\partial\varphi(\boldsymbol{R},t)}{\partial y} - \frac{\partial\varphi(\boldsymbol{R},t)}{\partial x}\frac{\partial\psi(\boldsymbol{R},t)}{\partial y},$$

and function $h(\boldsymbol{R}) = f_0\widetilde{h}(\boldsymbol{R})/H_0$ is the deviation of bottom topography $\widetilde{h}(\boldsymbol{R})$ relative to its average thickness H_0 (Fig. 1.26*a*).

The velocity field is expressed in terms of the stream function by the relationship

$$\boldsymbol{v}(\boldsymbol{R},t) = \left(-\frac{\partial\psi(\boldsymbol{R},t)}{\partial y}, \frac{\partial\psi(\boldsymbol{R},t)}{\partial x}\right).$$

Note that, under the neglect of Earth's rotation and effects of underlying surface topography, Eq. (1.188) reduces to the standard equation of two-dimensional hydrodynamics (see, e.g., [83])

$$\frac{\partial}{\partial t}\Delta\psi(\boldsymbol{R},t) + J\left\{\Delta\psi(\boldsymbol{R},t));\psi(\boldsymbol{R},t)\right\}, \quad \psi(\boldsymbol{R},0) = \psi_0(\boldsymbol{R}). \tag{1.189}$$

Equation (1.188) describes the *barotropic* motion of a fluid. In the more general case of *baroclinic* motions, investigation is usually carried out within the framework of the two-layer model of hydrodynamic flows described by the system of equations [120]

$$\begin{aligned}\frac{\partial}{\partial t}\left[\Delta\psi_1 - \alpha_1 F(\psi_1 - \psi_2)\right] + \beta_0\frac{\partial}{\partial x}\psi_1 = J\left\{\Delta\psi_1 - \alpha_1 F(\psi_1 - \psi_2);\psi_1\right\},\\ \frac{\partial}{\partial t}\left[\Delta\psi_2 - \alpha_2 F(\psi_2 - \psi_1)\right] + \beta_0\frac{\partial}{\partial x}\psi_2 = J\left\{\Delta\psi_2 - \alpha_2 F(\psi_2 - \psi_1) + f_0\alpha_2 h;\psi_2\right\},\end{aligned} \tag{1.190}$$

where additional parameters

$$F = f_0^2\rho/g(\Delta\rho), \quad \Delta\rho/\rho = (\rho_2 - \rho_1)/\rho_0 > 0$$

are introduced and $\alpha_1 = 1/H_1$ and $\alpha_2 = 1/H_2$ are the inverse thicknesses of layers (Fig. 1.26*b*).

Among the particular cases of Eqs. (1.188), (1.190) are the equations obtained by neglecting Earth's rotation (two-dimensional hydrodynamics) but with allowance for bottom topography and the linearized quasi-geostrophic equations similar to Eq. (1.55) that describe the effect of topography on propagation of the Rossby waves.

Chapter 2
Solution Dependence on Problem Type, Medium Parameters, and Initial Data

Below, we considered a number of dynamic systems described by both ordinary and partial differential equations. Many applications concerning research of statistical characteristics of the solutions to these equations require the knowledge of the solution dependence (generally, in the functional form) on the medium parameters appeared in the equation as coefficients and initial values. Some properties appear common of all such dependencies, and two of them are of special interest in the context of statistical descriptions. We illustrate these dependencies by the example of the simplest problem, namely, the system of ordinary differential equations (1.1) that describes particle dynamics under random velocity field and which we reformulate in the form of the nonlinear integral equation

$$\boldsymbol{r}(t) = \boldsymbol{r}_0 + \int_{t_0}^{t} d\tau \boldsymbol{U}(\boldsymbol{r}(\tau), \tau). \tag{2.1}$$

The solution to Eq. (2.1) functionally depends on vector field $\boldsymbol{U}(\boldsymbol{r}', \tau)$ and initial values $\boldsymbol{r}_0$, t_0.

2.1 Functional Representation of Problem Solution

2.1.1 Variational (Functional) Derivatives

Recall first the general definition of a functional. One says that a functional is given if a rule is fixed that associates a number to every function from certain function family. Below, we give some examples of functionals:

V.I. Klyatskin, *Stochastic Equations: Theory and Applications in Acoustics, Hydrodyn., Magnetohydrodyn., and Radiophys., Vol. 1*, Understanding Complex Systems,
DOI: 10.1007/978-3-319-07587-7_2,

$$\text{(a)} \qquad F[\varphi(\tau)] = \int_{t_1}^{t_2} d\tau a(\tau)\varphi(\tau),$$

where $a(t)$ is the given (fixed) function and limits t_1 and t_2 can be both finite and infinite. This is the linear functional.

$$\text{(b)} \qquad F[\varphi(\tau)] = \int_{t_1}^{t_2}\int_{t_1}^{t_2} d\tau_1 d\tau_2 B(\tau_1, \tau_2)\varphi(\tau_1)\varphi(\tau_2),$$

where $B(t_1, t_2)$ is the given (fixed) function. This is the quadratic functional.

$$\text{(c)} \qquad F[\varphi(\tau)] = f\left(\Phi[\varphi(\tau)]\right),$$

where $f(x)$ is the given function and quantity $\Phi[\varphi(\tau)]$ is the functional.

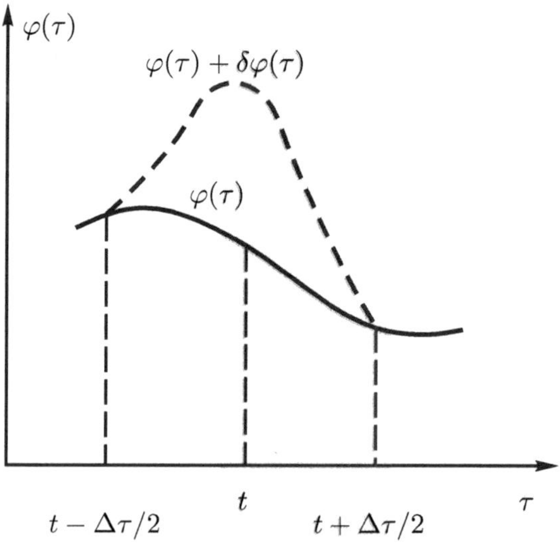

Fig. 2.1 To definition of variational derivative

Estimate the difference between the values of a functional calculated for functions $\varphi(\tau)$ and $\varphi(\tau) + \delta\varphi(\tau)$ for $t - \dfrac{\Delta\tau}{2} < \tau < t + \dfrac{\Delta\tau}{2}$ (see Fig. 2.1).

The variation of a functional is defined as the linear (in $\delta\varphi(\tau)$) portion of the difference

$$\delta F[\varphi(\tau)] = \{F[\varphi(\tau) + \delta\varphi(\tau)] - F[\varphi(\tau)]\}.$$

The limit

$$\frac{\delta F[\varphi(\tau)]}{\delta\varphi(t)dt} = \lim_{\Delta\tau\to 0} \frac{\delta F[\varphi(\tau)]}{\int\limits_{\Delta\tau} d\tau\delta\varphi(\tau)} \tag{2.2}$$

is called the *variational* (or *functional*) *derivative* (see, e.g., [112]).

For short, we will use notation $\dfrac{\delta F[\varphi(\tau)]}{\delta\varphi(t)}$ instead of $\dfrac{\delta F[\varphi(\tau)]}{\delta\varphi(t)dt}$.

Note that, if we use function $\delta\varphi(\tau) = \alpha\delta(\tau)$, where $\delta(\tau)$ is the Dirac delta function, then Eq. (2.2) can be represented in the form of the ordinary derivative

$$\frac{\delta F[\varphi(\tau)]}{\delta\varphi(t)} = \lim_{\alpha\to 0} \frac{d}{d\alpha} F[\varphi(\tau) + \alpha\delta(\tau - t)].$$

The variational derivative of functional $F[\varphi(\tau)]$ is again the functional of $\varphi(\tau)$, which depends additionally on point t as a parameter. As a result, this variational derivative will have two types of derivatives; one can differentiate it in the ordinary sense with respect to parameter t and in the functional sense with respect to $\varphi(\tau)$ at point $\tau = t'$ thus obtaining the second variational derivative of the initial functional

$$\frac{\delta^2 F[\varphi(\tau)]}{\delta\varphi(t')\delta\varphi(t)} = \frac{\delta}{\delta\varphi(t')}\left[\frac{\delta F[\varphi(\tau)]}{\delta\varphi(t)}\right].$$

The second variational derivative will now be the functional of $\varphi(\tau)$ dependent on two points t and t', and so forth.

Determine the variational derivatives of functionals (**a**), (**b**), and (**c**).

In the case (**a**), we have

$$\delta F[\varphi(\tau)] = F[\varphi(\tau) + \delta\varphi(\tau)] - F[\varphi(\tau)] = \int\limits_{t-\frac{\Delta\tau}{2}}^{t+\frac{\Delta\tau}{2}} d\tau a(\tau)\delta\varphi(\tau).$$

If function $a(t)$ is continuous on segment $\Delta\tau$, then, by the average theorem,

$$\delta F[\varphi(\tau)] = a(t') \int\limits_{\Delta\tau} d\tau\delta\varphi(\tau),$$

where point t' belongs to segment $\left[t - \dfrac{\Delta\tau}{2}, t + \dfrac{\Delta\tau}{2}\right]$. Consequently,

$$\frac{\delta F[\varphi(\tau)]}{\delta\varphi(t)} = \lim_{\Delta\tau\to 0} a(t') = a(t). \tag{2.3}$$

In the case (**b**), we obtain similarly

$$\frac{\delta F[\varphi(\tau)]}{\delta\varphi(t)} = \int\limits_{t_1}^{t_2} d\tau\,[B(\tau,t)+B(t,\tau)]\,\varphi(\tau) \quad (t_1 < t < t_2).$$

Note that function $B(\tau_1, \tau_2)$ can always be assumed a symmetric function of its arguments here.

In the case (**c**), we have

$$F[\varphi(\tau)+\delta\varphi(\tau)] = f\left(\Phi[\varphi(\tau)]\right) + \frac{\partial f(\Phi[\varphi(\tau)])}{\partial \Phi}\delta\Phi[\varphi(\tau)] + \cdots$$
$$= F[\varphi(\tau)] + \frac{\partial f(\Phi[\varphi(\tau)])}{\partial \Phi}\delta\Phi[\varphi(\tau)] + \cdots$$

and, consequently,

$$\frac{\delta}{\delta\varphi(t)} f\left(\Phi[\varphi(\tau)]\right) = \frac{\partial f(\Phi[\varphi(\tau)])}{\partial \Phi}\frac{\delta}{\delta\varphi(t)}\Phi[\varphi(\tau)]. \tag{2.4}$$

Consider now functional $\Phi[\varphi(\tau)] = F_1[\varphi(\tau)]F_2[\varphi(\tau)]$. We have

$$\delta\Phi[\varphi(\tau)] = F_1[\varphi(\tau)+\delta\varphi(\tau)]F_2[\varphi(\tau)+\delta\varphi(\tau)] - F_1[\varphi(\tau)]F_2[\varphi(\tau)]$$
$$= F_1[\varphi(\tau)]\delta F_2[\varphi(\tau)] + F_2[\varphi(\tau)]\delta F_1[\varphi(\tau)]$$

and, consequently,

$$\frac{\delta}{\delta\varphi(t)} F_1[\varphi(\tau)]F_2[\varphi(\tau)] = F_1[\varphi(\tau)]\frac{\delta}{\delta\varphi(t)}F_2[\varphi(\tau)] + F_2[\varphi(\tau)]\frac{\delta}{\delta\varphi(t)}F_1[\varphi(\tau)]. \tag{2.5}$$

We can define the expression for the variational derivative of functional $\varphi(\tau_0)$ with respect to function $\varphi(t)$ by the formal relationship

$$\frac{\delta\varphi(\tau_0)}{\delta\varphi(t)} = \delta(\tau_0 - t). \tag{2.6}$$

Formula (2.6) can be proved, for example, by considering the linear functional of the form

$$F[\varphi(\tau)] = \frac{1}{\sqrt{2\pi}\sigma}\int\limits_{-\infty}^{\infty} d\tau\varphi(\tau)\exp\left\{-\frac{(\tau-\tau_0)^2}{2\sigma^2}\right\}. \tag{2.7}$$

According to Eq. (2.3), the variational derivative of this functional has the form

$$\frac{\delta}{\delta\varphi(t)}F[\varphi(\tau)] = \frac{1}{\sqrt{2\pi}\sigma}\exp\left\{-\frac{(t-\tau_0)^2}{2\sigma^2}\right\}. \tag{2.8}$$

Performing now formal limit process $\sigma \to 0$ in Eq. (2.7) and (2.8), we obtain the desired formula (2.6). Moreover,

$$\frac{\delta F[\varphi(\tau)]}{\delta\varphi(t)} = \frac{\partial F[\varphi(\tau)]}{\partial\varphi(\tau)} \frac{\delta\varphi(\tau)}{\delta\varphi(t)} = \frac{\partial F[\varphi(\tau)]}{\partial\varphi(\tau)} \delta(\tau - t).$$

Formula (2.6) is very convenient for functional differentiation of functionals explicitly dependent on $\varphi(\tau)$. Indeed, for the quadratic functional (**b**), we have

$$\frac{\delta}{\delta\varphi(t)} \int\limits_{t_1}^{t_2}\int\limits_{t_1}^{t_2} d\tau_1 d\tau_2 B(\tau_1,\tau_2)\varphi(\tau_1)\varphi(\tau_2)$$

$$\overset{(2.5)}{=} \int\limits_{t_1}^{t_2}\int\limits_{t_1}^{t_2} d\tau_1 d\tau_2 B(\tau_1,\tau_2) \left[\frac{\delta\varphi(\tau_1)}{\delta\varphi(t)}\varphi(\tau_2) + \varphi(\tau_1)\frac{\delta\varphi(\tau_2)}{\delta\varphi(t)}\right]$$

$$\overset{(2.6)}{=} \int\limits_{t_1}^{t_2} d\tau \left[B(t,\tau)+B(\tau,t)\right]\varphi(\tau) \quad (t_1 < t < t_2).$$

Consider the functional

$$F[\varphi(\tau)] = \int\limits_{t_1}^{t_2} d\tau L\left(\tau, \varphi(\tau), \frac{d\varphi(\tau)}{d\tau}\right)$$

as another example. In this case,

$$\frac{\delta}{\delta\varphi(t)} F[\varphi(\tau)] \overset{(2.4)}{=} \int\limits_{t_1}^{t_2} d\tau \left[\frac{\partial L\left(\tau, \varphi(\tau), \frac{d\varphi(\tau)}{d\tau}\right)}{\partial\varphi(\tau)} + \frac{\partial L\left(\tau, \varphi(\tau), \frac{d\varphi(\tau)}{d\tau}\right)}{\partial\dot{\varphi}(\tau)} \frac{d}{d\tau}\right] \frac{\delta\varphi(\tau)}{\delta\varphi(t)}$$

$$\overset{(2.6)}{=} \int\limits_{t_1}^{t_2} d\tau \left[\frac{\partial L\left(\tau, \varphi(\tau), \frac{d\varphi(\tau)}{d\tau}\right)}{\partial\varphi(\tau)} + \frac{\partial L\left(\tau, \varphi(\tau), \frac{d\varphi(\tau)}{d\tau}\right)}{\partial\dot{\varphi}(\tau)} \frac{d}{d\tau}\right] \delta(\tau - t)$$

$$= \left(-\frac{d}{dt}\frac{\partial}{\partial\dot{\varphi}(t)} + \frac{\partial}{\partial\varphi(t)}\right) L\left(t, \varphi(t), \frac{d\varphi(t)}{dt}\right),$$

where $\dot{\varphi}(t) = \dfrac{d}{dt}\varphi(t)$ if point t belongs to interval (t_1, t_2).

Just as a function can be expanded in the Taylor series, a functional $F[\varphi(\tau) + \eta(\tau)]$ can be expanded in the functional Taylor series in function $\eta(\tau)$

$$F[\varphi(\tau)+\eta(\tau)] = F[\varphi(\tau)] + \int\limits_{-\infty}^{\infty} dt \frac{\delta F[\varphi(\tau)]}{\delta\varphi(t)}\eta(t)$$
$$+ \frac{1}{2!}\int\limits_{-\infty}^{\infty}\int\limits_{-\infty}^{\infty} dt_1 dt_2 \frac{\delta^2 F[\varphi(\tau)]}{\delta\varphi(t_1)\delta\varphi(t_2)}\eta(t_1)\eta(t_2) + \cdots . \quad (2.9)$$

Note that the operator expression

$$1 + \int\limits_{-\infty}^{\infty} dt\eta(t)\frac{\delta}{\delta\varphi(t)} + \frac{1}{2!}\int\limits_{-\infty}^{\infty}\int\limits_{-\infty}^{\infty} dt_1 dt_2 \eta(t_1)\eta(t_2)\frac{\delta^2}{\delta\varphi(t_1)\delta\varphi(t_2)} + \cdots$$
$$= 1 + \int\limits_{-\infty}^{\infty} dt\eta(t)\frac{\delta}{\delta\varphi(t)} + \frac{1}{2!}\left[\int\limits_{-\infty}^{\infty} dt\eta(t)\frac{\delta}{\delta\varphi(t)}\right]^2 + \cdots \quad (2.10)$$

can be written shortly as the operator

$$\exp\left\{\int\limits_{-\infty}^{\infty} dt\eta(t)\frac{\delta}{\delta\varphi(t)}\right\}, \quad (2.11)$$

whose action should be treated precisely in the sense of expansion (2.10). Using this operator, we can rewrite Eq. (2.9) in the form

$$F[\varphi(\tau)+\eta(\tau)] = e^{\int_{-\infty}^{\infty} dt\eta(t)\frac{\delta}{\delta\varphi(t)}} F[\varphi(\tau)], \quad (2.12)$$

which enables us to interpret operator (2.11) as the functional shift operator.

Consider now functional $F[t;\varphi(\tau)]$ dependent on parameter t. We can differentiate this functional with respect to t and determine its variational derivative with respect to $\varphi(t')$, as well. One can easily see that these operations commute, i.e., the equality

$$\frac{\partial}{\partial t}\frac{\delta F[t;\varphi(\tau)]}{\delta\varphi(t')} = \frac{\delta}{\delta\varphi(t')}\frac{\partial F[t;\varphi(\tau)]}{\partial t} \quad (2.13)$$

holds. If the domain of τ is independent of t, the validity of Eq. (2.13) is obvious. Otherwise, for example, for functionals $F[t;\varphi(\tau)]$ with $0 \le \tau \le t$, the validity of Eq. (2.13) can be checked on by expanding functional $F[t;\varphi(\tau)]$ in the functional Taylor series.

2.1.2 Principle of Dynamic Causality

Vary Eq. (2.1) with respect to field $\boldsymbol{U}(\boldsymbol{r},t)$. Assuming that the initial position $\boldsymbol{r}_0$ is independent of field $\boldsymbol{U}$, we obtain the equation linear in variational derivative (the linear variational differential equation)

$$\frac{\delta r_i(t)}{\delta U_j(\boldsymbol{r},t')} = \delta_{ij}\delta(\boldsymbol{r}-\boldsymbol{r}(t'))\,\theta(t'-t_0)\,\theta(t-t') + \int_{t_0}^{t} d\tau \frac{\partial U_i(\boldsymbol{r}(\tau),\tau)}{\partial r_k}\frac{\delta r_k(\tau)}{\delta U_j(\boldsymbol{r},t')}, \tag{2.14}$$

where $\delta(\boldsymbol{r}-\boldsymbol{r}')$ is the Dirac delta function, and $\theta(z)$ is the Heaviside step function. From Eq. (2.14) follows that

$$\frac{\delta r_i(t)}{\delta U_j(\boldsymbol{r},t')} = 0 \quad \text{for } t' > t \text{ or } t' < t_0, \tag{2.15}$$

which means that solution $\boldsymbol{r}(t)$ to the dynamic problem (2.1) as a functional of field $\boldsymbol{U}(\boldsymbol{r},t')$ depends on $\boldsymbol{U}(\boldsymbol{r},t')$ only for $t_0 < t' < t$. Consequently, function $\boldsymbol{r}(t)$ will remain unchanged if field $\boldsymbol{U}(\boldsymbol{r},t')$ varies outside the interval (t_0,t), i.e., for $t' < t_0$ or $t' > t$. We will call condition (2.15) the *dynamic causality condition.*

Taking this condition into account, we can rewrite Eq. (2.14) in the form

$$\frac{\delta r_i(t)}{\delta U_j(\boldsymbol{r},t')} = \delta_{ij}\delta(\boldsymbol{r}-\boldsymbol{r}(t'))\,\theta(t'-t_0)\,\theta(t-t') + \int_{t'}^{t} d\tau \frac{\partial U_i(\boldsymbol{r}(\tau),\tau)}{\partial r_k}\frac{\delta r_k(\tau)}{\delta U_j(\boldsymbol{r},t')}. \tag{2.16}$$

As a consequence, limit $t \to t'+0$ yields the equality

$$\left.\frac{\delta r_i(t)}{\delta U_j(\boldsymbol{r},t')}\right|_{t=t'+0} = \delta_{ij}\delta(\boldsymbol{r}-\boldsymbol{r}(t')). \tag{2.17}$$

Integral equation (2.16) in variational derivative is obviously equivalent to the linear differential equation with the initial value

$$\begin{gathered}\frac{\partial}{\partial t}\left(\frac{\delta r_i(t)}{\delta U_j(\boldsymbol{r},t')}\right) = \frac{\partial U_i(\boldsymbol{r}(t),t)}{\partial r_k}\left(\frac{\delta r_k(t)}{\delta U_j(\boldsymbol{r},t')}\right),\\ \left.\frac{\delta r_i(t)}{\delta U_j(\boldsymbol{r},t')}\right|_{t=t'} = \delta_{ij}\delta(\boldsymbol{r}-\boldsymbol{r}(t')).\end{gathered} \tag{2.18}$$

The dynamic causality condition is the general property of problems described by differential equations with initial values. The boundary-value problems possess no such property. Indeed, in the case of problem (1.33), (1.34), page 23 that describes propagation of a plane wave in a layer of inhomogeneous medium, wavefield $u(x)$ at point x and reflection and transmission coefficients depend functionally on function $\varepsilon(x)$ for all x of layer (L_0, L). However, using the imbedding method, we can convert this problem into the initial-value problem with respect to an auxiliary parameter L and make use the causality property in terms of the equations of the imbedding method.

2.2 Solution Dependence on Problem's Parameters

2.2.1 Solution Dependence on Initial Data

Here, we will use the vertical bar symbol to isolate the dependence of solution $\boldsymbol{r}(t)$ to Eq. (2.1) on the initial parameters $\boldsymbol{r}_0$ and t_0:

$$\boldsymbol{r}(t) = \boldsymbol{r}(t|\boldsymbol{r}_0, t_0), \quad \boldsymbol{r}_0 = \boldsymbol{r}(t_0|\boldsymbol{r}_0, t_0).$$

Let us differentiate Eq. (2.1) with respect to parameters r_{0k} and t_0. As a result, we obtain linear equations for Jacobi's matrix $\frac{\partial}{\partial r_{0k}} r_i(t|\boldsymbol{r}_0, t_0)$ and quantity $\frac{\partial}{\partial t_0} r_i(t|\boldsymbol{r}_0, t_0)$

$$\begin{aligned}
\frac{\partial r_i(t|\boldsymbol{r}_0, t_0)}{\partial r_{0k}} &= \delta_{ik} + \int\limits_{t_0}^{t} d\tau \frac{\partial U_i(\boldsymbol{r}(\tau), \tau)}{\partial r_j} \frac{\partial r_j(\tau|\boldsymbol{r}_0, t_0)}{\partial r_{0k}}, \\
\frac{\partial r_i(t|\boldsymbol{r}_0, t_0)}{\partial t_0} &= -U_i(\boldsymbol{r}_0(t_0), t_0) + \int\limits_{t_0}^{t} d\tau \frac{\partial U_i(\boldsymbol{r}(\tau), \tau)}{\partial r_j} \frac{\partial r_j(\tau|\boldsymbol{r}_0, t_0)}{\partial t_0}.
\end{aligned} \tag{2.19}$$

Multiplying now the first of these equations by $U_k(\boldsymbol{r}_0(t), t)$, summing over index k, adding the result to the second equation, and introducing the vector function

$$F_i(t|\boldsymbol{r}_0, t_0) = \left(\frac{\partial}{\partial t_0} + \boldsymbol{U}(\boldsymbol{r}_0, t_0)\frac{\partial}{\partial \boldsymbol{r}_0}\right) r_i(t|\boldsymbol{r}_0, t_0),$$

we obtain that this function satisfies the linear homogeneous equation

$$F_i(t|\boldsymbol{r}_0, t_0) = \int\limits_{t_0}^{t} d\tau \frac{\partial U_i(\boldsymbol{r}(\tau), \tau)}{\partial r_k} F_k(\tau|\boldsymbol{r}_0, t_0). \tag{2.20}$$

Differentiating this equation with respect to time, we arrive at the ordinary differential equation

$$\frac{\partial}{\partial t} F_i(t|\boldsymbol{r}_0, t_0) = \frac{\partial U_i(\boldsymbol{r}(t), t)}{\partial r_k} F_k(t|\boldsymbol{r}_0, t_0)$$

with the initial condition $F_i(t_0|\boldsymbol{r}_0, t_0) = 0$ at $t = t_0$, which follows from Eq. (2.20); as a consequence, we have $F_i(t|\boldsymbol{r}_0, t_0) \equiv 0$. Therefore, we obtain the equality

$$\left(\frac{\partial}{\partial t_0} + \boldsymbol{U}(\boldsymbol{r}_0, t_0)\frac{\partial}{\partial \boldsymbol{r}_0}\right) r_i(t|\boldsymbol{r}_0, t_0) = 0, \tag{2.21}$$

which can be considered as the linear partial differential equation with the derivatives with respect to variables $\boldsymbol{r}_0$, t_0 and the initial value at $t_0 = t$

$$\boldsymbol{r}(t|\boldsymbol{r}_0, t) = \boldsymbol{r}_0. \tag{2.22}$$

The variable t appears now in problem (2.21), (2.22) as a parameter.

Equation (2.21) is solved using the time direction inverse to that used in solving problem (1.1), page 3; for this reason, we will call it the *backward equation* .

Equation (2.21) with the initial condition (2.22) can be rewritten as the integral equation

$$\boldsymbol{r}(t|\boldsymbol{r}_0, t_0) = \boldsymbol{r}_0 + \int\limits_{t_0}^{t} d\tau \left(\boldsymbol{U}(\boldsymbol{r}_0, \tau)\frac{\partial}{\partial \boldsymbol{r}_0}\right)\boldsymbol{r}(t|\boldsymbol{r}_0, \tau). \tag{2.23}$$

Varying now Eq. (2.23) with respect to function $U_j(\boldsymbol{r}', t')$, we obtain the integral equation

$$\frac{\delta r_i(t|\boldsymbol{r}_0, t_0)}{\delta U_j(\boldsymbol{r}', t')} = \delta(\boldsymbol{r}_0 - \boldsymbol{r}')\theta(t' - t_0)\theta(t - t')\frac{\partial r_i(t|\boldsymbol{r}_0, t')}{\partial r_{j0}}$$
$$+ \int\limits_{t_0}^{t} d\tau \left(\boldsymbol{U}(\boldsymbol{r}_0, \tau)\frac{\partial}{\partial \boldsymbol{r}_0}\right)\frac{\delta r_i(t|\boldsymbol{r}_0, \tau)}{\delta U_j(\boldsymbol{r}', t')}, \tag{2.24}$$

from which follows that

$$\frac{\delta r_i(t|\boldsymbol{r}_0, t_0)}{\delta U_j(\boldsymbol{r}', t')} = 0, \quad \text{if } t' > t \quad \text{or} \quad t' < t_0,$$

which means that function $\boldsymbol{r}(t|\boldsymbol{r}_0, t_0)$ also possesses the property of dynamic causality with respect to parameter t_0 (2.15)—which is quite natural—and Eq. (2.24) can be rewritten in the form (for $t_0 < t' < t$)

$$\frac{\delta r_i(t|\boldsymbol{r}_0, t_0)}{\delta U_j(\boldsymbol{r}', t')} = \delta(\boldsymbol{r}_0 - \boldsymbol{r}')\frac{\partial r_i(t|\boldsymbol{r}_0, t')}{\partial r_{j0}} + \int\limits_{t_0}^{t'} d\tau \left(\boldsymbol{U}(\boldsymbol{r}_0, \tau)\frac{\partial}{\partial \boldsymbol{r}_0}\right)\frac{\delta r_i(t|\boldsymbol{r}_0, \tau)}{\delta U_j(\boldsymbol{r}', t')}. \tag{2.25}$$

Setting now $t' \to t_0 + 0$, we obtain the equality

$$\left.\frac{\delta r_i\left(t|\boldsymbol{r}_0,t_0\right)}{\delta U_j\left(\boldsymbol{r},t'\right)}\right|_{t'=t_0+0}=\delta\left(\boldsymbol{r}_0-\boldsymbol{r}\right)\frac{\partial r_i\left(t|\boldsymbol{r}_0,t_0\right)}{\partial r_{0j}}. \tag{2.26}$$

2.2.2 Imbedding Method for Boundary-Value Problems

Consider first boundary-value problems formulated in terms of ordinary differential equations. The *imbedding method* (or *invariant imbedding method*, as it is usually called in mathematical literature) offers a possibility of reducing boundary-value problems at hand to the evolution-type initial-value problems possessing the property of dynamic causality with respect to an auxiliary parameter.

The idea of this method was first suggested by V.A. Ambartsumyan (the so-called *Ambartsumyan invariance principle*) [2–4] for solving the equations of linear theory of radiative transfer. Further, mathematicians grasped this idea and used it to convert boundary-value (nonlinear, in the general case) problems into evolution-type initial-value problems that are more convenient for simulations. Several monographs (see, e.g., [8, 10, 38]) deal with this method and consider both physical and computational aspects.

Consider the dynamic system described in terms of the system of ordinary differential equations

$$\frac{d}{dt}\boldsymbol{x}(t)=\boldsymbol{F}\left(t,\boldsymbol{x}(t)\right), \tag{2.27}$$

defined on segment $t\in[0,T]$ with the boundary conditions

$$g\boldsymbol{x}(0)+h\boldsymbol{x}(T)=\boldsymbol{v}, \tag{2.28}$$

where g and h are the constant matrixes.

Dynamic problem (2.27), (2.28) possesses no dynamic causality property, which means that the solution $\boldsymbol{x}(t)$ to this problem at instant t functionally depends on external forces $\boldsymbol{F}\left(\tau,\boldsymbol{x}(\tau)\right)$ for all $0\le\tau\le T$. Moreover, even boundary values $\boldsymbol{x}(0)$ and $\boldsymbol{x}(T)$ are functionals of field $\boldsymbol{F}\left(\tau,\boldsymbol{x}(\tau)\right)$. The absence of dynamic causality in problem (2.27), (2.28) prevents us from using the known statistical methods of analyzing statistical characteristics of the solution to Eq. (2.27) if external force functional $\boldsymbol{F}\left(t,\boldsymbol{x}\right)$ is the random space- and time-domain field. Introducing the one-time probability density $P(t;\boldsymbol{x})$ of the solution to Eq. (2.27), we can easily see that condition (2.28) is insufficient for determining the value of this probability at any point. The boundary condition imposes only certain functional restriction.

Note that the solution to problem (2.27), (2.28) parametrically depends on T and $\boldsymbol{v}$, i.e., $\boldsymbol{x}(t)=\boldsymbol{x}(t;T,\boldsymbol{v})$. Abiding by paper [26], we introduce functions

$$\boldsymbol{R}(T,\boldsymbol{v})=\boldsymbol{x}(T;T,\boldsymbol{v}),\quad \boldsymbol{S}(T,\boldsymbol{v})=\boldsymbol{x}(0;T,\boldsymbol{v})$$

that describe the boundary values of the solution to Eq. (2.27).

Differentiate Eq. (2.27) with respect to T and $\boldsymbol{v}$. We obtain two linear equations in the corresponding derivatives

$$\begin{aligned} \frac{d}{dt}\frac{\partial x_i(t;T,\boldsymbol{v})}{\partial T} &= \frac{\partial F_i(t,\boldsymbol{x})}{\partial x_l}\frac{\partial x_l(t;T,\boldsymbol{v})}{\partial T}, \\ \frac{d}{dt}\frac{\partial x_i(t;T,\boldsymbol{v})}{\partial v_k} &= \frac{\partial F_i(t,\boldsymbol{x})}{\partial x_l}\frac{\partial x_l(t;T,\boldsymbol{v})}{\partial v_k}. \end{aligned} \tag{2.29}$$

These equations are identical in form; consequently, we can expect that their solutions are related by the linear expression

$$\frac{\partial x_i(t;T,\boldsymbol{v})}{\partial T} = \lambda_k(T,\boldsymbol{v})\frac{\partial x_i(t;T,\boldsymbol{v})}{\partial v_k} \tag{2.30}$$

if vector quantity $\boldsymbol{\lambda}(T,\boldsymbol{v})$ is such that boundary conditions (2.28) are satisfied and the solution is unique. To determine vector quantity $\boldsymbol{\lambda}(T,\boldsymbol{v})$, we first set $t=0$ in Eq. (2.30) and multiply the result by matrix g; then, we set $t=T$ and multiply the result by matrix h; and, finally, we combine the obtained expressions. Taking into account Eq. (2.28), we obtain

$$g\frac{\partial \boldsymbol{x}(0;T,\boldsymbol{v})}{\partial T} + h\left.\frac{\partial \boldsymbol{x}(t;T,\boldsymbol{v})}{\partial T}\right|_{t=T} = \boldsymbol{\lambda}(T,\boldsymbol{v}).$$

In view of the fact that

$$\left.\frac{\partial \boldsymbol{x}(t;T,\boldsymbol{v})}{\partial T}\right|_{t=T} = \frac{\partial \boldsymbol{x}(T;T,\boldsymbol{v})}{\partial T} - \left.\frac{\partial \boldsymbol{x}(t;T,\boldsymbol{v})}{\partial t}\right|_{t=T} = \frac{\partial \boldsymbol{R}(T,\boldsymbol{v})}{\partial T} - \boldsymbol{F}\left(T,\boldsymbol{R}(T,\boldsymbol{v})\right)$$

(with allowance for Eq. (2.27)), we obtain the desired expression for quantity $\boldsymbol{\lambda}(T,\boldsymbol{v})$,

$$\boldsymbol{\lambda}(T,\boldsymbol{v}) = -h\boldsymbol{F}\left(T,\boldsymbol{R}(T,\boldsymbol{v})\right). \tag{2.31}$$

Expression (2.30) with parameter $\boldsymbol{\lambda}(T,\boldsymbol{v})$ defined by Eq. (2.31), i.e., the expression

$$\frac{\partial x_i(t;T,\boldsymbol{v})}{\partial T} = -h_{kl}F_l\left(T,\boldsymbol{R}(T,\boldsymbol{v})\right)\frac{\partial x_i(t;T,\boldsymbol{v})}{\partial v_k}, \tag{2.32}$$

can be considered as the linear differential equation; one needs only to supplement it with the corresponding initial condition

$$\boldsymbol{x}(t;T,\boldsymbol{v})|_{T=t} = \boldsymbol{R}(t,\boldsymbol{v})$$

assuming that function $\boldsymbol{R}(T,\boldsymbol{v})$ is known.

The equation for this function can be obtained from the equality

$$\frac{\partial \boldsymbol{R}(T,\boldsymbol{v})}{\partial T} = \left.\frac{\partial \boldsymbol{x}(t;T,\boldsymbol{v})}{\partial t}\right|_{t=T} + \left.\frac{\partial \boldsymbol{x}(t;T,\boldsymbol{v})}{\partial T}\right|_{t=T}. \tag{2.33}$$

The right-hand side of Eq. (2.33) is the sum of the right-hand sides of Eq. (2.27) and (2.30) at $t = T$. As a result, we obtain the closed nonlinear (quasilinear) equation

$$\frac{\partial \boldsymbol{R}(T,\boldsymbol{v})}{\partial T} = -h_{kl}F_l\left(T,\boldsymbol{R}(T,\boldsymbol{v})\right)\frac{\partial \boldsymbol{R}(T,\boldsymbol{v})}{\partial v_k} + \boldsymbol{F}\left(T,\boldsymbol{R}(T,\boldsymbol{v})\right). \tag{2.34}$$

The initial condition for Eq. (2.34) follows from Eq. (2.28) for $T \to 0$

$$\boldsymbol{R}(T,\boldsymbol{v})|_{T=0} = (g+h)^{-1}\boldsymbol{v}. \tag{2.35}$$

Setting now $t = 0$ in Eq. (2.29), we obtain for the secondary boundary quantity $\boldsymbol{S}(T,\boldsymbol{v}) = \boldsymbol{x}(0;T,\boldsymbol{v})$ the equation

$$\frac{\partial \boldsymbol{S}(T,\boldsymbol{v})}{\partial T} = -h_{kl}F_l\left(T,\boldsymbol{R}(T,\boldsymbol{v})\right)\frac{\partial \boldsymbol{S}(T,\boldsymbol{v})}{\partial v_k} \tag{2.36}$$

with the initial condition

$$\boldsymbol{S}(T,\boldsymbol{v})|_{T=0} = (g+h)^{-1}\boldsymbol{v}$$

following from Eq. (2.35).

Thus, the problem reduces to the closed quasilinear equation (2.34) with initial value (2.35) and linear equation (2.30) whose coefficients and initial value are determined by the solution of Eq. (2.34).

In the problem under consideration, input 0 and output T are symmetric. For this reason, one can solve it not only from T to 0, but also from 0 to T. In the latter case, functions $\boldsymbol{R}(T,\boldsymbol{v})$ and $\boldsymbol{S}(T,\boldsymbol{v})$ switch the places.

An important point consists in the fact that, despite the initial problem (2.27) is nonlinear, Eq. (2.30) is the linear equation, because it is essentially the equation in variations. It is Eq. (2.34) that is responsible for nonlinearity.

Note that the above technique of deriving imbedding equations for Eq. (2.27) can be easily extended to the boundary condition of the form [26]

$$\boldsymbol{g}\left(\boldsymbol{x}(0)\right) + \boldsymbol{h}\left(\boldsymbol{x}(T)\right) + \int\limits_0^T d\tau \boldsymbol{K}\left(\tau,\boldsymbol{x}(\tau)\right) = \boldsymbol{v},$$

where $\boldsymbol{g}(\boldsymbol{x})$, $\boldsymbol{h}(\boldsymbol{x})$ and $\boldsymbol{K}(T,\boldsymbol{x})$ are arbitrary given vector functions.

If function $\boldsymbol{F}(t,\boldsymbol{x})$ is linear in $\boldsymbol{x}$, $F_i(t,\boldsymbol{x}) = A_{ij}(t)x_j(t)$, then boundary-value problem (2.27), (2.28) assumes the simpler form

$$\frac{d}{dt}\boldsymbol{x}(t) = A(t)\,\boldsymbol{x}(t), \quad g\boldsymbol{x}(0) + h\boldsymbol{x}(T) = \boldsymbol{v},$$

and the solution of Eq. (2.30), (2.34) and (2.36) will be the function linear in $\boldsymbol{v}$

$$\boldsymbol{x}(t; T, \boldsymbol{v}) = X(t; T)\boldsymbol{v}. \tag{2.37}$$

As a result, we arrive at the closed matrix Riccati equation for matrix $R(T) = X(T;T)$

$$\frac{d}{dT} R(T) = A(T)R(T) - R(T)hA(T)R(T), \quad R(0) = (g+h)^{-1}. \tag{2.38}$$

As regards matrix $X(t,T)$, it satisfies the linear matrix equation with the initial condition

$$\frac{\partial}{\partial T} X(t;T) = -X(t;T)hA(T)R(T), \quad X(t;T)_{T=t} = R(t). \tag{2.39}$$

Note that, for particular linear boundary problems of wave propagation, the direct derivation of imbedding equations from the specific problem statement is usually more convenient (see Appendix to Volume 2).

Chapter 3
Indicator Function and Liouville Equation

Modern apparatus of the theory of random processes is able of constructing closed descriptions of dynamic systems if these systems meet the condition of dynamic causality and are formulated in terms of linear partial differential equations or certain types of integral equations (see Chapter 5). One can use indicator functions to perform the transition from the initial, generally nonlinear system to the equivalent description in terms of the linear partial differential equations. However, this approach results in increasing the dimension of the space of variables. Consider such a transition using the dynamic systems described in Chapter 1.

3.1 Ordinary Differential Equations

Assume that a stochastic problem is described by the system of equations (1.1), page 3

$$\frac{d}{dt}\boldsymbol{r}(t) = \boldsymbol{U}\left(\boldsymbol{r}(t), t\right), \quad \boldsymbol{r}(t_0) = \boldsymbol{r}_0. \tag{3.1}$$

We introduce the scalar function

$$\varphi(\boldsymbol{r}, t) = \delta(\boldsymbol{r}(t) - \boldsymbol{r}), \tag{3.2}$$

which is concentrated on the section of the random process $\boldsymbol{r}(t)$ by a given plane $\boldsymbol{r}(t) = \boldsymbol{r}$ and is usually called the *indicator function*.

Differentiating Eq. (3.2) with respect to time t, we obtain, using Eq. (3.1), the equality

$$\frac{\partial}{\partial t}\varphi(\boldsymbol{r}, t) = -\frac{d\boldsymbol{r}(t)}{dt}\frac{\partial}{\partial \boldsymbol{r}}\delta(\boldsymbol{r}(t) - \boldsymbol{r}) = -\boldsymbol{U}\left(\boldsymbol{r}(t), t\right)\frac{\partial}{\partial \boldsymbol{r}}\delta(\boldsymbol{r}(t) - \boldsymbol{r})$$
$$= -\frac{\partial}{\partial \boldsymbol{r}}\left[\boldsymbol{U}\left(\boldsymbol{r}(t), t\right)\delta(\boldsymbol{r}(t) - \boldsymbol{r})\right].$$

V.I. Klyatskin, *Stochastic Equations: Theory and Applications in Acoustics, Hydrodyn., Magnetohydrodyn., and Radiophys., Vol. 1*, Understanding Complex Systems, DOI: 10.1007/978-3-319-07587-7_3, © Springer International Publishing Switzerland 2015

Using then the *probing property* of the delta-function

$$\boldsymbol{U}\left(\boldsymbol{r}(t), t\right) \delta(\boldsymbol{r}(t)-\boldsymbol{r})=\boldsymbol{U}\left(\boldsymbol{r}, t\right) \delta(\boldsymbol{r}(t)-\boldsymbol{r}),$$

we obtain the linear partial differential equation

$$\left(\frac{\partial}{\partial t}+\frac{\partial}{\partial \boldsymbol{r}} \boldsymbol{U}(\boldsymbol{r}, t)\right) \varphi(\boldsymbol{r}, t)=0, \quad \varphi(\boldsymbol{r}, t_0)=\delta(\boldsymbol{r}_0-\boldsymbol{r}) \tag{3.3}$$

equivalent to the initial system. This equation is called the *Liouville equation.* In terms of phase points moving in phase space $\{\boldsymbol{r}, t\}$, this equation corresponds to the equation of continuity.

Remark 3.1. ***Specifics of Working with Delta-Function***

The above derivation procedure of the Liouville equation clearly shows the necessity of distinguishing the function $\boldsymbol{r}(t)$ and the parameter $\boldsymbol{r}$. In this connection, the correct working with delta-function assumes that the argument of function $\boldsymbol{r}(t)$ was explicitly written in all intermediate calculations (cf. the forms of Eqs. (3.1) and (1.1), page 3). Neglecting this rule results usually in errors. ♦

The transition from system (3.1) to Liouville equation (3.3) entails enlargement of the phase space ($\boldsymbol{r}$, t), which, however, has the finite dimension. Note that Eq. (3.3) coincides in form with the equation of tracer transfer by the velocity field $\boldsymbol{U}(\boldsymbol{r}, t)$ (1.56); the only difference consists in the initial values.

Equation (3.3) can be rewritten in the form of the linear integral equation

$$\varphi(\boldsymbol{r}, t)=\delta(\boldsymbol{r}-\boldsymbol{r}_0)-\int_{t_0}^{t} d\tau \frac{\partial}{\partial \boldsymbol{r}}\left\{\boldsymbol{U}(\boldsymbol{r}, \tau) \varphi(\boldsymbol{r}, \tau)\right\}.$$

Varying this equation with respect to function $U_i(\boldsymbol{r}', t')$ with the use of Eq. (2.6), page 84, we obtain the integral equation

$$\begin{aligned}
\frac{\delta \varphi(\boldsymbol{r}, t)}{\delta U_j(\boldsymbol{r}', t')} &= -\int_{t_0}^{t} d\tau \frac{\partial}{\partial r_i}\left\{\frac{\delta U_i(\boldsymbol{r}, \tau)}{\delta U_j(\boldsymbol{r}', t')} \varphi(\boldsymbol{r}, \tau)+U_i(\boldsymbol{r}, \tau) \frac{\delta \varphi(\boldsymbol{r}, \tau)}{\delta U_j(\boldsymbol{r}', t')}\right\} \\
&= -\theta(t-t')\theta(t'-t_0) \frac{\partial}{\partial r_j}\left\{\delta\left(\boldsymbol{r}-\boldsymbol{r}'\right) \varphi(\boldsymbol{r}, t')\right\}-\int_{t_0}^{t} d\tau \frac{\partial}{\partial \boldsymbol{r}}\left\{\boldsymbol{U}(\boldsymbol{r}, \tau) \frac{\delta \varphi(\boldsymbol{r}, \tau)}{\delta U_j(\boldsymbol{r}', t')}\right\},
\end{aligned} \tag{3.4}$$

where $\theta(t)$ is, as earlier, the Heaviside step function. As a result, we have

$$\frac{\delta \varphi(\boldsymbol{r}, t)}{\delta U_j(\boldsymbol{r}', t')} \sim \theta(t-t')\theta(t'-t_0),$$

and the variational derivative satisfies the condition

$$\frac{\delta\varphi(\boldsymbol{r},t)}{\delta U_j(\boldsymbol{r}',t')} = 0 \quad \text{if} \quad t' < t_0 \quad \text{or} \quad t' > t,$$

which expresses the condition of dynamic causality for the Liouville equation (3.3).

For $t_0 < t' < t$, Eq. (3.4) can be written in the form

$$\frac{\delta\varphi(\boldsymbol{r},t)}{\delta U_j(\boldsymbol{r}',t')} = -\frac{\partial}{\partial r_j}\left\{\delta\left(\boldsymbol{r}-\boldsymbol{r}'\right)\varphi(\boldsymbol{r},t')\right\} - \int\limits_{t'}^{t} d\tau \frac{\partial}{\partial \boldsymbol{r}}\left\{\boldsymbol{U}(\boldsymbol{r},\tau)\frac{\delta\varphi(\boldsymbol{r},\tau)}{\delta U_j(\boldsymbol{r}',t')}\right\}, \tag{3.5}$$

from which follows that

$$\left.\frac{\delta\varphi(\boldsymbol{r},t)}{\delta U_i(\boldsymbol{r}',t')}\right|_{t=t'+0} = -\frac{\partial}{\partial r_i}\left\{\delta\left(\boldsymbol{r}-\boldsymbol{r}'\right)\varphi(\boldsymbol{r},t')\right\}. \tag{3.6}$$

Note that equality (3.6) can be derived immediately from Eq. (2.17), page 87. Indeed, by definition of function $\varphi(\boldsymbol{r},t)$, variational derivative $\delta\varphi(\boldsymbol{r},t)/\delta U_i(\boldsymbol{r}',t')$ has the form

$$\frac{\delta\varphi(\boldsymbol{r},t)}{\delta U_j(\boldsymbol{r}',t')} = \frac{\delta}{\delta U_j(\boldsymbol{r}',t')}\delta(\boldsymbol{r}(t)-\boldsymbol{r}) = -\frac{\partial}{\partial r_i}\delta(\boldsymbol{r}(t)-\boldsymbol{r})\frac{\delta r_i(t)}{\delta U_j(\boldsymbol{r}',t')}. \tag{3.7}$$

As a consequence, for $t = t' + 0$, we have the equality

$$\left.\frac{\delta\varphi(\boldsymbol{r},t)}{\delta U_j(\boldsymbol{r}',t')}\right|_{t=t'+0} = -\frac{\partial}{\partial r_i}\delta(\boldsymbol{r}(t')-\boldsymbol{r})\left.\frac{\delta r_i(t)}{\delta U_j(\boldsymbol{r}',t')}\right|_{t=t'+0}$$

$$= -\frac{\partial}{\partial r_j}\left\{\delta(\boldsymbol{r}(t')-\boldsymbol{r})\delta\left(\boldsymbol{r}'-\boldsymbol{r}\right)\right\} = -\frac{\partial}{\partial r_j}\left\{\varphi(\boldsymbol{r},t')\delta\left(\boldsymbol{r}'-\boldsymbol{r}\right)\right\}. \tag{3.8}$$

The solution to Eq. (3.1) and, consequently, function (3.2) depends on the initial values t_0, $\boldsymbol{r}_0$. Indeed, function $\boldsymbol{r}(t) = \boldsymbol{r}(t|\boldsymbol{r}_0,t_0)$ as a function of variables $\boldsymbol{r}_0$ and t_0 satisfies the linear first-order partial differential equation (2.21). The equations of such type also allow the transition to the equations in the indicator function $\varphi(t;\boldsymbol{r}|t_0,\boldsymbol{r}_0)$ (see the next section); in the case under consideration, this equation is again the linear first-order partial differential equation including the derivatives with respect to variables $\boldsymbol{r}_0$ and t_0,

$$\left(\frac{\partial}{\partial t_0} + \boldsymbol{U}(\boldsymbol{r}_0,t_0)\frac{\partial}{\partial \boldsymbol{r}_0}\right)\varphi(\boldsymbol{r},t|\boldsymbol{r}_0,t_0) = 0, \quad \varphi(\boldsymbol{r},t|\boldsymbol{r}_0,t) = \delta(\boldsymbol{r}_0-\boldsymbol{r}). \tag{3.9}$$

Equation (3.9) can be called the *backward Liouville equation*. At the same time, as a consequence of the dynamic causality, the variational derivative of indicator function $\varphi(\boldsymbol{r},t|\boldsymbol{r}_0,t_0)$,

$$\frac{\delta\varphi(\boldsymbol{r},t|\boldsymbol{r}_0,t_0)}{\delta U_j(\boldsymbol{r}',t')} = \frac{\delta}{\delta U_j(\boldsymbol{r}',t')}\delta(\boldsymbol{r}(t|\boldsymbol{r}_0,t_0)-\boldsymbol{r}) = -\frac{\partial}{\partial r_i}\delta(\boldsymbol{r}(t|\boldsymbol{r}_0,t_0)-\boldsymbol{r})\frac{\delta r_i(t|\boldsymbol{r}_0,t_0)}{\delta U_j(\boldsymbol{r}',t')},$$

can be expressed in view of Eq. (2.26), page 90 as follows

$$\left.\frac{\delta\varphi(\boldsymbol{r},t|\boldsymbol{r}_0,t_0)}{\delta U_j(\boldsymbol{r}',t')}\right|_{t'=t_0+0} = -\frac{\partial}{\partial r_i}\delta(\boldsymbol{r}(t|\boldsymbol{r}_0,t_0)-\boldsymbol{r})\left.\frac{\delta r_i(t|\boldsymbol{r}_0,t_0)}{\delta U_j(\boldsymbol{r}',t')}\right|_{t'=t_0+0}$$

$$= -\delta(\boldsymbol{r}_0-\boldsymbol{r}')\frac{\partial r_i(t|\boldsymbol{r}_0,t_0)}{\partial r_{0j}}\frac{\partial\varphi(\boldsymbol{r},t|\boldsymbol{r}_0,t_0)}{\partial r_i}. \qquad (3.10)$$

On the other hand,

$$\frac{\partial\varphi(\boldsymbol{r},t|\boldsymbol{r}_0,t_0)}{\partial r_{0j}} = -\frac{\partial r_i(t|\boldsymbol{r}_0,t_0)}{\partial r_{0j}}\frac{\partial\varphi(\boldsymbol{r},t|\boldsymbol{r}_0,t_0)}{\partial r_i},$$

and Eq. (3.10) assumes the final closed form

$$\left.\frac{\delta\varphi(\boldsymbol{r},t|\boldsymbol{r}_0,t_0)}{\delta U_j(\boldsymbol{r}',t')}\right|_{t'=t_0+0} = \delta(\boldsymbol{r}_0-\boldsymbol{r}')\frac{\partial\varphi(\boldsymbol{r},t|\boldsymbol{r}_0,t_0)}{\partial r_{0j}}. \qquad (3.11)$$

Note that, in the general case of the Hamilton systems described by Eq. (1.16), page 10, the indicator function $\varphi(\boldsymbol{r},\boldsymbol{v},t) = \delta(\boldsymbol{r}(t)-\boldsymbol{r})\delta(\boldsymbol{v}(t)-\boldsymbol{v})$ satisfies the Liouville equation (equation of continuity in $\{\boldsymbol{r},\,\boldsymbol{v}\}$-space)

$$\frac{\partial\varphi(\boldsymbol{r},\boldsymbol{v},t)}{\partial t} = \left\{-\frac{\partial}{\partial r_k}\frac{\partial H(\boldsymbol{r},\boldsymbol{v},t)}{\partial v_k} + \frac{\partial}{\partial v_k}\frac{\partial H(\boldsymbol{r},\boldsymbol{v},t)}{\partial r_k}\right\}\varphi(\boldsymbol{r},\boldsymbol{v},t)$$

that can be rewritten in the form

$$\frac{\partial\varphi(\boldsymbol{r},\boldsymbol{v},t)}{\partial t} = \{H(\boldsymbol{r},\boldsymbol{v},t),\varphi(\boldsymbol{r},\boldsymbol{v},t)\}, \qquad (3.12)$$

where

$$\{H,\varphi\} = \left(\frac{\partial H}{\partial r_k}\frac{\partial\varphi}{\partial v_k} - \frac{\partial H}{\partial v_k}\frac{\partial\varphi}{\partial r_k}\right)$$

is the Poisson bracket of functions $H(\boldsymbol{r},\boldsymbol{v},t)$ and $\varphi(\boldsymbol{r},\boldsymbol{v},t)$.

3.2 First-Order Partial Differential Equations

If the initial-value problem is formulated in terms of partial differential equations, we always can convert it to the equivalent formulation in terms of the linear variational differential equation in the infinite-dimensional space (the *Hopf equation*) [33–35] (see also [48, 49, 112]). For some particular types of problems, this approach is simplified. Indeed, if the initial dynamic system satisfies the first-order partial differential equation (either *linear* as Eq. (1.56), page 34, or *quasilinear* as Eq. (1.84), page 44, or in the general case *nonlinear* as Eq. (1.101), page 50), then the phase space of the corresponding indicator function will be the finite-dimension space [48, 49], which follows from the fact that first-order partial differential equations are equivalent to systems of ordinary (characteristic) differential equations. Consider these cases in more details.

3.2.1 Linear Equations

Backward Liouville Equation

First of all, we consider the derivation of the backward Liouville equation in indicator function

$$\varphi(\boldsymbol{r},t|\boldsymbol{r}_0,t_0) = \delta\left(\boldsymbol{r}\left(t|\boldsymbol{r}_0,t_0\right) - \boldsymbol{r}\right)$$

considered as a function of parameters $(\boldsymbol{r}_0,\, t_0)$. Differentiating function $\varphi(\boldsymbol{r},t|\boldsymbol{r}_0,t_0)$ with respect to t_0 and r_{0j} and taking into account Eq. (2.21), page 89, we obtain the expressions

$$\begin{aligned}\frac{\partial\varphi(\boldsymbol{r},t|\boldsymbol{r}_0,t_0)}{\partial t_0} &= \frac{\partial\varphi(\boldsymbol{r},t|\boldsymbol{r}_0,t_0)}{\partial r_i}\boldsymbol{U}\left(\boldsymbol{r}_0,t_0\right)\frac{\partial r_i\left(t|\boldsymbol{r}_0,t_0\right)}{\partial \boldsymbol{r}_0},\\ \frac{\partial\varphi(\boldsymbol{r},t|\boldsymbol{r}_0,t_0)}{\partial \boldsymbol{r}_0} &= -\frac{\partial\varphi(\boldsymbol{r},t|\boldsymbol{r}_0,t_0)}{\partial r_i}\frac{\partial r_i(t|\boldsymbol{r}_0,t_0)}{\partial \boldsymbol{r}_0}.\end{aligned} \tag{3.13}$$

Taking scalar product of the second equation in (3.13) with $\boldsymbol{U}\left(\boldsymbol{r}_0,t_0\right)$ and summing the result with the first one, we arrive at the closed equation

$$\left(\frac{\partial}{\partial t_0} + \boldsymbol{U}\left(\boldsymbol{r}_0,t_0\right)\frac{\partial}{\partial \boldsymbol{r}_0}\right)\varphi(\boldsymbol{r},t|\boldsymbol{r}_0,t_0) = 0, \quad \varphi(\boldsymbol{r},t|\boldsymbol{r}_0,t) = \delta(\boldsymbol{r}-\boldsymbol{r}_0), \tag{3.14}$$

which is just the backward Liouville equation (3.9) given earlier without derivation.

Rewrite Eq. (3.14) as the linear integral equation

$$\varphi(\boldsymbol{r},t|\boldsymbol{r}_0,t_0) = \delta(\boldsymbol{r}-\boldsymbol{r}_0) + \int\limits_{t_0}^{t} d\tau \boldsymbol{U}(\boldsymbol{r}_0,\tau)\frac{\partial}{\partial \boldsymbol{r}_0}\varphi(\boldsymbol{r},t|\boldsymbol{r}_0,\tau). \tag{3.15}$$

According to this equation, variational derivative $\dfrac{\delta\varphi(\boldsymbol{r},t|\boldsymbol{r}_0,t_0)}{\delta U_j(\boldsymbol{r}',t')}$ satisfies the linear integral equation

$$\frac{\delta\varphi(\boldsymbol{r},t|\boldsymbol{r}_0,t_0)}{\delta U_j(\boldsymbol{r}',t')} = \delta(\boldsymbol{r}_0-\boldsymbol{r}')\theta(t'-t_0)\theta(t-t')\frac{\partial\varphi(\boldsymbol{r},t|\boldsymbol{r}_0,t')}{\partial r_{0j}}$$

$$+\int\limits_{t_0}^{t} d\tau \boldsymbol{U}(\boldsymbol{r}_0,\tau)\frac{\partial}{\partial \boldsymbol{r}_0}\frac{\delta\varphi(\boldsymbol{r},t|\boldsymbol{r}_0,\tau)}{\delta U_j(\boldsymbol{r}',t')},$$

from which follows that

$$\frac{\delta\varphi(\boldsymbol{r},t|\boldsymbol{r}_0,t_0)}{\delta U_j(\boldsymbol{r}',t')} \sim \theta(t'-t_0)\theta(t-t'),$$

i.e., the condition of dynamic causality holds. For $t_0 < t' < t$, this equation is simplified to the form

$$\frac{\delta\varphi(\boldsymbol{r},t|\boldsymbol{r}_0,t_0)}{\delta U_j(\boldsymbol{r}',t')} = \delta(\boldsymbol{r}_0-\boldsymbol{r}')\frac{\partial\varphi(\boldsymbol{r},t|\boldsymbol{r}_0,t')}{\partial r_{0j}} + \int\limits_{t_0}^{t'} d\tau \boldsymbol{U}(\boldsymbol{r}_0,\tau)\frac{\partial}{\partial \boldsymbol{r}_0}\frac{\delta\varphi(\boldsymbol{r},t|\boldsymbol{r}_0,\tau)}{\delta U_j(\boldsymbol{r}',t')},$$

and we arrive, in the limit $t' = t_0+0$, at Eq. (3.11) derived earlier in a different way.

Density Field of Passive Tracer under Random Velocity Field

Consider the problem on tracer transfer by random velocity field in more details. The problem is formulated in terms of Eq. (1.56), page 34 that we rewrite in the form

$$\left(\frac{\partial}{\partial t} + \boldsymbol{U}(\boldsymbol{r},t)\frac{\partial}{\partial \boldsymbol{r}}\right)\rho(\boldsymbol{r},t) + \frac{\partial \boldsymbol{U}(\boldsymbol{r},t)}{\partial \boldsymbol{r}}\rho(\boldsymbol{r},t) = 0, \quad \rho(\boldsymbol{r},0) = \rho_0(\boldsymbol{r}). \quad (3.16)$$

To describe the density field in the Eulerian description, we introduce the indicator function

$$\varphi(\boldsymbol{r},t;\rho) = \delta(\rho(t,\boldsymbol{r}) - \rho), \quad (3.17)$$

which is similar to function (3.2) and is localized on surface $\rho(\boldsymbol{r},t) = \rho =$ const in the three-dimensional case or on a contour in the two-dimensional case. An equation for this function can be easily obtained either immediately from Eq. (3.16), or from the Liouville equation in the Lagrangian description. Indeed, differentiating Eq. (3.17) with respect to time and using dynamic equation (3.16) and probing property of the delta-function, we obtain the equation

$$\frac{\partial}{\partial t}\varphi(\boldsymbol{r},t;\rho) = \frac{\partial \boldsymbol{U}(\boldsymbol{r},t)}{\partial \boldsymbol{r}}\frac{\partial}{\partial \rho}\rho\varphi(\boldsymbol{r},t;\rho) + \boldsymbol{U}(\boldsymbol{r},t)\frac{\partial \rho(\boldsymbol{r},t)}{\partial \boldsymbol{r}}\frac{\partial}{\partial \rho}\varphi(\boldsymbol{r},t;\rho). \quad (3.18)$$

However, this equation is not closed because the right-hand side includes term $\partial\rho(\boldsymbol{r},t)/\partial\boldsymbol{r}$ that cannot be explicitly expressed through $\rho(\boldsymbol{r},t)$.

On the other hand, differentiating function (3.17) with respect to $\boldsymbol{r}$, we obtain the equality

$$\frac{\partial}{\partial \boldsymbol{r}}\varphi(\boldsymbol{r},t;\rho) = -\frac{\partial\rho(\boldsymbol{r},t)}{\partial \boldsymbol{r}}\frac{\partial}{\partial\rho}\varphi(\boldsymbol{r},t;\rho). \tag{3.19}$$

Eliminating now the last term in Eq. (3.18) with the use of (3.19), we obtain the closed Liouville equation in the Eulerian description

$$\begin{gathered}\left(\frac{\partial}{\partial t}+\boldsymbol{U}(\boldsymbol{r},t)\frac{\partial}{\partial \boldsymbol{r}}\right)\varphi(\boldsymbol{r},t;\rho) = \frac{\partial \boldsymbol{U}(\boldsymbol{r},t)}{\partial \boldsymbol{r}}\frac{\partial}{\partial\rho}\left[\rho\varphi(\boldsymbol{r},t;\rho)\right]\\ \varphi(\boldsymbol{r},0;\rho) = \delta(\rho_0(\boldsymbol{r})-\rho).\end{gathered} \tag{3.20}$$

Remark 3.2. ***The simplest model of velocity field***

Note that the Liouville equation (3.20) becomes the one-dimensional equation in the case of velocity field described by the simplest model (1.17), page 10 – $\boldsymbol{U}(x,t) = \boldsymbol{v}(t)f(x)$ (x is the dimensionless variable),

$$\begin{gathered}\left(\frac{\partial}{\partial t}+v_x(t)f(x)\frac{\partial}{\partial x}\right)\varphi(x,t;\rho) = v_x(t)\frac{\partial f(x)}{\partial x}\frac{\partial}{\partial\rho}\left[\rho\varphi(x,t;\rho)\right],\\ \varphi(x,0;\rho) = \delta(\rho_0(x)-\rho).\end{gathered} \tag{3.21}$$

♦

To obtain more complete description, we consider the extended indicator function including both density field $\rho(\boldsymbol{r},t)$ and its spatial gradient $\mathbf{p}(\boldsymbol{r},t) = \boldsymbol{\nabla}\rho(\boldsymbol{r},t)$

$$\varphi(\boldsymbol{r},t;\rho,\mathbf{p}) = \delta\left(\rho(\boldsymbol{r},t)-\rho\right)\delta\left(\mathbf{p}(\boldsymbol{r},t)-\mathbf{p}\right). \tag{3.22}$$

Differentiating Eq. (3.22) with respect to time, we obtain the equality

$$\frac{\partial}{\partial t}\varphi(\boldsymbol{r},t;\rho,\mathbf{p}) = -\left[\frac{\partial}{\partial\rho}\frac{\partial\rho(\boldsymbol{r},t)}{\partial t}+\frac{\partial}{\partial p_i}\frac{\partial p_i(\boldsymbol{r},t)}{\partial t}\right]\varphi(\boldsymbol{r},t;\rho,\mathbf{p}). \tag{3.23}$$

Using now dynamic equation (3.16) for density and Eq. (1.69), page 38 for the density spatial gradient, we can rewrite Eq. (3.23) as the equation

$$\begin{aligned}\frac{\partial}{\partial t}\varphi(\boldsymbol{r},t;\rho,\mathbf{p}) &= \frac{\partial}{\partial\rho}\left[\frac{\partial \boldsymbol{U}(\boldsymbol{r},t)}{\partial \boldsymbol{r}}\rho+\boldsymbol{U}(\boldsymbol{r},t)\mathbf{p}\right]\varphi(\boldsymbol{r},t;\rho,\mathbf{p})\\ &+\frac{\partial}{\partial p_i}\left[\boldsymbol{U}(\boldsymbol{r},t)\frac{\partial p_i(\boldsymbol{r},t)}{\partial \boldsymbol{r}}+\frac{\partial \boldsymbol{U}(\boldsymbol{r},t)}{\partial \boldsymbol{r}}p_i+p_k\frac{\partial U_k(\boldsymbol{r},t)}{\partial r_i}+\rho\frac{\partial^2 \boldsymbol{U}(\boldsymbol{r},t)}{\partial r_i\partial \boldsymbol{r}}\right]\varphi(\boldsymbol{r},t;\rho,\mathbf{p}),\end{aligned} \tag{3.24}$$

which is not closed because of the term $\partial p_i(\boldsymbol{r},t)/\partial\boldsymbol{r}$ in the right-hand side.

Differentiating function (3.22) with respect to $\boldsymbol{r}$, we obtain the equality

$$\frac{\partial}{\partial \boldsymbol{r}}\varphi(\boldsymbol{r},t;\rho,\boldsymbol{p}) = -\left[\boldsymbol{p}\frac{\partial}{\partial \rho} + \frac{\partial p_i(\boldsymbol{r},t)}{\partial \boldsymbol{r}}\frac{\partial}{\partial p_i}\right]\varphi(\boldsymbol{r},t;\rho,\boldsymbol{p}). \tag{3.25}$$

Now, multiplying Eq. (3.25) by $\boldsymbol{U}(\boldsymbol{r},t)$ and adding the result to Eq. (3.24), we obtain the closed Liouville equation for the extended indicator function

$$\begin{aligned}&\left(\frac{\partial}{\partial t} + \boldsymbol{U}(\boldsymbol{r},t)\frac{\partial}{\partial \boldsymbol{r}}\right)\varphi(\boldsymbol{r},t;\rho,\boldsymbol{p})\\ &= \left[\frac{\partial U_k(\boldsymbol{r},t)}{\partial r_i}\frac{\partial}{\partial p_i}p_k + \frac{\partial \boldsymbol{U}(\boldsymbol{r},t)}{\partial \boldsymbol{r}}\left(\frac{\partial}{\partial \rho}\rho + \frac{\partial}{\partial \boldsymbol{p}}\boldsymbol{p}\right) + \frac{\partial^2 \boldsymbol{U}(\boldsymbol{r},t)}{\partial r_i \partial \boldsymbol{r}}\frac{\partial}{\partial p_i}\rho\right]\varphi(\boldsymbol{r},t;\rho,\boldsymbol{p}),\end{aligned} \tag{3.26}$$

with the initial value

$$\varphi(\boldsymbol{r},0;\rho,\boldsymbol{p}) = \delta\left(\rho_0(\boldsymbol{r}) - \rho\right)\delta\left(\boldsymbol{p}_0(\boldsymbol{r}) - \boldsymbol{p}\right).$$

In the case of nondivergent velocity field ($\operatorname{div}\boldsymbol{U}(\boldsymbol{r},t) = 0$), the Liouville equation (3.26) is simplified to the form

$$\begin{aligned}&\left(\frac{\partial}{\partial t} + \boldsymbol{U}(\boldsymbol{r},t)\frac{\partial}{\partial \boldsymbol{r}}\right)\varphi(\boldsymbol{r},t;\rho,\boldsymbol{p}) = \frac{\partial U_k(\boldsymbol{r},t)}{\partial r_i}\frac{\partial}{\partial p_i}p_k\varphi(\boldsymbol{r},t;\rho,\boldsymbol{p}),\\ &\varphi(\boldsymbol{r},0;\rho,\boldsymbol{p}) = \delta\left(\rho_0(\boldsymbol{r}) - \rho\right)\delta\left(\boldsymbol{p}_0(\boldsymbol{r}) - \boldsymbol{p}\right).\end{aligned} \tag{3.27}$$

Derive now Eq. (3.20) starting from the Lagrangian description of the dynamic system. In the Lagrangian representation, the behavior of a passive tracer is described in terms of ordinary differential equations (1.59), (1.60), and (1.62), page 35. Using these equations, one can easily derive the linear Liouville equation in the corresponding phase space for the function

$$\varphi_{\text{Lag}}(\boldsymbol{r},\rho,j,t|\boldsymbol{r}_0) = \delta(\boldsymbol{r}(t|\boldsymbol{r}_0) - \boldsymbol{r})\delta(\rho(t|\boldsymbol{r}_0) - \rho)\delta(j(t|\boldsymbol{r}_0) - j), \tag{3.28}$$

which explicitly assumes that the solution to the initial dynamic equations is a function of the Lagrangian coordinates $\boldsymbol{r}_0$. This equation has the form

$$\left(\frac{\partial}{\partial t} + \frac{\partial}{\partial \boldsymbol{r}}\boldsymbol{U}(\boldsymbol{r},t)\right)\varphi_{\text{Lag}}(\boldsymbol{r},\rho,j,t|\boldsymbol{r}_0) = \frac{\partial \boldsymbol{U}(\boldsymbol{r},t)}{\partial \boldsymbol{r}}\left(\frac{\partial}{\partial \rho}\rho - \frac{\partial}{\partial j}j\right)\varphi_{\text{Lag}}(\boldsymbol{r},\rho,j,t|\boldsymbol{r}_0) \tag{3.29}$$

with the initial value

$$\varphi_{\text{Lag}}(\boldsymbol{r},\rho,j,0|\boldsymbol{r}_0) = \delta(\boldsymbol{r}_0 - \boldsymbol{r})\delta(\rho_0(\boldsymbol{r}_0) - \rho)\delta(j - 1).$$

Taking into account the equality

$$\delta(\boldsymbol{r}(t|\boldsymbol{r}_0) - \boldsymbol{r}) = \frac{1}{\|\partial r_\alpha / \partial r_{0\beta}\|}\delta(\boldsymbol{r}_0 - \boldsymbol{r}_0(t,\boldsymbol{r})) = \frac{1}{j(t|\boldsymbol{r}_0)}\delta(\boldsymbol{r}_0 - \boldsymbol{r}_0(t,\boldsymbol{r})),$$

we can rewrite function (3.28) in the form

$$\varphi_{\text{Lag}}(\boldsymbol{r},\rho,j,t|\boldsymbol{r}_0) = \frac{1}{j}\delta(\boldsymbol{r}_0 - \boldsymbol{r}_0(t,\boldsymbol{r}))\delta(j(t|\boldsymbol{r}_0) - j)\varphi(\boldsymbol{r},t;\rho),$$

where $\varphi(\boldsymbol{r},t;\rho)$ is the indicator function (3.17). Consequently,

$$\varphi(\boldsymbol{r},t;\rho) = \int d\boldsymbol{r}_0 \int_0^\infty j dj \varphi_{\text{Lag}}(\boldsymbol{r},\rho,j,t|\boldsymbol{r}_0). \tag{3.30}$$

Multiplying Eq. (3.29) by j and integrating the result over j and $\boldsymbol{r}_0$, we obtain the corresponding Liouville equation in the Eulerian representation, which coincides with Eq. (3.20).

In the case of nondivergent velocity field, Eqs. (3.16), (3.20), and (3.29) coincide. Fundamental differences appear only if the potential component is available in the velocity field.

Note that solutions to dynamic problems have the one-time and one-point probability densities that coincide with the corresponding indicator functions averaged over an ensemble of realizations

$$P(\boldsymbol{r},\rho,j,t|\boldsymbol{r}_0) = \left\langle \varphi_{\text{Lag}}\left(\boldsymbol{r},\rho,j,t|\boldsymbol{r}_0\right)\right\rangle,$$

$$P(\boldsymbol{r},t;\rho) = \left\langle \varphi\left(\boldsymbol{r},t;\rho\right)\right\rangle, \quad P(\boldsymbol{r},t;\rho,\boldsymbol{p}) = \left\langle \varphi\left(\boldsymbol{r},t;\rho,\boldsymbol{p}\right)\right\rangle.$$

This point explains the special interest to indicator functions in the statistical analyzing of dynamics of systems. In addition, indicator functions provide a good deal of data on geometric structure of random fields, which can be obtained using statistical topography of random fields (see Sect. 6.2, page 160).

Magnetic Field Diffusion under Random Velocity Field

Let us introduce the indicator function of magnetic field $\boldsymbol{H}(\boldsymbol{r},t)$,

$$\varphi(\boldsymbol{r},t;\boldsymbol{H}) = \delta(\boldsymbol{H}(\boldsymbol{r},t) - \boldsymbol{H}).$$

Differentiating this function with respect to time with the use of dynamic equation (1.78), page 40 and probing property of the delta function, we arrive at the equation

$$\frac{\partial}{\partial t}\varphi(\boldsymbol{r},t;\boldsymbol{H}) = -\frac{\partial H_i(\boldsymbol{r},t)}{\partial t}\frac{\partial}{\partial H_i}\varphi(\boldsymbol{r},t;\boldsymbol{H})$$

$$= -\left[\boldsymbol{H}(\boldsymbol{r},t)\frac{\partial u_i(\boldsymbol{r},t)}{\partial \boldsymbol{r}} - H_i(\boldsymbol{r},t)\frac{\partial \boldsymbol{u}(\boldsymbol{r},t)}{\partial \boldsymbol{r}}\right]\frac{\partial}{\partial H_i}\varphi(\boldsymbol{r},t;\boldsymbol{H})$$

$$+\boldsymbol{u}(\boldsymbol{r},t)\frac{\partial H_i(\boldsymbol{r},t)}{\partial \boldsymbol{r}}\frac{\partial}{\partial H_i}\varphi(\boldsymbol{r},t;\boldsymbol{H}) = -\frac{\partial}{\partial H_i}\left[\boldsymbol{H}\frac{\partial u_i(\boldsymbol{r},t)}{\partial \boldsymbol{r}} - H_i\frac{\partial \boldsymbol{u}(\boldsymbol{r},t)}{\partial \boldsymbol{r}}\right]\varphi(\boldsymbol{r},t;\boldsymbol{H})$$

$$+\boldsymbol{u}(\boldsymbol{r},t)\frac{\partial H_i(\boldsymbol{r},t)}{\partial \boldsymbol{r}}\frac{\partial}{\partial H_i}\varphi(\boldsymbol{r},t;\boldsymbol{H}),$$

which is unclosed because of the last term. However, since

$$\frac{\partial}{\partial \boldsymbol{r}}\varphi(\boldsymbol{r},t;\boldsymbol{H}) = -\frac{\partial H_i(\boldsymbol{r},t)}{\partial \boldsymbol{r}}\frac{\partial}{\partial H_i}\delta(\boldsymbol{H}(\boldsymbol{r},t)-\boldsymbol{H})$$

and, hence,

$$-\boldsymbol{u}(\boldsymbol{r},t)\frac{\partial}{\partial \boldsymbol{r}}\varphi(\boldsymbol{r},t;\boldsymbol{H}) = \boldsymbol{u}(\boldsymbol{r},t)\frac{\partial H_i(\boldsymbol{r},t)}{\partial \boldsymbol{r}}\frac{\partial}{\partial H_i}\delta(\boldsymbol{H}(\boldsymbol{r},t)-\boldsymbol{H}),$$

we obtain the desired closed Liouville equation [67]

$$\left(\frac{\partial}{\partial t} + \boldsymbol{u}(\boldsymbol{r},t)\frac{\partial}{\partial \boldsymbol{r}}\right)\varphi(\boldsymbol{r},t;\boldsymbol{H}) = -\frac{\partial}{\partial H_i}\left[\boldsymbol{H}\frac{\partial u_i(\boldsymbol{r},t)}{\partial \boldsymbol{r}} - H_i\frac{\partial \boldsymbol{u}(\boldsymbol{r},t)}{\partial \boldsymbol{r}}\right]\varphi(\boldsymbol{r},t;\boldsymbol{H}) \tag{3.31}$$

with the initial condition $\varphi(\boldsymbol{r},0;\boldsymbol{H}) = \delta(\boldsymbol{H}_0(\boldsymbol{r})-\boldsymbol{H})$.

Remark 3.3. ***The Simplest Model of Velocity Field***

Note that, in the case of velocity field described by the simplest model (1.17), page 10 – $\boldsymbol{u}(x,t) = \boldsymbol{v}(t)f(x)$ (x is the dimensionless variable), the Liouville equation (3.31) with homogeneous initial conditions becomes the equation

$$\frac{\partial}{\partial t}\varphi(x,t;\boldsymbol{H}) = \left[-\frac{\partial}{\partial x}f(x) + f'(x)\left(1+\frac{\partial}{\partial \boldsymbol{H}}\boldsymbol{H}\right)\right]v_x(t)\varphi(x,t;\boldsymbol{H})$$
$$- f'(x)H_{x0}v_i(t)\frac{\partial}{\partial H_i}\varphi(x,t;\boldsymbol{H}) \tag{3.32}$$

with the initial condition $\varphi(x,0;\boldsymbol{H}) = \delta(\boldsymbol{H}_0-\boldsymbol{H})$. ♦

3.2.2 Quasilinear Equations

Consider now the simplest quasilinear equation for scalar quantity $q(\boldsymbol{r},t)$ (1.84), page 44

$$\left(\frac{\partial}{\partial t}+\boldsymbol{U}(t,q)\frac{\partial}{\partial \boldsymbol{r}}\right)q(\boldsymbol{r},t)=Q\left(t,q\right),\qquad q(\boldsymbol{r},0)=q_0(\boldsymbol{r}). \tag{3.33}$$

In this case, an attempt of deriving a closed equation for the indicator function $\varphi(\boldsymbol{r},t;q)=\delta(q(\boldsymbol{r},t)-q)$ on the analogy of the linear problem will fail. Here, we must supplement Eq. (3.33) with Eq. (1.85), page 45 for the gradient field $\boldsymbol{p}(\boldsymbol{r},t)=\boldsymbol{\nabla}q(\boldsymbol{r},t)$

$$\left(\frac{\partial}{\partial t}+\boldsymbol{U}(t,q)\frac{\partial}{\partial \boldsymbol{r}}\right)\boldsymbol{p}(\boldsymbol{r},t)+\frac{\partial\left\{\boldsymbol{U}(t,q)\boldsymbol{p}(\boldsymbol{r},t)\right\}}{\partial q}\boldsymbol{p}(\boldsymbol{r},t)=\frac{\partial Q(t,q)}{\partial q}\boldsymbol{p}(\boldsymbol{r},t) \tag{3.34}$$

and consider the extended indicator function

$$\varphi(\boldsymbol{r},t;q,\boldsymbol{p})=\delta(q(\boldsymbol{r},t)-q)\delta(\boldsymbol{p}(\boldsymbol{r},t)-\boldsymbol{p}). \tag{3.35}$$

Differentiating Eq. (3.35) with respect to time and using Eqs. (3.33) and (3.34), we obtain the equation

$$\begin{aligned}\frac{\partial}{\partial t}\varphi(\boldsymbol{r},t;q,\boldsymbol{p})&=-\left[\frac{\partial}{\partial q}\frac{\partial q(\boldsymbol{r},t)}{\partial t}+\frac{\partial}{\partial p_k}\frac{\partial p_k(\boldsymbol{r},t)}{\partial t}\right]\varphi(\boldsymbol{r},t;q,\boldsymbol{p})\\&=\left\{\frac{\partial}{\partial q}\left[\boldsymbol{p}\boldsymbol{U}(t,q)-Q(t,q)\right]+\frac{\partial}{\partial p_k}\boldsymbol{U}(t,q)\frac{\partial p_k(\boldsymbol{r},t)}{\partial \boldsymbol{r}}\right\}\varphi(\boldsymbol{r},t;q,\boldsymbol{p})\\&\quad+\frac{\partial}{\partial p_k}\left[p_k\left(\boldsymbol{p}\frac{\partial \boldsymbol{U}(t,q)}{\partial q}-\frac{\partial Q(t,q)}{\partial q}\right)\right]\varphi(\boldsymbol{r},t;q,\boldsymbol{p}),\end{aligned} \tag{3.36}$$

which is not closed, however, because of the term $\partial p_k(\boldsymbol{r},t)/\partial\boldsymbol{r}$ in the right-hand side.

Differentiating function (3.35) with respect to $\boldsymbol{r}$, we obtain the equality

$$\frac{\partial}{\partial \boldsymbol{r}}\varphi(\boldsymbol{r},t;q,\boldsymbol{p})=-\left[\boldsymbol{p}\frac{\partial}{\partial q}+\frac{\partial p_k(\boldsymbol{r},t)}{\partial \boldsymbol{r}}\frac{\partial}{\partial p_k}\right]\varphi(\boldsymbol{r},t;q,\boldsymbol{p}), \tag{3.37}$$

from which follows that

$$\frac{\partial}{\partial p_k}\frac{\partial p_k(\boldsymbol{r},t)}{\partial \boldsymbol{r}}\varphi(\boldsymbol{r},t;q,\boldsymbol{p})=-\left[\frac{\partial}{\partial \boldsymbol{r}}+\boldsymbol{p}\frac{\partial}{\partial q}\right]\varphi(\boldsymbol{r},t;q,\boldsymbol{p}).$$

Consequently, Eq. (3.36) can be rewritten in the closed form

$$\begin{aligned}\left(\frac{\partial}{\partial t}+\boldsymbol{U}(t,q)\frac{\partial}{\partial \boldsymbol{r}}\right)\varphi(\boldsymbol{r},t;q,\boldsymbol{p})&=\left[\boldsymbol{p}\frac{\partial \boldsymbol{U}(t,q)}{\partial q}-\frac{\partial}{\partial q}Q(t,q)\right]\varphi(\boldsymbol{r},t;q,\boldsymbol{p})\\&\quad+\frac{\partial}{\partial p_k}\left[p_k\left(\boldsymbol{p}\frac{\partial \boldsymbol{U}(t,q)}{\partial q}-\frac{\partial Q(t,q)}{\partial q}\right)\right]\varphi(\boldsymbol{r},t;q,\boldsymbol{p}),\end{aligned} \tag{3.38}$$

which is just the desired Liouville equation for the quasilinear equation (3.33) in the extended phase space $\{q, \boldsymbol{p}\}$ with the initial value

$$\varphi(\boldsymbol{r}, 0; q, \boldsymbol{p}) = \delta(q_0(\boldsymbol{r}) - q)\delta(\boldsymbol{p}_0(\boldsymbol{r}) - \boldsymbol{p}).$$

Note that the equation of continuity for conserved quantity $I(\boldsymbol{r}, t)$

$$\frac{\partial}{\partial t} I(\boldsymbol{r}, t) + \frac{\partial}{\partial \boldsymbol{r}} \{\boldsymbol{U}(t, q) I(\boldsymbol{r}, t)\} = 0, \qquad I(\boldsymbol{r}, 0) = I_0(\boldsymbol{r})$$

can be combined with Eqs. (3.33), (3.34). In this case, the indicator function has the form

$$\varphi(\boldsymbol{r}, t; q, \boldsymbol{p}, I) = \delta(q(\boldsymbol{r}, t) - q)\delta(\boldsymbol{p}(\boldsymbol{r}, t) - \boldsymbol{p})\delta(I(\boldsymbol{r}, t) - I)$$

and derivation of the closed equation for this function appears possible in space $\{q, \boldsymbol{p}, I\}$, which follows from the fact that, in the Lagrangian description, quantity inverse to $I(\boldsymbol{r}, t)$ coincides with the divergence.

Note that the Liouville equation in indicator function

$$\varphi(x, t; q, p, \chi) = \delta\left(q(x, t) - u\right)\delta\left(p(x, t) - p\right)\delta\left(\chi(x, t) - \chi\right)$$

in the case of the one-dimensional Riemann equation (1.96), page 47 in function $q(x, t)$ complemented with the equations for the gradient field $p(x, t) = \dfrac{\partial q(x, t)}{\partial x}$ — (1.98) and the logarithm of the density field $\chi(x, t) = \ln \rho(x, t)$ — (1.100) assumes the form of the equation

$$\left(\frac{\partial}{\partial t} + q\frac{\partial}{\partial x}\right)\varphi(x, t; q, p, \chi) = \left(p + \frac{\partial}{\partial p}p^2 + p\frac{\partial}{\partial \chi}\right)\varphi(x, t; q, p, \chi) \tag{3.39}$$

with the initial condition

$$\varphi(x, 0; q, p, \chi) = \delta(q_0(x) - q)\delta(p_0(x) - p)\delta\left(\chi_0(x) - \chi\right).$$

The solution to Eq. (3.39) can be represented in the form

$$\varphi(x, t; q, p, \chi) = \exp\left\{-t\left(q\frac{\partial}{\partial x} - p\frac{\partial}{\partial \chi}\right)\right\}\widetilde{\varphi}(x, t; q, p, \chi),$$

where function $\widetilde{\varphi}(x, t; q, p)$ satisfies the equation

$$\left(\frac{\partial}{\partial t} - p^2\frac{\partial}{\partial p}\right)\widetilde{\varphi}(x, t; q, p, \chi) = 3p\widetilde{\varphi}(x, t; q, p, \chi) \tag{3.40}$$

with the initial condition

$$\varphi(x, 0; q, p, \chi) = \delta(q_0(x) - q)\delta(p_0(x) - p)\delta\left(\chi_0(x) - \chi\right).$$

Solving now Eq. (3.40) by the method of characteristics, we obtain the expression

$$\widetilde{\varphi}(x,t;q,p,\chi) = \frac{1}{(1-pt)^3}\delta(q_0(x)-q)\delta\left(p_0(x)-\frac{p}{1-pt}\right) \\ \times\delta\left(\chi_0(x)-\ln\left(1+tp_0(x)\right)-\chi\right),$$

that can be rewritten in the form

$$\widetilde{\varphi}(x,t;q,p,\chi) = [1+tp_0(x)]\,\delta(q_0(x)-q)\delta\left(\frac{p_0(x)}{1+tp_0(x)}-p\right) \\ \times\delta\left(\chi_0(x)-\ln\left(1+tp_0(x)\right)-\chi\right). \quad (3.41)$$

Thus, we obtain the solution to Eq. (3.39) in the final form

$$\varphi(x,t;q,p,\chi) = \exp\left\{-qt\frac{\partial}{\partial x}\right\}[1+tp_0(x)]\,\delta(q_0(x)-q)\delta\left(\frac{p_0(x)}{1+tp_0(x)}-p\right) \\ \times\delta\left(\chi_0(x)-\ln\left(1+tp_0(x)\right)-\chi\right), \quad (3.42)$$

in agreement with solutions on page 47.

Waves on Sea Surface

Consider now problem (1.174), page 71 that describes generation of sea surface displacements under the effect of random vertical velocity of the sea medium.

Introduce the indicator function of surface displacement

$$\varphi(\boldsymbol{R},t;\xi,\boldsymbol{p}) = \delta\left(\xi(\boldsymbol{R},t)-\xi\right) \quad (3.43)$$

and the joined indicator function of surface displacement and gradient

$$\varphi(\boldsymbol{R},t;\xi,\boldsymbol{p}) = \delta\left(\xi(\boldsymbol{R},t)-\xi\right)\delta\left(\boldsymbol{p}(\boldsymbol{R},t)-\boldsymbol{p}\right). \quad (3.44)$$

Then, using equality

$$\frac{\partial\varphi(\boldsymbol{R},t;\xi)}{\partial t} = -\frac{\partial\xi(\boldsymbol{R},t)}{\partial t}\frac{\partial}{\partial\xi}\varphi(\boldsymbol{R},t;\xi)$$

and taking into account the equation for the field of water surface displacement (Eq. (1.174), page 71), we obtain an equation in function (3.43)

$$\frac{\partial\varphi(\boldsymbol{R},t;\xi)}{\partial t} - \frac{\partial\xi(\boldsymbol{R},t)}{\partial R_i}\frac{\partial}{\partial\xi}u_i(\boldsymbol{R},\xi;t)\varphi(\boldsymbol{R},t;\xi) = -\frac{\partial}{\partial\xi}w(\boldsymbol{R},\xi;t)\varphi(\boldsymbol{R},t;\xi), \quad (3.45)$$

which is nonclosed because it contains an additional unknown function, namely, the gradient of the water surface displacement.

Equation for the joined indicator function $\varphi(\boldsymbol{R},t;\xi,\boldsymbol{p})$ has the form

$$\frac{\partial\varphi(\boldsymbol{R},t;\xi,\boldsymbol{p})}{\partial t}=-\frac{\partial\xi(\boldsymbol{R},t)}{\partial t}\frac{\partial\varphi(\boldsymbol{R},t;\xi,\boldsymbol{p})}{\partial\xi}-\frac{\partial p_k(\boldsymbol{R},t)}{\partial t}\frac{\partial\varphi(\boldsymbol{R},t;\xi,\boldsymbol{p})}{\partial p_k}. \quad (3.46)$$

Using Eq. (1.175), page (71) that describes the gradient of surface displacement $p_k(\boldsymbol{R},t)=\dfrac{\partial\xi(\boldsymbol{R},t)}{\partial R_k}$, we can rewrite Eq. (3.46) in the form

$$\begin{aligned}\frac{\partial\varphi(\boldsymbol{R},t;\xi,\boldsymbol{p})}{\partial t}&=\frac{\partial}{\partial\xi}\left[u_i(\boldsymbol{R},\xi,t)p_i-w(\boldsymbol{R},\xi;t)\right]\varphi(\boldsymbol{R},t;\xi,\boldsymbol{p})\\&+\frac{\partial}{\partial p_k}\left[\frac{\partial u_i(\boldsymbol{R},\xi;t)}{\partial R_k}p_i+\frac{\partial u_i(\boldsymbol{R},\xi;t)}{\partial z}p_kp_i\right]\varphi(\boldsymbol{R},t;\xi,\boldsymbol{p})\\&-\frac{\partial}{\partial p_k}\left[\frac{\partial w(\boldsymbol{R},\xi;t)}{\partial R_k}+\frac{\partial w(\boldsymbol{R},\xi;t)}{\partial z}p_k\right]\varphi(\boldsymbol{R},t;\xi,\boldsymbol{p})\\&+\frac{\partial^2\xi(\boldsymbol{R},t)}{\partial R_k\partial R_i}u_i(\boldsymbol{R},\xi(\boldsymbol{R},t);t)\frac{\partial}{\partial p_k}\varphi(\boldsymbol{R},t;\xi,\boldsymbol{p}), \quad (3.47)\end{aligned}$$

which is nonclosed again, but containes now the second spatial derivative of water surface displacement.

In order to close Eq. (3.47), we calculate spatial derivative

$$\frac{\partial}{\partial R_i}\varphi(\boldsymbol{R},t;\xi,\boldsymbol{p})=-\frac{\partial\xi(\boldsymbol{R},t)}{\partial R_i}\frac{\partial}{\partial\xi}\varphi(\boldsymbol{R},t;\xi,\boldsymbol{p})-\frac{\partial^2\xi(\boldsymbol{R},t)}{\partial R_i\partial R_k}\frac{\partial}{\partial p_k}\varphi(\boldsymbol{R},t;\xi,\boldsymbol{p}) \quad (3.48)$$

and use this expression to eliminate the last term in Eq. (3.47). As a result, we obtain the closed Liouville equation of the form

$$\begin{aligned}&\frac{\partial\varphi(\boldsymbol{R},t;\xi,\boldsymbol{p})}{\partial t}+u_i(\boldsymbol{R},\xi;t)\left(\frac{\partial}{\partial R_i}+p_i\frac{\partial}{\partial\xi}\right)\varphi(\boldsymbol{R},t;\xi,\boldsymbol{p})\\&\quad=\frac{\partial}{\partial\xi}\left[p_iu_i(\boldsymbol{R},\xi;t)-w(\boldsymbol{R},\xi;t)\right]\varphi(\boldsymbol{R},t;\xi,\boldsymbol{p})\\&\quad+\frac{\partial}{\partial p_k}\left[\frac{\partial u_i(\boldsymbol{R},\xi;t)}{\partial R_k}p_i+\frac{\partial u_i(\boldsymbol{R},\xi;t)}{\partial\xi}p_kp_i\right]\varphi(\boldsymbol{R},t;\xi,\boldsymbol{p})\\&\quad-\frac{\partial}{\partial p_k}\left[\frac{\partial w(\boldsymbol{R},\xi;t)}{\partial R_k}+\frac{\partial w(\boldsymbol{R},\xi;t)}{\partial\xi}p_k\right]\varphi(\boldsymbol{R},t;\xi,\boldsymbol{p}). \quad (3.49)\end{aligned}$$

3.2.3 General-Form Nonlinear Equations

Consider now the scalar nonlinear first-order partial differential equation in the general form (1.101), page 50

$$\frac{\partial}{\partial t}q(\boldsymbol{r},t)+H\left(\boldsymbol{r},t,q,\boldsymbol{p}\right)=0,\quad q\left(\boldsymbol{r},0\right)=q_0\left(\boldsymbol{r}\right),\tag{3.50}$$

where $\boldsymbol{p}(\boldsymbol{r},t)=\partial q(\boldsymbol{r},t)/\partial\boldsymbol{r}$. In order to derive the closed Liouville equation in this case, we must supplement Eq. (3.50) with equations for vector $\boldsymbol{p}(\boldsymbol{r},t)$ and second derivative matrix $U_{ik}(\boldsymbol{r},t)=\partial^2 q(\boldsymbol{r},t)/\partial r_i\partial r_k$.

Introduce now the extended indicator function

$$\varphi(\boldsymbol{r},t;q,\boldsymbol{p},U,I)=\delta(q(\boldsymbol{r},t)-q)\delta(\boldsymbol{p}(\boldsymbol{r},t)-\boldsymbol{p})\delta(U(\boldsymbol{r},t)-U)\delta(I(\boldsymbol{r},t)-I),\tag{3.51}$$

where we included for generality an additional conserved variable $I(\boldsymbol{r},t)$ satisfying the equation of continuity (1.103)

$$\frac{\partial}{\partial t}I(\boldsymbol{r},t)+\frac{\partial}{\partial\boldsymbol{r}}\left\{\frac{\partial H\left(\boldsymbol{r},t,q,\boldsymbol{p}\right)}{\partial\boldsymbol{p}}I(\boldsymbol{r},t)\right\}=0,\quad I\left(\boldsymbol{r},0\right)=I_0\left(\boldsymbol{r}\right).\tag{3.52}$$

Equations (3.50), (3.52) describe, for example, wave propagation in inhomogeneous media within the frames of the geometrical optics approximation of the parabolic equation of quasi-optics. Differentiating function (3.51) with respect to time and using dynamic equations for functions $q(\boldsymbol{r},t)$, $\boldsymbol{p}(\boldsymbol{r},t)$, $U(\boldsymbol{r},t)$ and $I(\boldsymbol{r},t)$, we generally obtain an unclosed equation containing third-order derivatives of function $q(\boldsymbol{r},t)$ with respect to spatial variable $\boldsymbol{r}$. However, the combination

$$\left(\frac{\partial}{\partial t}+\frac{\partial H\left(\boldsymbol{r},t,q,\boldsymbol{p}\right)}{\partial\boldsymbol{p}}\frac{\partial}{\partial\boldsymbol{r}}\right)\varphi(\boldsymbol{r},t;q,\boldsymbol{p},U,I)$$

will not include the third-order derivatives; as a result, we obtain the closed Liouville equation in space $\{q,\boldsymbol{r},U,I\}$ [48, 49].

3.3 Higher-Order Partial Differential Equations

If the initial dynamic system includes higher-order derivatives (e.g., Laplace operator), derivation of a closed equation for the corresponding indicator functionbecomes impossible. In this case, only the variational differential equation (the *Hopf equation*) can be derived in the closed form for the functional whose average over an ensemble of realizations coincides with the *characteristic functional* of the solution to the corresponding dynamic equation.

Consider such a transition using the partial differential equations considered in Chapter 1 as examples.

3.3.1 Parabolic Equation of Quasi-Optics

The first example concerns wave propagation in a random medium within the frames of the linear parabolic equation (1.153), page 63

$$\frac{\partial}{\partial x}u(x,\boldsymbol{R}) = \frac{i}{2k}\Delta_{\boldsymbol{R}}u(x,\boldsymbol{R}) + \frac{ik}{2}\varepsilon(x,\boldsymbol{R})u(x,\boldsymbol{R}), \qquad u(0,\boldsymbol{R}) = u_0(\boldsymbol{R}). \quad (3.53)$$

Equation (3.53) as the initial-value problem possesses the property of dynamic causality with respect to parameter x. Indeed, rewrite this equation in the integral form

$$u(x,\boldsymbol{R}) = u_0(\boldsymbol{R}) + \frac{i}{2k}\int\limits_0^x d\xi \Delta_{\boldsymbol{R}}u(\xi,\boldsymbol{R}) + \frac{ik}{2}\int\limits_0^x d\xi\varepsilon(\xi,\boldsymbol{R})u(\xi,\boldsymbol{R}). \quad (3.54)$$

Function $u(x,\boldsymbol{R})$ is a functional of field $\varepsilon(x,\boldsymbol{R})$, i.e.

$$u(x,\boldsymbol{R}) = u\left[x,\boldsymbol{R};\varepsilon\left(\widetilde{x},\widetilde{\boldsymbol{R}}\right)\right].$$

Varying Eq. (3.54) with respect to function $\varepsilon(x',\boldsymbol{R}')$, we obtain the integral equation in variational derivative $\dfrac{\delta u(x,\boldsymbol{R})}{\delta\varepsilon(x',\boldsymbol{R}')}$

$$\frac{\delta u(x,\boldsymbol{R})}{\delta\varepsilon(x',\boldsymbol{R}')} = \frac{ik}{2}\theta(x-x')\theta(x')\delta(\boldsymbol{R}-\boldsymbol{R}')u(x',\boldsymbol{R})$$

$$+\frac{ik}{2}\int\limits_0^x d\xi\varepsilon(\xi,\boldsymbol{R})\frac{\delta u(\xi,\boldsymbol{R})}{\delta\varepsilon(x',\boldsymbol{R}')} + \frac{i}{2k}\int\limits_0^x d\xi\Delta_{\boldsymbol{R}}\frac{\delta u(\xi,\boldsymbol{R})}{\delta\varepsilon(x',\boldsymbol{R}')}. \quad (3.55)$$

From this equation follows that

$$\frac{\delta u(x,\boldsymbol{R})}{\delta\varepsilon(x',\boldsymbol{R}')} = 0 \quad \text{if} \quad x < x' \quad \text{or} \quad x' < 0, \quad (3.56)$$

which just expresses the dynamic causality property. As a consequence of this fact, we can rewrite Eq. (3.55) in the form (for $0 < x' < x$)

$$\frac{\delta u(x,\boldsymbol{R})}{\delta\varepsilon(x',\boldsymbol{R}')} = \frac{ik}{2}\delta(\boldsymbol{R}-\boldsymbol{R}')u(x',\boldsymbol{R})$$

$$+\frac{ik}{2}\int\limits_{x'}^{x} d\xi\varepsilon(\xi,\boldsymbol{R})\frac{\delta u(\xi,\boldsymbol{R})}{\delta\varepsilon(x',\boldsymbol{R}')} + \frac{i}{2k}\int\limits_{x'}^{x} d\xi\Delta_{\boldsymbol{R}}\frac{\delta u(\xi,\boldsymbol{R})}{\delta\varepsilon(x',\boldsymbol{R}')} \tag{3.57}$$

from which follows the equality

$$\frac{\delta u(x,\boldsymbol{R})}{\delta\varepsilon(x-0,\boldsymbol{R}')} = \frac{ik}{2}\delta(\boldsymbol{R}-\boldsymbol{R}')u(x,\boldsymbol{R}). \tag{3.58}$$

Consider now the functional

$$\varphi[x;v(\boldsymbol{R}'),v^*(\boldsymbol{R}')] = \exp\left\{i\int d\boldsymbol{R}'\left[u(x,\boldsymbol{R}')v(\boldsymbol{R}') + u^*(x,\boldsymbol{R}')v^*(\boldsymbol{R}')\right]\right\}, \tag{3.59}$$

where wavefield $u(x,\boldsymbol{R})$ satisfies Eq. (3.53) and $u^*(x,\boldsymbol{R})$ is the complex conjugated function. Differentiating (3.59) with respect to x and using dynamic equation (3.53) and its complex conjugate, we obtain the equality

$$\frac{\partial}{\partial x}\varphi[x;v(\boldsymbol{R}'),v^*(\boldsymbol{R}')] = -\frac{1}{2k}\int d\boldsymbol{R}\left[v(\boldsymbol{R})\Delta_{\boldsymbol{R}}u(x,\boldsymbol{R})\right.$$

$$\left.-v^*(\boldsymbol{R})\Delta_{\boldsymbol{R}}u^*(x,\boldsymbol{R})\right]\varphi[x;v(\boldsymbol{R}'),v^*(\boldsymbol{R}')]$$

$$-\frac{k}{2}\int d\boldsymbol{R}\varepsilon(x,\boldsymbol{R})\left[v(\boldsymbol{R})u(x,\boldsymbol{R}) - v^*(\boldsymbol{R})u^*(x,\boldsymbol{R})\right]\varphi[x;v(\boldsymbol{R}'),v^*(\boldsymbol{R}')],$$

which can be written as the variational differential equation

$$\frac{\partial}{\partial x}\varphi[x;v(\boldsymbol{R}'),v^*(\boldsymbol{R}')] = \frac{ik}{2}\int d\boldsymbol{R}\varepsilon(x,\boldsymbol{R})\widehat{M}(\boldsymbol{R})\varphi[x;v(\boldsymbol{R}'),v^*(\boldsymbol{R}')]$$

$$+\frac{i}{2k}\int d\boldsymbol{R}\left[v(\boldsymbol{R})\Delta_{\boldsymbol{R}}\frac{\delta}{\delta v(\boldsymbol{R})} - v^*(\boldsymbol{R})\Delta_{\boldsymbol{R}}\frac{\delta}{\delta v^*(\boldsymbol{R})}\right]\varphi[x;v(\boldsymbol{R}'),v^*(\boldsymbol{R}')] \tag{3.60}$$

with the initial value

$$\varphi[0;v(\boldsymbol{R}'),v^*(\boldsymbol{R}')] = \exp\left\{i\int d\boldsymbol{R}'\left[u_0(\boldsymbol{R}')v(\boldsymbol{R}') + u_0^*(\boldsymbol{R}')v^*(\boldsymbol{R}')\right]\right\}, \tag{3.61}$$

and with the Hermitian operator

$$\widehat{M}(\boldsymbol{R}) = v(\boldsymbol{R})\frac{\delta}{\delta v(\boldsymbol{R})} - v^*(\boldsymbol{R})\frac{\delta}{\delta v^*(\boldsymbol{R})}.$$

Equation (3.60) is equivalent to the input equation (3.53).

Rewrite problem (3.60) with the initial condition (3.61) in the integral form

$$\varphi[x; v(\widetilde{\boldsymbol{R}}), v^*(\widetilde{\boldsymbol{R}})] = \varphi[0; v(\widetilde{\boldsymbol{R}}), v^*(\widetilde{\boldsymbol{R}})] + \frac{ik}{2}\int_0^x d\xi \int d\boldsymbol{R}\varepsilon(\xi,\boldsymbol{R})\widehat{M}(\boldsymbol{R})\varphi[\xi; v(\widetilde{\boldsymbol{R}}), v^*(\widetilde{\boldsymbol{R}})]$$
$$+\frac{i}{2k}\int_0^x d\xi \int d\boldsymbol{R}\left[v(\boldsymbol{R})\Delta_{\boldsymbol{R}}\frac{\delta}{\delta v(\boldsymbol{R})} - v^*(\boldsymbol{R})\Delta_{\boldsymbol{R}}\frac{\delta}{\delta v^*(\boldsymbol{R})}\right]\varphi[\xi; v(\widetilde{\boldsymbol{R}}), v^*(\widetilde{\boldsymbol{R}})] \tag{3.62}$$

and vary the resulting equation with respect to function $\varepsilon(x', \boldsymbol{R}')$. Then we obtain that variational derivative $\dfrac{\delta\varphi[x; v(\widetilde{\boldsymbol{R}}), v^*(\widetilde{\boldsymbol{R}})]}{\delta\varepsilon(x', \boldsymbol{R}')}$ satisfies the integral equation

$$\frac{\delta\varphi[x; v(\widetilde{\boldsymbol{R}}), v^*(\widetilde{\boldsymbol{R}})]}{\delta\varepsilon(x', \boldsymbol{R}')} = \theta(x-x')\theta(x')\frac{ik}{2}\widehat{M}(\boldsymbol{R}')\varphi[x'; v(\widetilde{\boldsymbol{R}}), v^*(\widetilde{\boldsymbol{R}})] + \frac{ik}{2}\int_0^x d\xi \int d\boldsymbol{R}\varepsilon(\xi,\boldsymbol{R})\widehat{M}(\boldsymbol{R})\frac{\delta\varphi[\xi; v(\widetilde{\boldsymbol{R}}), v^*(\widetilde{\boldsymbol{R}})]}{\delta\varepsilon(x', \boldsymbol{R}')}$$
$$+\frac{i}{2k}\int_0^x d\xi \int d\boldsymbol{R}\left[v(\boldsymbol{R})\Delta_{\boldsymbol{R}}\frac{\delta}{\delta v(\boldsymbol{R})} - v^*(\boldsymbol{R})\Delta_{\boldsymbol{R}}\frac{\delta}{\delta v^*(\boldsymbol{R})}\right]\frac{\delta\varphi[\xi; v(\widetilde{\boldsymbol{R}}), v^*(\widetilde{\boldsymbol{R}})]}{\delta\varepsilon(x', \boldsymbol{R}')}. \tag{3.63}$$

In view of the dynamic causality property, we can rewrite it in the form of the equation (for $0 < x' < x$)

$$\frac{\delta\varphi[x; v(\widetilde{\boldsymbol{R}}), v^*(\widetilde{\boldsymbol{R}})]}{\delta\varepsilon(x', \boldsymbol{R}')} = \frac{ik}{2}\widehat{M}(\boldsymbol{R}')\varphi[x'; v(\widetilde{\boldsymbol{R}}), v^*(\widetilde{\boldsymbol{R}})] + \frac{ik}{2}\int_{x'}^x d\xi \int d\boldsymbol{R}\varepsilon(\xi,\boldsymbol{R})\widehat{M}(\boldsymbol{R})\frac{\delta\varphi[\xi; v(\widetilde{\boldsymbol{R}}), v^*(\widetilde{\boldsymbol{R}})]}{\delta\varepsilon(x', \boldsymbol{R}')}$$
$$+\frac{i}{2k}\int_{x'}^x d\xi \int d\boldsymbol{R}\left[v(\boldsymbol{R})\Delta_{\boldsymbol{R}}\frac{\delta}{\delta v(\boldsymbol{R})} - v^*(\boldsymbol{R})\Delta_{\boldsymbol{R}}\frac{\delta}{\delta v^*(\boldsymbol{R})}\right]\frac{\delta\varphi[\xi; v(\widetilde{\boldsymbol{R}}), v^*(\widetilde{\boldsymbol{R}})]}{\delta\varepsilon(x', \boldsymbol{R}')} \tag{3.64}$$

from which follows the equality

$$\frac{\delta}{\delta\varepsilon(x-0,\boldsymbol{R})}\varphi[x;v(\widetilde{\boldsymbol{R}}),v^{*}(\widetilde{\boldsymbol{R}})]=\frac{ik}{2}\widehat{M}(\boldsymbol{R})\varphi[x;v(\widetilde{\boldsymbol{R}}),v^{*}(\widetilde{\boldsymbol{R}})]. \tag{3.65}$$

3.3.2 Random Forces in Hydrodynamic Theory of Turbulence

Consider now integro-differential equation (1.171), page 69 for the Fourier harmonics $\widehat{\boldsymbol{u}}(\boldsymbol{k},t)$ of the solution to the Navier–Stokes equation (1.164)

$$\frac{\partial}{\partial t}\widehat{u}_i(\boldsymbol{k},t)+\frac{i}{2}\int d\boldsymbol{k}_1\int d\boldsymbol{k}_2\Lambda_i^{\alpha\beta}(\boldsymbol{k}_1,\boldsymbol{k}_2,\boldsymbol{k})\,\widehat{u}_\alpha(\boldsymbol{k}_1,t)\,\widehat{u}_\beta(\boldsymbol{k}_2,t)$$
$$-\nu k^2\widehat{u}_i(\boldsymbol{k},t)=\widehat{f}_i(\boldsymbol{k},t), \tag{3.66}$$

where

$$\Lambda_i^{\alpha\beta}(\boldsymbol{k}_1,\boldsymbol{k}_2,\boldsymbol{k})=\frac{1}{(2\pi)^3}\left\{k_\alpha\Delta_{i\beta}(\boldsymbol{k})+k_\beta\Delta_{i\alpha}(\boldsymbol{k})\right\}\delta(\boldsymbol{k}_1+\boldsymbol{k}_2-\boldsymbol{k}),$$

$$\Delta_{ij}(\boldsymbol{k})=\delta_{ij}-\frac{k_ik_j}{\boldsymbol{k}^2}\quad(i,\,\alpha,\,\beta\,=\,1,\,2,\,3),$$

and $\widehat{\boldsymbol{f}}(\boldsymbol{k},t)$ is the spatial Fourier harmonic of external forces.

Function $\widehat{u}_i(\boldsymbol{k},t)$ satisfies the principle of dynamic causality, which means that the variational derivative of function

$$\widehat{u}_i(\boldsymbol{k},t)=\widehat{u}_i\left[\boldsymbol{k},t;\widehat{f}_\alpha\left(\widetilde{\boldsymbol{k}},\widetilde{t}\right)\right]$$

(considered as a functional of field $\widehat{\boldsymbol{f}}(\boldsymbol{k},t)$) with respect to field $\widehat{\boldsymbol{f}}(\boldsymbol{k}',t')$ vanishes for $t'>t$ and $t'<0$,

$$\frac{\delta\widehat{u}_i(\boldsymbol{k},t)}{\delta\widehat{f}_j(\boldsymbol{k}',t')}=0,\quad\text{if}\quad t'>t\quad\text{or}\quad t'<0,$$

and, moreover,

$$\frac{\delta\widehat{u}_i(\boldsymbol{k},t)}{\delta\widehat{f}_j(\boldsymbol{k}',t-0)}=\delta_{ij}\delta(\boldsymbol{k}-\boldsymbol{k}'). \tag{3.67}$$

Consider the functional

$$\varphi[t;\boldsymbol{z}]=\varphi[t;\boldsymbol{z}(\boldsymbol{k}')]=\exp\left\{i\int d\boldsymbol{k}'\widehat{\boldsymbol{u}}(\boldsymbol{k}',t)\boldsymbol{z}(\boldsymbol{k}')\right\}. \tag{3.68}$$

Differentiating this functional with respect to time t and using dynamic equation (3.66), we obtain the equality

$$\frac{\partial}{\partial t}\varphi[t;\boldsymbol{z}] = \frac{1}{2}\int d\boldsymbol{k}z_i(\boldsymbol{k})\int d\boldsymbol{k}_1\int d\boldsymbol{k}_2\Lambda_i^{\alpha\beta}(\boldsymbol{k}_1,\boldsymbol{k}_2,\boldsymbol{k})\,\widehat{u}_\alpha(\boldsymbol{k}_1,t)\,\widehat{u}_\beta(\boldsymbol{k}_2,t)\,\varphi[t;\boldsymbol{z}]$$
$$-i\int d\boldsymbol{k}z_i(\boldsymbol{k})\left\{\nu k^2\widehat{u}_i(\boldsymbol{k},t)-\widehat{f}_i(\boldsymbol{k},t)\right\}\varphi[t;\boldsymbol{z}],$$

which can be rewritten in the functional space as the linear *Hopf equation* containing variational derivatives

$$\frac{\partial}{\partial t}\varphi[t;\boldsymbol{z}] = -\frac{1}{2}\int d\boldsymbol{k}z_i(\boldsymbol{k})\int d\boldsymbol{k}_1\int d\boldsymbol{k}_2\Lambda_i^{\alpha\beta}(\boldsymbol{k}_1,\boldsymbol{k}_2,\boldsymbol{k})\,\frac{\delta^2\varphi[t;\boldsymbol{z}]}{\delta z_\alpha(\boldsymbol{k}_1)\,\delta z_\beta(\boldsymbol{k}_2)}$$
$$-\int d\boldsymbol{k}z_i(\boldsymbol{k})\left\{\nu k^2\frac{\delta}{\delta z_i(\boldsymbol{k})}-i\widehat{f}_i(\boldsymbol{k},t)\right\}\varphi[t;\boldsymbol{z}] \quad (3.69)$$

with the given initial condition at $t=0$.

This means that Eq. (3.69) satisfy the condition of dynamic causality; a consequence of this condition is the equality

$$\frac{\delta}{\delta\widehat{\boldsymbol{f}}(\boldsymbol{k},t-0)}\varphi[t;\boldsymbol{z}] = i\boldsymbol{z}(\boldsymbol{k})\varphi[t;\boldsymbol{z}]. \quad (3.70)$$

In a similar way, considering the space-time harmonics of the velocity field $\widehat{u}_i(\boldsymbol{K})$, where $\boldsymbol{K}$ is the four-dimensional wave vector $\{\boldsymbol{k},\omega\}$ and $\widehat{u}_i^*(\boldsymbol{K}) = \widehat{u}_i(-\boldsymbol{K})$ because field $u_i(\boldsymbol{r},t)$ is real, we obtain the nonlinear integral equation (1.172), page 70

$$(i\omega+\nu\boldsymbol{k}^2)\widehat{u}_i(\boldsymbol{K})+\frac{i}{2}\int d^4\boldsymbol{K}_1\int d^4\boldsymbol{K}_2\Lambda_i^{\alpha\beta}(\boldsymbol{K}_1,\boldsymbol{K}_2,\boldsymbol{K})\,\widehat{u}_\alpha(\boldsymbol{K}_1)\,\widehat{u}_\beta(\boldsymbol{K}) = \widehat{f}_i(\boldsymbol{K}), \quad (3.71)$$

where

$$\Lambda_i^{\alpha\beta}(\boldsymbol{K}_1,\boldsymbol{K}_2,\boldsymbol{K}) = \frac{1}{2\pi}\Lambda_i^{\alpha\beta}(\boldsymbol{k}_1,\boldsymbol{k}_2,\boldsymbol{k})\,\delta(\omega_1+\omega_2-\omega),$$

and $\widehat{f}_i(\boldsymbol{K})$ are the space-time Fourier harmonics of external forces. In this case, dealing with the functional

$$\varphi[\boldsymbol{z}] = \varphi[\boldsymbol{z}(\boldsymbol{K}')] = \exp\left\{i\int d^4\boldsymbol{K}'\widehat{\boldsymbol{u}}(\boldsymbol{K}')\boldsymbol{z}(\boldsymbol{K}')\right\}, \quad (3.72)$$

we derive the linear variational integro-differential equation of the form

$$(i\omega+\nu\boldsymbol{k}^2)\frac{\delta\varphi[\boldsymbol{z}]}{\delta z_i(\boldsymbol{K})} = -\frac{1}{2}\int d^4\boldsymbol{K}_1\int d^4\boldsymbol{K}_2\Lambda_i^{\alpha\beta}(\boldsymbol{K}_1,\boldsymbol{K}_2,\boldsymbol{K})\,\frac{\delta\varphi[\boldsymbol{z}]}{\delta z_\alpha(\boldsymbol{K}_1)\,\delta z_\beta(\boldsymbol{K}_2)}$$
$$+i\widehat{f}_i(\boldsymbol{K})\,\varphi[\boldsymbol{z}]. \quad (3.73)$$

Part II
Random Quantities, Processes, and Fields

Chapter 4
Random Quantities and Their Characteristics

Prior to consider statistical descriptions of the problems mentioned in Part 1, we discuss basic concepts of the theory of random quantities, processes, and fields.

This chapter concerns those basic properties of random quantities, processes, and fields that are widely used in analyzing dynamic systems with fluctuating parameters, but only slightly elucidated in textbooks. Here, we will follow monographs [46, 48, 49], in which one can also found the fundamental references concerning the problem.

The probability for a random quantity ξ to fall in interval$-\infty < \xi < z$ is the monotonic function

$$F(z) = \mathrm{P}(-\infty < \xi < z) = \langle \theta(z - \xi) \rangle_{\xi}\,, \quad F(\infty) = 1, \tag{4.1}$$

where

$$\theta(z) = \begin{cases} 1, & \text{if } z > 0, \\ 0, & \text{if } z < 0 \end{cases}$$

is the Heaviside step function and $\langle \cdots \rangle_{\xi}$ denotes averaging over anensemble of realizations of random quantity ξ. This function is called the *probability distribution function* or the *integral distribution function*. Definition (4.1) reflects the real-world procedure of finding the probability according to the rule

$$\mathrm{P}(-\infty < \xi < z) = \lim_{N\to\infty} \frac{n}{N},$$

where n is the integer equal to the number of realizations of event $\xi < z$ in N independent trials. Consequently, the probability for a random quantity ξ to fall into interval $z < \xi < z + dz$, where dz is the infinitesimal increment, can be written in the form

$$\mathrm{P}(z < \xi < z + dz) = p(z)dz,$$

V.I. Klyatskin, *Stochastic Equations: Theory and Applications in Acoustics, Hydrodyn., Magnetohydrodyn., and Radiophys., Vol. 1*, Understanding Complex Systems, DOI: 10.1007/978-3-319-07587-7_4, © Springer International Publishing Switzerland 2015

where function $p(z)$ called the *probability density* is represented by the formula

$$p(z) = \frac{d}{dz}\mathrm{P}(-\infty < \xi < z) = \langle \delta(z-\xi) \rangle_{\xi}\,, \tag{4.2}$$

where $\delta(z)$ is the Dirac delta function. In terms of probability density $p(z)$, the integral distribution function is expressed by the formula

$$F(z) = \mathrm{P}(-\infty < \xi < z) = \int\limits_{-\infty}^{z} d\xi p(\xi), \tag{4.3}$$

so that

$$p(z) > 0, \qquad \int\limits_{-\infty}^{\infty} dz p(z) = 1.$$

Multiplying Eq. (4.2) by arbitrary function $f(z)$ and integrating over the whole domain of variable z, we express the mean value of the arbitrary function of random quantity in the following form

$$\langle f(\xi) \rangle_{\xi} = \int\limits_{-\infty}^{\infty} dz p(z) f(z). \tag{4.4}$$

The *characteristic function* defined by the equality

$$\Phi(v) = \left\langle e^{iv\xi} \right\rangle_{\xi} = \int\limits_{-\infty}^{\infty} dz e^{ivz} p(z)$$

is a very important quantity that exhaustively describes all characteristics of random quantity ξ. The characteristic function being known, we can obtain the probability density (via the Fourier transform)

$$p(x) = \frac{1}{2\pi} \int\limits_{-\infty}^{\infty} dv \Phi(v) e^{-ivx},$$

moments

$$M_n = \langle \xi^n \rangle = \int\limits_{-\infty}^{\infty} dz p(z) z^n = \left(\frac{d}{idv} \right)^n \Phi(v) \bigg|_{v=0},$$

cumulants (or semi-invariants)

$$K_n = \left(\frac{d}{idv} \right)^n \Theta(v) \bigg|_{v=0},$$

where $\Theta(v) = \ln \Phi(v)$, and other statistical characteristics. In terms of moments and cumulants of random quantity ξ, functions $\Theta(v)$ and $\Phi(v)$ are the Taylor series

$$\Phi(v) = \sum_{n=0}^{\infty} \frac{i^n}{n!} M_n v^n, \quad \Theta(v) = \sum_{n=1}^{\infty} \frac{i^n}{n!} K_n v^n. \tag{4.5}$$

In the case of multidimensional random quantity $\boldsymbol{\xi} = \{z_1, \cdots, z_n\}$, the exhaustive statistical description assumes the multidimensional characteristic function

$$\Phi(\boldsymbol{v}) = \left\langle e^{i\boldsymbol{v}\boldsymbol{\xi}} \right\rangle_{\boldsymbol{\xi}}, \quad \boldsymbol{v} = \{v_1, \cdots, v_n\}. \tag{4.6}$$

The corresponding joined probability density for quantities $\xi_1, \cdots, \xi_n$ is the Fourier transform of characteristic function $\Phi(\boldsymbol{v})$, i.e.,

$$P(\boldsymbol{x}) = \frac{1}{(2\pi)^n} \int d\boldsymbol{v} \Phi(\boldsymbol{v}) e^{-i\boldsymbol{v}\boldsymbol{x}}, \quad \boldsymbol{x} = \{x_1, \cdots, x_n\}. \tag{4.7}$$

Substituting function $\Phi(\boldsymbol{v})$ defined by Eq. (4.6) in Eq. (4.7) and integrating the result over $\boldsymbol{v}$, we obtain the obvious equality

$$P(\boldsymbol{x}) = \langle \delta(\boldsymbol{\xi} - \boldsymbol{x}) \rangle_{\boldsymbol{\xi}} = \langle \delta(\xi_1 - x_1) \cdots \delta(\xi_n - x_n) \rangle \tag{4.8}$$

that can serve the definition of the probability density of random vector quantity $\boldsymbol{\xi}$.

In this case, the moments and cumulants of random quantity $\boldsymbol{\xi}$ are defined by the expressions

$$M_{i_1,\cdots,i_n} = \left. \frac{\partial^n}{i^n \partial v_{i_1} \cdots \partial v_{i_n}} \Phi(\boldsymbol{v}) \right|_{\boldsymbol{v}=0}, \quad K_{i_1,\cdots,i_n} = \left. \frac{\partial^n}{i^n \partial v_{i_1} \cdots \partial v_{i_n}} \Theta(\boldsymbol{v}) \right|_{\boldsymbol{v}=0},$$

where $\Theta(\boldsymbol{v}) = \ln \Phi(\boldsymbol{v})$, and functions $\Theta(\boldsymbol{v})$ and $\Phi(\boldsymbol{v})$ are expressed in terms of moments $M_{i_1,\cdots,i_n}$ and cumulants $K_{i_1,\cdots,i_n}$ via the Taylor series

$$\Phi(\boldsymbol{v}) = \sum_{n=0}^{\infty} \frac{i^n}{n!} M_{i_1,\cdots,i_n} v_{i_1} \cdots v_{i_n}, \quad \Theta(\boldsymbol{v}) = \sum_{n=1}^{\infty} \frac{i^n}{n!} K_{i_1,\cdots,i_n} v_{i_1} \cdots v_{i_n}. \tag{4.9}$$

Note that, for quantities ξ assuming only discrete values $\xi_i (i = 1, 2, \cdots)$ with probabilities p_i, formula (4.8) is replaced with its discrete analog

$$p_k = \left\langle \delta_{z,\xi_k} \right\rangle,$$

where $\delta_{i,k}$ is the Kronecker delta ($\delta_{i,k} = 1$ for $i = k$ and 0 otherwise).

Consider now statistical average $\langle \xi f(\xi) \rangle_{\xi}$, where $f(z)$ is arbitrary deterministic function such that the above average exists. We calculate this average using the procedure that will be widely used in what follows. Instead of $f(\xi)$, we consider function $f(\xi + \eta)$, where η is arbitrary deterministic quantity.

Expand function $f(\xi+\eta)$ in the Taylor series in powers of ξ,, i.e., represent it in the form

$$f(\xi+\eta)=\sum_{n=0}^{\infty}\frac{1}{n!}f^{(n)}(\eta)\xi^{n}=\exp\left\{\xi\frac{d}{d\eta}\right\}f(\eta),$$

where we introduced the shift operator with respect to η. Then we can write the equality

$$\langle\xi f(\xi+\eta)\rangle_{\xi}=\left\langle\xi\exp\left\{\xi\frac{d}{d\eta}\right\}\right\rangle_{\xi}f(\eta)$$

$$=\frac{\left\langle\xi\exp\left\{\xi\frac{d}{d\eta}\right\}\right\rangle_{\xi}}{\left\langle\exp\left\{\xi\frac{d}{d\eta}\right\}\right\rangle_{\xi}}\left\langle\exp\left\{\xi\frac{d}{d\eta}\right\}\right\rangle_{\xi}f(\eta)=\Omega\left(\frac{d}{id\eta}\right)\langle f(\xi+\eta)\rangle_{\xi}, \quad (4.10)$$

where function

$$\Omega(v)=\frac{\langle\xi e^{i\xi v}\rangle_{\xi}}{\langle e^{i\xi v}\rangle_{\xi}}=\frac{d}{idv}\ln\Phi(v)=\frac{d}{idv}\Theta(v),$$

and $\Phi(v)$ is the characteristic function of random quantity ξ. Using now the Taylor series (4.5) for function $\Theta(v)$, we obtain function $\Omega(v)$ in the form of the series

$$\Omega(v)=\sum_{n=0}^{\infty}\frac{i^{n}}{n!}K_{n+1}v^{n}. \quad (4.11)$$

Because variable η appears in the right-hand side of Eq. (4.10) only as the term of sum $\xi+\eta$, we can replace differentiation with respect to η with differentiation with respect to ξ (in so doing, operator $\Omega(d/id\xi)$ should be introduced into averaging brackets) and set $\eta=0$. As a result, we obtain the equality

$$\langle\xi f(\xi)\rangle_{\xi}=\left\langle\Omega\left(\frac{d}{id\xi}\right)f(\xi)\right\rangle_{\xi},$$

which can be rewritten, using expansion (4.11) for $\Omega(v)$, as the series in cumulants K_n

$$\langle\xi f(\xi)\rangle_{\xi}=\sum_{n=0}^{\infty}\frac{1}{n!}K_{n+1}\left\langle\frac{d^{n}f(\xi)}{d\xi^{n}}\right\rangle_{\xi}. \quad (4.12)$$

Note that, setting $f(\xi)=\xi^{n-1}$ in Eq. (4.12), we obtain the recurrence formula that relates moments and cumulants of random quantity ξ in the form

$$M_n = \sum_{k=1}^{n} \frac{(n-1)!}{(k-1)!(n-k)!} K_k M_{n-k} \quad (M_0 = 1, \quad n = 1, 2, \cdots). \tag{4.13}$$

In a similar way, we can obtain the following operator expression for statistical average $\langle g(\xi) f(\xi) \rangle_\xi$

$$\langle g(\xi+\eta_1) f(\xi+\eta_2) \rangle_\xi = \exp\left\{ \Theta\left(\frac{d}{id\eta_1} + \frac{d}{id\eta_2}\right) - \Theta\left(\frac{d}{id\eta_1}\right) - \Theta\left(\frac{d}{id\eta_2}\right) \right\}$$
$$\times \langle g(\xi+\eta_1) \rangle_\xi \langle f(\xi+\eta_2) \rangle_\xi . \tag{4.14}$$

In the particular case $g(z) = e^{\omega z}$, where parameter ω assumes complex values too, we obtain the expression

$$\left\langle e^{\omega\xi} f(\xi+\eta) \right\rangle_\xi = \exp\left\{ \Theta\left(\frac{1}{i}\left(\omega + \frac{d}{d\eta}\right)\right) - \Theta\left(\frac{d}{id\eta}\right) \right\} \langle f(\xi+\eta) \rangle_\xi . \tag{4.15}$$

To illustrate practicability of the above formulas, we consider two types of random quantities ξ as examples.

1. Let ξ be the Gaussian random quantity with the probability density

$$p(z) = \frac{1}{\sqrt{2\pi}\sigma} \exp\left\{-\frac{z^2}{2\sigma^2}\right\}.$$

Then, we have

$$\Phi(v) = \exp\left\{-\frac{v^2\sigma^2}{2}\right\}, \quad \Theta(v) = -\frac{v^2\sigma^2}{2},$$

so that

$$M_1 = K_1 = \langle \xi \rangle = 0, \quad M_2 = K_2 = \sigma^2 = \left\langle \xi^2 \right\rangle, \quad K_{n>2} = 0.$$

In this case, the recurrence formula (4.13) assumes the form

$$M_n = (n-1)\sigma^2 M_{n-2}, \quad n = 2, \cdots, \tag{4.16}$$

from which follows that

$$M_{2n+1} = 0, \quad M_{2n} = (2n-1)!!\sigma^{2n}.$$

For averages (4.12) and (4.15), we obtain the expressions

$$\langle \xi f(\xi) \rangle_\xi = \sigma^2 \left\langle \frac{df(\xi)}{d\xi} \right\rangle_\xi , \quad \left\langle e^{\omega\xi} f(\xi) \right\rangle_\xi = \exp\left\{\frac{\omega^2\sigma^2}{2}\right\} \left\langle f(\xi + \omega\sigma^2) \right\rangle_\xi . \tag{4.17}$$

Additional useful formulas can be derived from Eqs. (4.17); for example,

$$\left\langle e^{\omega\xi}\right\rangle_{\xi} = \exp\left\{\frac{\omega^2\sigma^2}{2}\right\}, \quad \left\langle \xi e^{\omega\xi}\right\rangle_{\xi} = \omega\sigma^2 \exp\left\{\frac{\omega^2\sigma^2}{2}\right\},$$

and so on.

If random quantity ξ has a nonzero mean value, $\langle\xi\rangle \neq 0$, the probability density is given by the expression

$$P(z) = \frac{1}{\sqrt{2\pi\sigma^2}} \exp\left\{-\frac{(\xi - \langle\xi\rangle)^2}{2\sigma^2}\right\}. \tag{4.18}$$

In this case, the characteristic function assumes the form

$$\Phi(v) = \exp\left\{iv\left\langle\xi\right\rangle - \frac{v^2\sigma^2}{2}\right\}, \quad \Theta(v) = iv\left\langle\xi\right\rangle - \frac{v^2\sigma^2}{2},$$

and, consequently, moments and cumulants are given by the expressions

$$M_1 = K_1 = \langle\xi\rangle, \;\; M_2 = \langle\xi\rangle^2 + \sigma^2, \;\; K_2 = \sigma^2, \;\; K_{n>2} = 0.$$

The corresponding integral distribution function (4.3) assumes the form

$$F(z) = \frac{1}{\sqrt{2\pi\sigma^2}} \int\limits_{-\infty}^{z} d\xi \, \exp\left\{-\frac{(\xi - \langle\xi\rangle)^2}{2\sigma^2}\right\}$$

$$= \frac{1}{\sqrt{2\pi}} \int\limits_{-\infty}^{\frac{z-\langle\xi\rangle}{\sigma}} dy \, \exp\left\{-\frac{y^2}{2}\right\} = \Pr\left(\frac{z - \langle\xi\rangle}{\sigma}\right), \tag{4.19}$$

where function

$$\Pr(z) = \frac{1}{\sqrt{2\pi}} \int\limits_{-\infty}^{z} dx \, \exp\left\{-\frac{x^2}{2}\right\} \tag{4.20}$$

is known as the *probability integral.* It is clear that

$$\Pr(\infty) = 1 \quad \text{and} \quad \Pr(0) = \frac{1}{2}. \tag{4.21}$$

For negative z ($z < 0$), this function satisfies the relationship

$$\Pr(-|z|) = \frac{1}{\sqrt{2\pi}} \int\limits_{|z|}^{\infty} dx \, \exp\left\{-\frac{x^2}{2}\right\} = 1 - \Pr(|z|). \tag{4.22}$$

Asymptotic behavior of probability integral for $z \to \pm\infty$ can be easily derived from Eqs. (4.20) and (4.22); namely,

$$\Pr(z) \approx \begin{cases} 1 - \dfrac{1}{z\sqrt{2\pi}} \exp\left\{-\dfrac{z^2}{2}\right\}, & z \to \infty, \\ \dfrac{1}{|z|\sqrt{2\pi}} \exp\left\{-\dfrac{z^2}{2}\right\} & z \to -\infty. \end{cases} \tag{4.23}$$

If we deal with the random Gaussian vector whose components are ξ_i ($\langle \xi_i \rangle_{\boldsymbol{\xi}} = 0$, $i = 1, \cdots, n$), then the characteristic function is given by the equality

$$\Phi(\boldsymbol{v}) = \exp\left\{-\frac{1}{2} B_{ij} v_i v_j\right\}, \quad \Theta(\boldsymbol{v}) = -\frac{1}{2} B_{ij} v_i v_j,$$

where matrix $B_{ij} = \langle \xi_i \xi_j \rangle$ and repeated indices assume summation. In this case, Eq. (4.17) is replaced with the equality

$$\langle \boldsymbol{\xi} f(\boldsymbol{\xi}) \rangle_{\boldsymbol{\xi}} = B \left\langle \frac{df(\boldsymbol{\xi})}{d\boldsymbol{\xi}} \right\rangle_{\boldsymbol{\xi}}. \tag{4.24}$$

2. Let $\xi \equiv n$ be the integer random quantity governed by Poisson distribution

$$p_n = \frac{\bar{n}^n}{n!} e^{-\bar{n}},$$

where $\bar{n}$ is the average value of quantity n. In this case, we have

$$\Phi(v) = \exp\left\{\bar{n}\left(e^{iv} - 1\right)\right\}, \quad \Theta(v) = \bar{n}\left(e^{iv} - 1\right).$$

The recurrence formula (4.13) and Eq. (4.12) assume for this random quantity the forms

$$M_l = \bar{n} \sum_{k=0}^{l-1} \frac{(l-1)!}{k!(l-1-k)!} M_k \equiv \bar{n} \left\langle (n+1)^{l-1} \right\rangle, \ \langle n f(n) \rangle = \bar{n} \langle f(n+1) \rangle. \tag{4.25}$$

Chapter 5
Random Processes and Their Characteristics

5.1 General Remarks

If we deal with random function (random process) $z(t)$, then all statistical characteristics of this function at any fixed instant t are exhaustively described in terms of the one-time probability density

$$P(z,t) = \langle \delta\left(z(t) - z\right) \rangle \tag{5.1}$$

dependent parametrically on time t by the following relationship

$$\langle f(z(t)) \rangle = \int\limits_{-\infty}^{\infty} dz f(z) P(z,t).$$

Note that the singular Dirac delta function

$$\varphi(z,t) = \delta\left(z(t) - z\right)$$

appeared in Eq. (5.1) in angle brackets of averaging is called the *indicator function*.

The integral distribution function for this process, i.e. the probability of the event that process $z(t) < Z$ at instant t, is calculated by the formula

$$F(t,Z) = \mathrm{P}(z(t) < Z) = \int\limits_{-\infty}^{Z} dz P(z,t)$$

from which follows that

$$F(t,Z) = \langle \theta(Z - z(t)) \rangle, \quad F(t,\infty) = 1, \tag{5.2}$$

V.I. Klyatskin, *Stochastic Equations: Theory and Applications in Acoustics, Hydrodyn., Magnetohydrodyn., and Radiophys., Vol. 1*, Understanding Complex Systems,
DOI: 10.1007/978-3-319-07587-7_5,

where $\theta(z)$ is the Heaviside step function equal to zero for $z < 0$ and unity for $z > 0$.

Similar definitions hold for the two-time probability density

$$P(z_1, t_1; z_2, t_2) = \langle \varphi(z_1, t_1; z_2, t_2) \rangle$$

and for the general case of the n-time probability density

$$P(z_1, t_1; \cdots ; z_n, t_n) = \langle \varphi(z_1, t_1; \cdots ; z_n, t_n) \rangle ,$$

where

$$\varphi(z_1, t_1; \cdots ; z_n, t_n) = \delta(z(t_1) - z_1) \cdots \delta(z(t_n) - z_n).$$

is the n-time indicator function.

Process $z(t)$ is called *stationary* if all its statistical characteristics are invariant with respect to arbitrary temporal shift, i.e., if

$$P(z_1, t_1 + \tau; \cdots ; z_n, t_n + \tau) = P(z_1, t_1; \cdots ; z_n, t_n). \tag{5.3}$$

In particular, the one-time probability density of stationary process is at all independent of time, and the correlation function depends only on difference of times,

$$B_z(t_1, t_2) = \langle z(t_1) z(t_2) \rangle = B_z(t_1 - t_2).$$

Temporal scale τ_0 characteristic of correlation function $B_z(t)$ is called the temporal correlation radius of process $z(t)$. We can determine this scale, say, by the equality

$$\int_0^\infty \langle z(t + \tau) z(t) \rangle \, d\tau = \tau_0 \left\langle z^2(t) \right\rangle . \tag{5.4}$$

Note that the Fourier transform of stationary process correlation function

$$\Phi_z(\omega) = \int_{-\infty}^{\infty} dt B_z(t) e^{i\omega t}$$

is called the *temporal spectral function* (or simply *temporal spectrum*).

An exhaustive description of random function $z(t)$ can be given in terms of the characteristic functional

$$\Phi[v(\tau)] = \left\langle \exp\left\{ i \int_{-\infty}^{\infty} d\tau v(\tau) z(\tau) \right\} \right\rangle , \tag{5.5}$$

where $v(t)$ is arbitrary (but sufficiently smooth) function. Functional $\Phi[v(\tau)]$ being known, one can determine such characteristics of random function $z(t)$

as mean value $\langle z(t)\rangle$, correlation function $\langle z(t_1)z(t_2)\rangle$, n-time moment function $\langle z(t_1)\cdots z(t_n)\rangle$, etc.

Indeed, expanding functional $\Phi[v(\tau)]$ in the functional Taylor series, we obtain the representation of characteristic functional in terms of the moment functions of process $z(t)$:

$$\Phi[v(\tau)] = \sum_{n=0}^{\infty} \frac{i^n}{n!} \int_{-\infty}^{\infty} dt_1 \cdots \int_{-\infty}^{\infty} dt_n M_n(t_1, \cdots, t_n) v(t_1) \cdots v(t_n),$$

$$M_n(t_1, \cdots, t_n) = \langle z(t_1) \cdots z(t_n)\rangle = \frac{1}{i^n} \frac{\delta^n}{\delta v(t_1) \cdots \delta v(t_n)} \Phi[v(\tau)]\bigg|_{v=0}.$$

Consequently, the moment functions of random process $z(t)$ are expressed in terms of variational derivatives of the characteristic functional. See Sect. 2.1.1, page 81 for variational derivative definitions and the corresponding operation rules.

Represent functional $\Phi[v(\tau)]$ in the form $\Phi[v(\tau)] = \exp\{\Theta[v(\tau)]\}$. Functional $\Theta[v(\tau)]$ also can be expanded in the functional Taylor series

$$\Theta[v(\tau)] = \sum_{n=1}^{\infty} \frac{i^n}{n!} \int_{-\infty}^{\infty} dt_1 \cdots \int_{-\infty}^{\infty} dt_n K_n(t_1, \cdots, t_n) v(t_1) \cdots v(t_n), \tag{5.6}$$

where function

$$K_n(t_1, \cdots, t_n) = \frac{1}{i^n} \frac{\delta^n}{\delta v(t_1) \cdots \delta v(t_n)} \Theta[v(\tau)]\bigg|_{v=0}$$

is called the n-th order *cumulant function* of random process $z(t)$.

Characteristic functionals ensure an exhaustive description of random processes. However, even one-time probability densities provide certain information about temporal behavior of random processes for arbitrary long temporal intervals.

5.1.1 Random Process Typical Realization Curve

Following works [53, 71], we discuss first the concept of *typical realization curve* of random process $z(t)$. This concept concerns the fundamental features of the behavior of a separate process realization as a whole for temporal intervals of arbitrary duration.

A feature common of all random functions (processes) $z(t)$ consists in the fact that their particular realizations contain irregular spikes (*fluctuations*) both positive and negative relative certain deterministic curve called the *typical realization curve*

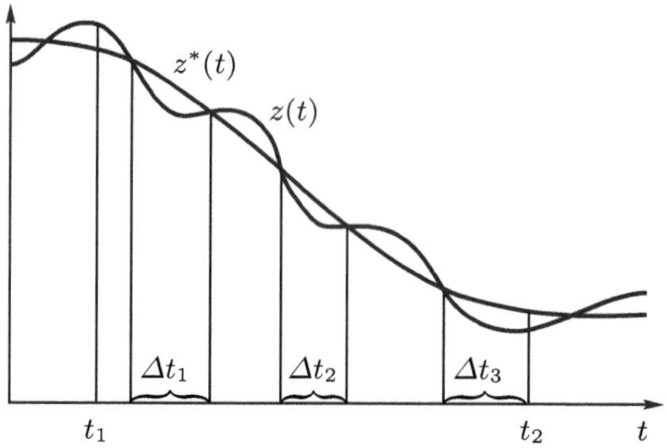

Fig. 5.1 To the definition of the typical realization curve of a random process

The typical realization curve of random process $z(t)$ is defined as the deterministic curve $z^*(t)$, which is the *median of the integral distribution function* (5.2) and is determined as the solution to the algebraic equation

$$F(t, z^*(t)) = \int_{-\infty}^{z^*(t)} dz\, P(z,t) = \frac{1}{2}. \tag{5.7}$$

This means, on the one hand, that for any t probabilities

$$\mathrm{P}\{z(t) > z^*(t)\} = \mathrm{P}\{z(t) < z^*(t)\} = 1/2.$$

On the other hand, a specific property of this curve as the median is that, for any temporal interval (t_1, t_2), random process $z(t)$ entwines about curve $z^*(t)$ in a way to force the identity of average times during which the inequalities $z(t) > z^*(t)$ and $z(t) < z^*(t)$ hold (Fig. 5.1):

$$\left\langle T_{z(t)>z^*(t)} \right\rangle = \left\langle T_{z(t)<z^*(t)} \right\rangle = \frac{1}{2}(t_2 - t_1). \tag{5.8}$$

Indeed, integrating Eq. (5.7) over temporal interval (t_1, t_2), we obtain

$$\int_{t_1}^{t_2} dt F(t, z^*(t)) = \frac{1}{2}(t_2 - t_1). \tag{5.9}$$

On the other hand, in view of definition of the integral distribution function (5.2), the integral in the right-hand side of Eq. (5.9) can be represented as

$$\int_{t_1}^{t_2} dtF\left(t, z^*(t)\right) = \left\langle T(t_1, t_2)\right\rangle, \tag{5.10}$$

where $T(t_1, t_2) = \sum_1^N \Delta t_k$ is the combined time during which the realization of process $z(t)$ appears below curve $z^*(t)$ in interval (t_1, t_2). Combining Eqs. (5.9) and (5.10), we obtain Eq. (5.8).

Curve $z^*(t)$ can significantly differ from any particular realization of process $z(t)$ and cannot describe possible magnitudes of spikes. Nevertheless, the definitional domain of typical realization curve $z^*(t)$ of random process $z(t)$ derived from the one-time probability density is the whole temporal axis $t \in (0, \infty)$. This property common of all random processes is called *intermittency*. For more detailed discussion of this concept, see Sect. 10.4.4, page 320.

Consideration of specific random processes allow obtaining an additional information concerning the realization spikes relative to this curve.

5.1.2 Statistics of Random Process Cross Points with a Line

The one-time probability density (5.1) of random process $z(t)$ is a result of averaging the singular indicator function over an ensemble of realizations of this process. This function is concentrated at points at which process $z(t)$ crosses line $z =$ const. Because the cross points are determined as roots of algebraic equation

$$z(t_n) = z \quad (n = 0, 1, \cdots, \infty),$$

we can rewrite the indicator function in the following form

$$\varphi(z, t) = \sum_{k=1}^{n} \frac{1}{|p(t_k)|} \delta(t - t_k),$$

where $p(t) = \dfrac{d}{dt} z(t)$.

The number of cross points by itself is obviously a random quantity described by the formula

$$n(t, z) = \int_{-\infty}^{t} d\tau |p(\tau)| \varphi(\tau; z). \tag{5.11}$$

As a consequence, the average number of points where process $z(t)$ crosses line $z =$ const can be described in terms of the correlation between the process derivative with respect to time and the process indicator function, or in terms of joint one-time probability density of process $z(t)$ and its derivative with respect to time $\frac{d}{dt}z(t)$.

In a similar way, we can determine certain elements of statistics related to some other special points (such as points of maxima or minima) of random process $z(t)$.

Consider several examples of random processes.

5.2 Gaussian Random Process

We start the discussion with the continuous processes; namely, we consider the Gaussian random process $z(t)$ with zero-valued mean ($\langle z(t)\rangle = 0$) and correlation function $B(t_1,t_2) = \langle z(t_1)z(t_2)\rangle$. The corresponding characteristic functional assumes the form

$$\Phi[v(\tau)] = \exp\left\{-\frac{1}{2}\int\limits_{-\infty}^{\infty}\int\limits_{-\infty}^{\infty} dt_1 dt_2 B(t_1,t_2)v(t_1)v(t_2)\right\}. \tag{5.12}$$

Only one cumulant function (the correlation function)

$$K_2(t_1,t_2) = B(t_1,t_2)$$

is different from zero for this process, so that

$$\Theta[v(\tau)] = -\frac{1}{2}\int\limits_{-\infty}^{\infty}\int\limits_{-\infty}^{\infty} dt_1 dt_2 B(t_1,t_2)v(t_1)v(t_2). \tag{5.13}$$

Consider the nth-order variational derivative of functional $\Phi[v(\tau)]$. It satisfies the following line of equalities:

$$\frac{\delta^n}{\delta v(t_1)\cdots\delta v(t_n)}\Phi[v(\tau)] = \frac{\delta^{n-1}}{\delta v(t_2)\cdots\delta v(t_n)}\frac{\delta\Theta[v(\tau)]}{\delta v(t_1)}\Phi[v(\tau)]$$

$$= \frac{\delta^2\Theta[v(\tau)]}{\delta v(t_1)\delta v(t_2)}\frac{\delta^{n-2}}{\delta v(t_3)\cdots\delta v(t_n)}\Phi[v(\tau)] + \frac{\delta^{n-2}}{\delta v(t_3)\cdots\delta v(t_n)}\frac{\delta\Theta[v(\tau)]}{\delta v(t_1)}\frac{\delta\Phi[v(\tau)]}{\delta v(t_2)}.$$

Setting now $v = 0$, we obtain that moment functions of the Gaussian process $z(t)$ satisfy the recurrence formula

$$M_n(t_1,\cdots,t_n) = \sum_{k=2}^{n} B(t_1,t_2)M_{n-2}(t_2,\cdots,t_{k-1},t_{k+1},\cdots,t_n). \tag{5.14}$$

From this formula follows that, for the Gaussian process with zero-valued mean, all moment functions of odd orders are identically equal to zero and the moment functions of even orders are represented as sums of terms which are the products of averages of all possible pairs $z(t_i)z(t_k)$.

If we assume that function $v(\tau)$ in Eq. (5.13) is different from zero only in interval $0 < \tau < t$, the characteristic functional

$$\Phi[t; v(\tau)] = \left\langle \exp\left(i\int\limits_0^t d\tau z(\tau)v(\tau)\right)\right\rangle = \exp\left\{-\int\limits_0^t d\tau_1 \int\limits_0^{\tau_1} d\tau_2 B(\tau_1,\tau_2)v(\tau_1)v(\tau_2)\right\} \tag{5.15}$$

becomes a function of time t and satisfies the ordinary differential equation

$$\frac{d}{dt}\Phi[t; v(\tau)] = -v(t)\int\limits_0^t dt_1 B(t,t_1)v(t_1)\Phi[t; v(\tau)], \quad \Phi[0; v(\tau)] = 1. \tag{5.16}$$

To obtain the one-time characteristic function of the Gaussian random process at instant t, we specify function $v(\tau)$ in Eq. (5.12) in the form

$$v(\tau) = v\delta(\tau - t).$$

Then, we obtain

$$\Phi(v,t) = \left\langle e^{ivz(t)}\right\rangle = \int\limits_{-\infty}^{\infty} dz\, P(z,t)e^{ivz} = \exp\left\{-\frac{1}{2}\sigma^2(t)v^2\right\}, \tag{5.17}$$

where $\sigma^2(t) = B(t,t)$. Using the inverse Fourier transform of (5.17), we obtain the one-time probability density of the Gaussian random process

$$P(z,t) = \frac{1}{2\pi}\int\limits_{-\infty}^{\infty} dv\,\Phi(v,t)e^{-ivz} = \frac{1}{\sqrt{2\pi\sigma^2(t)}}\exp\left\{-\frac{z^2}{2\sigma^2(t)}\right\}. \tag{5.18}$$

Note that, in the case of stationary random process $z(t)$, variance $\sigma^2(t)$ is independent of time t, i.e., $\sigma^2(t) = \sigma^2 = \text{const}$.

Density $P(z,t)$ as a function of z is symmetric relative to point $z = 0$,

$$P(z,t) = P(-z,t). \tag{5.19}$$

If mean value of the Gaussian random process is different from zero, then we can consider process $z(t) - \langle z(t)\rangle$ to obtain instead of Eq. (5.18) the expression

$$P(z,t) = \frac{1}{\sqrt{2\pi\sigma^2(t)}}\exp\left\{-\frac{(z - \langle z(t)\rangle)^2}{2\sigma^2(t)}\right\} \tag{5.20}$$

and the corresponding integral distribution function assumes the form

$$F(z,t) = \frac{1}{\sqrt{2\pi\sigma^2(t)}} \int\limits_{-\infty}^{z} dz \exp\left\{-\frac{(z - \langle z(t)\rangle)^2}{2\sigma^2(t)}\right\} = \Pr\left(\frac{z - \langle z(t)\rangle}{\sigma(t)}\right),$$

where probability integral $\Pr(z)$ is defined by Eq. (4.20), page 122.

In view of Eq. (4.21), page 122, the typical realization curve (5.7) of the Gaussian random process $z(t)$ in this case coincides with the mean value of process $z(t)$

$$z^*(t) = \langle z(t)\rangle . \tag{5.21}$$

5.3 Logarithmically Normal Random Process

The so-called *logarithmically normal* (*lognormal*) random process $y(t)$

$$y(t) = e^{z(t)},$$

whose logarithm is the Gaussian random process, is described by the one-time probability density $P(y,t)$ of the form

$$P(y,t) = \frac{1}{y} P(z = \ln y, t) = \frac{1}{y\sqrt{2\pi\sigma^2(t)}} \exp\left\{-\frac{\ln^2\left[e^{-\langle z(t)\rangle} y\right]}{2\sigma^2(t)}\right\}$$

and, consequently, the corresponding integral distribution function assumes the form

$$F(y,t) = \int\limits_{-\infty}^{y} dy' \, P(y',t) = \frac{1}{\sqrt{2\pi\sigma^2(t)}} \int\limits_{-\infty}^{y} \frac{dy'}{y'} \exp\left\{-\frac{\ln^2\left[e^{-\langle z(t)\rangle} y'\right]}{2\sigma^2(t)}\right\}$$

$$= \frac{1}{\sqrt{2\pi\sigma^2(t)}} \int\limits_{-\infty}^{\ln\left[e^{-\langle z(t)\rangle} y\right]} dz \exp\left\{-\frac{z^2}{2\sigma^2(t)}\right\} = \Pr\left(\frac{\ln\left[e^{-\langle z(t)\rangle} y\right]}{\sigma(t)}\right),$$

where probability integral $\Pr(z)$ is defined by Eq. (4.20), page 122.

In view of Eq. (4.21), page 122, the right-hand side of this equality turns into 1/2 if

$$e^{-\langle z(t)\rangle} y(t) = 1,$$

which means that the typical realization curve $y^*(t)$ of the lognormal random process $y(t)$ is given by the equality

$$y^*(t) = e^{\langle z(t)\rangle} = e^{\langle \ln y(t)\rangle}. \tag{5.22}$$

Note that, if we know the moment functions of random process $y(t)$ as functions of time, i.e., if we know functions $\langle y^n(t)\rangle$ $(n = 1, 2, \cdots)$, we know also the statistical characteristics of random process $z(t) = \ln y(t)$. Indeed,

$$\langle y^n(t)\rangle = \left\langle e^{n\ln y(t)}\right\rangle = \exp\left\{n\langle \ln y(t)\rangle + \frac{n^2}{2}\sigma^2_{\ln y}(t)\right\}$$

and, consequently,

$$\langle \ln y(t)\rangle = \lim_{n\to 0}\frac{1}{n}\ln\langle y^n(t)\rangle\,, \quad \sigma^2_{\ln y}(t) = \lim_{n\to\infty}\frac{2}{n^2}\ln\langle y^n(t)\rangle\,. \tag{5.23}$$

5.4 Discontinuous Random Processes

Consider now some examples of discontinuous processes. The discontinuous processes are the random functions that change their time-dependent behavior at discrete instants $t_1, t_2, \cdots$ given statistically. The description of discontinuous processes requires first of all either the knowledge of the statistics of these instants, or the knowledge of the statistics of number $n(0,t)$ of instants t_i falling in time interval $(0,t)$. In the latter case, we have the equality

$$n(0,t) = n(0,t') + n(t',t), \quad 0 \le t' \le t.$$

The quantity $n(0,t)$ by itself is a random process, and Fig. 5.2 shows its possible realization.

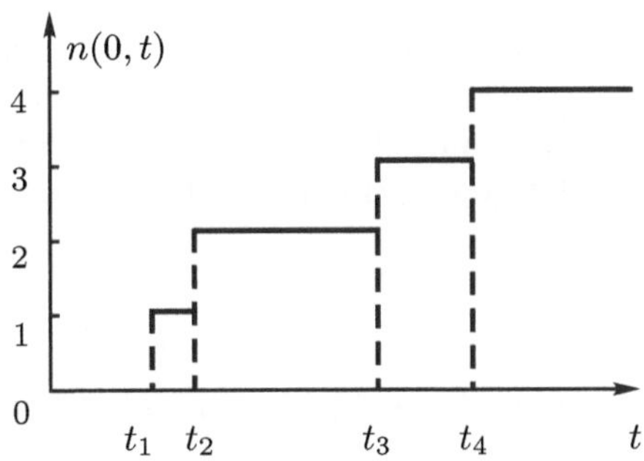

Fig. 5.2 A possible realization of process $n(0,t)$

The set of points of discontinuity t_1, $t_2, \cdots$ of process $z(t)$ is called the *stream of points*. In what follows, we will consider *the Poisson stationary stream of points* in which the probability of falling n points in interval (t_1, t_2) is specified by the Poisson formula

$$P_{n(t_1,t_2)=n} = \frac{\left[\overline{n(t_1,t_2)}\right]^n}{n!} e^{-\overline{n(t_1,t_2)}} \tag{5.24}$$

with the mean number of points in interval (t_1, t_2) given by the formula

$$\overline{n(t_1,t_2)} = \nu|t_1 - t_2|,$$

where ν is the mean number of points per unit time. It is assumed here that the numbers of points falling in nonoverlapping intervals are statistically independent and the instants at which points were fallen in interval (t_1, t_2) under the condition that their total number was n are also statistically independent and uniformly distributed over the interval (t_1, t_2). The length of the interval between adjacent points of discontinuity satisfies the exponential distribution.

The Poisson stream of points is an example of the Markovian processes (see Sect. 5.5).

Note that quantity (5.24)

$$P(n,t) = \langle \delta\left(n(0,t) - n\right)\rangle,$$

which is the probability density of falling n points in time interval $(0, t)$, satisfies as a function of parameter t the recurrence equations

$$\frac{d}{dt}P(n,t) = -\nu\left[P(n-1,t) - P(n,t)\right], \quad P(n,0) = 0 \quad (n = 1, 2, \cdots),$$

$$\frac{d}{dt}P(0,t) = -\nu P(0,t), \quad P(0,0) = 1. \tag{5.25}$$

Consider now random processes whose points of discontinuity form Poisson streams of points. Currently, three types of such processes — *Poisson process*, *telegrapher's process*, and *generalized telegrapher's process* — are mainly used in the model problems of physics. Below, we focus our attention on these processes.

5.4.1 Poisson (Impulse) Random Process

The Poisson (impulse) random process $z(t)$ is the process described by the formula

$$z(t) = \sum_{i=1}^{n} \xi_i g(t - t_i), \tag{5.26}$$

where random quantities ξ_i are statistically independent and distributed with probability density $p(\xi)$; random points t_k are uniformly distributed on interval $(0,T)$, so that their number n obeys the Poisson law with parameter $\bar{n}=\nu T$; and function $g(t)$ is the deterministic function that describes the pulse envelope.

Calculate the characteristic functional of this process

$$\Phi[t;v(\tau)]=\left\langle \exp\left\{i\int\limits_0^t d\tau\, v(\tau)z(\tau)\right\}\right\rangle_z . \tag{5.27}$$

We will average Eq. (5.27) in two stages (see, e.g., [46]). At first, we perform averaging over random quantity ξ and positions of random points t_k:

$$\begin{aligned}
\Phi[t;v(\tau)] &= \left\langle\left\langle \exp\left\{i\int\limits_0^t d\tau\, v(\tau)\sum_{k=1}^{n}\xi_k g(\tau-t_k)\right\}\right\rangle_{\xi_i,t_k}\right\rangle_n \\
&= \left\langle\left[\frac{1}{T}\int\limits_0^T dt'\int\limits_{-\infty}^{\infty} d\xi p(\xi)\exp\left\{i\xi\int\limits_0^t d\tau\, v(\tau)g(\tau-t')\right\}\right]^n\right\rangle_n \\
&= \left\langle\left[\frac{1}{T}\int\limits_0^T dt'\, W\left(\int\limits_{t'}^t d\tau\, v(\tau)g(\tau-t')\right)\right]^n\right\rangle_n ,
\end{aligned} \tag{5.28}$$

where $W(v)=\displaystyle\int_{-\infty}^{\infty} d\xi\, p(\xi)e^{i\xi v}$ is the characteristic function of random quantity ξ. Let $t<T$. Then Eq. (5.28) can be rewritten in the form

$$\Phi[t;v(\tau)]=\left\langle\left[\frac{1}{T}\int\limits_0^t dt'\, W\left(\int\limits_{t'}^t d\tau\, v(\tau)g(\tau-t')\right)+\frac{T-t}{T}\right]^n\right\rangle_n . \tag{5.29}$$

Average now Eq. (5.29) over Poisson's distribution $p_n=\dfrac{e^{-\overline{n}}}{n!}(\overline{n})^n$ of random quantity n:

$$\begin{aligned}
\Phi[t;v(\tau)] &= \sum_0^{\infty}\frac{e^{-\overline{n}}}{n!}(\overline{n})^n\left\langle\left[\frac{1}{T}\int\limits_0^t dt'\, W\left(\int\limits_{t'}^t d\tau\, v(\tau)g(\tau-t')\right)+\frac{T-t}{T}\right]^n\right\rangle_n \\
&= \exp\left\{-\overline{n}+\frac{\overline{n}}{T}\left[\int\limits_0^t dt'\, W\left(\int\limits_{t'}^t d\tau\, v(\tau)g(\tau-t')\right)+T-t\right]\right\}.
\end{aligned}$$

As a result, we obtain the following expression for the characteristic functional of Poisson's random process

$$\Phi[t;v(\tau)] = \exp\left\{\nu\int_0^t dt'\left[W\left(\int_{t'}^t d\tau v(\tau)g(t-t')\right)-1\right]\right\}. \tag{5.30}$$

Consequently, functional $\Theta[t;v(\tau)]$ is given by the formula

$$\Theta[t;v(\tau)] = \ln\Phi[t;v(\tau)] = \nu\int_0^t dt'\int_{-\infty}^{\infty} d\xi p(\xi)\left\{\exp\left[i\xi\int_{t'}^t d\tau v(\tau)g(t-t')\right]-1\right\} \tag{5.31}$$

and cumulant functions assume the form

$$K_n(t_1,\cdots,t_n) = \nu\langle\xi^n\rangle\int_0^{\min\{t_1,\cdots,t_n\}} dt' g(t-t')\cdots g(t_n-t').$$

We consider two types of Poisson's processes important for applications.

1. Let $g(t)=\theta(t)=\begin{cases}1, & t>0,\\ 0, & t<0,\end{cases}$ i.e., $z(t)=\sum_{i=1}^n \xi_i\theta(t-t_i)$. In this case,

$$K_n(t_1,\cdots,t_n) = \nu\langle\xi^n\rangle\min\{t_1,\cdots,t_n\}.$$

If additionally $\xi=1$, then process $z(t)\equiv n(0,t)$, and we have

$$K_n(t_1,\cdots,t_n) = \nu\min\{t_1,\cdots,t_n\},$$

$$\Theta[t;v(\tau)] = \nu\int_0^t dt'\left\{\exp\left[i\int_{t'}^t d\tau v(\tau)\right]-1\right\}. \tag{5.32}$$

2. Let now $g(t)=\delta(t)$. In this case, process

$$z(t)=\sum_{i=1}^n \xi_i\delta(t-t_i)$$

is usually called the *shot noise process.* This process is a particular case of the *delta-correlated processes* (see Sect. 7.8, page 181). For such a process, functional $\Theta[t;v(\tau)]$ is

$$\Theta[t;v(\tau)] = \nu\int_0^t d\tau\int_{-\infty}^{\infty} d\xi p(\xi)\left\{e^{i\xi v(\tau)}-1\right\} \tag{5.33}$$

and cumulant functions assume the forms

$$K_n(t_1,\cdots,t_n) = \nu\langle\xi^n\rangle\delta(t_1-t_2)\delta(t_2-t_3)\cdots\delta(t_{n-1}-t_n).$$

Note that the Poisson process $z(t)$ with arbitrary given impulse function $g(t)$ can be always expressed in terms of the Poisson delta-correlated random process $z_\delta(t)$ by the formula

$$z(t) = \int_0^t d\tau g(t-\tau) z_\delta(\tau). \tag{5.34}$$

5.4.2 Telegrapher's Random Process

Consider now statistical characteristics of telegrapher's random process (Fig. 5.3) defined by the formula

$$z(t) = a(-1)^{n(0,t)} \quad (z(0) = a, \quad z^2(t) \equiv a^2), \tag{5.35}$$

where $n(t_1, t_2)$ is the random sequence of integers equal to the number of points of discontinuity in interval (t_1, t_2).

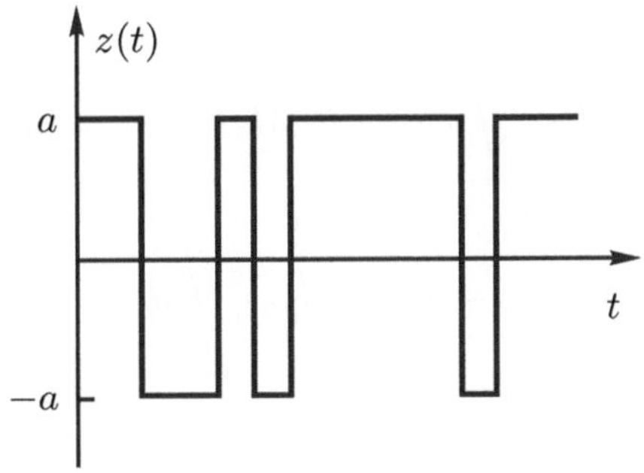

Fig. 5.3 A possible realization of telegrapher's random process

We consider two cases.

1. We will assume first that amplitude a is a deterministic quantity.

For the two first moment functions of process $z(t)$, we have the expressions

$$\langle z(t) \rangle = a \sum_{n(0,t)=0}^{\infty} (-1)^{n(0,t)} P_{n(0,t)} = a e^{-2\overline{n}(0,t)} = a e^{-2\nu t},$$

$$\langle z(t_1) z(t_2) \rangle = a^2 \left\langle (-1)^{n(0,t_1)+n(0,t_2)} \right\rangle = a^2 \left\langle (-1)^{n(t_2,t_1)} \right\rangle$$
$$= a^2 e^{-2\overline{n}(t_2,t_1)} = a^2 e^{-2\nu(t_1-t_2)} \quad (t_1 \geq t_2).$$

The higher moment functions for $t_1 \geqslant t_2 \geqslant \cdots \geqslant t_n$ satisfy the recurrence relationship

$$M_n(t_1,\cdots,t_n) = \langle z(t_1)\cdots z(t_n)\rangle = a^2\left\langle(-1)^{n(0,t_1)+n(0,t_2)+n(0,t_3)+\cdots+n(0,t_n)}\right\rangle$$

$$= a^2\left\langle(-1)^{n(t_2,t_1)}\right\rangle\left\langle(-1)^{n(0,t_3)+\cdots+n(0,t_n)}\right\rangle = \langle z(t_1)z(t_2)\rangle\, M_{n-2}(t_3,\cdots,t_n). \tag{5.36}$$

This relationship is very similar to Eq. (5.14) for the Gaussian process with correlation function $B(t_1,t_2)$. The only difference is that the right-hand side of Eq. (5.36) coincides with only one term of the sum in Eq. (5.14), namely, with the term that corresponds to the above order of times.

Consider now the characteristic functional of this process

$$\Phi_a[t;v(\tau)] = \left\langle \exp\left\{i\int\limits_0^t d\tau z(\tau)v(\tau)\right\}\right\rangle,$$

where index a means that amplitude a is the deterministic quantity. Expanding the characteristic functional in the functional Taylor series and using recurrence formula (5.36), we obtain the expansion

$$\Phi_a[t;v(\tau)] = \sum_{n=0}^{\infty}\frac{i^n}{n!}\int\limits_0^t\cdots\int\limits_0^t dt_1\cdots dt_n M_n(t_1,\cdots,t_n)v(t_1)\cdots v(t_n)$$

$$= 1 + ia\int\limits_0^t dt_1 e^{-2\nu t_1}v(t_1) - a^2\int\limits_0^t dt_1\int\limits_0^{t_1} dt_2 e^{-2\nu(t_1-t_2)}v(t_1)v(t_2)$$

$$\times\sum_{n=2}^{\infty} i^n\int\limits_0^{t_2} dt_3\cdots\int\limits_0^{t_{n-1}} dt_n M_n(t_3,\cdots,t_n)v(t_3)\cdots v(t_n). \tag{5.37}$$

The sum in the right-hand side of Eq. (5.37) can be expressed in terms of the characteristic functional; as a result, we obtain the integral equation

$$\Phi_a[t;v(\tau)] = 1 + ia\int\limits_0^t dt_1 e^{-2\nu t_1}v(t_1) - a^2\int\limits_0^t dt_1\int\limits_0^{t_1} dt_2 e^{-2\nu(t_1-t_2)}v(t_1)v(t_2)\Phi_a[t_2;v(\tau)]. \tag{5.38}$$

Differentiating Eq. (5.38) with respect to t, we obtain the integro-differential equation

$$\frac{d}{dt}\Phi_a[t;v(\tau)] = iae^{-2\nu t}v(t) - a^2 v(t)\int\limits_0^t dt_1 e^{-2\nu(t-t_1)}v(t_1)\Phi_a[t_1;v(\tau)]. \tag{5.39}$$

No general solution to Eq. (5.39) is known. It can be shown that this equation is equivalent to the second-order differential equation

$$\left\{\frac{d^2}{dt^2}+\left[2\nu+\frac{d\ln v(t)}{dt}\right]\frac{d}{dt}+a^2v(t)\right\}\Phi_a[t;v(\tau)]=0,$$

$$\Phi_a[0;v(\tau)]=1,\qquad \left.\frac{d}{dt}\Phi_a[t;v(\tau)]\right|_{t=0}=iav(0).$$

2. Let now amplitude a be the random quantity with probability density $p(a)$. To obtain the characteristic functional of process $z(t)$ in this case, we should average Eq. (5.39) with respect to random amplitude a. In the general case, such averaging cannot be performed analytically. Analytical averaging of Eq. (5.39) appears possible only if probability density of random amplitude a has the form

$$p(a)=\frac{1}{2}\left[\delta\left(a-a_0\right)+\delta\left(a+a_0\right)\right]$$

with $\langle a\rangle=0$ and $\langle a^2\rangle=a_0^2$ (in fact, this very case is what is called usually telegrapher's process). As a result, we obtain the integro-differential equation

$$\frac{d}{dt}\Phi[t;v(\tau)]=-a_0^2v(t)\int\limits_0^t dt_1e^{-2\nu(t-t_1)}v(t_1)\Phi[t_1;v(\tau)] \tag{5.40}$$

equivalent to the second-order equation

$$\begin{gathered}\left\{\frac{d^2}{dt^2}+\left[2\nu+\frac{d\ln v(t)}{dt}\right]\frac{d}{dt}+a_0^2v(t)\right\}\Phi[t;v(\tau)]=0,\\ \left.\frac{d}{dt}\Phi[t;v(\tau)]\right|_{t=0}=0.\qquad \Phi[0;v(\tau)]=1,\end{gathered} \tag{5.41}$$

Note that in the special case of $v(t)\equiv v$, Eq. (5.41) can be solved analytically, and the solution has the form

$$\Phi[t;v]=\left\langle\exp\left\{iv\int\limits_0^t d\tau z(\tau)\right\}\right\rangle=e^{-\nu t}\left\{\cosh\sqrt{\nu^2-a_0^2v^2}t+\frac{\nu}{\sqrt{\nu^2-a_0^2v^2}}\sinh\sqrt{\nu^2-a_0^2v^2}t\right\}. \tag{5.42}$$

One can easily see that this expression is the one-time characteristic function of random process $\xi(t)=\int_0^t d\tau z(\tau)$.

Now, we dwell on an important limit theorem concerning telegrapher's random processes.

Remark 5.4. ***Limiting Theorem for Telegrapher'S Random Processes.***

Consider the random process

$$\xi_N(t) = z_1(t) + \cdots + z_N(t),$$

where all $z_k(t)$ are statistically independent telegrapher's processes with zero-valued means and correlation functions

$$\langle z(t)z(t+\tau)\rangle = \frac{\sigma^2}{N} e^{-\alpha|\tau|}.$$

In this case, the characteristic functional of process $z_k(t)$ satisfies Eq. (5.40)

$$\frac{d}{dt}\Phi[t; v(\tau)] = -\frac{\sigma^2}{N} v(t) \int\limits_0^t dt_1 e^{-\alpha(t-t_1)} v(t_1)\Phi[t_1; v(\tau)],$$

from which follows that $\Phi[t; v(\tau)] \to 1$ for $N \to \infty$.

For the characteristic functional of random process $\xi_N(t)$, we have the expression

$$\Phi_N[t; v(\tau)] = \left\langle \exp\left\{ i \int\limits_0^t d\tau \xi_N(\tau) v(\tau) \right\} \right\rangle = \{\Phi[t; v(\tau)]\}^N.$$

Consequently, it satisfies the equation

$$\frac{d}{dt} \ln \Phi_N[t; v(\tau)] = -\sigma^2 v(t) \int\limits_0^t dt_1 e^{-\alpha(t-t_1)} v(t_1) \frac{\Phi[t_1; v(\tau)]}{\Phi[t; v(\tau)]}.$$

In the limit $N \to \infty$, we obtain the equation

$$\frac{d}{dt} \ln \Phi_\infty[t; v(\tau)] = -\sigma^2 v(t) \int\limits_0^t dt_1 e^{-\alpha(t-t_1)} v(t_1),$$

which means that process $\xi(t) = \lim_{N\to\infty} \xi_N(t)$ is the Gaussian random process with the exponential correlation function

$$\langle \xi(t)\xi(t+\tau)\rangle = \sigma^2 e^{-\alpha|\tau|},$$

i.e., the Gaussian Markovian process (see Sect. 4.3). Thus, process $\xi_N(t)$ for finite N is the finite-number-of-states process approximating the Gaussian Markovian process. This approximation appears practicable for studying various functions of the Gaussian Markovian processes rather than only the Gaussian Markovian processes by themselves. As an example, for process

$$z(t) = x^2(t) - \langle x^2(t)\rangle,$$

where $x(t)$ is the Gaussian Markovian process with the exponential correlation function, the finite-series approximation assumes the form

$$z_N(t) = \sum_{i \neq j=1}^{N} z_i(t) z_j(t), \quad z(t) = \lim_{N \to \infty} z_N(t).$$

This representation is much more convenient for analyzing stochastic equations than the immediate use of processes $x(t)$ and $z(t)$. ♦

5.4.3 Generalized Telegrapher'S Random Process

Consider now generalized telegrapher's process defined by the formula

$$z(t) = a_{n(0,t)}. \tag{5.43}$$

Here, $n(0,t)$ is the sequence of integers described above and quantities a_k are assumed statistically independent with distribution function $p(a)$. Figure 5.4 shows a possible realization of such a process.

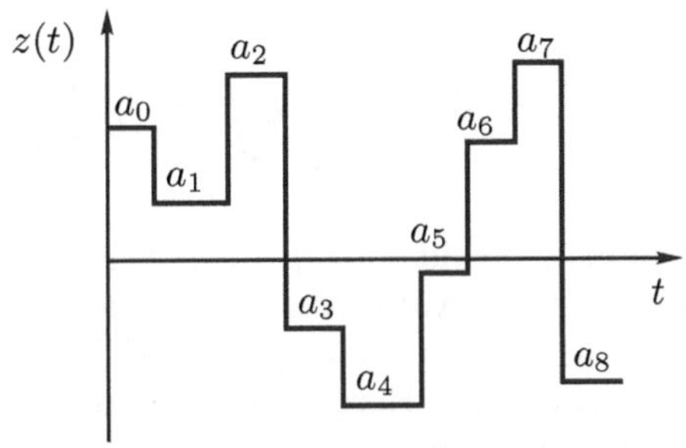

Fig. 5.4 A possible realization of generalized telegrapher's random process

For process $z(t)$, we have

$$\langle z(t) \rangle = \sum_{k=0}^{\infty} \left\langle a_k \delta_{k,n(0,t)} \right\rangle = \langle a \rangle \,,$$

$$\langle z(t_1)z(t_2)\rangle = \sum_{k=0}^{\infty}\sum_{k=0}^{\infty} \langle a_k a_l\rangle \left\langle \delta_{k,n(0,t_1)}\delta_{l,n(0,t_2)}\right\rangle$$

$$= \left\langle a^2\right\rangle \left\{\sum_{k=0}^{\infty} \left\langle \delta_{k,n(0,t_1)}\right\rangle \left\langle \delta_{0,n(t_2,t_1)}\right\rangle + 1 - \sum_{k=0}^{\infty} \left\langle \delta_{k,n(0,t_2)}\right\rangle \left\langle \delta_{0,n(t_2,t_1)}\right\rangle\right\}$$

$$= \left\langle a^2\right\rangle e^{-\nu(t_1-t_2)} + \left\langle a^2\right\rangle \left(1 - e^{-\nu(t_1-t_2)}\right) \quad (t_1 \geqslant t_2),$$

and so on. In addition, the probability of absence of points of discontinuity in interval (t_2, t_1) is given by the formula

$$P_{n(t_2,t_1)=0} = \left\langle \delta_{0,n(t_2,t_1)}\right\rangle = e^{-\nu|t_1-t_2|}.$$

For such a process, no relationship similar to Eq. (5.37) can be obtained, and derivation of the equation for the characteristic functional is essentially based on the fact tat this process is the Markovian process. The resulting equation is the integro-differential equation

$$\Phi[t, v(\tau)] = \left\langle \exp\left\{ia \int_0^t d\tau v(\tau)\right\}\right\rangle e^{-\nu t}$$
$$+ \nu \int_0^t dt_1 e^{-\nu(t-t_1)} \left\langle \exp\left\{ia \int_{t_1}^t d\tau v(\tau)\right\}\right\rangle \Phi[t_1, v(\tau)]. \quad (5.44)$$

The first term in the right-hand side of Eq. (5.44) corresponds to the absence of points of discontinuity in interval $(0, t)$, and the second term corresponds to the situations in which the number of points of discontinuity in interval $(0, t)$ can vary from one to infinity. Here, time t_1 is the instance at which the last point of discontinuity appears.

Note that, for probability density of the form

$$p(a) = \frac{1}{2}[\delta(a - a_0) + \delta(a + a_0)],$$

Eq. (5.44) coincides (after replacing ν with $\nu/2$) with the equation for telegrapher's process. This fact is quite expectable, because, unlike telegrapher's process, the process $z(t)$ considered here can change the sign at a point of discontinuity with a probability of 1/2, which just results in doubling the mean time between the discontinuities.

Earlier, we noted that the Poisson stream of points and processes based on these streams are the Markovian processes. Below, we consider this important class of random processes in detail.

5.5 Markovian Processes

5.5.1 General Properties

In the foregoing section, we considered the characteristic functional that describes all statistical characteristics of random process $z(t)$. Specification of the argument of the functional in the form

$$v(t) = \sum_{k=1}^{n} v_k \delta(t - t_k)$$

transforms the characteristic functional into the joint characteristic function of random quantities $z_k = z(t_k)$

$$\Phi_n(v_1, \cdots, v_n) = \left\langle \exp\left\{ i \sum_{k=1}^{n} v_k z(t_k) \right\} \right\rangle,$$

whose Fourier transform is the joint probability density of process $z(t)$ at discrete instants

$$P_n(z_1, t_1; \cdots; z_n, t_n) = \langle \delta(z(t_1) - z_1) \cdots \delta(z(t_n) - z_n) \rangle . \tag{5.45}$$

Assume that the above instants are ordered according to the line of inequalities

$$t_1 \geq t_2 \geq \cdots \geq t_n.$$

Then, by definition of the conditional probability, we have

$$P_n(z_1, t_1; \cdots; z_n, t_n) = p_n(z_1, t_1 | z_2, t_2; \cdots; z_n, t_n) P_{n-1}(z_2, t_2; \cdots; z_n, t_n), \tag{5.46}$$

where p_n is the conditional probability density of the value of process $z(t)$ at instant t_1 under the condition that function $z(t)$ was equal to z_k at instants t_k for $k = 2, \cdots, n$ ($z(t_k) = z_k$, $k = 2, \cdots, n$). If process $z(t)$ is such that the conditional probability density for all $t_1 > t_2$ is unambiguously determined by the value z_2 of the process at instant t_2 and is independent of the previous history, i.e., if

$$p_n(z_1, t_1 | z_2, t_2; \cdots; z_n, t_n) = p(z_1, t_1 | z_2, t_2), \tag{5.47}$$

then this process is called the Markovian process, or the memoryless process. In this case, function

$$p(z, t | z_0, t_0) = \langle \delta(z(t) - z) | z(t_0) = z_0 \rangle \quad (t > t_0) \tag{5.48}$$

is called the *transition probability density*. Setting $t = t_0$ in Eq. (5.48), we obtain the equality

$$p(z, t_0|z_0, t_0) = \delta(z - z_0).$$

Substituting expression (5.46) in Eq. (5.45), we obtain the recurrence formula for the n-time probability density of process $z(t)$. Iterating this formula, we find the relationship of probability density P_n with the one-time probability density ($t_1 \geq t_2 \geq \cdots \geq t_n$)

$$P_n(z_1, t_1; \cdots; z_n, t_n) = p(z_1, t_1|z_2, t_2) \cdots p(z_{n-1}, t_{n-1}|z_n, t_n)P(t_n, z_n). \tag{5.49}$$

Thus, only two functions — transition probability density $p(z, t|z_0, t_0)$ and one-time probability density $P(t, z)$ — are sufficient to exhaustively describe all statistical characteristics of the Markovian process $z(t)$. It appears that transition probability density as a function of its arguments satisfies the nonlinear integral equation called the *Smolukhovsky equation* (or the *Kolmogorov–Chapman equation*). In the context of the derivation of this equation, we note the following fact: if process $z(t)$ assumes values $z(t_0) = z_0$, $z(t_1) = z_1$, $z(t) = z$ at fixed instants $t_0 < t_1 < t$, then the coordination condition

$$\int_{-\infty}^{\infty} dz_1 P_3(z, t; z_1, t_1; z_0, t_0) = P_2(z, t; z_0, t_0). \tag{5.50}$$

holds. Substituting now P_3 and P_2 expressed in the form of Eq. (5.49) in Eq. (5.50), we obtain the desired equation

$$p(z, t; z_0, t_0) = \int_{-\infty}^{\infty} dz_1 p(z, t|z_1, t_1)p(z_1, t_1; z_0, t_0). \tag{5.51}$$

Integrating Eq. (5.51) over z_0, we obtain the linear integral equation for the one-time probability density $P(z, t)$

$$P(z, t) = \int_{-\infty}^{\infty} dz_1 p(z, t|z_1, t_1)P(z_1, t_1). \tag{5.52}$$

Integral equations (5.51) and (5.52) offer a possibility of deriving differential or integro-differential equation for simple Markovian processes. The simplest Markovian processes with continuous time can be classified as follows:

1) discrete processes,
2) continuous processes, and
3) discrete-continuous processes that can undergo discontinuous variations at certain instants and behave as continuous processes between these instants.

Discrete Markovian Process

Consider the discrete Markovian process $z(t)$. This assumes that the process can take on only discrete values $z_1, \cdots, z_n$ and switching between the values occurs at random time instants. We introduce the transition probability density

$$p_{ij}(t, t_0) = \langle \delta(z(t) - z_i) | z(t_0) = z_j \rangle, \quad \sum_i p_{ij}(t, t_0) = 1 \quad (t_0 < t), \quad (5.53)$$

which is the conditional probability of the event that process $z(t)$ assumes value z_i at instant t under the condition that its value at instant t_0 was z_j. It is obvious that

$$p_{ij}(t_0, t_0) = \delta_{ij}. \quad (5.54)$$

For short temporal intervals $\Delta t \to 0$, we have

$$p_{ij}(t + \Delta t, t) = \delta_{ij} + a_{ij}(t)\Delta t + o(\Delta t), \quad (5.55)$$

where $a_{ij}(t)\Delta t$ is the transition probability from state z_j at instant t to state z_i during time Δt. It is assumed that

$$a_{ij}(t) > 0 \quad (i \neq j), \quad a_{jj}(t) = -\sum_{i(i \neq j)} a_{ij}(t), \quad (5.56)$$

because normalization condition (5.53) must hold.

Using Eq. (5.55), one can easily show from the Smolukhovsky equation (5.51) that probability $p_{ij}(t, t_0)$ satisfies the system of linear differential equations

$$\frac{d}{dt} p_{ij}(t, t_0) = \sum_{k=1}^{n} a_{ik}(t) p_{kj}(t, t_0) \quad (i, j = 1, \cdots, n). \quad (5.57)$$

Representing the one-time probability $P_i(t)$ in the form

$$P_i(t) = \sum_j p_{ij}(t, t_0) p_j^0, \quad (5.58)$$

where p_j^0 are the initial probabilities of states ($p_j^0 = P_j(t_0)$), we obtain that this one-time probability satisfies the system of equations

$$\frac{d}{dt} P_i(t) = \sum_{k=1}^{n} a_{ik}(t) P_k(t), \quad P_i(t_0) = p_i^0. \quad (5.59)$$

Consider three examples as illustrations of the above consideration.

1. Let random process $z(t) = n(0, t)$ represents the number of discontinuities occurred in interval $(0, t)$ at random instants (see Fig. 5.2 for a possible realization of this process). It is assumed that process $z(t)$ takes on only integer values 0, 1, 2, $\cdots$, and it is obvious that

$$p_{ij}(t, t_0) = 0 \quad \text{for} \quad i < j, \quad t \geq t_0.$$

Assuming additionally that, in temporal interval $(t, t + \Delta t)$, the probability of one change of state is $\nu \Delta t + o(\Delta t)$ and the probability of the absence of discontinuities is $1 - \nu \Delta t + o(\Delta t)$ and neglecting the possibility of two and more changes of state in this interval (these assumptions are just the assumptions that govern the Poisson stream of instants at which the discontinuities appear), we can write the system of equations (5.59) for this process. In the case under consideration, this system assumes the form

$$\begin{aligned} \frac{d}{dt} P_0(t) &= -\nu P_0(t), \quad P_0(0) = 1, \\ \frac{d}{dt} P_i(t) &= -\nu \left[P_i(t) - P_{i-1}(t) \right], \quad P_{i \neq 0}(0) = 0 \end{aligned} \tag{5.60}$$

and coincides with the system of equations (5.25) at $i = n$.

2. As the second example, we consider the simplest Markovian process with the finite number of states, namely, telegrapher's random process that can take on only two values $z(t) = \pm a$. In the foregoing section, we considered this process from another viewpoint. Here, we assume that the probabilities of transitions $(a \to -a)$ and $(-a \to a)$ during short interval Δt coincide and are $\nu \Delta t + o(\Delta t)$, the corresponding probabilities of state preservation during interval Δt are $1 - \nu \Delta t + o(\Delta t)$, and probabilities of initial states are p_a^0 and $p_{-a}^0 = 1 - p_a^0$. In this case, the transition probabilities satisfy the system of equations (5.57) with parameters

$$a_{11} = a_{22} = -\nu, \quad a_{12} = a_{21} = \nu.$$

The solution of this system is $(\tau = t - t_0)$

$$p_{11}(\tau) = p_{22}(\tau) = \frac{1}{2} \left[1 + e^{-2\nu\tau} \right], \quad p_{12}(\tau) = p_{21}(\tau) = \frac{1}{2} \left[1 - e^{-2\nu\tau} \right]. \tag{5.61}$$

Expressions for the one-time probabilities are obtained similarly:

$$P_1(\tau) = \frac{1}{2} + \left[p_a^0 - \frac{1}{2} \right] e^{-2\nu\tau}, \quad P_2(\tau) = \frac{1}{2} - \left[p_a^0 - \frac{1}{2} \right] e^{-2\nu\tau}. \tag{5.62}$$

If process $z(t)$ had at the initial instant the fixed value $z(t_0) = a$, then $p_a^0 = 1$ and Eqs. (5.62) assume the form

$$P_1(\tau) = \frac{1}{2} \left[1 + e^{-2\nu\tau} \right], \quad P_2(\tau) = \frac{1}{2} \left[1 - e^{-2\nu\tau} \right]. \tag{5.63}$$

For $t \to \infty$, these probability distributions tend to steady-state values $P_{1,2}(\infty) = 1/2$ and the process behavior tends to steady-state regime.

If $p_a^0 = p_{-a}^0 = 1/2$ at the initial instant, the process $z(t)$ is always steady.

Note that, in the case of telegrapher's process, formulas (5.61) can be combined in one formula; namely,

$$p(z,t|z_0,t_0) = \delta(z-z_0)P_1(\tau) + \delta(z+z_0)P_2(\tau), \tag{5.64}$$

where $P_1(\tau)$ and $P_2(\tau)$ are given by Eqs. (5.63) and $\tau = t - t_0$. Differentiating Eq. (5.64) with respect to time, we obtain the equation for the transition probability density $p(z,t|z_0,t_0)$

$$\frac{\partial}{\partial t}p(z,t|z_0,t_0) = -\nu\left\{p(z,t|z_0,t_0) - p(-z,t|z_0,t_0)\right\} \tag{5.65}$$

with the initial value $p(z,t_0|z_0,t_0) = \delta(z-z_0)$.

Thus, the transition probability density $p(z,t|z_0,t_0)$ of telegrapher's process satisfies the linear operator equation

$$\frac{\partial}{\partial t}p(z,t|z_0,t_0) = \widehat{L}(z)p(z,t|z_0,t_0), \tag{5.66}$$

where operator $\widehat{L}(z)$ is defined by the equality

$$\widehat{L}(z)f(z) = -\nu\left\{f(z) - f(-z)\right\}. \tag{5.67}$$

Note that this is the property characteristic of all Markovian processes. However, the equation for the transition probability density not always allows the compact representation such as (5.66). In the general case of arbitrary Markovian process with a finite number of states, operator $\widehat{L}(z)$ is matrix $||a_{ij}||$ appeared in Eq. (5.57) and probability density $p(z,t|z_0,t_0)$ itself is the matrix function. In this case, any realization of process $z(t)$ satisfies the identity

$$[z(t)-z_1][z(t)-z_2]\cdots[z(t)-z_n] = 0. \tag{5.68}$$

Opening the brackets in Eq. (5.68), we see that different powers of process $z(t)$ satisfy the algebraic relationship

$$z^n(t) = (z_1 + \cdots + z_n)z^{n-1}(t) + \cdots + (-1)^{n+1}z_1z_2\cdots z_n. \tag{5.69}$$

In the case of telegrapher's random process, i.e., the process with two possible states $z(t) = \pm a$, identity (5.69) reduces to $z^2(t) = a^2$, which appears to be very useful for analyzing stochastic equations whose parameters fluctuate by the law of telegrapher's process.

3. Consider now generalized telegrapher's process as an example of the spasmodic process. This process is defined by Eq. (5.43), and its transition probability density has the form

$$p(z,t|z_0,t_0) = \langle \delta(z(t)-z)|z(t_0)=z_0 \rangle$$
$$= \delta(z-z_0)P_{n(0,t)=0} + \langle \delta(z-a) \rangle_a \left\{ P_{n(0,t)=1} + P_{n(0,t)=2} + \cdots \right\}. \quad (5.70)$$

Taking into account the normalization condition $\sum\limits_{n=0}^{\infty} P_{n(0,t)=n} = 1$, we obtain the final expression in the form

$$p(z,t|z_0,t_0) = \delta(z-z_0)P_0(t,t_0) + p_a(z)\left\{1 - P_0(t,t_0)\right\}, \quad (5.71)$$

where $P_0(t,t_0) = e^{-\nu(t-t_0)}$ is the probability of absence of jumps within temporal interval (t_0,t) and $p_a(z)$ is the probability of the event that random quantity a assumes value z.

The one-time probability distribution of process $z(t)$ is obviously the steady-state distribution

$$P(z,t) = p_a(z). \quad (5.72)$$

It is obvious that quantity (5.71) satisfies, as a function of variable t, the differential equation

$$\frac{\partial}{\partial t}p(z,t|z_0,t_0) = -\nu\left\{p(z,t|z_0,t_0) - p_a(z)\right\}, \quad (5.73)$$

which can be rewritten in the operator form

$$\frac{\partial}{\partial t}p(z,t|z_0,t_0) = \widehat{L}(z)p(z,t|z_0,t_0), \quad (5.74)$$

where $\widehat{L}(z)$ in this particular case is the integral operator

$$\widehat{L}(z)f(z) = -\nu\left\{ f(z) - p_a(z)\int\limits_{-\infty}^{\infty} dz' f(z') \right\}. \quad (5.75)$$

Continuous Markovian Processes

Consider now continuous Markovian processes. In this case, the transition probability density $p(z,t|z_0,t_0)$ satisfies the operator equation (this equation is a consequence of the Smolukhovsky equation (5.51))

$$\frac{\partial}{\partial t}p(z,t|z_0,t_0) = \sum_{n=1}^{\infty}\frac{(-1)^n}{n!}\frac{\partial^n}{\partial z^n}\left[B_n(z,t)p(z,t|z_0,t_0)\right], \quad (5.76)$$

where functions $B_n(z,t)$ are determined by the equalities

$$B_n(z,t) = \lim_{\Delta t \to 0} \frac{1}{\Delta t} \left\langle \{z(t+\Delta t) - z(t)\}^n \left| z(t) \right.\right\rangle$$

$$= \lim_{\Delta t \to 0} \frac{1}{\Delta t} \int\limits_{-\infty}^{\infty} dz \, \{z(t+\Delta t) - z(t)\}^n \, p(z, t+\Delta t | z, t). \quad (5.77)$$

A consequence of Eq. (5.76) is the similar equation for the one-time probability density $P(z,t) = \langle \delta(z(t) - z) \rangle$:

$$\frac{\partial}{\partial t} P(t,z) = \sum_{n=1}^{\infty} \frac{(-1)^n}{n!} \frac{\partial^n}{\partial z^n} \left[B_n(z,t) P(t,z) \right].$$

Continuous processes for which all coefficients B_n in Eq. (5.76) with $n \geq 3$ vanish form an important particular class. Markovian processes having this property are called the *diffusion processes*. In the context of such processes, Eq. (5.76) assumes the form

$$\frac{\partial}{\partial t} p(z,t|z_0,t_0) = -\frac{\partial}{\partial z} \left[B_1(z,t) p(z,t|z_0,t_0) \right] + \frac{1}{2} \frac{\partial^2}{\partial z^2} \left[B_2(z,t) p(z,t|z_0,t_0) \right]. \quad (5.78)$$

This equation is called the *Fokker–Planck equation* and functions $B_1(z,t)$ and $B_2(z,t)$ are called the *drift* and *diffusion coefficients*, respectively.

In the particular case of $B_2(z,t) = \text{const}$ and $B_1(z,t) = -B_1 z$, Markovian process $z(t)$ is the Gaussian process with the exponential correlation function

$$\langle z(t) z(t+\tau) \rangle = \sigma^2(t) e^{-B_1|\tau|}.$$

In this case Eq. (5.78) is replaced with the equation

$$\frac{\partial}{\partial t} p(z,t|z_0,t_0) = \widehat{L}(z) p(z,t|z_0,t_0),$$

where operator

$$\widehat{L}(z) = B_1 \frac{\partial}{\partial z} z + \frac{1}{2} B_2 \frac{\partial^2}{\partial z^2}. \quad (5.79)$$

Note that the converse is also valid; namely, any Gaussian process with the exponential correlation function is the Markovian process.

Discrete-Continuous Markovian Processes

Consider now the one-dimensional discrete-continuous Markovian process. Two cases are possible here: the case of a purely discontinuous (spasmodic) process and the case of a process varying both continuously and discontinuously. In the first case, two functions — $q(z,t)$ and $u(z,z',t)$ — characterize random process $z(t)$. The meaning of these functions is as follows: within the short temporal interval $(t, t+\Delta t)$, the probability for the process to preserve its previous value is $1-q(z,t)\Delta t$ and the probability for the process to change its value from z to z'' is $u(z,z',t)\Delta t\Delta z'$ (here, $z' < z'' < z'+\Delta z'$). Of course, the normalization condition

$$\int_{-\infty}^{\infty} dz' u(z,z',t) = q(z,t). \tag{5.80}$$

is assumed additionally. For this process, a consequence of the Smolukhovsky equation (5.51) is the integro-differential equation

$$\frac{\partial}{\partial t}p(z,t|z_0,t_0) = -q(z,t)p(z,t|z_0,t_0) + \int_{-\infty}^{\infty} dz' u(z,z',t)p(z',t|z_0,t_0), \tag{5.81}$$

which is called the *Kolmogorov–Feller equation*. The equation for the one-time probability density has the similar form.

If, in addition to jumps, the process allows continuous variation, then the right-hand side of Eq. (5.81) is added with the right-hand side of Eq. (5.78). Note that random process $z(t) = n(0,t)$ (the number of jumps within temporal interval $(0,t)$) that we considered earlier is a special case of spasmodic processes; correspondingly, difference-differential equations (5.60) are special cases of the integro-differential equation (5.81).

It is obvious that Eq. (5.74) for generalized telegrapher's process is the Kolmogorov–Feller equation (5.81) with specially defined parameters $q(z,t) = \nu$ and $u(z,z',t) = \nu p_a(z)$.

Up to this point, we dealt with the one-dimensional processes; it is clear however that all results remain valid for multidimensional processes, i.e., for vector random functions $\boldsymbol{z}(t)$. In particular, the transition probability density

$$p(\boldsymbol{z},t|\boldsymbol{z}_0,t_0) = \langle \delta(\boldsymbol{z}(t)-\boldsymbol{z})|\boldsymbol{z}(t_0)=\boldsymbol{z}_0)\rangle$$

will satisfy the linear operator equation

$$\frac{\partial}{\partial t}p(\boldsymbol{z},t|\boldsymbol{z}_0,t_0) = \widehat{L}(\boldsymbol{z})p(\boldsymbol{z},t|\boldsymbol{z}_0,t_0). \tag{5.82}$$

Note additionally that transition probability density $p(\boldsymbol{z},t|\boldsymbol{z}_0,t_0)$ satisfies, as a function of its arguments, not only Eq. (5.82) (we will call this equation the *forward equation*), but also the equation with respect to variable t_0

$$\frac{\partial}{\partial t_0}p(\boldsymbol{z},t|\boldsymbol{z}_0,t_0) = \widehat{L}^{+}(\boldsymbol{z}_0)p(\boldsymbol{z},t|\boldsymbol{z}_0,t_0), \tag{5.83}$$

which we will call the *backward equation*. Here, $\widehat{L}^{+}(\boldsymbol{z}_0)$ is the operator conjugated to operator $\widehat{L}(\boldsymbol{z})$. Equation (5.83) is convenient for analyzing the problems that deal with dependencies on initial locations of space-time points.

We mentioned earlier that two functions — transition probability density $p(z,t|z_0,t_0)$ and one-time probability density $P(z,t)$ — are sufficient to exhaustively describe all statistical characteristics of the Markovian process $z(t)$. Nevertheless, statistical analysis of stochastic equations requires additionally the knowledge of the characteristic functional of random process $z(t)$.

5.5.2 Characteristic Functional of the Markovian Process

For the Markovian process $z(t)$, no closed equation can be derived in the general case for the characteristic functional $\Phi[t;v(\tau)] = \langle\varphi[t;v(\tau)]\rangle$, where

$$\varphi[t;v(\tau)] = \exp\left\{i\int\limits_0^t d\tau z(\tau)v(\tau)\right\}.$$

Instead, we can derive the closed equation for the functional

$$\Psi[z,t;v(\tau)] = \langle\delta(z(t)-z)\varphi[t;v(\tau)]\rangle \tag{5.84}$$

describing correlations of process $z(t)$ with its prehistory. The characteristic functional $\Phi[t;v(\tau)]$ can be obtained from functional $\Psi[z,t;v(\tau)]$ by the formula

$$\Phi[t;v(\tau)] = \int\limits_{-\infty}^{\infty} dz\Psi[z,t;v(\tau)]. \tag{5.85}$$

To derive the equation for functional $\Psi[z,t;v(\tau)]$, we note that the following equality

$$\varphi[t;v(\tau)] = 1 + i\int\limits_0^t dt_1 z(t_1)v(t_1)\varphi[t_1;v(\tau)] \tag{5.86}$$

holds. Substituting Eq. (5.86) in Eq. (5.84), we obtain the expression

$$\Psi[z,t;v(\tau)] = P(z,t) + i\int\limits_0^t dt_1 v(t_1)\left\langle\delta(z(t)-z)z(t_1)\varphi[t_1;v(\tau)]\right\rangle, \tag{5.87}$$

where $P(z,t) = \langle\delta(z(t)-z)\rangle$ is the one-time probability density of random quantity $z(t)$.

We rewrite Eq. (5.87) in the form

$$\Psi[z,t;v(\tau)] = P(z,t) + i\int\limits_0^t dt_1 v(t_1) \int\limits_{-\infty}^{\infty} dz_1 z_1 \left\langle \delta(z(t)-z)\delta(z(t_1)-z_1)\varphi[t_1;v(\tau)]\right\rangle . \tag{5.88}$$

Taking into account the fact that process $z(t)$ is the Markovian process, we can perform averaging in (5.88) to obtain the closed integral equation

$$\Psi[z,t;v(\tau)] = P(z,t) + i\int\limits_0^t dt_1 v(t_1) \int\limits_{-\infty}^{\infty} dz_1 z_1 p(z,t|z_1,t_1)\Psi[z_1,t_1;v(\tau)], \tag{5.89}$$

where $p(z,t|z_1,t_1)$ is the transition probability density.

We note that the integral equation similar to Eq. (5.89) can be derived also for the functional

$$\Psi[z,t',t;v(\tau)] = \langle \delta(z(t')-z)\varphi[t;v(\tau)]\rangle \quad (t' \geq t). \tag{5.90}$$

This equation has the form

$$\Psi[z,t',t;v(\tau)] = P(z,t') + i\int\limits_0^t dt_1 v(t_1) \int\limits_{-\infty}^{\infty} dz_1 z_1 p(z,t'|z_1,t_1)\Psi[z_1,t_1;v(\tau)]. \tag{5.91}$$

Integrating Eq. (5.89) with respect to z, we obtain an additional relationship between the characteristic functional $\Phi[t;v(\tau)]$ and functional $\Psi[z,t;v(\tau)]$. This relationship has the form

$$\frac{1}{iv(t)}\frac{d}{dt}\Phi[t;v(\tau)] = \int\limits_{-\infty}^{\infty} dz_1 z_1 \Psi[z_1,t;v(\tau)] = \Psi[t;v(\tau)]. \tag{5.92}$$

Multiplying Eq. (5.89) by z and integrating the result over z, we obtain the relationship between functionals $\Psi[t;v(\tau)]$ and $\Psi[z,t;v(\tau)]$

$$\Psi[t;v(\tau)] = \langle z(t)\rangle + i\int\limits_0^t dt_1 v(t_1) \int\limits_{-\infty}^{\infty} dz_1 \left\langle z(t)|z_1,t_1\right\rangle \Psi[z_1,t_1;v(\tau)]. \tag{5.93}$$

Equation (5.89) is in the general case a complicated integral equation whose explicit form depends on functions $P(z,t)$ and $p(z,t|z_1,t_1)$, i.e., on parameters of the Markovian process. Preliminarily differentiating this equation with respect to t and using Eq. (5.66), we can convert it into the integro-differential equation

$$\frac{d}{dt}\Psi[z,t;v(\tau)] = izv(t)\Psi[z,t;v(\tau)] + \widehat{L}(z)\Psi[z,t;v(\tau)],$$
$$\Psi[z,0;v(\tau)] = P(z,0). \tag{5.94}$$

In this case, functional $\Psi[z,t',t;v(\tau)]$ (5.90) as a function of variable t' satisfies the equation with the initial value at $t' = t$

$$\frac{d}{dt'}\Psi[z,t',t;v(\tau)] = \widehat{L}(z)\Psi[z,t',t;v(\tau)] \quad (t' > t),$$
$$\Psi[z,t,t;v(\tau)] = \Psi[z,t;v(\tau)]. \tag{5.95}$$

Thus, Eq. (5.94) together with Eqs. (5.92) and (5.93) forms the starting point for the determination of the characteristic functional of the Markovian process.

We demonstrate this fact using the processes considered earlier as examples.

For telegrapher's process, Eq. (5.64) gives

$$\langle z(t)|z_1,t_1\rangle = z_1 e^{-2\nu(t-t_1)}, \quad \langle z(t)\rangle = 0,$$

and we obtain Eq. (4.24).

Consider now generalized telegrapher's process. By virtue of Eq. (5.75), Eq. (5.94) for functional $\Psi[t,z;v(\tau)]$ assumes in this case the form

$$\frac{d}{dt}\Psi[z,t;v(\tau)] = \{izv(t) - \nu\}\,\Psi[z,t;v(\tau)] + \nu p_a(z)\Phi[t;v(\tau)],$$
$$\Psi[z,0;v(\tau)] = p_a(z). \tag{5.96}$$

Deriving Eq (5.96), we used equality (5.85). Solving Eq. (5.96) in functional $\Psi[z,t;v(\tau)]$, we relate it to the characteristic functional

$$\Psi[z,t;v(\tau)] = p_a(z)\exp\left\{-\nu t + iz\int\limits_0^t d\tau v(\tau)\right\}$$
$$+ \nu p_a(z)\int\limits_0^t dt_1 \Phi[t_1;v(\tau)]\exp\left\{-\nu(t-t_1) + iz\int\limits_{t_1}^t d\tau v(\tau)\right\}. \tag{5.97}$$

Integrating Eq. (5.97) over z, we obtain the closed integral equation for the characteristic functional $\Phi[t,v(\tau)]$

$$\Phi[t, v(\tau)] = \left\langle \exp\left\{ia\int_0^t d\tau v(\tau)\right\}\right\rangle_a e^{-\nu t}$$

$$+\nu\int_0^t dt_1 e^{-\nu(t-t_1)}\left\langle \exp\left\{ia\int_{t_1}^t d\tau v(\tau)\right\}\right\rangle_a \Phi[t_1, v(\tau)], \quad (5.98)$$

which coincides with Eq. (5.44).

Multiplying Eq. (5.97) by arbitrary function $F(z)$ and integrating the result over z, we obtain the equality

$$\left\langle F\left(z(t)\right)\exp\left\{i\int_0^t d\tau z(\tau)v(\tau)\right\}\right\rangle = \left\langle F(a)\exp\left\{ia\int_0^t d\tau v(\tau)\right\}\right\rangle_a e^{-\nu t}$$

$$+\nu\int_0^t dt_1 e^{-\nu(t-t_1)}\left\langle F(a)\exp\left\{ia\int_{t_1}^t d\tau v(\tau)\right\}\right\rangle_a \Phi[t_1, v(\tau)]. \quad (5.99)$$

In the particular case of $F(z) = z$, Eq. (5.99) can be reduced to the integro-differential equation for the characteristic functional $\Phi[t, v(\tau)]$

$$\frac{d}{iv(t)dt}\Phi[t, v(\tau)] = \left\langle a\exp\left\{ia\int_0^t d\tau v(\tau)\right\}\right\rangle_a e^{-\nu t}$$

$$+\nu\int_0^t dt_1 e^{-\nu(t-t_1)}\left\langle a\exp\left\{ia\int_{t_1}^t d\tau v(\tau)\right\}\right\rangle_a \Phi[t_1, v(\tau)], \quad (5.100)$$

which is equivalent to Eq. (5.98).

In the case of generalized telegrapher's process, we can additionally establish the relationship between functionals $\Psi[z, t', t; v(\tau)]$ and $\Psi[z, t; v(\tau)]$. This relationship has the form ($t' \geq t$)

$$\Psi[z, t', t; v(\tau)] = \Psi[z, t; v(\tau)]e^{-\nu(t'-t)} + p_a(z)\Phi[t; v(\tau)][1 - e^{-\nu(t'-t)}]. \quad (5.101)$$

Chapter 6
Random Fields

For random field $f(\boldsymbol{x},t)$, the one- and n-point probability densities are defined similarly to those for random processes

$$\begin{aligned} P(\boldsymbol{x},t;f) &= \langle\varphi(\boldsymbol{x},t;f)\rangle\,, \\ P(\boldsymbol{x}_1,t_1,f_1;\cdots;\boldsymbol{x}_n,t_n,f_n) &= \langle\varphi(\boldsymbol{x}_1,t_1,f_1;\cdots;\boldsymbol{x}_n,t_n,f_n)\rangle\,, \end{aligned} \tag{6.1}$$

where the indicator functions are defined as follows:

$$\begin{aligned} \varphi(\boldsymbol{x},t;f) &= \delta(f(\boldsymbol{x},t)-f), \\ \varphi(\boldsymbol{x}_1,t_1,f_1;\cdots;\boldsymbol{x}_n,t_n,f_n) &= \delta\left(f(\boldsymbol{x}_1,t_1)-f_1\right)\cdots\delta\left(f(\boldsymbol{x}_n,t_n)-f_n\right). \end{aligned} \tag{6.2}$$

For clarity, we use here variables $\boldsymbol{x}$ and t as spatial and temporal coordinates; however, in many physical problems, some preferred spatial coordinate can play the role of the temporal coordinate.

Random field $f(\boldsymbol{x},t)$ is called the spatially homogeneous field if all its statistical characteristics are invariant relative to spatial translations by arbitrary vector $\boldsymbol{a}$, i.e., if

$$P(\boldsymbol{x}_1+\boldsymbol{a},t_1,f_1;\cdots;\boldsymbol{x}_n+\boldsymbol{a},t_n,f_n) = P(\boldsymbol{x}_1,t_1,f_1;\cdots;\boldsymbol{x}_n,t_n,f_n).$$

In this case, the one-point probability density $P(\boldsymbol{x},t;f)=P(t;f)$ is independent of $\boldsymbol{x}$, and the spatial correlation function $B_f(\boldsymbol{x}_1,t_1;\boldsymbol{x}_2,t_2)$ depends on the difference $\boldsymbol{x}_1-\boldsymbol{x}_2$:

$$B_f(\boldsymbol{x}_1,t_1;\boldsymbol{x}_2,t_2) = \langle f(\boldsymbol{x}_1,t_1)f(\boldsymbol{x}_2,t_2)\rangle = B_f(\boldsymbol{x}_1-\boldsymbol{x}_2;t_1,t_2).$$

If random field $f(\boldsymbol{x},t)$ is additionally invariant with respect to rotation of all vectors $\boldsymbol{x}_i$ by arbitrary angle, i.e., with respect to rotations of the reference system, then field $f(\boldsymbol{x},t)$ is called the homogeneous isotropic random field. In this case, the correlation function depends only on length $|\boldsymbol{x}_1-\boldsymbol{x}_2|$:

$$B_f(\boldsymbol{x}_1,t_1;\boldsymbol{x}_2,t_2) = \langle f(\boldsymbol{x}_1,t_1)f(\boldsymbol{x}_2,t_2)\rangle = B_f(|\boldsymbol{x}_1-\boldsymbol{x}_2|;t_1,t_2).$$

V.I. Klyatskin, *Stochastic Equations: Theory and Applications in Acoustics, Hydrodyn., Magnetohydrodyn., and Radiophys., Vol. 1*, Understanding Complex Systems, DOI: 10.1007/978-3-319-07587-7_6,

The corresponding Fourier transform of the correlation function with respect to the spatial variable defines the spatial spectral function (called also the angular spectrum)

$$\Phi_f(\boldsymbol{k},t) = \int d\boldsymbol{x} B_f(\boldsymbol{x},t) e^{i\boldsymbol{k}\boldsymbol{x}},$$

and the Fourier transform of the correlation function of random field $f(\boldsymbol{x},t)$ stationary in time and homogeneous in space defines the space-time spectrum

$$\Phi_f(\boldsymbol{k},\omega) = \int d\boldsymbol{x} \int\limits_{-\infty}^{\infty} dt B_f(\boldsymbol{x},t) e^{i(\boldsymbol{k}\boldsymbol{x}+\omega t)}.$$

In the case of isotropic random field $f(\boldsymbol{x},t)$, the space-time spectrum appears isotropic in the $\boldsymbol{k}$-space:

$$\Phi_f(\boldsymbol{k},\omega) = \Phi_f(k,\omega).$$

The characteristic functional and the n-th order cumulant functions of scalar random field $f(\boldsymbol{x},t)$ are defined similarly

$$\Phi[v(\boldsymbol{x}',\tau)] = \left\langle \exp\left\{ i\int d\boldsymbol{x} \int\limits_{-\infty}^{\infty} dt v(\boldsymbol{x},t) f(\boldsymbol{x},t) \right\}\right\rangle = \exp\left\{\Theta[v(\boldsymbol{x}',\tau)]\right\},$$

$$M_n(\boldsymbol{x}_1,t_1,\cdots,\boldsymbol{x}_n,t_n) = \frac{1}{i^n}\frac{\delta^n}{\delta v(\boldsymbol{x}_1,t_1)\cdots\delta v(\boldsymbol{x}_n,t_n)}\Phi[v(\boldsymbol{x}',\tau)]\bigg|_{v=0},$$

$$K_n(\boldsymbol{x}_1,t_1,\cdots,\boldsymbol{x}_n,t_n) = \frac{1}{i^n}\frac{\delta^n}{\delta v(\boldsymbol{x}_1,t_1)\cdots\delta v(\boldsymbol{x}_n,t_n)}\Theta[v(\boldsymbol{x}',\tau)]\bigg|_{v=0}.$$

In the case of vector random field $\boldsymbol{f}(\boldsymbol{x},t)$, we must assume that $\boldsymbol{v}(\boldsymbol{x},t)$ is the vector function.

6.1 Gaussian Vector Random Field

Consider now the basic space–time statistical characteristics of the vector Gaussian random field $\boldsymbol{u}(\boldsymbol{r},t)$ with zero-valued mean and correlation tensor

$$B_{ij}(\boldsymbol{r},t;\boldsymbol{r}',t') = \langle u_i(\boldsymbol{r},t) u_j(\boldsymbol{r}',t')\rangle.$$

For definiteness, we will identify this vector field with the velocity field.

In the general case, random field $\boldsymbol{u}(\boldsymbol{r},t)$ is assumed to be the divergent ($\operatorname{div}\boldsymbol{u}(\boldsymbol{r},t) \neq 0$) Gaussian field statistically homogeneous and possessing spherical symmetry (but not possessing reflection symmetry) in space and stationary in time with correlation and spectral tensors ($\tau = t - t_1$)

$$B_{ij}(\boldsymbol{r} - \boldsymbol{r}_1, \tau) = \langle u_i(\boldsymbol{r}, t) u_j(\boldsymbol{r}_1, t_1)\rangle = \int d\boldsymbol{k}\, E_{ij}(\boldsymbol{k}, \tau) e^{i\boldsymbol{k}(\boldsymbol{r}-\boldsymbol{r}_1)},$$

$$E_{ij}(\boldsymbol{k}, \tau) = \frac{1}{(2\pi)^d} \int d\boldsymbol{r}\, B_{ij}(\boldsymbol{r}, \tau) e^{-i\boldsymbol{k}\boldsymbol{r}},$$

where d is the dimension of space. In view of the assumed symmetry conditions, correlation tensor $B_{ij}(\boldsymbol{r} - \boldsymbol{r}_1, \tau)$ has the structure [7] ($\boldsymbol{r} - \boldsymbol{r}_1 \to \boldsymbol{r}$)

$$B_{ij}(\boldsymbol{r}, \tau) = B_{ij}^{\text{iso}}(\boldsymbol{r}, \tau) + C(r, \tau)\varepsilon_{ijk} r_k, \tag{6.3}$$

where the isotropic portion of the correlation tensor is expressed as follows

$$B_{ij}^{\text{iso}}(\boldsymbol{r}, \tau) = A(r, \tau) r_i r_j + B(r, \tau)\delta_{ij}.$$

Here, ε_{ijk} is the pseudotensor described in Sect. 1.3.1, page 39. This pseudotensor is used, for example, to determine the *velocity vortex field*

$$\boldsymbol{\omega}(\boldsymbol{r}, t) = \operatorname{curl} \boldsymbol{u}(\boldsymbol{r}, t) = [\boldsymbol{\nabla} \times \boldsymbol{u}(\boldsymbol{r}, t)] \tag{6.4}$$

whose components have the form

$$\omega_i(\boldsymbol{r}, t) = \varepsilon_{ijk} \frac{\partial u_k(\boldsymbol{r}, t)}{\partial r_j}.$$

Note that, in the two-dimensional case, vector $\boldsymbol{\omega}(\boldsymbol{r}, t)$ has a single component orthogonal to velocity field $\boldsymbol{u}(\boldsymbol{r}, t)$; as a result, quantity $\boldsymbol{u}(\boldsymbol{r}, t) \cdot \boldsymbol{\omega}(\boldsymbol{r}, t)$ called *velocity field helicity* vanishes,

$$\boldsymbol{u}(\boldsymbol{r}, t) \cdot \boldsymbol{\omega}(\boldsymbol{r}, t) = 0.$$

Expression (1.72), page 39 allows obtaining the following representations for the vortex field correlation and spectral tensors [7]:

$$\langle \omega_i(\boldsymbol{r}, t)\omega_j(\boldsymbol{r}', t')\rangle = -\delta_{ij}\Delta_{\boldsymbol{r}} B_{ll}(\boldsymbol{r} - \boldsymbol{r}', t - t') + \frac{\partial^2 B_{ll}(\boldsymbol{r} - \boldsymbol{r}', t - t')}{\partial r_i \partial r_j} + \Delta_{\boldsymbol{r}} B_{ij}(\boldsymbol{r} - \boldsymbol{r}', t - t'),$$

$$\Omega_{ij}(\boldsymbol{k}, t - t') = (\delta_{ij}\boldsymbol{k}^2 - k_i k_j) E_{ll}(\boldsymbol{k}, t - t') - \boldsymbol{k}^2 E_{ij}(\boldsymbol{k}, t - t').$$

Convolving them with respect to indices i and j and setting $\boldsymbol{r} = \boldsymbol{r}'$, we obtain the expressions for the correlation and spectral functions of the vortex field

$$\langle \omega_i(\boldsymbol{r},t)\omega_i(\boldsymbol{r},t')\rangle = -\Delta_{\boldsymbol{r}} B_{ll}(\boldsymbol{0},t-t') = -\langle \boldsymbol{u}(\boldsymbol{r},t)\Delta_{\boldsymbol{r}}\boldsymbol{u}(\boldsymbol{r},t')\rangle, \qquad (6.5)$$
$$\Omega_{ii}(\boldsymbol{k},t-t') = \boldsymbol{k}^2 E_{ii}(\boldsymbol{k},t-t').$$

Note that the procedure similar to that used to calculate vortex field variance and spectrum allows additionally calculating average helicity of the velocity field

$$\chi(\boldsymbol{r},t) = \boldsymbol{u}(\boldsymbol{r},t)\cdot\boldsymbol{\omega}(\boldsymbol{r},t) = \boldsymbol{u}(\boldsymbol{r},t)\cdot \operatorname{curl}\boldsymbol{u}(\boldsymbol{r},t)$$

in the three-dimensional case in the absence of reflection symmetry. Indeed, the use of Eq. (1.73), page 39 results in the following expression for statistically averaged quantity $\langle \omega_i(\boldsymbol{r},t)u_j(\boldsymbol{r},t)\rangle$

$$\left\langle \omega_i(\boldsymbol{r},t)u_p(\boldsymbol{r}',t)\right\rangle_{\boldsymbol{r}=\boldsymbol{r}'} = \varepsilon_{ijk}\left\langle \frac{\partial u_k(\boldsymbol{r},t)}{\partial r_j}u_p(\boldsymbol{r}',t)\right\rangle_{\boldsymbol{r}=\boldsymbol{r}'} = \varepsilon_{ijk}\frac{\partial}{\partial r_j}B_{kp}(\boldsymbol{r},0)_{\boldsymbol{r}=\boldsymbol{0}}$$
$$= \lim_{\boldsymbol{r}\to\boldsymbol{0}} \varepsilon_{ijk}\frac{\partial}{\partial r_j}C(r,0)\varepsilon_{kpm}r_m = C(0,0)\varepsilon_{ijm}\varepsilon_{mpj} = 2C(0,0)\delta_{ip}$$

and, hence, the average helicity of the Gaussian random field in the three-dimensional case is

$$\langle \chi(\boldsymbol{r},t)\rangle = 6C(0,0). \qquad (6.6)$$

The isotropic portion of correlation tensor corresponds to the spatial spectral tensor of the form

$$E_{ij}(\boldsymbol{k},\tau) = E^{\mathrm{s}}_{ij}(\boldsymbol{k},\tau) + E^{\mathrm{p}}_{ij}(\boldsymbol{k},\tau),$$

where the spectral components of the tensor of velocity field have the following structure

$$E^{\mathrm{s}}_{ij}(\boldsymbol{k},\tau) = E^{\mathrm{s}}(k,\tau)\left(\delta_{ij} - \frac{k_ik_j}{k^2}\right), \quad E^{\mathrm{p}}_{ij}(\boldsymbol{k},\tau) = E^{\mathrm{p}}(k,\tau)\frac{k_ik_j}{k^2}.$$

Here, $E^{\mathrm{s}}(k,\tau)$ and $E^{\mathrm{p}}(k,\tau)$ are the solenoidal and potential components of the spectral density of velocity field, respectively.

Define now function $B_{ij}(\boldsymbol{r})$ as the integral of correlation function (6.3) over time, i.e.,

$$B_{ij}(\boldsymbol{r}) = \int\limits_0^\infty d\tau B_{ij}(\boldsymbol{r},\tau) = B^{\mathrm{iso}}_{ij}(\boldsymbol{r}) + C(r)\varepsilon_{ijk}r_k. \qquad (6.7)$$

Then, $B_{ij}(\boldsymbol{0}) = D_0\delta_{ij}$ and, hence, quantity

$$B_{ii}(\boldsymbol{0}) = D_0 d = \tau_0\sigma^2_{\boldsymbol{u}} = \int d\boldsymbol{k}\ \left[(d-1)\,E^{\mathrm{s}}(k) + E^{\mathrm{p}}(k)\right] \qquad (6.8)$$

defines temporal correlation radius of the velocity field τ_0. Here, $\sigma_{\boldsymbol{u}}^2 = B_{ii}(\mathbf{0},0) = \langle \boldsymbol{u}^2(\boldsymbol{r},t)\rangle$ is the variance of the velocity field, and functions $E^{\rm s}(k)$ and $E^{\rm p}(k)$ are defined as follows

$$E^{\rm s}(k) = \int_0^\infty d\tau\; E^{\rm s}(k,\tau), \quad E^{\rm p}(k) = \int_0^\infty d\tau\; E^{\rm p}(k,\tau). \tag{6.9}$$

The use of correlation function (6.7) appreciably simplifies calculations dealing with spatial derivatives of the velocity field. Indeed, we have

$$\frac{\partial B_{ij}(\mathbf{0})}{\partial r_k} = C(0)\varepsilon_{ijk}, \tag{6.10}$$

$$\begin{aligned}-\frac{\partial^2 B_{ij}(0)}{\partial r_k \partial r_l} &= \frac{D^{\rm s}}{d(d+2)}\left[(d+1)\delta_{kl}\delta_{ij} - \delta_{ki}\delta_{lj} - \delta_{kj}\delta_{li}\right] \\ &\quad + \frac{D^{\rm p}}{d(d+2)}\left[\delta_{kl}\delta_{ij} + \delta_{ki}\delta_{lj} + \delta_{kj}\delta_{li}\right],\end{aligned} \tag{6.11}$$

$$\frac{\partial^3 B_{kp}(\mathbf{0})}{\partial r_n \partial r_m \partial r_j} = -2\alpha\left(\varepsilon_{kpj}\delta_{nm} + \varepsilon_{kpm}\delta_{nj} + \varepsilon_{kpn}\delta_{mj}\right), \tag{6.12}$$

and, consequently,

$$\frac{\partial^3 B_{kp}(\mathbf{0})}{\partial \boldsymbol{r}^2 \partial r_j} = -2\alpha(d+2)\varepsilon_{kpj},$$

where

$$\begin{aligned} D^{\rm s} &= \int d\boldsymbol{k}\, k^2 E^{\rm s}(k) = \frac{1}{d-1}\int_0^\infty d\tau\, \langle \boldsymbol{\omega}(\boldsymbol{r},t+\tau)\boldsymbol{\omega}(\boldsymbol{r},t)\rangle, \\ D^{\rm p} &= \int d\boldsymbol{k}\, k^2 E^{\rm p}(k) = \int_0^\infty d\tau \left\langle \frac{\partial \boldsymbol{u}(\boldsymbol{r},t+\tau)}{\partial \boldsymbol{r}}\frac{\partial \boldsymbol{u}(\boldsymbol{r},t)}{\partial \boldsymbol{r}}\right\rangle, \\ C(r) &= C(0) - \alpha r^2, \end{aligned} \tag{6.13}$$

and $\boldsymbol{\omega}(\boldsymbol{r},t)$ is the velocity field vortex.

As we noted earlier, characteristic functionals ensure the exhaustive description of random processes and fields. However, even one-time probability densities provide certain information about temporal behavior and spatial structure of random processes for arbitrary long temporal intervals. The ideas of statistical topography of random processes and fields can assist in obtaining this information.

6.2 Statistical Topography of Random Processes and Fields

The term *statistical topography* was seemingly for the first time introduced in book [141], though the underlying ideas of this approach can be traced back to much earlier works (see, e.g., books [1, 11] and review [36] with detailed reference lists on the problem).

Similarly to common topography of mountain ranges, the statistical topography studies the systems of contours (level lines in the two-dimensional case and surfaces of constant values in the three-dimensional case) specified by the equality $f(\boldsymbol{r}, t) = f = \text{const}$. Figure 6.1 shows examples of realizations of two random fields characterized by different statistical structures.

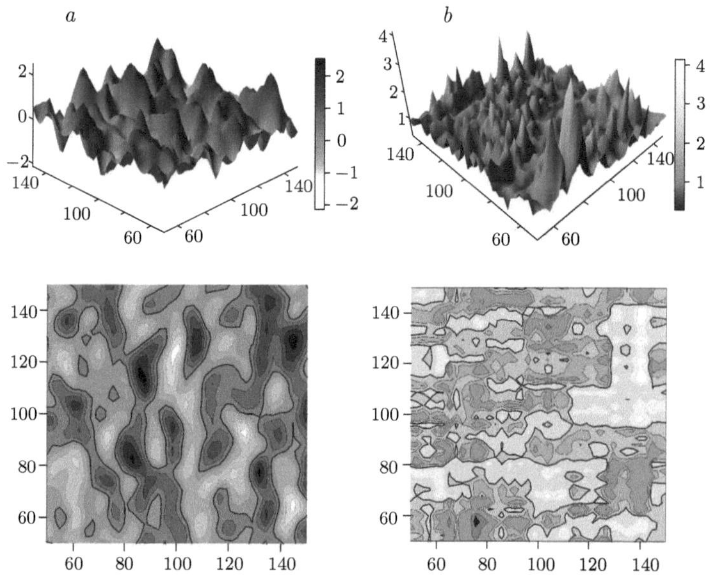

Fig. 6.1 Realizations of the fields governed by (a) Gaussian and (b) lognormal distributions(a) and the corresponding topographic level lines. The bold curves in the bottom patterns show level lines corresponding to levels 0 (a) and 1 (b).

For analyzing a system of contours (in this section, we will deal for simplicity with the two-dimensional case and assume $\boldsymbol{r} = \boldsymbol{R}$), we introduce the singular indicator function (6.2) concentrated on these contours.

The convenience of function (6.2) consists, in particular, in the fact that it allows simple expressions for quantities such as the total area of regions where $f(\boldsymbol{R}, t) > f$ (i.e., within level lines $f(\boldsymbol{R}, t) = f$) and the total mass of the field within these regions [72]

$$S(t;f) = \int \theta(f(\boldsymbol{R},t) - f)d\boldsymbol{R} = \int_f^{\infty} df' \int d\boldsymbol{R}\varphi(t,\boldsymbol{R};f'),$$

$$M(t;f) = \int f(\boldsymbol{R},t)\theta(f(\boldsymbol{R},t) - f)d\boldsymbol{R} = \int_f^{\infty} f'df' \int d\boldsymbol{R}\varphi(t,\boldsymbol{R};f').$$

As we mentioned earlier, the mean value of indicator function (6.2) over an ensemble of realizations determines the one-time (in time) and one-point (in space) probability density

$$P(\boldsymbol{R},t;f) = \langle \varphi(\boldsymbol{R},t;f)\rangle = \langle \delta\left(f(\boldsymbol{R},t) - f\right)\rangle .$$

Consequently, this probability density immediately determines ensemble-averaged values of the above expressions.

If we include into consideration the spatial gradient

$$\boldsymbol{p}(\boldsymbol{R},t) = \boldsymbol{\nabla} f(\boldsymbol{R},t),$$

we can obtain additional information on details of the structure of field $f(\boldsymbol{R},t)$. For example, quantity

$$l(t;f) = \int d\boldsymbol{R}\, |\boldsymbol{p}(\boldsymbol{R},t)|\, \delta(f(\boldsymbol{R},t) - f) = \oint dl \tag{6.14}$$

is the total length of contours [84 – 88] and extends formula (5.11) to random fields.

The integrand in Eq. (6.14) is described in terms of the extended indicator function

$$\varphi(\boldsymbol{R},t;f,\boldsymbol{p}) = \delta\left(f(\boldsymbol{R},t) - f\right)\delta\left(\boldsymbol{p}(\boldsymbol{R},t) - \boldsymbol{p}\right), \tag{6.15}$$

so that the average value of total length (6.14) is related to the joint one-time probability density of field $f(\boldsymbol{R},t)$ and its gradient $\boldsymbol{p}(\boldsymbol{R},t)$, which is defined as the ensemble average of indicator function (6.15), i.e., as the function

$$P(\boldsymbol{R},t;f,\boldsymbol{p}) = \langle \delta\left(f(\boldsymbol{R},t) - f\right)\delta\left(\boldsymbol{p}(\boldsymbol{R},t) - \boldsymbol{p}\right)\rangle .$$

Inclusion of second-order spatial derivatives into consideration allows estimating the total number of contours $f(\boldsymbol{R},t) = f = \text{const}$ by the approximate formula (neglecting unclosed lines) [127]

$$N(t;f) = N_{\text{in}}(t;f) - N_{\text{out}}(t;f) = \frac{1}{2\pi}\int d\boldsymbol{R}\kappa(t,\boldsymbol{R};f)\,|\boldsymbol{p}(\boldsymbol{R},t)|\,\delta\left(f(\boldsymbol{R},t) - f\right),$$

where $N_{\text{in}}(t;f)$ and $N_{\text{out}}(t;f)$ are the numbers of contours for which vector $\boldsymbol{p}$ is directed along internal and external normals, respectively; and $\kappa(\boldsymbol{R},t;f)$ is the curvature of the level line.

Recall that, in the case of the spatially homogeneous field $f(\boldsymbol{R},t)$, the corresponding probability densities $P(\boldsymbol{R},t;f)$ and $P(\boldsymbol{R},t;f,\boldsymbol{p})$ are independent of $\boldsymbol{R}$. In this case, statistical averages of the above expressions without integration over $\boldsymbol{R}$ characterize – if exits – the corresponding specific (per unit area) values of these quantities.

So, the specific mean area $\langle s_{\mathrm{hom}}(t;f)\rangle$ over which the random field $f(\boldsymbol{R},t)$ exceeds a given level f, coincides with the probability of the event $f(\boldsymbol{R},t) > f$ at any spatial point, i.e.,

$$\langle s_{\mathrm{hom}}(t;f)\rangle = \langle \theta(f(\boldsymbol{R},t) - f)\rangle = \mathrm{P}\{f(\boldsymbol{R},t) > f\}$$

and therefore the mean specific area offers a geometric interpretation of the probability of the event $f(\boldsymbol{R},t) > f$, which is apparently independent of point $\boldsymbol{R}$.

In Chapter 1 we noted that, in a number of cases, stochastic structures of the type of clustering can be formed in stochastic dynamic systems.

Clustering of a field is identified as the emergence of compact areas with large field values against the residual background of areas where these values are fairly low. *Naturally, statistical averaging completely destroys all data on clustering.*

The notion of clustering by itself is related to the spatial behavior of a random field *in separate realizations*! Clustering either exists or not exists.

Consider now the conditions of occurrence of stochastic structure formation. *Phenomenon of clustering* random fields can be detected and described only on the basis of the ideas of statistical topography.

6.3 On the Criterion of Stochastic Structure Formation in Random Media

It is clear that, for a *positive field* $f(\boldsymbol{R},t)$, the condition of clustering with a probability one, i.e., almost in all realizations, is formulated *in the general case* as simultaneous tendency of fulfillment of the following asymptotic equalities for $t \to \infty$

$$\langle S(t;f)\rangle \to 0, \quad \langle M(t;f)\rangle \to \int d\boldsymbol{R}\,\langle f(\boldsymbol{R},t)\rangle\,.$$

On the contrary, simultaneous tendency of fulfillment of the asymptotic equalities for $t \to \infty$

$$\langle S(t;f)\rangle \to \infty, \quad \langle M(t;f)\rangle \to \int d\boldsymbol{R}\,\langle f(\boldsymbol{R},t)\rangle$$

corresponds to the absence of structure formation.

In the case of a spatially homogeneous field $f(\boldsymbol{R}, t)$, conditions of clustering are reduced to the tendency of asymptotic equalities for $t \to \infty$

$$\langle s_{\text{hom}}(t; f)\rangle = \mathrm{P}\{f(\boldsymbol{R}, t) > f\} \to 0, \quad \langle m_{\text{hom}}(t; f)\rangle \to \langle f(t)\rangle .$$

Absence of clustering corresponds to the tendency of asymptotic equalities

$$\langle s_{\text{hom}}(t; f)\rangle = \mathrm{P}\{f(\boldsymbol{R}, t) > f\} \to 1, \quad \langle m_{\text{hom}}(t; f)\rangle \to \langle f(t)\rangle .$$

Thus, in spatially homogeneous problems, clustering is the physical phenomenon realized with probability one (i.e., occurred in almost all realizations of a positive random field) generated by a rare event whose probability tends to zero. Namely availability of these rare events is the trigger that starts the process of structure formation, and structure formation itself is the property of random medium [61–66].

In the conditions of developed clustering, the random field is simply absent in the most part of space!

As for setup time of such spatial structure formation, it depends on limiting behavior of the right-hand expressions in all above asymptotic equalities.

Stochastic structure formation is governed by diffusion of random field $f(\boldsymbol{R}, t)$ in its phase space $\{f\}$. This means that clustering of such a positive field $f(\boldsymbol{R}, t)$ is inherent in dynamic fields of arbitrary nature, and we can say that structure formation *is a law of nature* for such fields.

The above criterion of *'ideal'* clustering (analogously to *ideal hydrodynamic*) describes dynamics of cluster formation in the dynamic systems described in general by the first-order partial differential equations (*Eulerian description*). This *ideal* structure originates in the form of very *thin belts* (in the two-dimensional case) or very *thin tubes* (in the three-dimensional case).

We note that first-order partial differential equations can be solved in the general case by the method of characteristics. This corresponds to the *Lagrangian description* of dynamic systems. Moreover, the characteristic curves, which are described by ordinary differential equations, can of course have various peculiarities and even singularities. The conditions for such peculiarities to arise in the Lagrangian description are not directly related to the phenomenon of clustering in space and time, i.e., in the *Eulerian description.* A more detailed description of this fact is given in Volume 2, Part 2.

As for *actual* physical systems, various additional factors come to play with time; they are related to generation of random field spatial derivatives like spatial diffusion or diffraction, which *deform* the pattern of clustering, but not *dispose* it.

In particular, a possible situation can occur when the probability density rapidly approaches its steady-state regime $P(\boldsymbol{R}; f)$ for $t \to \infty$. In this case, functionals like $\langle S(f)\rangle$ and $\langle M(f)\rangle$ cease to describe further deformation of the clustering pattern, and we must study temporal evolution of functionals related to the spatial derivatives of field $f(\boldsymbol{R}, t)$, like the number of contours.

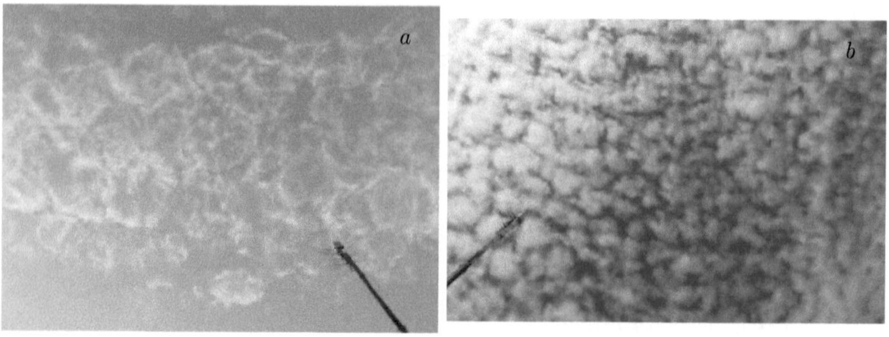

Fig. 6.2 Photo of cluster structure of cloudy sky: (*a*) nearly "ideal" clustering and (*b*) "deformed" clustering

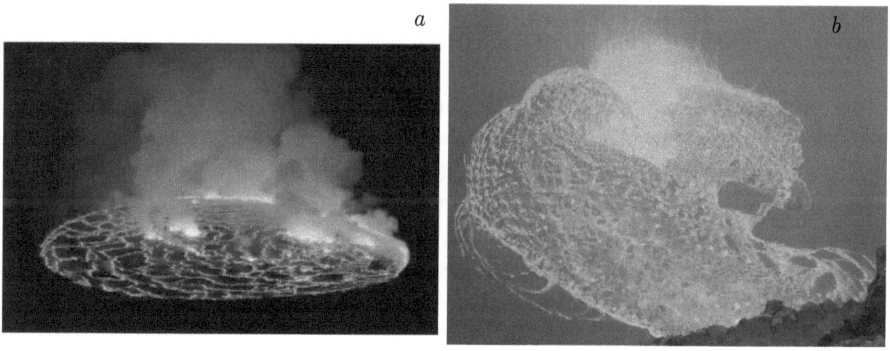

Fig. 6.3 The lava lakes in the depths of Nyiragongo Crater (*a*), and Kilauea Crater (*b*)

As illustrations of '*ideal*' and '*deformed*' clustering in Nature we cite photos of cluster structure of cloudy sky[1] (Fig. 6.2), and the lava lakes boiling in the depths of Nyiragongo Crater (in the heart of the Great Lakes region of Africa) and in the depth of Kilauea Crater (Hawaiian National Volcano Park) in Figs. 6.3[2]. The problem of *wave propagation in random media* yields other illustrations of the phenomenon of clustering (see Fig. 1.19, page 65 in Chapter 1, and similar patterns in Chapter 20 of Volume 2, which were obtained in both laboratory experiments and numerical simulations).

---

[1] These photos were obtained by V.A. Dovzhenko on June 15 and August 2, 2013 at the coast of the Azov Sea (Russia) at 21:00.

[2] Photos in Figs. 6.3 can be found at `http://bigpicture.ru/?p=128340`, `http://pacificislandparks.com/2010/01/20/more-amazing-lava-lake-photos/`

Chapter 7
Correlation Splitting

7.1 General Remarks

For simplicity, we content ourselves here with the one-dimensional random processes (extensions to multidimensional cases are obvious). We need the ability of calculating correlation $\langle F[z(\tau)]R[z(\tau)]\rangle$, where $F[z(\tau)]$ is the functional explicitly dependent on process $z(t)$ and $R[z(\tau)]$ is the functional that can depend on process $z(t)$ both explicitly and implicitly. To calculate this average, we consider auxiliary functionals $F[z(\tau)+\eta_1(\tau)]$ and $R[z(\tau)+\eta_2(\tau)]$, where $\eta_i(t)$ are arbitrary deterministic functions, and calculate the correlation $\langle F[z(\tau)+\eta_1(\tau)]R[z(\tau)+\eta_2(\tau)]\rangle$. The correlation of interest will be obtained by setting $\eta_i(\tau)=0$ in the final result.

We can expand the above auxiliary functionals in the functional Taylor series with respect to $z(\tau)$. The result can be represented in the form

$$F[z(\tau)+\eta_1(\tau)] = e^{\int\limits_{-\infty}^{\infty} d\tau z(\tau)\frac{\delta}{\delta\eta_1(\tau)}} F[\eta_1(\tau)],$$

$$R[z(\tau)+\eta_2(\tau)] = e^{\int\limits_{-\infty}^{\infty} d\tau z(\tau)\frac{\delta}{\delta\eta_2(\tau)}} R[\eta_2(\tau)],$$

where we introduced the functional shift operators. With this representation, we can obtain the following expression for the correlation

$$\begin{aligned} &\langle F[z(\tau)+\eta_1(\tau)]R[z(\tau)+\eta_2(\tau)]\rangle \\ &= \exp\left\{\Theta\left[\frac{1}{i}\left(\frac{\delta}{\delta\eta_1(\tau)}+\frac{\delta}{\delta\eta_2(\tau)}\right)\right] - \Theta\left[\frac{1}{i}\frac{\delta}{\delta\eta_1(\tau)}\right] - \Theta\left[\frac{1}{i}\frac{\delta}{\delta\eta_2(\tau)}\right]\right\} \\ &\qquad\times \langle F[z(\tau)+\eta_1(\tau)]\rangle\,\langle R[z(\tau)+\eta_2(\tau)]\rangle\,. \qquad (7.1)\end{aligned}$$

Here $\Theta[v(\tau)] = \ln\Phi[v(\tau)]$ and $\Phi[v(\tau)]$ is the characteristic functional of random process $z(t)$ (See Eq. (5.5)).

V.I. Klyatskin, *Stochastic Equations: Theory and Applications in Acoustics, Hydrodyn., Magnetohydrodyn., and Radiophys., Vol. 1*, Understanding Complex Systems,
DOI: 10.1007/978-3-319-07587-7_7, © Springer International Publishing Switzerland 2015

This formula expresses the average of the product of functionals through the product of averages of the functionals themselves. The main problem here consists in calculating the action of the functional operator

$$\exp\left\{\Theta\left[\frac{1}{i}\left(\frac{\delta}{\delta\eta_1(\tau)}+\frac{\delta}{\delta\eta_2(\tau)}\right)\right]-\Theta\left[\frac{1}{i}\frac{\delta}{\delta\eta_1(\tau)}\right]-\Theta\left[\frac{1}{i}\frac{\delta}{\delta\eta_2(\tau)}\right]\right\}$$

on the product of average functionals.

In a number of statistical problems, the intensity of parameter fluctuation can be considered small. In these situations, we can expand functional $F[z(\tau)]$ in the Taylor series with respect to process $z(\tau)$ and content themselves with the linear term of the expansion. In the case of the linear functional $F[z(\tau)] \equiv z(t')$, we obtain the following expression for the correlation

$$\langle z(t')R[z(\tau)+\eta(\tau)]\rangle = \Omega\left[t';\frac{\delta}{i\delta\eta(\tau)}\right]\langle R[z(\tau)+\eta(\tau)]\rangle\,,$$

where functional

$$\Omega[t';v(\tau)] = \frac{\left\langle z(t')\exp\left\{i\int\limits_{-\infty}^{\infty}d\tau z(\tau)v(\tau)\right\}\right\rangle}{\left\langle \exp\left\{i\int\limits_{-\infty}^{\infty}d\tau z(\tau)v(\tau)\right\}\right\rangle} = \frac{\delta}{i\delta v(t')}\Theta[v(\tau)].$$

Setting now $\eta(\tau)=0$, we obtain the expression

$$\langle z(t')R[z(\tau)\rangle = \left\langle \Omega\left[t';\frac{\delta}{i\delta z(\tau)}\right]R[z(\tau)]\right\rangle. \tag{7.2}$$

If we expand functional $\Theta[v(\tau)]$ in the functional Taylor series (5.6), page 127

$$\Omega[t';v(\tau)] = \sum_{n=0}^{\infty}\frac{i^n}{n!}\int\limits_{-\infty}^{\infty}dt_1\cdots\int\limits_{-\infty}^{\infty}dt_n K_{n+1}(t',t_1,\cdots,t_n)v(t_1)\cdots v(t_n),$$

then expression (7.2) assumes the form

$$\langle z(t')R[z(\tau)]\rangle = \sum_{n=0}^{\infty}\frac{1}{n!}\int\limits_{-\infty}^{\infty}dt_1\cdots\int\limits_{-\infty}^{\infty}dt_n K_{n+1}(t',t_1,\cdots,t_n)\left\langle\frac{\delta^n R[z(\tau)]}{\delta z(t_1)\cdots\delta z(t_n)}\right\rangle. \tag{7.3}$$

Note that, if functional $R[z(\tau)]$ has the form of the power monomial

$$R[z(\tau)] = z(t_1)\cdots z(t_n),$$

then Eq. (7.3) recursively relates the n-point moment of process $z(t)$ to its cumulants.

If process $z(t)$ is simply random quantity z, operator $\int dt \frac{\delta}{\delta z(t)}$ reduces to the ordinary derivative $\frac{d}{dz}$ and Eq. (7.3) grades into Eq. (4.12), page 120. Thus, Eq. (7.3) extends Eq. (4.12) to random processes.

In physical problems satisfying the condition of dynamical causality in time t, statistical characteristics of the solution at instant t depend on the statistical characteristics of process $z(\tau)$ for $0 \le \tau \le t$, which are completely described by the characteristic functional

$$\Phi[t; v(\tau)] = \exp\{\Theta[t; v(\tau)]\} = \left\langle \exp\left\{ i \int_0^t d\tau z(\tau) v(\tau) \right\} \right\rangle.$$

In this case, the obtained formulas hold also for calculating statistical averages $\langle z(t')R[t; z(\tau)]\rangle$ for $t' < t$, $\tau \le t$, i.e., we have the equality

$$\langle z(t')R[t; z(\tau)]\rangle = \left\langle \Omega\left[t', t; \frac{\delta}{i\delta z(\tau)}\right] R[t; z(\tau)] \right\rangle \quad (0 < t' < t), \tag{7.4}$$

where

$$\begin{aligned} \Omega[t', t; v(\tau)] &= \frac{\delta}{i\delta v(t')]} \Theta[t; v(\tau)] \\ &= \sum_{n=0}^{\infty} \frac{i^n}{n!} \int_0^t dt_1 \cdots \int_0^t dt_n K_{n+1}(t', t_1, \cdots, t_n) v(t_1) \cdots v(t_n). \end{aligned} \tag{7.5}$$

For $t' = t - 0$, formula (7.4) holds as before, i.e.

$$\langle z(t)R[t; z(\tau)]\rangle = \left\langle \Omega\left[t, t; \frac{\delta}{i\delta z(\tau)}\right] R[t; z(\tau)] \right\rangle. \tag{7.6}$$

However, expansion (7.5) not always gives the correct result in the limit $t' \to t - 0$ (which means that the limiting process and the procedure of expansion in the functional Taylor series can appear non-commutable). In this case,

$$\Omega[t, t; v(\tau)] = \frac{\left\langle z(t) \exp\left\{ i \int_0^t d\tau z(\tau) v(\tau) \right\} \right\rangle}{\left\langle \exp\left\{ i \int_0^t d\tau z(\tau) v(\tau) \right\} \right\rangle} = \frac{d}{iv(t)dt} \Theta[t; v(\tau)], \tag{7.7}$$

and statistical averages in Eqs. (7.4) and (7.6) can be discontinuous at $t' = t - 0$.

Consider several examples of random processes.

7.2 Gaussian Process

In the case of the Gaussian random process $z(t)$, all formulas obtained in the previous section become significantly simpler. In this case, the logarithm of characteristic functional $\Phi[v(\tau)]$ is given by Eq. (5.13), page 130 (we assume that the mean value of process $z(t)$ is zero); as a consequence, Eq. (7.1) assumes the form

$$\langle F[z(\tau)+\eta_1(\tau)]R[z(\tau)+\eta_2(\tau)]\rangle = \exp\left\{\int\limits_{-\infty}^{\infty}\int\limits_{-\infty}^{\infty} d\tau_1 d\tau_2 B(\tau_1,\tau_2)\frac{\delta^2}{\delta\eta_1(\tau_1)\delta\eta_2(\tau_2)}\right\}$$
$$\times \langle F[z(\tau)+\eta_1(\tau)]\rangle \langle R[z(\tau)+\eta_2(\tau)]\rangle . \quad (7.8)$$

We can easily calculate variational derivative of Eq. (7.8) with respect to $\eta_1(\tau)$ (this operation reduces to functional shift) and set $\eta_1(\tau) = 0$. As a result, we obtain the equality

$$\langle F[z(\tau)]R[z(\tau)+\eta(\tau)]\rangle = \left\langle F\left[z(\tau)+\int\limits_{-\infty}^{\infty} d\tau_1 B(\tau,\tau_1)\frac{\delta}{\delta\eta(\tau_1)}\right]\right\rangle \langle R[z(\tau)+\eta(\tau)]\rangle . \quad (7.9)$$

Let $F[z(\tau)] = z(t)$ for example. Then Eq. (7.9) assumes the form

$$\langle z(t)R[z(\tau)+\eta(\tau)]\rangle = \int\limits_{-\infty}^{\infty} d\tau_1 B(t,\tau_1)\frac{\delta}{\delta\eta(\tau_1)}\langle R[z(\tau)+\eta(\tau)]\rangle . \quad (7.10)$$

Replacing now differentiation with respect to $\eta(\tau)$ by differentiation with respect to $z(\tau)$ and setting $\eta(\tau) = 0$, we obtain the equality

$$\langle z(t)R[z(\tau)]\rangle = \int\limits_{-\infty}^{\infty} d\tau_1 B(t,\tau_1)\left\langle \frac{\delta}{\delta z(\tau_1)}R[z(\tau)]\right\rangle \quad (7.11)$$

commonly known in physics as the *Furutsu–Novikov formula* [20, 114]. Note that this formula can be obtained by partial integration in appropriate functional space [15].

One can easily obtain the multi-dimensional extension of Eq. (7.11); it can be written in the form

$$\langle z_{i_1,\cdots,i_n}(\boldsymbol{r})R[\boldsymbol{z}]\rangle = \int d\boldsymbol{r}' \langle z_{i_1,\cdots,i_n}(\boldsymbol{r})z_{j_1,\cdots,j_n}(\boldsymbol{r}')\rangle \left\langle \frac{\delta R[\boldsymbol{z}]}{\delta z_{j_1,\cdots,j_n}(\boldsymbol{r}')}\right\rangle , \quad (7.12)$$

where $\boldsymbol{r}$ stands for all continuous arguments of random vector field $\boldsymbol{z}(\boldsymbol{r})$ and $i_1, \cdots, i_n$ are the discrete (index) arguments. Repeated index arguments in the right-hand side of Eq. (7.12) assume summation.

If we set $F[z(\tau)] = \exp\left\{ i \int\limits_{-\infty}^{\infty} d\tau\, z(\tau)v(\tau) \right\}$ in Eq. (7.9), then we obtain, at $\eta(\tau) = 0$, the equality

$$\left\langle \exp\left\{ i \int\limits_{-\infty}^{\infty} d\tau\, z(\tau)v(\tau) \right\} R[z(\tau)] \right\rangle = \Phi[v(\tau)] \left\langle R\left[z(\tau) + i \int\limits_{-\infty}^{\infty} d\tau_1 B(\tau,\tau_1)v(\tau_1) \right] \right\rangle, \tag{7.13}$$

in which random process $z(\tau)$ within the averaging brackets in the right-hand side is added with the deterministic imaginary term. Formulas (7.11), (7.12) and (7.13) extend formulas (4.17), (4.24), page 123 to the Gaussian random processes.

If random process $z(\tau)$ is defined only on time interval $[0,t]$, then functional $\Theta[t, v(\tau)]$ will assume the form

$$\Theta[t, v(\tau)] = -\frac{1}{2} \int\limits_0^t \int\limits_0^t d\tau_1 d\tau_2 B(\tau_1,\tau_2) v(\tau_1) v(\tau_2), \tag{7.14}$$

and functionals $\Omega[t', t; v(\tau)]$ and $\Omega[t, t; v(\tau)]$ will be the linear functionals

$$\begin{aligned} \Omega[t', t; v(\tau)] &= \frac{\delta}{i\delta v(t')} \Theta[t, v(\tau)] = i \int\limits_0^t d\tau B(t',\tau) v(\tau), \\ \Omega[t, t; v(\tau)] &= \frac{d}{iv(t)dt} \Theta[t, v(\tau)] = i \int\limits_0^t d\tau B(t,\tau) v(\tau). \end{aligned} \tag{7.15}$$

As a consequence, Eqs. (7.4), (7.6) will assume the form

$$\langle z(t') R[t, z(\tau)] \rangle = \int\limits_0^t d\tau B(t',\tau) \left\langle \frac{\delta R[z(\tau)]}{\delta z(\tau)} \right\rangle \quad (t' \leqslant t) \tag{7.16}$$

that coincides with Eq. (7.11) if the condition

$$\frac{\delta R[t; z(\tau)]}{\delta z(\tau)} = 0 \quad \text{for} \quad \tau < 0, \quad \tau > t \tag{7.17}$$

holds. Note that Eq. (7.13) assumes in this case the form

$$\left\langle \exp\left\{ i\int\limits_0^t d\tau z(\tau)v(\tau) \right\} R[t;z(\tau)] \right\rangle =$$

$$\Phi[t;v(\tau)] \left\langle R\left[t; z(\tau) + i\int\limits_0^t d\tau_1 B(\tau,\tau_1)v(\tau_1) \right] \right\rangle, \tag{7.18}$$

where $\Phi[t;v(\tau)]$ is the characteristic functional of the Gaussian random process $z(t)$.

The corresponding extension of multidimensional formula (7.12) to the case of functional $R[t;\boldsymbol{z}(\widetilde{\boldsymbol{r}},\tau)]$ causal in time has the form

$$\langle z_{i_1,\cdots,i_n}(\boldsymbol{r},t)R[t;\boldsymbol{z}(\widetilde{\boldsymbol{r}},\tau)]\rangle =$$

$$\int d\boldsymbol{r}' \int\limits_0^t dt' \, \langle z_{i_1,\cdots,i_n}(\boldsymbol{r},t) z_{j_1,\cdots,j_n}(\boldsymbol{r}',t')\rangle \left\langle \frac{\delta R[t;\boldsymbol{z}(\widetilde{\boldsymbol{r}},\tau)]}{\delta z_{j_1,\cdots,j_n}(\boldsymbol{r}',t')} \right\rangle. \tag{7.19}$$

7.3 Gaussian Random Fields

We consider Gaussian random fields in more detail because they play very important role in physical applications.

A statistical quantity such as the joint characteristic functional (*generating functional*) of two correlated Gaussian fields $\boldsymbol{u}(\boldsymbol{r},t)$ and $\boldsymbol{v}(\boldsymbol{r},t)$ with zero means has the following structure

$$\left\langle e^{\int d\boldsymbol{r}' \int\limits_{-\infty}^{\infty} dt' \, [\boldsymbol{\kappa}_1(\boldsymbol{r}',t')\boldsymbol{v}(\boldsymbol{r}',t') + \boldsymbol{\kappa}(\boldsymbol{r}',t')\boldsymbol{u}(\boldsymbol{r}',t')]} \right\rangle$$

$$= \exp\left\{ \frac{1}{2}\int d\boldsymbol{r}'d\boldsymbol{r}'' \int\limits_{-\infty}^{\infty} dt'dt'' \, \left[\langle v_i(\boldsymbol{r}',t')v_j(\boldsymbol{r}'',t'')\rangle \varkappa_{1i}(\boldsymbol{r}',t')\varkappa_{1j}(\boldsymbol{r}'',t'') \right.\right.$$

$$\left.\left. +2\langle v_i(\boldsymbol{r}',t')u_j(\boldsymbol{r}'',t'')\rangle \kappa_{1i}(\boldsymbol{r}',t')\kappa_j(\boldsymbol{r}'',t'') + \langle u_i(\boldsymbol{r}',t')u_j(\boldsymbol{r}'',t'')\rangle \kappa_i(\boldsymbol{r}',t')\kappa_j(\boldsymbol{r}'',t'') \right] \right\}.$$

Varying this expression with respect to κ_{1l} at point $(\boldsymbol{r},t)$ (i.e., calculating variational derivative $\dfrac{\delta}{i\delta\kappa_{1l}(\boldsymbol{r},t)}$) and setting $\boldsymbol{\kappa}_1 = \boldsymbol{0}$, we obtain

$$\left\langle v_l(\boldsymbol{r},t) e^{\int d\boldsymbol{r}' \int_{-\infty}^{\infty} dt' \, \boldsymbol{u}(\boldsymbol{r}',t')\boldsymbol{\kappa}(\boldsymbol{r}',t')} \right\rangle$$

$$= \int d\boldsymbol{r}'' \int_{-\infty}^{\infty} dt'' \, \langle v_l(\boldsymbol{r},t) u_j(\boldsymbol{r}'',t'')\rangle \, \kappa_j(\boldsymbol{r}'',t'') \left\langle e^{\int d\boldsymbol{r}' \int_{-\infty}^{\infty} dt' \, \boldsymbol{u}(\boldsymbol{r}',t')\boldsymbol{\kappa}(\boldsymbol{r}',t')} \right\rangle.$$

Using this expression and the functional shift operator, we can represent correlation $\langle v_l(\boldsymbol{r},t) F[\boldsymbol{u}(\widetilde{\boldsymbol{r}},\tau) + \boldsymbol{\eta}(\widetilde{\boldsymbol{r}},\tau)]\rangle$ in the form

$$\langle v_l(\boldsymbol{r},t) F[\boldsymbol{u}(\widetilde{\boldsymbol{r}},\tau) + \boldsymbol{\eta}(\widetilde{\boldsymbol{r}},\tau)]\rangle = \left\langle v_l(\boldsymbol{r},t) e^{\int d\boldsymbol{r}' \int_{-\infty}^{\infty} dt' \, \boldsymbol{u}(\boldsymbol{r}',t') \frac{\delta}{\delta \boldsymbol{\eta}(\boldsymbol{r}',t')}} \right\rangle F[\boldsymbol{\eta}(\widetilde{\boldsymbol{r}},\tau)]$$

$$= \int d\boldsymbol{r}'' \int_{-\infty}^{\infty} dt'' \, \langle v_l(\boldsymbol{r},t) u_j(\boldsymbol{r}'',t'')\rangle \frac{\delta}{\delta \eta_j(\boldsymbol{r}'',t'')} \left\langle e^{\int d\boldsymbol{r}' \int_{-\infty}^{\infty} dt' \, \boldsymbol{u}(\boldsymbol{r}',t') \frac{\delta}{\delta \boldsymbol{\eta}(\boldsymbol{r}',t')}} \right\rangle F[\boldsymbol{\eta}(\widetilde{\boldsymbol{r}},\tau)]$$

$$= \int d\boldsymbol{r}' \int_{-\infty}^{\infty} dt' \, \langle v_l(\boldsymbol{r},t) u_j(\boldsymbol{r}',t')\rangle \frac{\delta}{\delta \eta_j(\boldsymbol{r}',t')} \langle F[\boldsymbol{u}(\widetilde{\boldsymbol{r}},\tau) + \boldsymbol{\eta}(\widetilde{\boldsymbol{r}},\tau)]\rangle .$$

Setting now $\boldsymbol{\eta}(\widetilde{\boldsymbol{r}},\tau) = 0$, we arrive at a formula that generalizes the Furutsu-Novikov (7.12) to two correlated Gaussian fields

$$\langle v_l(\boldsymbol{r},t) F[\boldsymbol{u}(\widetilde{\boldsymbol{r}},\tau)]\rangle = \int d\boldsymbol{r}' \int_{-\infty}^{\infty} dt' \, \langle v_l(\boldsymbol{r},t) u_j(\boldsymbol{r}',t')\rangle \left\langle \frac{\delta}{\delta u_j(\boldsymbol{r}',t')} F[\boldsymbol{u}(\widetilde{\boldsymbol{r}},\tau)] \right\rangle. \tag{7.20}$$

Here, field $\boldsymbol{v}(\boldsymbol{r},t)$ is, generally speaking, arbitrary. In particular, it can depend on field $\boldsymbol{u}(\boldsymbol{r},t)$. For example, if $\boldsymbol{v}(\boldsymbol{r},t) = \dfrac{\partial \boldsymbol{u}(\boldsymbol{r},t)}{\partial r_k}$, then we obtain the expression

$$\left\langle \frac{\partial u_l(\boldsymbol{r},t)}{\partial r_k} F[\boldsymbol{u}(\widetilde{\boldsymbol{r}},\tau)] \right\rangle = \int d\boldsymbol{r}' \int_{-\infty}^{\infty} dt' \, \frac{\partial \langle u_l(\boldsymbol{r},t) u_j(\boldsymbol{r}',t')\rangle}{\partial r_k} \left\langle \frac{\delta}{\delta u_j(\boldsymbol{r}',t')} F[\boldsymbol{u}(\widetilde{\boldsymbol{r}},\tau)] \right\rangle$$

that, in the case of the Gaussian field $\boldsymbol{u}(\boldsymbol{r},t)$ homogeneous in space and stationary in time reduces to the equality

$$\left\langle \frac{\partial u_l(\boldsymbol{r},t)}{\partial r_k} F[\boldsymbol{u}(\widetilde{\boldsymbol{r}},\tau)] \right\rangle = \int d\boldsymbol{r}' \int_{-\infty}^{\infty} dt' \, \frac{\partial B_{lj}(\boldsymbol{r}-\boldsymbol{r}',t-t')}{\partial r_k} \left\langle \frac{\delta}{\delta u_j(\boldsymbol{r}',t')} F[\boldsymbol{u}(\widetilde{\boldsymbol{r}},\tau)] \right\rangle, \tag{7.21}$$

where $B_{lj}(\boldsymbol{r}-\boldsymbol{r}',t-t') = \langle u_l(\boldsymbol{r},t) u_j(\boldsymbol{r}',t')\rangle$ is the correlation tensor of random vector field $\boldsymbol{u}(\boldsymbol{r},t)$.

Similar expressions can be derived for both other spatial derivatives of field $\boldsymbol{u}(\boldsymbol{r},t)$ and temporal derivatives of this field.

If condition of dynamic causality holds, formulas (7.20) and (7.21) are simplified and assume the form

$$\left\langle v_l(\boldsymbol{r},t)F[\boldsymbol{u}(\widetilde{\boldsymbol{r}},\tau)]\right\rangle = \int d\boldsymbol{r}'\int\limits_0^t dt'\ \left\langle v_l(\boldsymbol{r},t)u_j(\boldsymbol{r}',t')\right\rangle\left\langle\frac{\delta}{\delta u_j(\boldsymbol{r}',t')}F[\boldsymbol{u}(\widetilde{\boldsymbol{r}},\tau)]\right\rangle, \tag{7.22}$$

$$\left\langle\frac{\partial u_l(\boldsymbol{r},t)}{\partial r_k}F[\boldsymbol{u}(\widetilde{\boldsymbol{r}},\tau)]\right\rangle = \int d\boldsymbol{r}'\int\limits_0^t dt'\ \frac{\partial B_{lj}(\boldsymbol{r}-\boldsymbol{r}',t-t')}{\partial r_k}\left\langle\frac{\delta}{\delta u_j(\boldsymbol{r}',t')}F[\boldsymbol{u}(\widetilde{\boldsymbol{r}},\tau)]\right\rangle. \tag{7.23}$$

7.4 Poisson Process

The Poisson process is defined by Eq. (5.26), page 134, and its characteristic functional logarithm is given by Eq. (5.30). In this case, formulas (7.5) and (7.7) for functionals $\Omega[t',t;v(\tau)]$ and $\Omega[t,t;v(\tau)]$ assume the forms

$$\Omega[t',t;v(\tau)]=\frac{\delta}{i\delta v(t')}\Theta[t,v(\tau)] = -i\int\limits_0^{t'} d\tau g(t'-\tau)\dot{W}\left(\int\limits_{\tau}^{t} d\tau_1 v(\tau_1)g(\tau_1-\tau)\right),$$

$$\Omega[t,t;v(\tau)] = \frac{d}{iv(t)dt}\Theta[t,v(\tau)] = -i\int\limits_0^{t} d\tau g(t-\tau)\dot{W}\left(\int\limits_{\tau}^{t} d\tau_1 v(\tau_1)g(\tau_1-\tau)\right), \tag{7.24}$$

where $\dot{W}(v) = \dfrac{dW(v)}{dv} = i\displaystyle\int_{-\infty}^{\infty} d\xi\xi p(\xi)e^{i\xi v}$.

Changing the integration order, we can rewrite equalities (7.24) in the form

$$\Omega[t',t;v(\tau)] = i\int\limits_{-\infty}^{\infty} d\xi\xi p(\xi)\int\limits_0^{t'} d\tau g(t'-\tau)\exp\left\{i\xi\int\limits_{\tau}^{t} d\tau_1 v(\tau_1)g(\tau_1-\tau)\right\} \quad (t'\leqslant t). \tag{7.25}$$

As a result, we obtain that correlations of the Poisson random process $z(t)$ with functionals of this process are described by the expression

$$\left\langle z(t')R[t;z(\tau)]\right\rangle = \nu\int\limits_{-\infty}^{\infty} d\xi\xi p(\xi)\int\limits_0^{t'} d\tau' g(t'-\tau')\left\langle R[t;z(\tau)+\xi g(\tau-\tau')]\right\rangle \quad (t'\leqslant t). \tag{7.26}$$

As we mentioned earlier, random process $n(0,t)$ describing the number of jumps during temporal interval $(0,t)$ is the special case of the Poisson process. In this case, $p(\xi) = \delta(\xi-1)$ and $g(t) = \theta(t)$, so that Eq. (7.26) assumes the extra-simple form

$$\langle n(0,t)R[t;n(0,\tau)]\rangle = \nu \int_0^{t'} d\tau \, \langle R[t;n(0,\tau)+\theta(t'-\tau)]\rangle \quad (t' \leqslant t). \tag{7.27}$$

Equality (7.27) extends formula (4.25) for the Poisson random quantities to the Poisson random processes.

7.5 Telegrapher's Random Process

Now, we dwell on telegrapher's random process defined by formula (5.35), page 137

$$z(t) = a(-1)^{n(0,t)}, \tag{7.28}$$

where a is the deterministic quantity. The n-th order moment functions of this process satisfy recurrence equation (5.36) from which immediately follows the relationship

$$\langle z(t_1)z(t_2)R[z(\tau)]\rangle = \langle z(t_1)z(t_2)\rangle \langle R[z(\tau)]\rangle, \tag{7.29}$$

which holds for arbitrary functional $R[z(\tau)]$ under the condition that $\tau \leq t_2 \leq t_1$ [9]. The proof of Eq. (7.29) consists in expanding functional $R[z(\tau)]$ in the Taylor series in $z(\tau)$ and using formula (5.36).

Let now quantity a be the random quantity with probability distribution density

$$P(a) = \frac{1}{2}\left[\delta(a-a_0) + \delta(a+a_0)\right]. \tag{7.30}$$

In this case $M_{2k+1} = 0$ and, in addition to Eq. (7.29), the equality [9]

$$\begin{aligned} &\langle F[z(\tau_1)]z(t_1)z(t_2)R[z(\tau_2)]\rangle \\ &= \langle F[z(\tau_1)\rangle] \langle z(t_1)z(t_2)\rangle \langle R[z(\tau_2)]\rangle + \langle F[z(\tau_1)]z(t_1)\rangle \langle z(t_2)R[z(\tau_2)]\rangle \end{aligned} \tag{7.31}$$

holds for any $\tau_1 \leq t_1 \leq t_2 \leq \tau_2$ and arbitrary functionals $F[z(\tau_1)]$ and $R[z(\tau_2)]$. Indeed, in terms of the Taylor expansion in $z(\tau)$, functional $R[z(\tau_2)]$ can be represented in the form $R[z(\tau_2)] = \sum_{2k} + \sum_{2k+1}$, where the first sum consists of terms with the even number of co-factors $z(\tau)$ and the second sum consists of terms with the odd number of such co-factors. Then, taking into account Eq. (5.36) and equalities

$$\langle R[z(\tau_2)]\rangle = \left\langle \sum_{2k} \right\rangle, \quad \langle z(t_2)R[z(\tau_2)]\rangle = \left\langle z(t_2) \sum_{2k+1} \right\rangle,$$

we obtain Eq. (7.31).

Formula (7.31) allows another representation. Denote functional $F[z(\tau_1)]z(t_1)$ as $F[t_1; z(\tau_1)]$, where $\tau_1 \leq t_1$, and functional $z(t_2)R[z(\tau_2)]$ as $R[t_2; z(\tau_2)]$, where $t_1 \leq t_2 \leq \tau_2$. Then, Eq. (7.31) can be written in the form

$$\langle F[t_1; z(\tau_1)]R[t_2; z(\tau_2)]\rangle = \langle F[t_1; z(\tau_1)]\rangle \langle R[t_2; z(\tau_2)]\rangle$$
$$+ \frac{1}{a_0^2} e^{-2\nu(t_2-t_1)} \langle z(t_1)F[t_1; z(\tau_1)]\rangle \langle z(t_2)R[t_2; z(\tau_2)]\rangle . \quad (7.32)$$

Because functionals $F[z(\tau_1)]$ and $R[z(\tau_2)]$ in Eq. (7.31) are arbitrary functionals, functionals $F[t_1; z(\tau_1)]$ and $R[t_2; z(\tau_2)]$ in (7.32) are also arbitrary functionals.

Formulas (7.29) and (7.31) include bilinear combinations of process $z(t)$, which is not always practicable.

As we mentioned earlier, calculation of correlator $\langle z(t)R[t; z(\tau)]\rangle$ for $\tau \leq t$ assumes the knowledge of the characteristic functional of process $z(t)$; unfortunately, the characteristic functional is unavailable in this case. We know only Eqs. (5.39) and (5.40), page 139 that describe the relationship between the functional

$$\Psi[t; v(\tau)] = \frac{1}{iv(t)} \frac{d}{dt} \Phi[t; v(\tau)] = \left\langle z(t) e^{i\int_0^t d\tau z(\tau) v(\tau)} \right\rangle$$

and the characteristic functional $\Phi[t; v(\tau)]$ in the form

$$\left\langle z(t) e^{i\int_0^t d\tau z(\tau) v(\tau)} \right\rangle = ae^{-2\nu t} + ia^2 \int_0^t dt_1 e^{-2\nu(t-t_1)} v(t_1) \left\langle e^{i\int_0^{t_1} d\tau z(\tau) v(\tau)} \right\rangle .$$

This relationship is sufficient to split correlator $\langle z(t)R[t; z(\tau)]\rangle$, i.e., to express it immediately in terms of the average functional $\langle R[t; z(\tau)]\rangle$.

Indeed, the equality

$$\langle z(t)R[t; z(\tau) + \eta(\tau)]\rangle = \left\langle z(t) e^{\int_0^t d\tau z(\tau) \frac{\delta}{\delta\eta(\tau)}} \right\rangle R[t; \eta(\tau)]$$

$$= aR[t; \eta(\tau)]e^{-2\nu t} + a^2 \int_0^t dt_1 e^{-2\nu(t-t_1)} \frac{\delta}{\delta\eta(t_1)} \langle R[t; z(\tau)\theta(t_1 - \tau) + \eta(\tau)]\rangle ,$$
$$(7.33)$$

where $\eta(t)$ is arbitrary deterministic function, holds for any functional $R[t; z(\tau)]$ for $\tau \leq t$. The final result is obtained by the limit process $\eta \to 0$:

$$\langle z(t)R[t;z(\tau)]\rangle = aR[t;0]e^{-2\nu t} + a^2\int\limits_0^t dt_1 e^{-2\nu(t-t_1)}\left\langle \frac{\delta}{\delta z(t_1)}\widetilde{R}[t,t_1;z(\tau)]\right\rangle, \tag{7.34}$$

where functional $\widetilde{R}[t,t_1;z(\tau)]$ is defined by the formula

$$\widetilde{R}[t,t_1;z(\tau)] = R[t;z(\tau)\theta(t_1-\tau+0)]. \tag{7.35}$$

In the case of random quantity a distributed according to probability density distribution (7.30), additional averaging (7.34) over a results in the equality

$$\langle z(t)R[t;z(\tau)]\rangle = a_0^2\int\limits_0^t dt_1 e^{-2\nu(t-t_1)}\left\langle \frac{\delta}{\delta z(t_1)}\widetilde{R}[t,t_1;z(\tau)]\right\rangle. \tag{7.36}$$

Formula (7.36) is very similar to the formula for splitting the correlator of the Gaussian process $z(t)$ characterized by the exponential correlation function (i.e., the Gaussian Markovian process) with functional $R[t;z(\tau)]$. The difference consists in the fact that the right-hand side of Eq. (7.36) depends on the functional that is cut by the process $z(\tau)$ rather than on functional $R[t;z(\tau)]$ itself.

Note that correlation $\langle z(t')R[t;z(\tau)]\rangle$,where $t' \geq t$ and $\tau < t$, can be represented in the form

$$\langle z(t)R[t;z(\tau)]\rangle = \frac{1}{a_0^2}\langle z(t')z(t)z(t)R[t;z(\tau)]\rangle.$$

As a consequence, the equality

$$\langle z(t')R[t;z(\tau)]\rangle = e^{-2\nu(t'-t)}\langle z(t)R[t;z(\tau)]\rangle \quad (t' \geqslant t) \tag{7.37}$$

holds according to formula (7.29).

In a similar way, we can obtain the expression

$$\langle z(t')R[t_0,t;z(\tau)]\rangle = e^{-2\nu(t_0-t')}\langle z(t_0)R[t_0,t;z(\tau)]\rangle \quad (t' \geqslant t), \tag{7.38}$$

where $t' \leq t_0 \leq \tau \leq t$ and

$$\langle z(t_0)R[t_0,t;z(\tau)]\rangle = a_0^2\int\limits_{t_0}^t dt_1 e^{-2\nu(t_1-t_0)}\left\langle \frac{\delta}{\delta z(t_1)}R[t_0,t,;z(\tau)\theta(\tau-t_1+0]\right\rangle. \tag{7.39}$$

In the case of the general-form correlator $\langle z(\xi)R[t_0,t;z(\tau)]\rangle$, where $t_0 \leqslant \xi \leqslant t$ and $t_0 \leqslant \tau \leqslant t$, one can show [48, 49] the validity of the following equality:

$$\langle z(\xi)R[t_0,t;z(\tau)]\rangle = a_0^2\int\limits_{t_0}^{\xi} dt_1 e^{-2\nu(\xi-t_1)}\left\langle\frac{\delta R[t_0,t,;z_1(\tau)\theta(\tau-\xi)+z_2(\tau)\theta(t_1-\tau+0]}{\delta z_2(t_1)}\right\rangle$$

$$+a_0^2\int\limits_{\xi}^{t} dt_1 e^{-2\nu(t_1-\xi)}\left\langle\frac{\delta R[t_0,t,;z_1(\tau)\theta(\xi-\tau)+z_2(\tau)\theta(\tau-t_1+0]}{\delta z_2(t_1)}\right\rangle, \quad (7.40)$$

where $z_1(t)$ and $z_2(t)$ are statistically independent telegrapher's processes characterized by the same correlation function

$$\langle z(t)z(t')\rangle = a_0^2 e^{-2\nu|t-t'|}.$$

The limits of integration t_0 and t in Eq. (7.40) can assume arbitrary values from $-\infty$ to ∞. At $\xi = t$ or $\xi = t_0$, Eq. (7.40) grades into Eq. (7.36) or (7.39), respectively.

If we set $v(\tau) = v$ and $R[t_0,t;z(\tau)] = \exp\left\{iv\int\limits_{t_0}^{t} d\tau z(\tau)\right\}$ in Eq. (7.40) and take into account Eq. (5.42), page 139, we obtain the expression

$$\left\langle z(\xi)\exp\left\{iv\int\limits_{t_0}^{t} d\tau z(\tau)\right\}\right\rangle$$

$$= \frac{iva_0^2}{\lambda}e^{-\nu(t-t_0)}\left\{\sinh\lambda(t-t_0)+\frac{2\nu}{\lambda}\sinh\lambda(t-\xi)\sinh\lambda(\xi-t_0)\right\}, \quad (7.41)$$

where $\lambda = \sqrt{\nu^2 - a_0^2 v^2}$.

Let us differentiate Eq. (7.34) with respect to time t. Taking into account the fact that there is no need in differentiating with respect to the upper limit of the integral in the right-hand side of Eq. (7.34), we obtain the expression [124, 125]

$$\left(\frac{d}{dt}+2\nu\right)\langle z(t)R[t;z(\tau)]\rangle = \left\langle z(t)\frac{d}{dt}R[t;z(\tau)]\right\rangle \quad (7.42)$$

called usually the *differentiation formula*.

One essential point should be noticed. Functional $R[t;z(\tau)]$ in differentiation formula (7.42) is an arbitrary functional and can simply coincide with process $z(t-0)$. In the general case, the realization of telegrapher's process is a generalized function. The derivative of this process is also a generalized function (the sequence of delta-functions), so that

$$z(t)\frac{d}{dt}z(t) \neq \frac{1}{2}\frac{d}{dt}z^2(t) \equiv 0$$

in the general case. These generalized functions, as any generalized functions, are defined only in terms of functionals constructed on them. In the case of our interest, such functionals are the average quantities denoted by angle brackets $\langle\cdots\rangle$, and the above differentiation formula describes the differential constraint between different functionals related to random process $z(t)$ and its one-sided derivatives for $t \to t-0$, such as dz/dt, d^2z/dt^2. For example, formula (7.42) allows derivation of equalities, such as

$$\left\langle z(t)\frac{d}{d\tau}z(\tau)\right\rangle\Big|_{\tau=t-0} = 2\nu\left\langle z^2\right\rangle, \quad \left\langle z(t)\frac{d^2}{d\tau^2}z(\tau)\right\rangle\Big|_{\tau=t-0} = 4\nu^2\left\langle z^2\right\rangle.$$

It is clear that these formulas can be obtained immediately by differentiating the correlation function $\langle z(t)z(t')\rangle$ with respect to t' $(t' < t)$ for $t' \to t-0$.

Above, we considered the correlation of random process $z(t)$ with a functional of this process. If we deal with an arbitrary function of telegrapher's process $F(z(t))$, then, clearly, the equality

$$F\left(z(t)\right) = \frac{F(a)+F(-a)}{2} + \frac{F(a)-F(-a)}{2a}z(t) \tag{7.43}$$

will hold, and all results valid for telegrapher's process $z(t)$ will be valid (with small variations) for process $F(z(t))$.

7.6 Generalized Telegrapher's Random Process

Consider now generalized telegrapher's process described by Eq. (5.43), page 141. In this case, functional

$$\Psi[t; v(\tau)] = \frac{1}{iv(t)}\frac{d}{dt}\Phi[t; v(\tau)] = \left\langle z(t)\exp\left\{i\int\limits_0^t d\tau z(\tau)v(\tau)\right\}\right\rangle$$

is related to the characteristic functional $\Phi[t; v(\tau)]$ of process $z(t)$ by Eq. (5.100)

$$\left\langle z(t)\exp\left\{i\int\limits_0^t d\tau z(\tau)v(\tau)\right\}\right\rangle = \left\langle a\exp\left\{ia\int\limits_0^t d\tau v(\tau)\right\}\right\rangle_a e^{-\nu t}$$
$$+\nu\int\limits_0^t dt_1 e^{-\nu(t-t_1)}\left\langle a\exp\left\{ia\int\limits_{t_1}^t d\tau v(\tau)\right\}\right\rangle_a \Phi[t_1, v(\tau)],$$

As in the case of telegrapher's process, this formula allows one to express correlator $\langle z(t)R[t; z(\tau)]\rangle$, where $R[t; z(\tau)]$ is arbitrary functional, in terms of the mean value of the functional. Indeed, if we replicate operations used for telegrapher's process, we obtain the equality

$$\langle z(t)R[t;z(\tau)+\eta(\tau)]\rangle=\Psi\left[t;\frac{\delta}{i\delta\eta(\tau)}\right]R[t;\eta(\tau)]=\langle\widetilde{a}R[t;\eta(\tau)+\widetilde{a}]\rangle_{\widetilde{a}}\,e^{-\nu t}$$

$$+\nu\int_0^t dt_1 e^{-\nu(t-t_1)}\left\langle\widetilde{a}R\left[t;\eta(\tau)+\widetilde{a}\theta(\tau-t_1)+z\left(\tau\right)\theta(t_1-\tau)\right]\right\rangle_{\widetilde{a},z}, \quad (7.44)$$

where $\eta(\tau)$ is arbitrary function and random quantity $\widetilde{a}$ is distributed with probability density $p(\widetilde{a})$ and is statistically independent of process $z(t)$. Setting now $\eta(\tau)=0$, we obtain the final expression

$$\langle z(t)R[t;z(\tau)]\rangle=\langle\widetilde{a}R[t;\widetilde{a}]\rangle_{\widetilde{a}}\,e^{-\nu t}$$

$$+\nu\int_0^t dt_1 e^{-\nu(t-t_1)}\left\langle\widetilde{a}R\left[t;\widetilde{a}\theta(\tau-t_1)+z\left(\tau\right)\theta(t_1-\tau)\right]\right\rangle_{\widetilde{a},z}. \quad (7.45)$$

Note additionally that, in the case of generalized telegrapher's process, correlation $\langle F(z(t))R[t;z(\tau)]\rangle$, where $F(z)$ is arbitrary function, is described, in view of Eq. (5.99), by the formula similar to Eq. (7.45)

$$\langle F\left(z(t)\right)R[t;z(\tau)]\rangle=\langle F\left(\widetilde{a}\right)R[t;\widetilde{a}]\rangle_{\widetilde{a}}\,e^{-\nu t}$$

$$+\nu\int_0^t dt_1 e^{-\nu(t-t_1)}\left\langle F\left(\widetilde{a}\right)R\left[t;\widetilde{a}\theta(\tau-t_1)+z\left(\tau\right)\theta(t_1-\tau)\right]\right\rangle_{\widetilde{a},z}. \quad (7.46)$$

7.7 General-Form Markovian Processes

Processes such as mentioned telegrapher's processes are the simplest examples of the Markovian processes. Here, we consider what consequences can be derived for the correlation of functionals from the sole assumption that process $z(t)$ is the Markovian process.

In the case of the general-form Markovian process $z(t)$, we have no equation for the characteristic functional. We have only integral equation (5.89), page 152 for the functional

$$\Psi[t,z;v(\tau)]=\left\langle\delta\left(z(t)-z\right)\exp\left\{i\int_0^t d\tau z(\tau)v(\tau)\right\}\right\rangle$$

that describes statistical relationship of process $z(t)$ at instant t with its prehistory

$$\Psi[t,z;v(\tau)] = P(t,z) + i\int\limits_{0}^{t} dt_1 v(t_1) \int\limits_{-\infty}^{\infty} dz_1 z_1 p(z,t|z_1,t_1)\Psi[t_1,z_1;v(\tau)], \tag{7.47}$$

where $P(t,z)$ is the one-time probability density and $p(z,t|z_0,t_0)$ is the transition probability density of process $z(t)$. In this case, we can again use the method described above to obtain the expression for the correlator

$$\langle\delta\left(z(t)-z\right)R[t;z(\tau)+\eta(\tau)]\rangle \quad (\tau\leqslant t)$$

in the form of the integral equality with variational derivatives

$$\langle\delta\left(z(t)-z\right)R[t;z(\tau)+\eta(\tau)]\rangle = \Psi\left[t,z;\frac{\delta}{i\delta\eta(\tau)}\right]R[t;\eta(\tau)], \tag{7.48}$$

where $\eta(t)$ is arbitrary function.

For the Markovian processes $z(t)$, functions $P(t,z)$ and $p(z,t|z_0,t_0)$ satisfy linear operator equations (5.66), page 147

$$\frac{\partial}{\partial t}P(t,z) = \widehat{L}(z)P(t,z), \quad \frac{\partial}{\partial t}p(z,t|z_0,t_0) = \widehat{L}(z)p(z,t|z_0,t_0), \tag{7.49}$$

where $\widehat{L}(z)$ is the integro-differential operator with respect to variable z.

Let us differentiate Eq. (7.48) with respect to time t and take into account that variational derivative $\dfrac{\delta}{\delta\eta(t_1)}R[t;\eta(\tau)] \sim \theta(t-t_1)$ by the definition of functional $R[t;z(\tau)]$, so that we have no need in differentiating the integral in Eq. (7.47) with respect to the upper limit (we can set it to ∞). An additional point consists in the fact that the differentiation operation commutes with the variational differentiation operation (see Eq. (2.13), page 86):

$$\frac{\partial}{\partial t}\frac{\delta}{\delta\eta(t_1)}R[t;\eta(\tau)] = \frac{\delta}{\delta\eta(t_1)}\frac{\partial}{\partial t}R[t;\eta(\tau)].$$

Taking into account Eqs. (7.49), we obtain the formula for the derivative of the correlation of interest with respect to time (function $\eta(t)$ can be set to zero) [47–49]

$$\begin{aligned}\frac{\partial}{\partial t}\langle\delta\left(z(t)-z\right)R[t;z(\tau)]\rangle &= \left\langle\delta\left(z(t)-z\right)\frac{\partial}{\partial t}R[t;z(\tau)]\right\rangle\\ &\quad+\widehat{L}(z)\left\langle\delta\left(z(t)-z\right)R[t;z(\tau)]\right\rangle.\end{aligned} \tag{7.50}$$

Multiply now Eq. (7.50) by arbitrary function $f(z)$ and integrate the result over z. The result will be the differentiation formula

$$\frac{\partial}{\partial t}\langle f(z(t))\,R[t;z(\tau)]\rangle = \left\langle f(z(t))\,\frac{\partial}{\partial t}R[t;z(\tau)]\right\rangle + \int\limits_{-\infty}^{\infty} dz f(z)\widehat{L}(z)\,\langle\delta(z(t)-z)\,R[t;z(\tau)]\rangle\,, \tag{7.51}$$

that can be rewritten in the form

$$\frac{\partial}{\partial t}\langle f(z(t))\,R[t;z(\tau)]\rangle - \left\langle f(z(t))\,\frac{\partial}{\partial t}R[t;z(\tau)]\right\rangle = \left\langle R[t;z(\tau)]\left[\widehat{L}^{+}(z)f(z)\right]\right\rangle, \tag{7.52}$$

where we introduced operator $\widehat{L}^{+}(z)$ conjugated to operator $\widehat{L}(z)$.

Thus, Eqs. (7.50)–(7.52) govern the rules of differentiating with respect to time the correlators of functions of the Markovian process $z(t)$ with functionals of this process.

Note that, if the mean value of process $z(t)$ is equal to zero, the right-hand side of Eq. (7.52) can be expressed in terms of the desired correlation $\langle z(t)R[t;z(\tau)]\rangle$ for all Markovian processes considered earlier. This is most probably the reason of low practicability of this formula. However, for telegrapher's and generalized telegrapher's processes, Eq. (7.52) appears practicable for analyzing linear stochastic equations. Indeed, for telegrapher's process $z(t)$, the right-hand side of Eq. (7.52) for correlation $\langle z(t)R[t;z(\tau)]\rangle$ has the form

$$-2\nu\,\langle z(t)R[t;z(\tau)]\rangle\,. \tag{7.53}$$

For generalized telegrapher's process $z(t)$, the right-hand side of Eq. (7.52) under the condition that $\langle f(a)\rangle = 0$ assumes the form

$$-\nu\,\langle f(z(t))\,R[t;z(\tau)]\rangle\,. \tag{7.54}$$

In the special case $f(z) = z$, Eq. (7.54) reduces to

$$-\nu\,\langle z(t)R[t;z(\tau)]\rangle\,. \tag{7.55}$$

Now, we dwell on some extensions of the above formulas. First of all, we note that if we deal with the vector Markovian process $\boldsymbol{z}(t) = \{z_1(t),\cdots,z_N(t)\}$ described by operator $\widehat{L}(\boldsymbol{z})$, then functional

$$\Psi[t,\boldsymbol{z};\boldsymbol{v}(\tau)] = \left\langle \delta(\boldsymbol{z}(t)-\boldsymbol{z})\exp\left\{i\int\limits_0^t d\tau\boldsymbol{z}(\tau)\boldsymbol{v}(\tau)\right\}\right\rangle,$$

where $\boldsymbol{v}(t) = \{v_1(t),\cdots,v_N(t)\}$, satisfies the equation

$$\frac{\partial}{\partial t}\Psi[t,\boldsymbol{z};\boldsymbol{v}(\tau)] = \left\{\widehat{L}(\boldsymbol{z})+i\boldsymbol{z}\boldsymbol{v}\right\}\Psi[t,\boldsymbol{z};\boldsymbol{v}(\tau)] \tag{7.56}$$

with the initial value $\Psi[0, \boldsymbol{z}; \boldsymbol{v}(\tau)] = P_0(\boldsymbol{z})$. With this remark, we can easily derive the formula of differentiating the correlator $\langle F(\boldsymbol{z}(t))R[t; \boldsymbol{z}(\tau)]\rangle$ with respect to time; it assumes the form

$$\frac{\partial}{\partial t}\left\langle F\left(\boldsymbol{z}(t)\right)R[t;\boldsymbol{z}(\tau)]\right\rangle=\left\langle F\left(\boldsymbol{z}(t)\right)\frac{\partial}{\partial t}R[t;\boldsymbol{z}(\tau)]\right\rangle+\left\langle R[t;\boldsymbol{z}(\tau)]\left[\widehat{L}^{+}(\boldsymbol{z})F(\boldsymbol{z})\right]\right\rangle, \tag{7.57}$$

where $\widehat{L}^{+}(\boldsymbol{z})$ is the operator conjugated to operator $\widehat{L}(\boldsymbol{z})$.

An important special case corresponds to the situation in which all components of vector $\boldsymbol{z}(t)$ are statistically independent Markovian processes described by the same operator $\widehat{L}(\boldsymbol{z})$; in this case, Eq. (7.57) reduces to the form

$$\frac{\partial}{\partial t}\left\langle F\left(\boldsymbol{z}(t)\right)R[t;\boldsymbol{z}(\tau)]\right\rangle=\left\langle F\left(\boldsymbol{z}(t)\right)\frac{\partial}{\partial t}R[t;\boldsymbol{z}(\tau)]\right\rangle+\sum_{k=1}^{N}\left\langle R[t;\boldsymbol{z}(\tau)]\left[\widehat{L}^{+}(z_k)F(\boldsymbol{z})\right]\right\rangle. \tag{7.58}$$

For example, all above Markovian processes having the exponential correlation function $\langle z(t)z(t+\tau)\rangle = \langle z^2\rangle\, e^{-\alpha|\tau|}$ satisfy the equality

$$\left(\frac{\partial}{\partial t}+\alpha k\right)\left\langle z_1(t)\cdots z_k(t)R[t;\boldsymbol{z}(\tau)]\right\rangle=\left\langle z_1(t)\cdots z_k(t)\frac{\partial}{\partial t}R[t;\boldsymbol{z}(\tau)]\right\rangle. \tag{7.59}$$

Formula (7.59) defines the rule of factoring the operation of differentiating with respect to time out of angle brackets of averaging; in particular, we have

$$\left\langle z_1(t)\cdots z_k(t)\frac{\partial^n}{\partial t^n}R[t;\boldsymbol{z}(\tau)]\right\rangle=\left(\frac{\partial}{\partial t}+\alpha k\right)^n\left\langle z_1(t)\cdots z_k(t)R[t;\boldsymbol{z}(\tau)]\right\rangle, \tag{7.60}$$

where $k = 1, \cdots, N$.

7.8 Delta-Correlated Random Processes

7.8.1 General Remarks

Random processes $z(t)$ that can be treated as delta-correlated processes are of special interest in physics. The importance of this approximation follows first of all from the fact that it is physically obvious in the context of many physical problems; moreover, the corresponding dynamic systems always allow obtaining closed equations for probability densities of the solutions to these systems.

For the Gaussian delta-correlated (in time) process, correlation function has the form

$$B(t_1, t_2) = \langle z(t_1)z(t_2)\rangle = B(t_1)\delta(t_1 - t_2), \quad (\langle z(t)\rangle = 0).$$

In this case, functionals $\Theta[t;v(\tau)]$, $\Omega[t',t;v(\tau)]$, and $\Omega[t,t;v(\tau)]$ introduced earlier in Eqs. (7.14), (7.15), page 169 assume the forms

$$\Theta[t;v(\tau)] = -\frac{1}{2}\int\limits_0^t d\tau B(\tau)v^2(\tau), \quad \Omega[t',t;v(\tau)] = iB(t')v(t'),$$

$$\Omega[t,t;v(\tau)] = \frac{i}{2}B(t)v(t),$$

and Eqs. (7.16) appear significantly simpler

$$\begin{aligned} \langle z(t')R[t;v(\tau)]\rangle &= B(t')\left\langle \frac{\delta}{\delta z(t')}R[t;v(\tau)]\right\rangle \quad (0 < t' < t), \\ \langle z(t)R[t;v(\tau)]\rangle &= \frac{1}{2}B(t)\left\langle \frac{\delta}{\delta z(t)}R[t;v(\tau)]\right\rangle. \end{aligned} \tag{7.61}$$

Formulas (7.61) show that statistical averages of the Gaussian delta-correlated process considered here are discontinuous at $t' = t$. This discontinuity is completely dictated by the fact that this process is delta-correlated; if process is not delta-correlated, no discontinuity occurs (see Eq. (7.16), page 169).

The Poisson delta-correlated random process corresponds to the limit process

$$g(t) \to \delta(t).$$

In this case, the logarithm of the characteristic functional has the simple form (see (5.33), page 136); as a consequence, Eqs. (7.25), page 172 for functionals $\Omega[t',t;v(\tau)]$ and $\Omega[t,t;v(\tau)]$ assume the forms

$$\Omega[t',t;v(\tau)] = i\int\limits_{-\infty}^{\infty} d\xi\xi p(\xi)e^{i\xi v(t')} \quad (t' < t),$$

$$\Omega[t,t;v(\tau)] = \frac{\nu}{iv(t)}\int\limits_{-\infty}^{\infty} d\xi\xi p(\xi)\left[e^{i\xi v(t)} - 1\right] = \nu\int\limits_{-\infty}^{\infty} d\xi p(\xi)\int\limits_0^{\xi} d\eta e^{i\eta v(t)},$$

and we obtain the following expression for the correlation of the Poisson random process $z(t)$ with a functional of this process

$$\begin{aligned} \langle z(t')R[t;z(\tau)]\rangle &= \nu\int\limits_{-\infty}^{\infty} d\xi\xi p(\xi)\left\langle R[t;z(\tau)+\xi\delta(\tau-\tau')]\right\rangle \quad (t' \leqslant t), \\ \langle z(t)R[t;z(\tau)]\rangle &= \nu\int\limits_{-\infty}^{\infty} d\xi p(\xi)\int\limits_0^{\xi} d\eta\left\langle R[t;z(\tau)+\eta\delta(t-\tau)]\right\rangle. \end{aligned} \tag{7.62}$$

These expressions also show that statistical averages are discontinuous at $t' = t$. As in the case of the Gaussian process, this discontinuity is completely dictated by the fact that this process is delta-correlated.

In the general case of delta-correlated process $z(t)$, we can expand the logarithm of characteristic functional in the functional Taylor series

$$\Theta[t; v(\tau)] = \sum_{n=1}^{\infty} \frac{i^n}{n!} \int_0^t d\tau K_n(\tau) v^n(\tau), \tag{7.63}$$

where cumulant functions assume the form

$$K_n(t_1, \cdots, t_n) = K_n(t_1)\delta(t_1 - t_2) \cdots \delta(t_{n-1} - t_n).$$

As can be seen from Eq. (7.63), a characteristic feature of these processes consists in the validity of the equality

$$\dot{\Theta}[t; v(\tau)] = \dot{\Theta}[t; v(t)] \quad \left(\dot{\Theta}[t; v(\tau)] = \frac{d}{dt}\Theta[t; v(\tau)] \right), \tag{7.64}$$

which is of fundamental significance. This equality shows that, in the case of arbitrary delta-correlated process, quantity $\Theta[t; v(\tau)]$ appears not a functional, but simply a function of time t. In this case, functionals $\Omega[t', t; v(\tau)]$ and $\Omega[t, t; v(\tau)]$ are

$$\Omega[t', t; v(\tau)] = \sum_{n=0}^{\infty} \frac{i^n}{n!} K_{n+1}(t') v^n(t') \quad (t' < t),$$

$$\Omega[t, t; v(\tau)] = \sum_{n=0}^{\infty} \frac{i^n}{(n+1)!} K_{n+1}(t) v^n(t),$$

and formulas for correlation splitting assume the forms

$$\begin{aligned} \langle z(t') R[t; z(\tau)] \rangle &= \sum_{n=0}^{\infty} \frac{i^n}{n!} K_{n+1}(t') \left\langle \frac{\delta^n R[t; z(\tau)]}{\delta z^n(t')} \right\rangle \quad (t' < t), \\ \langle z(t) R[t; z(\tau)] \rangle &= \sum_{n=0}^{\infty} \frac{i^n}{(n+1)!} K_{n+1}(t) \left\langle \frac{\delta^n R[t; z(\tau)]}{\delta z^n(t)} \right\rangle. \end{aligned} \tag{7.65}$$

These formulas describe the discontinuity of statistical averages at $t' = t$ in the general case of delta-correlated processes.

Note that, for $t' > t$, delta-correlated processes satisfy the obvious equality

$$\langle z(t') R[t; z(\tau)] \rangle = \langle z(t') \rangle \, \langle R[t; z(\tau)] \rangle . \tag{7.66}$$

Now, we dwell on the concept of random delta-correlated (in time) fields.

We will deal with vector field $\boldsymbol{f}(\boldsymbol{x},t)$, where $\boldsymbol{x}$ describes the spatial coordinates and t is the temporal coordinate. In this case, the logarithm of characteristic functional can be expanded in the Taylor series with coefficients expressed in terms of cumulant functions of random field $\boldsymbol{f}(\boldsymbol{x},t)$ (see Sect. 3.2). In the special case of cumulant functions

$$K_n^{i_1,\cdots,i_m}(\boldsymbol{x}_1,t_1;\cdots,\boldsymbol{x}_n,t_n)=K_n^{i_1,\cdots,i_m}(\boldsymbol{x}_1,\cdots,\boldsymbol{x}_n;t_1)\delta(t_1-t_2)\cdots\delta(t_{n-1}-t_n), \tag{7.67}$$

we will call field $\boldsymbol{f}(\boldsymbol{x},t)$ the random field delta-correlated in time t. In this case, functional $\Theta[t;\boldsymbol{\psi}(\boldsymbol{x}',\tau)]$ assumes the form

$$\Theta[t;\boldsymbol{\psi}(\boldsymbol{x}',\tau)]=\sum_{n=1}^{\infty}\frac{i^n}{n!}\int\limits_0^t d\tau\int\cdots\int d\boldsymbol{x}_1\cdots d\boldsymbol{x}_n K_n^{i_1,\cdots,i_m}(\boldsymbol{x}_1,\cdots,\boldsymbol{x}_n;\tau)$$

$$\times\,\psi_{i_1}(\boldsymbol{x}_1,\tau)\cdots\psi_{im}(\boldsymbol{x}_n,\tau), \tag{7.68}$$

An important feature of this functional is the fact that it satisfies the equality similar to Eq. (7.64):

$$\dot{\Theta}[t;\boldsymbol{\psi}(\boldsymbol{x}',\tau)]=\dot{\Theta}[t;\boldsymbol{\psi}(\boldsymbol{x}',t)]. \tag{7.69}$$

7.8.2 Asymptotic Meaning of Delta-Correlated Processes and Fields

The nature knows no delta-correlated processes. All actual processes and fields are characterized by a finite temporal correlation radius, and delta-correlated processes and fields result from asymptotic expansions in terms of their temporal correlation radii.

We illustrate the appearance of delta-correlated processes using the stationary Gaussian process with correlation radius τ_0 as an example. In this case, the logarithm of the characteristic functional is described by the expression

$$\Theta[t;v(\tau)]=-\int\limits_0^t d\tau_1 v(\tau_1)\int\limits_0^{\tau_1} d\tau_2 B\left(\frac{\tau_1-\tau_2}{\tau_0}\right)v(\tau_2). \tag{7.70}$$

Setting $\tau_1-\tau_2=\xi\tau_0$, we transform Eq. (7.70) to the form

$$\Theta[t;v(\tau)]=-\tau_0\int\limits_0^t d\tau_1 v(\tau_1)\int\limits_0^{\tau_1/\tau_0} d\xi B\left(\xi\right)v(\tau_1-\xi\tau_0).$$

Assume now that $\tau_0\to 0$. In this case, the leading term of the asymptotic expansion in parameter τ_0 is given by the formula

$$\Theta[t; v(\tau)] = -\tau_0 \int_0^\infty d\xi B(\xi) \int_0^t d\tau_1 v^2(\tau_1)$$

that can be represented in the form

$$\Theta[t; v(\tau)] = -B^{\text{eff}} \int_0^t d\tau_1 v^2(\tau_1), \tag{7.71}$$

where

$$B^{\text{eff}} = \int_0^\infty d\tau B\left(\frac{\tau}{\tau_0}\right) = \frac{1}{2} \int_{-\infty}^\infty d\tau B\left(\frac{\tau}{\tau_0}\right). \tag{7.72}$$

Certainly, asymptotic expression (7.71) holds only for the functions $v(t)$ that vary during times about τ_0 only slightly, rather than for arbitrary functions $v(t)$. Indeed, if we specify this function as $v(t) = v\delta(t - t_0)$, asymptotic expression (7.71) fails; in this case, we must replace Eq. (7.70) with the expression

$$\Theta[t; v(\tau)] = -\frac{1}{2} B(0) v^2 \quad (t > t_0)$$

corresponding to the characteristic function of process $z(t)$ at a fixed time $t = t_0$.

Consider now correlation $\langle z(t) R[t; z(\tau)] \rangle$ given, according to the Furutsu–Novikov formula (7.16), page 169, by the expression

$$\langle z(t) R[t; z(\tau)] \rangle = \int_0^t dt_1 B\left(\frac{t - t_1}{\tau_0}\right) \left\langle \frac{\delta}{\delta z(t_1)} R[t; z(\tau)] \right\rangle.$$

The change of integration variable $t - t_1 \to \xi\tau_0$ transforms this expression to the form

$$\langle z(t) R[t; z(\tau)] \rangle = \tau_0 \int_0^{t/\tau_0} d\xi B(\xi) \left\langle \frac{\delta}{\delta z(t - \xi\tau_0)} R[t; z(\tau)] \right\rangle, \tag{7.73}$$

which grades for $\tau_0 \to 0$ into the equality obtained earlier for the Gaussian delta-correlated process

$$\langle z(t) R[t; z(\tau)] \rangle = B^{\text{eff}} \left\langle \frac{\delta}{\delta z(t)} R[t; z(\tau)] \right\rangle$$

provided that the variational derivative in Eq. (7.73) varies only slightly during times of about τ_0.

Thus, the approximation of process $z(t)$ by the delta-correlated one is conditioned by small variations of functionals of this process during times of about process' temporal correlation radius.

Consider now telegrapher's and generalized telegrapher's processes. In the case of telegrapher's process, the characteristic functional satisfies Eq. (5.40), page 139. The correlation radius of this process is $\tau_0 = 1/2\nu$, and, for $\nu \to \infty$ ($\tau_0 \to 0$), this equation grades for sufficiently smooth functions $v(t)$ into the equation

$$\frac{d}{dt}\Phi[t; v(\tau)] = -\frac{a_0^2}{2\nu} v^2(t)\Phi[t; v(\tau)], \tag{7.74}$$

which corresponds to the Gaussian delta-correlated process. If we additionally assume that $a_0^2 \to \infty$ and

$$\lim_{\nu\to\infty} a_0^2/2\nu = \sigma_0^2,$$

then Eq. (7.74) appears independent of parameter ν. Of cause, this fact does not mean that telegrapher's process looses its telegrapher's properties for $\nu \to \infty$. Indeed, for $\nu \to \infty$, the one-time probability distribution of process $z(t)$ will as before correspond to telegrapher's process, i.e., to the process with two possible states. As regards the correlation function and higher-order moment functions, they will possess for $\nu \to \infty$ all properties of delta-functions in view of the fact that

$$\lim_{\nu\to\infty} \left\{ 2\nu e^{-2\nu|\tau|} \right\} = \begin{cases} 0, & \text{if} \quad \tau \neq 0, \\ \infty, & \text{if} \quad \tau = 0. \end{cases}$$

Such functions should be considered the generalized functions; their delta-functional behavior will manifest itself in the integrals of them (see, e.g., [22]). As Eq. (7.74) shows, the limit process $\nu \to \infty$ is equivalent for these quantities to the replacement of process $z(t)$ by the Gaussian delta-correlated process. This situation is completely similar to the approximation of the Gaussian random process with a finite correlation radius τ_0 by the delta-correlated process for $\tau_0 \to 0$.

In a similar way, we obtain that generalized telegrapher's process whose characteristic functional satisfies integro-differential equation (5.100), page 154 is governed for $\nu \to \infty$ and sufficiently smooth functions $v(t)$ by the equation (here, we assume $\langle a \rangle = 0$ for simplicity)

$$\frac{d}{dt}\Phi[t; v(\tau)] = -\frac{\langle a^2 \rangle}{\nu} v^2(t)\Phi[t; v(\tau)],$$

which again corresponds to the Gaussian delta-correlated process.

Consider the square of the Gaussian stationary process, i.e., process $z(t) = \xi^2(t)$, where $\xi(t)$ is the Gaussian process with parameters

$$\langle \xi(t) \rangle = 0, \quad \langle \xi(t_1)\xi(t_2) \rangle = B(t_1 - t_2),$$

as a more complicated example.

Let as calculate the characteristic functional of this process

$$\Phi[t;v(\tau)] = \langle\varphi[t;\xi(\tau)]\rangle\,, \quad \varphi[t;\xi(\tau)] = \exp\left\{i\int\limits_0^t d\tau v(\tau)\xi^2(\tau)\right\}. \tag{7.75}$$

The characteristic functional of process $z(t)$ satisfies the stochastic equation

$$\frac{d}{dt}\Phi[t;v(\tau)] = iv(t)\left\langle\xi^2(t)\varphi[t;\xi(\tau)]\right\rangle. \tag{7.76}$$

Consider quantity $\Psi(t_1,t) = \langle\xi(t_1)\xi(t)\varphi[t;\xi(\tau)]\rangle$. According to the Furutsu–Novikov formula (7.11), page 168,

$$\Psi(t_1,t) = \int\limits_0^t dt' B(t_1-t')\left\langle\frac{\delta}{\delta\xi(t')}\xi(t)\varphi[t;\xi(\tau)]\right\rangle. \tag{7.77}$$

Calculating now the variational derivative in the right-hand side of Eq. (7.77) with the use of the explicit expression for functional $\varphi[t;\xi(\tau)]$, we obtain the integral equation for function $\Psi(t_1,t)$

$$\Psi(t_1,t) = B(t_1-t)\Phi[t;v(\tau)] + 2i\int\limits_0^t d\tau B(t_1-\tau)v(\tau)\Psi(\tau,t). \tag{7.78}$$

Function $\Psi(t_1,t)$ is representable in the form $\Psi(t_1,t) = S(t_1,t)\Phi[t;v(\tau)]$, where function $S(t_1,t)$ satisfies the linear integral equation

$$S(t_1,t) = B(t_1-t) + 2i\int\limits_0^t d\tau B(t_1-\tau)v(\tau)S(\tau,t). \tag{7.79}$$

As a consequence, characteristic functional $\Phi[t;v(\tau)]$ can be represented in the form

$$\Phi[t;v(\tau)] = \exp\left\{i\int\limits_0^t d\tau v(\tau)S(\tau,\tau)\right\}. \tag{7.80}$$

Thus, the expansion of quantity $S(t,t)$ in the functional Taylor series in $v(\tau)$ determines the cumulants of process $z(t) = \xi^2(t)$. Because Eq. (7.79) is the linear integral equation, we can represent its solution as the iterative series $S(t,t) = \sum\limits_{n=0}^{\infty} S^{(n)}(t,t)$, where

$$S^{(n)}(t,t) = (2i)^n \int\limits_0^t d\tau_1 \cdots \int\limits_0^t d\tau_n \, v(\tau_1)\cdots v(\tau_n) B(t-\tau_1) B(\tau_1-\tau_2)\cdots B(\tau_n-t). \tag{7.81}$$

If function $v(t)$ varies slowly during correlation time τ_0 of process $\xi(t)$ (which means that we omit from consideration the one-time characteristic functions of process $z(t) = \xi^2(t)$), we can proceed to the limit $\tau_0 \to 0$. As a result, we obtain the expressions

$$\begin{aligned} S^{(0)}(t,t) &= B(0), \\ S^{(n)}(t,t) &= (2i)^n v^n(t) \int\limits_0^\infty \cdots \int\limits_0^\infty d\tau_1 \cdots d\tau_n B(\tau_1) B(\tau_1 - \tau_2) \cdots B(\tau_n), \end{aligned} \tag{7.82}$$

from which follows that process $z(t) = \xi^2(t)$ in this limit can be considered the delta-correlated (in time t) random process. The effective expansion parameter of quantity $S(t,t)$ in series (7.81) is in this case $\beta = \tau_0 B(0) v(t)$. If $\beta \ll 1$, we can content ourselves with the first term of series (7.82), which corresponds to the standard perturbation theory. But if $\beta \sim 1$, one needs take into account the whole series for function $S(t,t)$.

The Gaussian Markovian process $\xi(t)$ with correlation function $B(\tau) = \sigma^2 e^{-\alpha|\tau|}$, where $\alpha = 1/\tau_0$, allows a more detailed analysis. In this case, integral equation (7.79) assumes for $t_1 < t$ the form

$$S(t_1,t) = \sigma^2 e^{-\alpha(t-t_1)} + 2i\sigma^2 \int\limits_0^t d\tau e^{-\alpha|t_1-\tau|} v(\tau) S(\tau,t). \tag{7.83}$$

The solution to this equation as a function of parameter t can be described as the initial-value problem

$$\begin{aligned} \frac{\partial}{\partial t} S(t_1,t) &= \{-\alpha + 2iv(t)S(t,t)\} S(t_1,t), \quad S(t_1,t)|_{t=t_1} = S(t_1,t_1), \\ \frac{d}{dt} S(t,t) &= -2\alpha \left[S(t,t) - \sigma^2\right] + 2iv(t) S^2(t,t), \quad S(t,t)|_{t=0} = \sigma^2. \end{aligned} \tag{7.84}$$

The asymptotic solution of the latter equation for $\alpha \to \infty$ $(\tau_0 \to 0)$ has the form

$$S(t,t) = \frac{1 - \sqrt{1 - 4i\sigma^2 \tau_0 v(t)}}{2i\tau_0 v(t)}, \quad S^{(n)}(t,t) = \left[2i\tau_0 v(t)\right]^n \sigma^{2(n+1)} \frac{(2n-1)!!}{(n+1)!},$$

which coincides with the solution of Eq. (7.84) under the assumption that function $v(t)$ is a constant. As a consequence, the logarithm of the characteristic functional in this asymptotic case is given by the formula $\Theta[t; v(\tau)] =$

$\frac{1}{2\tau_0}\int\limits_0^t d\tau\left(1-\sqrt{1-4i\sigma^2\tau_0 v(\tau)}\right)$, and cumulants of process $z(t)=\xi^2(t)$ have the forms ($K_1=\sigma^2$)

$$K_n(t_1,\cdots,t_n)=(2\tau_0)^{n-1}\sigma^{2n}(2n-3)!!\delta(t_1-t_2)\cdots\delta(t_{n-1}-t_n),$$

which correspond to Eqs. (7.82); in this case, we have

$$I_n=\int\limits_0^\infty d\tau_1\cdots\int\limits_0^\infty d\tau_n\exp\left\{-\tau_1-|\tau_1-\tau_2|-\cdots-|\tau_{n-1}-\tau_n|-\tau_n\right\}=\frac{(2n-1)!!}{(n+1)!}.$$

Process $z(t)=z\xi(t)$, where $\xi(t)$ is the Gaussian delta-correlated process with parameters $\langle\xi(t)\rangle=0$, $\langle\xi(t)\xi(t')\rangle=2\sigma^2\tau_0\delta(t-t')$, and quantity z is the random quantity with probability density $p(z)$, is an example of the process that cannot be considered the delta correlated process. In this case, the characteristic functional is given by the equality

$$\Phi[t;v(\tau)]=\left\langle\exp\left\{iz\int\limits_0^t d\tau v(\tau)\xi(\tau)\right\}\right\rangle_{z,\xi}=\int\limits_{-\infty}^{\infty}dzp(z)\exp\left\{-z^2\sigma^2\tau_0\int\limits_0^t d\tau v^2(\tau)\right\},\tag{7.85}$$

and process $z(t)$ cannot be considered the delta-correlated process, because it does not satisfy Eq. (7.64) despite its second cumulant has the form

$$K_2(t_1,t_2)=\langle z(t_1)z(t_2)\rangle=2\left\langle z^2\right\rangle\sigma^2\tau_0\delta(t_1-t_2).$$

This follows from the fact that process $z(t)$ is formed as the product of two processes — process z with the infinite correlation radius and process $\xi(t)$ with the zero-valued correlation radius.

Now, we proceed to the direct statistical analysis of stochastic dynamic systems.

Part III
Stochastic Equations

Chapter 8
General Approaches to Analyzing Stochastic Dynamic Systems

In this chapter, we will consider basic methods of determining statistical characteristics of solutions to stochastic equations.

Consider a linear (differential, integro-differential, or integral) stochastic equation. Averaging of such an equation over an ensemble of realizations of fluctuating parameters will not result generally in a closed equation for the corresponding average value. To obtain the closed equation, we must deal with an additional extended space whose dimension appears to be infinite in most cases. This approach makes it possible to derive the linear equation for average quantity of interest, but this equation will contain variational derivatives.

Consider some special types of dynamic systems.

8.1 Ordinary Differential Equations

Let dynamics of vector function $\boldsymbol{x}(t)$ is described by the ordinary differential equation

$$\frac{d}{dt}\boldsymbol{x}(t) = \boldsymbol{v}(\boldsymbol{x}, t) + \boldsymbol{f}(\boldsymbol{x}, t), \quad \boldsymbol{x}(t_0) = \boldsymbol{x}_0. \tag{8.1}$$

Here, function $\boldsymbol{v}(\boldsymbol{x}, t)$ is the deterministic function and $\boldsymbol{f}(\boldsymbol{x}, t)$ is the random function.

The solution to Eq. (8.1) is a functional of $\boldsymbol{f}(\boldsymbol{y}, \tau) + \boldsymbol{v}(\boldsymbol{y}, \tau)$ with $\tau \in (t_0, t)$, i.e.,

$$\boldsymbol{x}(t) = \boldsymbol{x}[t; \boldsymbol{f}(\boldsymbol{y}, \tau) + \boldsymbol{v}(\boldsymbol{y}, \tau)].$$

From this fact follows the equality

$$\frac{\delta}{\delta f_j(\boldsymbol{y}, \tau)} F\left(\boldsymbol{x}(t)\right) = \frac{\delta}{\delta v_j(\boldsymbol{y}, \tau)} F\left(\boldsymbol{x}(t)\right) = \frac{\partial F(\boldsymbol{x}(t))}{\partial x_i} \frac{\delta x_i(t)}{\delta f_j(\boldsymbol{y}, \tau)}$$

valid for arbitrary function $F(\boldsymbol{x})$. In addition, we have

V.I. Klyatskin, *Stochastic Equations: Theory and Applications in Acoustics, Hydrodyn., Magnetohydrodyn., and Radiophys., Vol. 1*, Understanding Complex Systems,
DOI: 10.1007/978-3-319-07587-7_8,

$$\frac{\delta}{\delta f_j(\boldsymbol{y},t-0)} x_i(t) = \frac{\delta}{\delta v_j(\boldsymbol{y},t-0)} x_i(t) = \delta_{ij}\delta\left(\boldsymbol{x}(t)-\boldsymbol{y}\right).$$

The corresponding Liouville equation for the indicator function

$$\varphi(\boldsymbol{x},t) = \delta(\boldsymbol{x}(t)-\boldsymbol{x})$$

follows from Eq. (8.1) and has the form

$$\frac{\partial}{\partial t}\varphi(\boldsymbol{x},t) = -\frac{\partial}{\partial \boldsymbol{x}}\left\{\left[\boldsymbol{v}(\boldsymbol{x},t)+\boldsymbol{f}(\boldsymbol{x},t)\right]\varphi(\boldsymbol{x},t)\right\}, \quad \varphi(\boldsymbol{x},t_0) = \delta(\boldsymbol{x}-\boldsymbol{x}_0), \quad (8.2)$$

from which follows the equality

$$\frac{\delta}{\delta \boldsymbol{f}(\boldsymbol{y},t-0)}\varphi(\boldsymbol{x},t) = \frac{\delta}{\delta \boldsymbol{v}(\boldsymbol{y},t-0)}\varphi(\boldsymbol{x},t) = -\frac{\partial}{\partial \boldsymbol{x}}\left\{\delta(\boldsymbol{x}-\boldsymbol{y})\varphi(\boldsymbol{x},t)\right\}. \quad (8.3)$$

Using this equality, we can rewrite Eq. (8.2) in the form, which may look at first sight more complicated

$$\left(\frac{\partial}{\partial t}+\frac{\partial}{\partial \boldsymbol{x}}\boldsymbol{v}(\boldsymbol{x},t)\right)\varphi(\boldsymbol{x},t) = \int d\boldsymbol{y}\boldsymbol{f}(\boldsymbol{y},t)\frac{\delta}{\delta \boldsymbol{v}(\boldsymbol{y},t)}\varphi(\boldsymbol{x},t). \quad (8.4)$$

Consider now the one-time probability density for solution $\boldsymbol{x}(t)$ of Eq. (8.1)

$$P(\boldsymbol{x},t) = \langle\varphi(\boldsymbol{x},t)\rangle = \langle\delta(\boldsymbol{x}(t)-\boldsymbol{x})\rangle\,.$$

Here, $\boldsymbol{x}(t)$ is the solution of Eq. (8.1) corresponding to a particular realization of random field $\boldsymbol{f}(\boldsymbol{x},t)$, and angle brackets $\langle\cdots\rangle$ denote averaging over an ensemble of realizations of field $\boldsymbol{f}(\boldsymbol{x},t)$.

Averaging Eq. (8.4) over an ensemble of realizations of field $\boldsymbol{f}(\boldsymbol{x},t)$, we obtain the expression

$$\left(\frac{\partial}{\partial t}+\frac{\partial}{\partial \boldsymbol{x}}\boldsymbol{v}(\boldsymbol{x},t)\right)P(\boldsymbol{x},t) = \int d\boldsymbol{y}\frac{\delta}{\delta \boldsymbol{v}(\boldsymbol{y},t)}\langle\boldsymbol{f}(\boldsymbol{y},t)\varphi(\boldsymbol{x},t)\rangle\,. \quad (8.5)$$

Quantity $\langle\boldsymbol{f}(\boldsymbol{y},t)\varphi(\boldsymbol{x},t)\rangle$ in the right-hand side of Eq. (8.5) is the correlator of random field $\boldsymbol{f}(\boldsymbol{y},t)$ with function $\varphi(\boldsymbol{x},t)$, which is a functional of random field $\boldsymbol{f}(\boldsymbol{y},\tau)$ and is given either by Eq. (8.2), or by Eq. (8.4).

The characteristic functional

$$\Phi[t,t_0;\boldsymbol{u}(\boldsymbol{y},\tau)] = \left\langle\exp\left\{i\int\limits_{t_0}^{t}d\tau\int d\boldsymbol{y}\boldsymbol{f}(\boldsymbol{y},\tau)\boldsymbol{u}(\boldsymbol{y},\tau)\right\}\right\rangle$$
$$= \exp\left\{\Theta[t,t_0;\boldsymbol{u}(\boldsymbol{y},\tau)]\right\}$$

exhaustively describes all statistical characteristics of random field $\boldsymbol{f}(\boldsymbol{y},\tau)$ for $\tau\in(t_0,t)$.

We split correlator $\langle \boldsymbol{f}(\boldsymbol{y},t)\varphi(\boldsymbol{x},t)\rangle$ using the technique of functionals. Introducing functional shift operator with respect to field $\boldsymbol{v}(\boldsymbol{y},\tau)$, we represent functional $\varphi[\boldsymbol{x},t;\boldsymbol{f}(\boldsymbol{y},\tau)+\boldsymbol{v}(\boldsymbol{y},\tau)]$ in the operator form

$$\varphi[\boldsymbol{x},t;\boldsymbol{f}(\boldsymbol{y},\tau)+\boldsymbol{v}(\boldsymbol{y},\tau)] = \exp\left[\int\limits_{t_0}^{t} d\tau \int d\boldsymbol{y}\boldsymbol{f}(\boldsymbol{y},\tau)\frac{\delta}{\delta\boldsymbol{v}(\boldsymbol{y},\tau)}\right]\varphi[\boldsymbol{x},t;\boldsymbol{v}(\boldsymbol{y},\tau)].$$

With this representation, the term in the right-hand side of Eq. (8.5) assumes the form

$$\int d\boldsymbol{y}\frac{\delta}{\delta v_j(\boldsymbol{y},t)}\frac{\left\langle f_j(\boldsymbol{y},t)\exp\left\{\int\limits_{t_0}^{t} d\tau \int d\boldsymbol{y}'\boldsymbol{f}(\boldsymbol{y}',\tau)\frac{\delta}{\delta\boldsymbol{v}(\boldsymbol{y}',\tau)}\right\}\right\rangle}{\left\langle \exp\left\{\int\limits_{t_0}^{t} d\tau \int d\boldsymbol{y}'\boldsymbol{f}(\boldsymbol{y}',\tau)\frac{\delta}{\delta\boldsymbol{v}(\boldsymbol{y}',\tau)}\right\}\right\rangle}P(\boldsymbol{x},t)$$

$$= \dot{\Theta}_t\left[t,t_0;\frac{\delta}{i\delta\boldsymbol{v}(\boldsymbol{y},\tau)}\right]P(\boldsymbol{x},t),$$

where we introduced the functional

$$\dot{\Theta}_t[t,t_0;\boldsymbol{u}(\boldsymbol{y},\tau)] = \frac{d}{dt}\ln\Phi[t,t_0;\boldsymbol{u}(\boldsymbol{y},\tau)].$$

Consequently, we can rewrite Eq. (8.5) in the form

$$\left(\frac{\partial}{\partial t}+\frac{\partial}{\partial\boldsymbol{x}}\boldsymbol{v}(\boldsymbol{x},t)\right)P(\boldsymbol{x},t) = \dot{\Theta}_t\left[t,t_0;\frac{\delta}{i\delta\boldsymbol{v}(\boldsymbol{y},\tau)}\right]P(\boldsymbol{x},t). \tag{8.6}$$

Equation (8.6) is the closed equation with variational derivatives in the functional space of all possible functions $\{\boldsymbol{v}(\boldsymbol{y}\ ,\tau)\}$. However, for a fixed function $\boldsymbol{v}(\boldsymbol{x},t)$, we arrive at unclosed equation [47–49]

$$\left(\frac{\partial}{\partial t}+\frac{\partial}{\partial\boldsymbol{x}}\boldsymbol{v}(\boldsymbol{x},t)\right)P(\boldsymbol{x},t) = \left\langle\dot{\Theta}_t\left[t,t_0;\frac{\delta}{i\delta\boldsymbol{f}(\boldsymbol{y},t)}\right]\varphi[\boldsymbol{x},t;\boldsymbol{f}(\boldsymbol{y},\tau)]\right\rangle,$$
$$P(\boldsymbol{x},t_0) = \delta(\boldsymbol{x}-\boldsymbol{x}_0). \tag{8.7}$$

Equation (8.7) is the exact consequence of the initial dynamic equation (8.1). Statistical characteristics of random field $\boldsymbol{f}(\boldsymbol{x},t)$ appear in this equation only through functional $\dot{\Theta}_t[t,t_0;\boldsymbol{u}(\boldsymbol{y},\tau)]$ whose expansion in the functional Taylor series in powers of $\boldsymbol{u}(\boldsymbol{x},\tau)$ depends on all space-time cumulant functions of random field $\boldsymbol{f}(\boldsymbol{x},t)$.

Note that equation for the one-time probability density $P(\boldsymbol{x}, t)$ preserve the form of Eq. (8.7) even for more general integro-differential equation

$$\frac{d}{dt}x_i(t) = v_i(\boldsymbol{x}, t) + \int d\boldsymbol{y} D_{ij}(\boldsymbol{x}, \boldsymbol{y}, t) f_j(\boldsymbol{y}, t), \quad \boldsymbol{x}(t_0) = \boldsymbol{x}_0,$$

in which case the variational derivative has the form

$$\frac{\delta}{\delta f_j(\boldsymbol{y}, t-0)} x_i(t) = D_{ij}(\boldsymbol{x}(t), \boldsymbol{y}, t).$$

As we mentioned earlier, Eq. (8.7) is not closed in the general case with respect to function $P(\boldsymbol{x}, t)$, because quantity

$$\dot{\Theta}_t \left[t, t_0; \frac{\delta}{i\delta \boldsymbol{f}(\boldsymbol{y}, \tau)}\right] \delta(\boldsymbol{x}(t) - \boldsymbol{x})$$

appeared in averaging brackets depends on the solution $\boldsymbol{x}(t)$ (which is a functional of random field $\boldsymbol{f}(\boldsymbol{y}, \tau)$) for all times $t_0 < \tau < t$. However, in some cases, the variational derivative in Eq. (8.7) can be expressed in terms of ordinary differential operators. In such conditions, equations like Eq. (8.7) will be the closed equations in the corresponding probability densities. The corresponding examples will be given below.

Note that Eq. (8.2) is the forward Liouville equation and describes the evolution of indicator function

$$\varphi(\boldsymbol{x}, t) = \varphi(\boldsymbol{x}, t|\boldsymbol{x}_0, t_0) = \delta\left(\boldsymbol{x}(t) - \boldsymbol{x}|\boldsymbol{x}(t_0) = \boldsymbol{x}_0\right)$$

in time t. For this reason, we can call Eq. (8.7) the *forward equation* for probability density.

In Chapter 3, we obtained the backward Liouville equation (3.9), page 97 for the indicator function, which describes the evolution of dynamic system (8.1) in terms of initial values t_0 and $\boldsymbol{x}_0$. In our case, this equation has the form,

$$\left(\frac{\partial}{\partial t_0} + \boldsymbol{v}(\boldsymbol{x}_0, t_0)\frac{\partial}{\partial \boldsymbol{x}_0}\right)\varphi(\boldsymbol{x}, t|\boldsymbol{x}_0, t_0) = -\boldsymbol{f}(\boldsymbol{x}_0, t_0)\frac{\partial}{\partial \boldsymbol{x}_0}\varphi(\boldsymbol{x}, t|\boldsymbol{x}_0, t_0),$$
$$\varphi(\boldsymbol{x}, t|\boldsymbol{x}_0, t) = \delta(\boldsymbol{x} - \boldsymbol{x}_0). \tag{8.8}$$

Averaging Eq. (8.8) over an ensemble of realizations of field $\boldsymbol{f}(\boldsymbol{x}_0, t_0)$ and performing the procedure similar to that used in derivation of Eq. (8.7), we obtain the equation for probability density

$$P(\boldsymbol{x}, t|\boldsymbol{x}_0, t_0) = \langle \delta(\boldsymbol{x}, t|\boldsymbol{x}_0, t_0)\rangle$$

in the form

$$\left(\frac{\partial}{\partial t_0} + \boldsymbol{v}(\boldsymbol{x}_0, t_0)\frac{\partial}{\partial \boldsymbol{x}_0}\right) P(\boldsymbol{x}, t|\boldsymbol{x}_0, t_0) = \left\langle \dot{\Theta}_{t_0}\left[t, t_0; \frac{\delta}{i\delta \boldsymbol{f}(\boldsymbol{y}, \tau)}\right] \delta\left(\boldsymbol{x}(t|\boldsymbol{x}_0, t_0) - \boldsymbol{x}\right)\right\rangle,$$

$$P(\boldsymbol{x}, t|\boldsymbol{x}_0, t) = \delta(\boldsymbol{x} - \boldsymbol{x}_0), \tag{8.9}$$

where

$$\dot{\Theta}_{t_0}[t, t_0; \boldsymbol{u}(\boldsymbol{y}, \tau)] = \frac{d}{dt_0} \ln \Phi[t, t_0; \boldsymbol{u}(\boldsymbol{y}, \tau)].$$

Equation (8.9) describes the evolution of the probability density as a function of initial parameters $\{\boldsymbol{x}_0, t_0\}$; for this reason, we can call it the *backward equation* .

The forward and backward equations are equivalent. The forward equation appears to be more convenient for studying the behavior of statistical characteristics of solutions to Eq. (8.1) in time domain. The backward equation is more convenient for studying the statistical characteristics that concern the residence of random process $\boldsymbol{x}(t)$ within certain region of space, such as residence duration within the region and time of arrival at region boundary. Indeed, the probability of residence of random process $\boldsymbol{x}(t)$ in spatial region V is given by the integral

$$G(t; \boldsymbol{x}_0, t_0) = \int\limits_V d\boldsymbol{x} P(x, t|\boldsymbol{x}_0, t_0),$$

which, according to Eq. (8.9), will satisfy the equation

$$\left(\frac{\partial}{\partial t_0} + \boldsymbol{v}(\boldsymbol{x}_0, t_0)\frac{\partial}{\partial \boldsymbol{x}_0}\right) G(t|\boldsymbol{x}_0, t_0) = \dot{\Theta}_{t_0}\left[t, t_0; \frac{\delta}{i\delta \boldsymbol{f}(\boldsymbol{y}, \tau)}\right] \int\limits_V d\boldsymbol{x}\delta\left(\boldsymbol{x}(t|\boldsymbol{x}_0, t_0) - \boldsymbol{x}\right),$$

$$G(t|\boldsymbol{x}_0, t_0) = \begin{cases} 1 \; (\boldsymbol{x}_0 \in V) \\ 0 \; (\boldsymbol{x}_0 \notin V) \end{cases}. \tag{8.10}$$

This equation must be supplemented with the boundary conditions following from the nature of each particular problem and depend on region V and region boundaries.

Proceeding in a similar way, one can obtain the equation similar to Eq. (8.7) for the m-time probability density that refers to m different instants $t_1 < t_2 < \cdots < t_m$

$$P_m(\boldsymbol{x}_1, t_1; \cdots; \boldsymbol{x}_m, t_m) = \langle \varphi_m(\boldsymbol{x}_1, t_1; \cdots; \boldsymbol{x}_m, t_m) \rangle, \tag{8.11}$$

where the indicator function is defined by the equality

$$\varphi_m(\boldsymbol{x}_1, t_1; \cdots; \boldsymbol{x}_m, t_m) = \delta(\boldsymbol{x}(t_1) - \boldsymbol{x}_1) \cdots \delta(\boldsymbol{x}(t_m) - \boldsymbol{x}_m).$$

Differentiating Eq. (8.11) with respect to time t_m and using then dynamic equation (8.1), one can obtain the equation

$$\left(\frac{\partial}{\partial t_m} + \frac{\partial}{\partial \boldsymbol{x}_m} \boldsymbol{v}(\boldsymbol{x}_m, t_m)\right) P_m(\boldsymbol{x}_1, t_1; \cdots; \boldsymbol{x}_m, t_m)$$
$$= \left\langle \dot{\Theta}_{t_m} \left[t_m, t_0; \frac{\delta}{i\delta \boldsymbol{f}(\boldsymbol{y}, \tau)}\right] \varphi_m(\boldsymbol{x}_1, t_1; \cdots; \boldsymbol{x}_m, t_m) \right\rangle. \quad (8.12)$$

No summation over index m is performed here. The initial value to Eq. (8.12) can be derived from Eq. (8.11). Setting $t_m = t_{m-1}$ in Eq. (8.11), we obtain the equality

$$P_m(\boldsymbol{x}_1, t_1; \cdots; \boldsymbol{x}_m, t_{m-1}) = \delta(\boldsymbol{x}_m - \boldsymbol{x}_{m-1}) P_{m-1}(\boldsymbol{x}_1, t_1; \cdots; \boldsymbol{x}_{m-1}, t_{m-1}),$$

which just determines the initial value for Eq. (8.12).

8.2 Partial Differential Equations

Above, we considered statistical description of dynamic systems starting from the Liouville equation (8.2) that matches the ordinary differential equation (8.1). It is quite obvious that the derivation procedure of Eqs. (8.6), (8.7), (8.12), and the like can be applied to other dynamic systems specified in terms of linear equations both in finite- and infinite-dimension spaces, i.e., in terms of partial differential equations of the first and higher orders. Below, we consider the use of these equations by the specific examples, such as passive tracer transfer in random field of velocities Eqs. (1.56), page 34, (3.16), and (3.20), page 101; wave propagation in random media described within the framework of the parabolic equation of quasi-optics (1.153), page 63; and hydrodynamic turbulence evolution described by the integro-differential equation (1.171), page 69.

8.2.1 Passive Tracer Transfer in Random Field of Velocities

The first example deals with Eq. (3.20), page 101

$$\left(\frac{\partial}{\partial t} + \boldsymbol{U}(\boldsymbol{r}, t)\frac{\partial}{\partial \boldsymbol{r}}\right) \varphi(\boldsymbol{r}, t; \rho) = \frac{\partial \boldsymbol{U}(\boldsymbol{r}, t)}{\partial \boldsymbol{r}} \frac{\partial}{\partial \rho} \rho \varphi(\boldsymbol{r}, t; \rho) \quad (8.13)$$

for indicator function

$$\varphi(\boldsymbol{r}, t; \rho) = \delta\left(\rho(\boldsymbol{r}, t) - \rho\right).$$

We assume that $\boldsymbol{U}(\boldsymbol{r},t) = \boldsymbol{u}_0(\boldsymbol{r}\ ,t) + \boldsymbol{u}(\boldsymbol{r},t)$, where $\boldsymbol{u}_0(\boldsymbol{r},t)$ is the deterministic component of the velocity field (mean flow) and $\boldsymbol{u}(\boldsymbol{r},t)$ is the random component.

A consequence of Eq. (8.13) is the equality

$$\frac{\delta\varphi(\boldsymbol{r},t;\rho)}{\delta U_j(\boldsymbol{r}',t-0)} = \frac{\delta\varphi(\boldsymbol{r},t;\rho)}{\delta u_j(\boldsymbol{r}',t-0)} = \frac{\delta\varphi(\boldsymbol{r},t;\rho)}{\delta u_{0j}(\boldsymbol{r}',t-0)}$$
$$= \left\{-\delta(\boldsymbol{r}-\boldsymbol{r}')\frac{\partial}{\partial r_j} + \frac{\partial\delta(\boldsymbol{r}-\boldsymbol{r}')}{\partial r_j}\frac{\partial}{\partial\rho}\rho\right\}\varphi(\boldsymbol{r},t;\rho). \quad (8.14)$$

Statistical characteristics of random field $\boldsymbol{u}(\boldsymbol{r},t)$ can be exhaustively described in terms of the characteristic functional

$$\Phi[t;\boldsymbol{\psi}(\boldsymbol{r}',\tau)] = \left\langle \exp\left\{i\int_0^t d\tau\int d\boldsymbol{r}'\boldsymbol{u}(\boldsymbol{r}',\tau)\boldsymbol{\psi}(\boldsymbol{r}',\tau)\right\}\right\rangle = \exp\left\{\Theta[t;\boldsymbol{\psi}(\boldsymbol{r}',\tau)]\right\}.$$

Now, we average Eq. (8.13) over an ensemble of realizations of random field $\boldsymbol{u}(\boldsymbol{r},t)$. Then, replicating derivation of Eq.(8.7) and taking into account Eq. (8.14), we obtain that the one-time probability density

$$P(\boldsymbol{r},t;\rho) = \langle\varphi(\boldsymbol{r},t;\rho)\rangle = \langle\delta\left(\rho(\boldsymbol{r},t)-\rho\right)\rangle$$

satisfies the equation

$$\left(\frac{\partial}{\partial t} + \boldsymbol{u}_0(\boldsymbol{r},t)\frac{\partial}{\partial\boldsymbol{r}}\right)P(\boldsymbol{r},t;\rho) = \frac{\partial\boldsymbol{u}_0(\boldsymbol{r},t)}{\partial\boldsymbol{r}}\frac{\partial}{\partial\rho}\rho P(\boldsymbol{r},t;\rho)$$
$$+\left\langle\int d\boldsymbol{r}\boldsymbol{u}(\boldsymbol{r},t)\frac{\delta}{\delta\boldsymbol{u}_0(\boldsymbol{r},t-0)}\varphi(\boldsymbol{r},t;\rho)\right\rangle, \quad (8.15)$$

whose last term can be rewritten in the form

$$\left\langle\int d\boldsymbol{r}\boldsymbol{u}(\boldsymbol{r},t)\frac{\delta}{\delta\boldsymbol{u}_0(\boldsymbol{r},t-0)}\varphi[\boldsymbol{r},t;\rho;\boldsymbol{u}+\boldsymbol{u}_0]\right\rangle$$
$$= \left\langle\int d\boldsymbol{r}\boldsymbol{u}(\boldsymbol{r},t)\frac{\delta}{\delta\boldsymbol{u}_0(\boldsymbol{r},t-0)}\exp\left\{\int_0^t d\tau\int d\boldsymbol{r}'\boldsymbol{u}(\boldsymbol{r}',\tau)\frac{\delta}{\delta\boldsymbol{u}_0(\boldsymbol{r}',\tau)}\right\}\right\rangle\varphi[\boldsymbol{r},t;\rho;\boldsymbol{u}_0]$$
$$= \left\langle\frac{d}{dt}\exp\left\{\int_0^t d\tau\int d\boldsymbol{r}'\boldsymbol{u}(\boldsymbol{r}',\tau)\frac{\delta}{\delta\boldsymbol{u}_0(\boldsymbol{r}',\tau)}\right\}\right\rangle\varphi[\boldsymbol{r},t;\rho;\boldsymbol{u}_0]$$
$$= \dot{\Theta}_t\left[t;\frac{\delta}{i\delta\boldsymbol{u}_0(\boldsymbol{r}',\tau)}\right]\langle\varphi[\boldsymbol{r},t;\rho;\boldsymbol{u}+\boldsymbol{u}_0]\rangle = \dot{\Theta}_t\left[t;\frac{\delta}{i\delta\boldsymbol{u}_0(\boldsymbol{r}',\tau)}\right]P(\boldsymbol{r},t;\rho),$$

where

$$\dot{\Theta}_t\left[t;\boldsymbol{\psi}(\boldsymbol{r}',\tau)\right]=\frac{d}{dt}\ln\left\langle\exp\left\{i\int\limits_0^t d\tau\int d\boldsymbol{r}'\boldsymbol{u}(\boldsymbol{r}',\tau)\boldsymbol{\psi}(\boldsymbol{r}',\tau)\right\}\right\rangle$$

is the derivative of the characteristic functional logarithm of random field $\boldsymbol{u}(\boldsymbol{r},t)$.

Thus, expression (8.15) can be represented as the functional linear variational differential equation in the functional space of functions $\{\boldsymbol{u}_0(\boldsymbol{r},t)\}$

$$\left(\frac{\partial}{\partial t}+\boldsymbol{u}_0(\boldsymbol{r},t)\frac{\partial}{\partial\boldsymbol{r}}\right)P(\boldsymbol{r},t;\rho)$$
$$=\frac{\partial\boldsymbol{u}_0(\boldsymbol{r},t)}{\partial\boldsymbol{r}}\frac{\partial}{\partial\rho}\rho P(\boldsymbol{r},t;\rho)+\dot{\Theta}_t\left[t;\frac{\delta}{i\delta\boldsymbol{u}_0(\boldsymbol{r}',\tau)}\right]P(\boldsymbol{r},t;\rho),\quad(8.16)$$

However, if we deal with a fixed mean flow $\boldsymbol{u}_0(\boldsymbol{r}\ ,t)$ (e.g., $\boldsymbol{u}_0(\boldsymbol{r},t)=0$), then Eq. (8.16) assumes the form of an unclosed equality

$$\left(\frac{\partial}{\partial t}+\boldsymbol{u}_0(\boldsymbol{r},t)\frac{\partial}{\partial\boldsymbol{r}}\right)P(\boldsymbol{r},t;\rho)$$
$$=\frac{\partial\boldsymbol{u}_0(\boldsymbol{r},t)}{\partial\boldsymbol{r}}\frac{\partial}{\partial\rho}\rho P(\boldsymbol{r},t;\rho)+\left\langle\dot{\Theta}_t\left[t;\frac{\delta}{i\delta\boldsymbol{u}(\boldsymbol{r}',\tau)}\right]\varphi(\boldsymbol{r},t;\rho)\right\rangle.\quad(8.17)$$

8.2.2 Parabolic Equation of Quasi-Optics

The second example deals with wave propagation in random medium within the framework of the linear parabolic equation (1.153), page 63

$$\frac{\partial}{\partial x}u(x,\boldsymbol{R})=\frac{i}{2k}\Delta_{\boldsymbol{R}}u(x,\boldsymbol{R})+i\frac{k}{2}\varepsilon(x,\boldsymbol{R})u(x,\boldsymbol{R}),\quad u(0,\boldsymbol{R})=u_0(\boldsymbol{R}).\ (8.18)$$

In this case, functional (3.59)

$$\varphi[x;v(\boldsymbol{R}'),v^*(\boldsymbol{R}')]=\varphi[x;v,v^*]=\exp\left\{i\int d\boldsymbol{R}'\left[u(x,\boldsymbol{R}')v(\boldsymbol{R}')+u^*(x,\boldsymbol{R}')v^*(\boldsymbol{R}')\right]\right\}$$

is described by the variational differential equation (the Hopf equation) (3.60), page 111

$$\frac{\partial}{\partial x}\varphi[x;v,v^*]=i\frac{k}{2}\int d\boldsymbol{R}'\varepsilon(x,\boldsymbol{R}')\widehat{M}(\boldsymbol{R}')\varphi[x;v,v^*]$$
$$+\frac{i}{2k}\left\{\int d\boldsymbol{R}'\left[v(\boldsymbol{R}')\Delta_{\boldsymbol{R}'}\frac{\delta}{\delta v(\boldsymbol{R}')}-v^*(\boldsymbol{R}')\Delta_{\boldsymbol{R}'}\frac{\delta}{\delta v^*(\boldsymbol{R}')}\right]\right\}\varphi[x;v,v^*]\quad(8.19)$$

equivalent to the initial equation (8.18). Here, $\widehat{M}(\boldsymbol{R}')$ is the Hermitian operator

$$\widehat{M}(\boldsymbol{R}') = v(\boldsymbol{R}')\frac{\delta}{\delta v(\boldsymbol{R}')} - v^*(\boldsymbol{R}')\frac{\delta}{\delta v^*(\boldsymbol{R}')}.$$

A consequence of Eq. (8.19) is the equality

$$\frac{\delta}{\delta\varepsilon(x-0,\boldsymbol{R}')}\varphi[x;v,v^*] = i\frac{k}{2}\widehat{M}(\boldsymbol{R}')\varphi[x;v,v^*]. \tag{8.20}$$

Averaging Eq. (8.19) over an ensemble of realizations of random field $\varepsilon(x,\boldsymbol{R})$ and replicating derivation of Eq. (8.7) and (8.17), we obtain that the characteristic functional of the solution to problem (8.18)

$$\Phi[x;v(\boldsymbol{R}'),v^*(\boldsymbol{R}')] = \Phi[x;v,v^*] = \langle\varphi[x;v(\boldsymbol{R}'),v^*(\boldsymbol{R}')]\rangle$$

satisfies the variational derivative equation

$$\frac{\partial}{\partial x}\Phi[x;v,v^*] = \left\langle \dot{\Theta}_x\left[x;\frac{\delta}{i\delta\varepsilon(\xi,\boldsymbol{R}')}\right]\varphi[x;v,v^*]\right\rangle + \frac{i}{2k}\left\{\int d\boldsymbol{R}'\left[v(\boldsymbol{R}')\Delta_{\boldsymbol{R}'}\frac{\delta}{\delta v(\boldsymbol{R}')} - v^*(\boldsymbol{R}')\Delta_{\boldsymbol{R}'}\frac{\delta}{\delta v^*(\boldsymbol{R}')}\right]\right\}\Phi[x;v,v^*], \tag{8.21}$$

where

$$\dot{\Theta}_x\left[x;\psi(\xi,\boldsymbol{R}')\right] = \frac{d}{dx}\ln\left\langle\exp\left\{i\int\limits_0^x d\xi\int d\boldsymbol{R}'\varepsilon(\xi,\boldsymbol{R}')\psi(\xi,\boldsymbol{R}')\right\}\right\rangle$$

is the derivative of the characteristic functional logarithm of random field $\varepsilon(x,\boldsymbol{R})$.

8.2.3 *Random Forces in the Theory of Hydrodynamic Turbulence*

In Chapter 1, page 69 we obtained that stationary and homogeneous hydrodynamic turbulence can be described in terms of the Fourier transform of the velocity field ($\widehat{u}_i^*(\boldsymbol{k},t) = \widehat{u}_i(-\boldsymbol{k},t)$)

$$\widehat{u}_i(\boldsymbol{k},t) = \int d\boldsymbol{r}u_i(\boldsymbol{r},t)e^{-i\boldsymbol{k}\boldsymbol{r}}, \quad u_i(\boldsymbol{r},t) = \frac{1}{(2\pi)^3}\int d\boldsymbol{k}\widehat{u}_i(\boldsymbol{k},t)e^{i\boldsymbol{k}\boldsymbol{r}},$$

which satisfies the nonlinear integro-differential equation (1.171)

$$\frac{\partial}{\partial t}\widehat{u}_i(\boldsymbol{k},t)+\frac{i}{2}\int d\boldsymbol{k}_1\int d\boldsymbol{k}_2\Lambda_i^{\alpha\beta}(\boldsymbol{k}_1,\boldsymbol{k}_2,\boldsymbol{k})\,\widehat{u}_\alpha(\boldsymbol{k}_1,t)\,\widehat{u}_\beta(\boldsymbol{k}_2,t)-\nu k^2\widehat{u}_i(\boldsymbol{k},t)=\widehat{f}_i(\boldsymbol{k},t),\tag{8.22}$$

where

$$\Lambda_i^{\alpha\beta}(\boldsymbol{k}_1,\boldsymbol{k}_2,\boldsymbol{k})=\frac{1}{(2\pi)^3}\left\{k_\alpha\Delta_{i\beta}(\boldsymbol{k})+k_\beta\Delta_{i\alpha}(\boldsymbol{k})\right\}\delta(\boldsymbol{k}_1+\boldsymbol{k}_2-\boldsymbol{k}),$$

$$\Delta_{ij}(\boldsymbol{k})=\delta_{ij}-\frac{k_ik_j}{\boldsymbol{k}^2}\quad(i,\,\alpha,\,\beta\,=\,1,\,2,\,3),$$

and $\widehat{\boldsymbol{f}}(\boldsymbol{k},t)$ is the spatial Fourier harmonics of the external force.

A specific feature of the three-dimensional hydrodynamic motions consists in the existence of the integral of energy under the condition that external forces and effects related to the molecular viscosity are absent.

Furthermore, in Chapter 2, page 114 we obtained that functional

$$\varphi[t;\boldsymbol{z}(\boldsymbol{k}')]=\varphi[t;\boldsymbol{z}]=\exp\left\{i\int d\boldsymbol{k}'\widehat{\boldsymbol{u}}(\boldsymbol{k}',t)\boldsymbol{z}(\boldsymbol{k}')\right\}$$

satisfies the linear variational differential Hopf equation in functional space (3.69)

$$\begin{aligned}\frac{\partial}{\partial t}\varphi[t;\boldsymbol{z}]&=-\int d\boldsymbol{k}z_i(\boldsymbol{k})\left\{\nu k^2\frac{\delta}{\delta z_i(\boldsymbol{k})}-i\widehat{f}_i(\boldsymbol{k},t)\right\}\varphi[t;\boldsymbol{z}]\\&-\frac{1}{2}\int d\boldsymbol{k}z_i(\boldsymbol{k})\int d\boldsymbol{k}_1\int d\boldsymbol{k}_2\Lambda_i^{\alpha\beta}(\boldsymbol{k}_1,\boldsymbol{k}_2,\boldsymbol{k})\frac{\delta^2\varphi[t;\boldsymbol{z}]}{\delta z_\alpha(\boldsymbol{k}_1)\,\delta z_\beta(\boldsymbol{k}_2)}.\end{aligned}\tag{8.23}$$

A consequence of Eq. (8.23) is the equality

$$\frac{\delta}{\delta\widehat{\boldsymbol{f}}(\boldsymbol{k},t-0)}\varphi[t;\boldsymbol{z}]=i\boldsymbol{z}(\boldsymbol{k})\varphi[t;\boldsymbol{z}].\tag{8.24}$$

Average Eq. (8.23) over an ensemble of realizations of random force $\widehat{\boldsymbol{f}}(\boldsymbol{k},t)$. Then, replicating derivation of Eq. (8.7), we obtain that characteristic functional of the velocity field

$$\Phi[t;\boldsymbol{z}(\boldsymbol{k}')]=\Phi[t;\boldsymbol{z}]=\left\langle\varphi[t;\boldsymbol{z}(\boldsymbol{k}')]\right\rangle,$$

satisfies the unclosed variational differential equation

$$\begin{aligned}\frac{\partial}{\partial t}\Phi[t;\boldsymbol{z}]&=\left\langle\dot{\Theta}_t\left[t;\frac{\delta}{i\delta\boldsymbol{f}(\boldsymbol{\kappa},\tau)}\right]\varphi[t;\boldsymbol{z}]\right\rangle\\&-\int d\boldsymbol{k}z_i(\boldsymbol{k})\left\{\frac{1}{2}\int d\boldsymbol{k}_1\int d\boldsymbol{k}_2\Lambda_{i,\alpha\beta}(\boldsymbol{k}_1,\boldsymbol{k}_2,\boldsymbol{k})\frac{\delta^2}{\delta z_\alpha(\boldsymbol{k}_1)\delta z_\beta(\boldsymbol{k}_2)}+\nu k^2\frac{\delta}{\delta z_i(\boldsymbol{k})}\right\}\Phi[t;\boldsymbol{z}],\end{aligned}\tag{8.25}$$

where

$$\dot{\Theta}_t\left[t;\boldsymbol{\psi}(\boldsymbol{\kappa},\tau)\right]=\frac{d}{dt}\ln\left\langle\exp\left\{i\int\limits_0^t d\tau\int d\boldsymbol{\kappa}\boldsymbol{f}(\boldsymbol{\kappa},\tau)\boldsymbol{\psi}(\boldsymbol{\kappa},\tau)\right\}\right\rangle$$

is the derivative of the characteristic functional logarithm of the random field of external force $\widehat{\boldsymbol{f}}(\boldsymbol{k},t)$. The equation for the characteristic functional $\Phi[t;\boldsymbol{z}(\boldsymbol{k}')]$ describes all one-time statistical characteristics of the velocity field.

8.3 Stochastic Integral Equations (Methods of Quantum Field Theory)

Problems discussed in the above sections allow deriving the closed (or unclosed) statistical description in functional space due to the fact that every of these problems can be formulated in terms of some system of differential equations of the first-order with respect to time and initial values at $t=0$. Such systems satisfy the causality condition formulated in Sect. 2.1.2, page 87 which reads as follows: problem solution at instant t depends only on fluctuations of system parameters for times preceding this instant and is independent of fluctuations for posterior times.

Problems formulated in terms of integral equations that cannot be generally reduced to the system of differential equations also can satisfy the causality condition.

However, prior to consider this class of stochastic problems, we dwell on general methods of statistical description of dynamic systems, which are borrowed from the quantum field theory. The essence of these methods consists in constructing a perturbation series for statistical characteristics of quantity of interest and analyzing the result with the use of the methods developed in the quantum field theory. It appears convenient to represent each term of these perturbation series diagrammatically (in the form of the so-called *Feynman diagrams*) and associate every diagram element with certain function or operator; as a result, each diagram corresponds to certain analytical expression. We will not consider the diagram technique as such (for its exhaustive description in the context of statistical problems, see, e.g., monographs [112, 130]); instead, we derive the basic results directly, using the functional methods described above [48, 49].

8.3.1 Linear Integral Equations

The input stochastic equation is the linear integral (or integro-differential) equation for Green's function

$$S(\boldsymbol{r},\boldsymbol{r}') = S_0(\boldsymbol{r},\boldsymbol{r}') + \int d\boldsymbol{r}_1 \int d\boldsymbol{r}_2 \int d\boldsymbol{r}_3 S_0(\boldsymbol{r},\boldsymbol{r}_1)\Lambda(\boldsymbol{r}_1,\boldsymbol{r}_2,\boldsymbol{r}_3)f(\boldsymbol{r}_2)S(\boldsymbol{r}_3,\boldsymbol{r}'), \tag{8.26}$$

where $\boldsymbol{r}$ denotes all arguments of functions S and f, including the index arguments that assume summation instead of integration. It is assumed that function $f(\boldsymbol{r})$ is the random field and function $S_0(\boldsymbol{r},\boldsymbol{r}')$ is Green's function of the problem without parameter fluctuations, i.e., at $f(\boldsymbol{r}) = 0$. In some problems, quantity $\Lambda(\boldsymbol{r}_1,\boldsymbol{r}_2,\boldsymbol{r}_3)$ can be an operator; the notation of Eq. (8.26) assumes that this operator acts on all factors appeared to the right from it. For example, the nonlinear system of ordinary differential equations can be reduced to the equation like Eq. (8.26) by constructing the equivalent linear stochastic partial differential equation (the Liouville equation) whose characteristics correspond to the solution of the nonlinear system. In this case, function S is Green's function of the stochastic Liouville equation and quantity Λ is the differential operator. For problems formulated in terms of systems of linear equations, quantity $\Lambda(\boldsymbol{r}_1,\boldsymbol{r}_2,\boldsymbol{r}_3)$ is a function.

Below, we will assume for simplicity that quantity $\Lambda(\boldsymbol{r}_1,\boldsymbol{r}_2,\boldsymbol{r}_3)$ is not an operator, but a function. The consideration of operator quantity $\Lambda(\boldsymbol{r}_1,\boldsymbol{r}_2,\boldsymbol{r}_3)$ involves only insignificant differences. Indeed, if quantity $\Lambda(\boldsymbol{r}_1,\boldsymbol{r}_2,\boldsymbol{r}_3)$ is an operator, we can reduce the problem to the problem under consideration by introducing delta-functions with arguments coinciding with variables on which this operator is acting and adding the corresponding integrations.

Equation (8.26) can be represented in the symbolic form

$$S = S_0 + S_0 \Lambda f S, \tag{8.27}$$

where integration is assumed with respect to all arguments of function $\Lambda(\{\boldsymbol{r}_i\})$.

We can solve Eq. (8.27) by the iteration method with function $S_0(\boldsymbol{r},\boldsymbol{r}')$ as the zero-order approximation. As a result, we obtain the solution in the form of a series that we represent in the symbolic form again

$$S = \{1 + S_0\Lambda f + \; S_0\Lambda f S_0 \Lambda f + \cdots\}\, S_0 = \sum_{n=0}^{\infty} (S_0\Lambda f)^n S_0. \tag{8.28}$$

The same iteration series represents the solution of the integral equation

$$S = S_0 + S\Lambda f S_0.$$

Consequently Eq. (8.27) is equivalent to the equation

$$S(\boldsymbol{r},\boldsymbol{r}') = S_0(\boldsymbol{r},\boldsymbol{r}') + \int d\boldsymbol{r}_1 \int d\boldsymbol{r}_2 \int d\boldsymbol{r}_3 S(\boldsymbol{r},\boldsymbol{r}_1)\Lambda(\boldsymbol{r}_1,\boldsymbol{r}_2,\boldsymbol{r}_3)f(\boldsymbol{r}_2)S_0(\boldsymbol{r}_3,\boldsymbol{r}'). \tag{8.29}$$

The solution to Eqs. (8.26) and (8.29) is the functional of field $f(\boldsymbol{r})$, i.e.,

$$S(\boldsymbol{r},\boldsymbol{r}') = S[\boldsymbol{r},\boldsymbol{r}'; f(\widetilde{\boldsymbol{r}})].$$

It is not difficult to show that Eqs. (8.26) and (8.29) are equivalent to the variational differential equation in functional space $\{f(\widetilde{\boldsymbol{r}})\}$

$$\frac{\delta}{\delta f(\boldsymbol{r}_0)} S[\boldsymbol{r},\boldsymbol{r}'; f(\widetilde{\boldsymbol{r}})] = \int d\boldsymbol{r}_1 \int d\boldsymbol{r}_2 S[\boldsymbol{r},\boldsymbol{r}_1; f(\widetilde{\boldsymbol{r}})] \Lambda(\boldsymbol{r}_1,\boldsymbol{r}_0,\boldsymbol{r}_2) S[\boldsymbol{r}_2,\boldsymbol{r}'; f(\widetilde{\boldsymbol{r}})] \tag{8.30}$$

with the initial value

$$S[\boldsymbol{r},\boldsymbol{r}'; f(\widetilde{\boldsymbol{r}})]_{f=0} = S_0(\boldsymbol{r},\boldsymbol{r}').$$

Indeed, varying Eq. (8.27) for $S(\boldsymbol{r},\boldsymbol{r}')$ with respect to function $f(\boldsymbol{r}_0)$, we obtain the equation

$$\frac{\delta}{\delta f} S = S_0 \Lambda \delta S + S_0 \Lambda f \frac{\delta}{\delta f} S,$$

where δ denotes the delta-function of the corresponding arguments. The solution to this equation can be represented as the iteration series

$$\frac{\delta}{\delta f} S = \left\{1 + S_0 \Lambda f + (S_0 \Lambda f)^2 + \ \ldots \right\} S_0 \Lambda \delta S.$$

Taking into account the iteration series (8.28) for S, we obtain the desired formula (8.30).

Averaging now the obtained iteration series (8.28) over an ensemble of realizations of field $f(\boldsymbol{r})$, we obtain function $\langle S(\boldsymbol{r},\boldsymbol{r}')\rangle$ in the form of the iteration series dependent on all moment functions of field $f(\boldsymbol{r})$. Rearranging the terms of this series, we can then express the right-hand side of the expansion in terms of function $\langle S(\boldsymbol{r},\boldsymbol{r}')\rangle$ itself. This rearrangement produces new unknown functions specified by the corresponding iteration series and called, by analogy with the quantum field theory, the *mass* and *vertex functions.*

Consider instead of Eq. (8.26) the auxiliary equation

$$\begin{aligned} S[\boldsymbol{r},\boldsymbol{r}'; f+\eta] &= S_0(\boldsymbol{r},\boldsymbol{r}') \\ &+ \int d\boldsymbol{r}_1 \int d\boldsymbol{r}_2 \int d\boldsymbol{r}_3 S_0(\boldsymbol{r},\boldsymbol{r}_1) \Lambda(\boldsymbol{r}_1,\boldsymbol{r}_2,\boldsymbol{r}_3)[f(\boldsymbol{r}_2)+\eta(\boldsymbol{r}_2)] S[\boldsymbol{r}_3,\boldsymbol{r}'; f+\eta], \end{aligned} \tag{8.31}$$

where $\eta(\boldsymbol{r})$ is arbitrary deterministic function. We can find the desired function $S(\boldsymbol{r},\boldsymbol{r}')$ by setting $\eta(\boldsymbol{r}) = 0$ in Eq. (8.31), i.e.,

$$S(\boldsymbol{r},\boldsymbol{r}') = S[\boldsymbol{r},\boldsymbol{r}'; f(\boldsymbol{r})] = S[\boldsymbol{r},\boldsymbol{r}'; f(\boldsymbol{r}) + \eta(\boldsymbol{r})]_{\eta=0}.$$

Let us average Eq. (8.31). Splitting the correlator $\langle fS\rangle$ by formula (7.2), page 153, we obtain the equation

$$G[\boldsymbol{r},\boldsymbol{r}';\eta] = S_0(\boldsymbol{r},\boldsymbol{r}') + \int d\boldsymbol{r}_1 \int d\boldsymbol{r}_2 \int d\boldsymbol{r}_3 S_0(\boldsymbol{r},\boldsymbol{r}_1)\Lambda(\boldsymbol{r}_1,\boldsymbol{r}_2,\boldsymbol{r}_3)\eta(\boldsymbol{r}_2)G[\boldsymbol{r}_3,\boldsymbol{r}';\eta]$$
$$+ \int d\boldsymbol{r}_1 \int d\boldsymbol{r}_2 \int d\boldsymbol{r}_3 S_0(\boldsymbol{r},\boldsymbol{r}_1)\Lambda(\boldsymbol{r}_1,\boldsymbol{r}_2,\boldsymbol{r}_3) \left\langle \Omega_{\boldsymbol{r}_2}\left[\frac{\delta}{i\delta f(\boldsymbol{r})}\right] S[\boldsymbol{r}_3,\boldsymbol{r}';f+\eta]\right\rangle . \tag{8.32}$$

Here, functional $\Omega_{\boldsymbol{r}}\left[v(\boldsymbol{r})\right]$ is given by the formula

$$\Omega_{\boldsymbol{r}} = \frac{\delta}{i\delta v(\boldsymbol{r})}\Theta\left[v(\boldsymbol{r})\right],$$

functional

$$\Theta\left[v(\boldsymbol{r})\right] = \ln\left\langle \exp\left\{i\int d\boldsymbol{r} f(\boldsymbol{r})v(\boldsymbol{r})\right\}\right\rangle$$

is the logarithm of the characteristic functional of random field $f(\boldsymbol{r})$, and

$$G[\boldsymbol{r},\boldsymbol{r}';\eta(\boldsymbol{r})] = \langle S[\boldsymbol{r},\boldsymbol{r}';f(\boldsymbol{r})+\eta(\boldsymbol{r})]\rangle .$$

Taking into account the fact that functional $S[\boldsymbol{r},\boldsymbol{r}';f(\boldsymbol{r})+\eta(\boldsymbol{r})]$ is the functional of argument $f(\boldsymbol{r})+\eta(\boldsymbol{r})$, we can replace variational differentiation with respect to $f(\boldsymbol{r})$ by differentiation with respect to $\eta(\boldsymbol{r})$ and rewrite Eq. (8.32) in the form of the closed variational differential equation similar to the *Schwinger equation* of the quantum field theory

$$G[\boldsymbol{r},\boldsymbol{r}';\eta] = S_0(\boldsymbol{r},\boldsymbol{r}') + \int d\boldsymbol{r}_1 \int d\boldsymbol{r}_2 \int d\boldsymbol{r}_3 S_0(\boldsymbol{r},\boldsymbol{r}_1)\Lambda(\boldsymbol{r}_1,\boldsymbol{r}_2,\boldsymbol{r}_3)\eta(\boldsymbol{r}_2)G[\boldsymbol{r}_3,\boldsymbol{r}';\eta]$$
$$+ \int d\boldsymbol{r}_1 \int d\boldsymbol{r}_2 \int d\boldsymbol{r}_3 S_0(\boldsymbol{r},\boldsymbol{r}_1)\Lambda(\boldsymbol{r}_1,\boldsymbol{r}_2,\boldsymbol{r}_3)\Omega_{\boldsymbol{r}_2}\left[\frac{\delta}{i\delta\eta(\boldsymbol{r})}\right] G[\boldsymbol{r}_3,\boldsymbol{r}';\eta]. \tag{8.33}$$

We can solve Eq. (8.33) for functional $G[\boldsymbol{r},\boldsymbol{r}';\eta(\boldsymbol{r})]$ by the iteration method with function $S_0(\boldsymbol{r},\boldsymbol{r}')$ as the zero-order approximation. Setting $\eta(\boldsymbol{r}) = 0$ in the resulting expansion, we obtain the iteration series for function $\langle S(\boldsymbol{r},\boldsymbol{r}')\rangle$.

To simplify further presentation, we rewrite Eq. (8.33) in the symbolic form (the corresponding complete expressions can be easily restored at every step)

$$G = S_0 + S_0\Lambda\left(\eta + \Omega\left[\frac{\delta}{i\delta\eta}\right]\right)G. \tag{8.34}$$

We introduce the inverse functional G^{-1}, such that

$$G^{-1}G = 1, \quad GG^{-1} = 1. \tag{8.35}$$

Here, the unity is understood as the corresponding delta-function. In addition, we introduce the functional

$$\Gamma = -\frac{\delta G^{-1}}{\delta\eta}, \tag{8.36}$$

which we call the *vertex functional.*

Varying Eq. (8.35) with respect to field η, we obtain the equality

$$\frac{\delta G}{\delta \eta} = G\Gamma G, \tag{8.37}$$

whose substitution in Eq. (8.34) results in the equation

$$G = S_0 + S_0 \Lambda \eta G + S_0 Q G. \tag{8.38}$$

We call the quantity

$$Q = \Lambda \left(\Omega \left[\frac{\delta}{i\delta\eta} \right] G \right) G^{-1} \tag{8.39}$$

the *mass functional.*

Multiplying now Eq. (8.38) by G^{-1} from the right and by S_0^{-1} from the left (and integrating over the corresponding arguments), we obtain the equation for functional G^{-1}

$$S_0^{-1} - G^{-1} = \Lambda \eta + Q. \tag{8.40}$$

Varying Eq. (8.40) with respect to field η, we obtain the equation for functional Γ

$$\Gamma = \Lambda + \frac{\delta}{\delta \eta} Q. \tag{8.41}$$

The system of functional equations (8.38) and (8.41) is closed in functionals G and Γ. An additional point is Eq. (8.37) that relates the solutions to these equations. We can solve Eq. (8.41) for Γ by the iteration method with quantity Λ as the zero-order approximation. If we use formula (8.37) to express variational derivatives of functional G with respect to η, we obtain the integral equations for Γ and G with infinite number of terms every of which includes no functionals other than G and Γ. Setting now $\eta(\boldsymbol{r}) = 0$, we can obtain the closed system of integral equations. In particular, Eq. (8.38) assumes the following form

$$\langle S \rangle = S_0 + S_0 Q \langle S \rangle, \quad \langle S \rangle = S_0 + \langle S \rangle Q S_0, \tag{8.42}$$

and is called the *Dyson equation.*

Now, we consider in more detail the case of the Gaussian random field $f(\boldsymbol{r})$ with correlation function $B(\boldsymbol{r}, \boldsymbol{r}') = \langle f(\boldsymbol{r}) f(\boldsymbol{r}') \rangle$. In this case

$$\Omega_{\boldsymbol{r}} [v(\boldsymbol{r})] = i \int d\boldsymbol{r}' B(\boldsymbol{r}, \boldsymbol{r}') v(\boldsymbol{r}'),$$

the mass functional assumes the form

$$Q = \Lambda B G \Gamma, \tag{8.43}$$

and Eqs. (8.38) and (8.41) assume the form

$$G = S_0 + S_0 \Lambda \eta G + S_0 \Lambda B G \Gamma G,$$
$$\Gamma = \Lambda + \Lambda B G \Gamma G \Gamma + \Lambda B G \frac{\delta \Gamma}{\delta \eta}. \tag{8.44}$$

Setting now $\eta = 0$, we obtain the closed system of equations

$$\langle S \rangle = S_0 + S_0 Q \langle S \rangle \quad \text{(Dyson equation)},$$
$$Q = \Lambda B \langle S \rangle \widetilde{\Gamma}, \tag{8.45}$$
$$\widetilde{\Gamma} = \Lambda + \Lambda B \langle S \rangle \widetilde{\Gamma} \langle S \rangle \widetilde{\Gamma} + \cdots \quad (\widetilde{\Gamma} = \Gamma|_{\eta=0}).$$

System of equations (8.45) is very complicated and low understood. The simplest way of its simplification consists in cutting the infinite series in the equation for $\widetilde{\Gamma}$. If we content ourselves with the first term, we obtain the closed nonlinear equation (the *Kraichnan approximation*)

$$\langle S \rangle = S_0 + S_0 Q_{\mathrm{Kr}} \langle S \rangle , \quad Q_{\mathrm{Kr}} = \Lambda B \langle S \rangle \Lambda. \tag{8.46}$$

Further, if we replace $\langle S \rangle$ in the expression for the mass function Q_{Kr} with S_0, we obtain the linear equation (the *Bourret approximation*)

$$\langle S \rangle = S_0 + S_0 \Lambda B S_0 \Lambda \langle S \rangle . \tag{8.47}$$

Functional Γ and, consequently, function $\widetilde{\Gamma}$ are closely related to quantity $\langle S \Lambda S \rangle$. Indeed, in view of formula (8.30), we can rewrite expression (8.37) at $\eta = 0$ in the form

$$\langle S \Lambda S \rangle = \langle S \rangle \widetilde{\Gamma} \langle S \rangle . \tag{8.48}$$

Thus, different approximations for function $\widetilde{\Gamma}$ are equivalent to certain hypotheses about splitting the correlation $\langle S \Lambda S \rangle$. The Kraichnan approximation (8.46) corresponds to the equality

$$\langle S \Lambda S \rangle = \langle S \rangle \Lambda \langle S \rangle ,$$

while the Bourret approximation (8.47) is equivalent to the requirement that

$$\langle S \Lambda S \rangle = S_0 \Lambda \langle S \rangle .$$

Splitting the correlation $\langle S \Lambda S \rangle$ by formula (7.9), page 168, we obtain the general operator expression

$$\langle S \Lambda S \rangle = G \left[i B \frac{\delta}{\delta \eta} \right] \Lambda G[\eta]|_{\eta=0},$$

which is in essence equivalent to the introduction of the vertex function.

Note that the knowledge of functional $G[\boldsymbol{r},\boldsymbol{r}';\eta(\boldsymbol{r})]$ is equivalent, in the case of the Gaussian field $f(\boldsymbol{r})$, to the knowledge of the functional

$$\Phi[\boldsymbol{r},\boldsymbol{r}';v(\boldsymbol{r})] = \left\langle S(\boldsymbol{r},\boldsymbol{r}')e^{i\int d\boldsymbol{r} f(\boldsymbol{r})v(\boldsymbol{r})}\right\rangle$$

describing all statistical correlations between function $S(\boldsymbol{r},\boldsymbol{r}')$ and field $f(\boldsymbol{r})$. Indeed, according to formula (7.13), page 169, we can rewrite functional $\Phi[\boldsymbol{r},\boldsymbol{r}';v(\boldsymbol{r})]$ in the form

$$\Phi[\boldsymbol{r},\boldsymbol{r}';v(\boldsymbol{r})] = \left\langle e^{i\int d\boldsymbol{r} f(\boldsymbol{r})v(\boldsymbol{r})}\right\rangle \left\langle S\left[\boldsymbol{r},\boldsymbol{r}';f(\boldsymbol{r})+i\int d\boldsymbol{r}_1 B(\boldsymbol{r},\boldsymbol{r}_1)\right]\right\rangle,$$

wherefrom we obtain the equality

$$\Phi[\boldsymbol{r},\boldsymbol{r}';v(\boldsymbol{r})] = \left\langle e^{i\int d\boldsymbol{r} f(\boldsymbol{r})v(\boldsymbol{r})}\right\rangle G\left[\boldsymbol{r},\boldsymbol{r}';i\int d\boldsymbol{r}_1 B(\boldsymbol{r},\boldsymbol{r}_1)\right].$$

To complete the picture, we dwell now on the so-called renormalization method. The point is that, even if the mass function is known, the Dyson equation (8.45) is a very complicated integral equation, which only rarely can be solved analytically. At the same time, the Dyson equation with the simplified mass function can be easily solved in a number of cases. The renormalization method lies in rearranging the Dyson equation into the integral equation in which function $S_0(\boldsymbol{r},\boldsymbol{r}')$ is replaced with the solution to the simplified problem.

Denote $\widetilde{S}$ the solution of the Dyson equation (8.45) with the simplified mass function $\widetilde{Q}$. Then, function $\widetilde{S}$ will satisfy the equation

$$\widetilde{S} = S_0 + S_0\widetilde{Q}\widetilde{S}. \tag{8.49}$$

In view of the fact that Eq. (8.49) is linear in $\widetilde{S}$, it is obvious that it can be rewritten in the form

$$\widetilde{S} = S_0 + \widetilde{S}\widetilde{Q}S_0 = (1+\widetilde{S}\widetilde{Q})S_0, \tag{8.50}$$

where 1 is the unit operator.

To exclude function $S_0(\boldsymbol{r},\boldsymbol{r}')$ from the Dyson equation (8.45), we apply operator $(1+\widetilde{S}\widetilde{Q})$ to it. Then, we obtain, in view of Eq. (8.50), the equation

$$\langle S\rangle = \widetilde{S} + \widetilde{S}\{Q-\widetilde{Q}\}\langle S\rangle. \tag{8.51}$$

Now, we can solve Eq. (8.51) by the iteration method with function $\widetilde{S}$ as the zero-order approximation.

At $\widetilde{Q} = 0$, function $\widetilde{S} = S_0$, and we turn back to the Dyson equation (8.45). It is obvious that the above derivation of Eq. (8.51) holds not only for the Gaussian field $f(\boldsymbol{r})$, but for any arbitrary field $f(\boldsymbol{r})$, because the form of the Dyson equation is independent of the type of field $f(\boldsymbol{r})$.

Now, we dwell on the general-form Dyson equation (8.38). Note that we can represent functional $\Omega[v]$ in terms of the Taylor series in powers of v

$$\Omega[v] = \sum_{n=1}^{\infty} \frac{1}{n!} K_{n+1} v^n,$$

where K_n are the cumulant functions of random field $f(\boldsymbol{r})$. As a consequence, we can represent the mass functional (8.39) in the form

$$Q = \Lambda \sum_{n=0}^{\infty} \frac{1}{n!} K_{n+1} \left\{ \frac{\delta^n}{\delta \eta^n} G \right\} G^{-1},$$

where variational derivatives of functional G with respect to η are calculated by formula (8.37). In this case, the Dyson equation has very complicated structure. The standard ways of simplifying this equation are quite similar to those used in the case of the Gaussian parameter fluctuations.

If we set $\Gamma = \Lambda$, expression (8.37) assumes the form

$$\frac{\delta G}{\delta \eta} = G \Lambda G$$

and, consequently,

$$\frac{\delta^n G}{\delta \eta^n} = n!(G\Lambda)^n G = \int_0^{\infty} d\lambda e^{-\lambda} (G\lambda\Lambda)^n G.$$

In this case, we arrive at the generalized Kraichnan equation

$$\langle S \rangle = S_0 + S_0 Q_{\text{Kr}} \langle S \rangle ,$$

$$Q_{\text{Kr}} = \Lambda \sum_{n=0}^{\infty} K_{n+1} \{\langle S \rangle \Lambda\}^n = \Lambda \int_0^{\infty} d\lambda e^{-\lambda} \Omega[\langle S \rangle \lambda\Lambda]. \tag{8.52}$$

If we replace $\langle S \rangle$ in operator Q_{Kr} in Eq. (8.52) with S_0, we obtain the generalized Bourret equation

$$Q_{\text{B}} = \Lambda \int_0^{\infty} d\lambda e^{-\lambda} \Omega[S_0 \lambda\Lambda],$$

which coincides with the so-called *one-group approximation for the Dyson equation.*

Above, we considered the derivation of the equation for averaged Green's function (the Dyson equation). In a similar way, we can derive the equation for the correlation function

$$\Gamma(\boldsymbol{r},\boldsymbol{r}';\boldsymbol{r}_1,\boldsymbol{r}_1') = \langle S(\boldsymbol{r},\boldsymbol{r}')S^*(\boldsymbol{r}_1,\boldsymbol{r}_1')\rangle\,.$$

With this goal in view, we multiply Eq. (8.26) by $S^*(\boldsymbol{r}_1,\boldsymbol{r}_1')$ and average the result over an ensemble of realizations of random field $f(\boldsymbol{r})$. In this way, we obtain the equation

$$\Gamma = S_0\,\langle S^*\rangle + S_0\Lambda\,\langle fSS^*\rangle\,. \tag{8.53}$$

Taking into account the Dyson equation (8.42)

$$\langle S\rangle = 1+\langle S\rangle\,QS_0,$$

we apply operator $\{1+\langle S\rangle\,Q\}$ to Eq. (8.53). As a result, we obtain the equation

$$\Gamma = \langle S\rangle\,\langle S^*\rangle + \langle S\rangle\,\{\Lambda\,\langle fSS^*\rangle - Q\Gamma\}\,. \tag{8.54}$$

In standard notation, Eq. (8.54) assumes the form

$$\begin{aligned}\Gamma(\boldsymbol{r},\boldsymbol{r}';\boldsymbol{r}_1,\boldsymbol{r}_1') &= \langle S(\boldsymbol{r},\boldsymbol{r}')\rangle\,\langle S^*(\boldsymbol{r}_1,\boldsymbol{r}_1')\rangle\\ &+\int d\boldsymbol{r}_2\int d\boldsymbol{r}_2'\int d\boldsymbol{r}_3\int d\boldsymbol{r}_3'\,\langle S(\boldsymbol{r},\boldsymbol{r}_2)\rangle\,\langle S^*(\boldsymbol{r}_1,\boldsymbol{r}_2')\rangle\,K(\boldsymbol{r}_2,\boldsymbol{r}_2';\boldsymbol{r}_3,\boldsymbol{r}_3')\Gamma(\boldsymbol{r}_3,\boldsymbol{r}';\boldsymbol{r}_3',\boldsymbol{r}_1'),\end{aligned} \tag{8.55}$$

and is called the *Bete–Salpeter equation* . Function $K(\boldsymbol{r}_2,\boldsymbol{r}_2';\boldsymbol{r}_3,\boldsymbol{r}_3')$ is called the *kernel of the intensity operator.*

The simplest approximation to this equation — the so-called *ladder approximation* — corresponds to function $K(\boldsymbol{r}_2,\boldsymbol{r}_2';\boldsymbol{r}_3,\boldsymbol{r}_3')$ of the form

$$K(\boldsymbol{r}_2,\boldsymbol{r}_2';\boldsymbol{r}_3,\boldsymbol{r}_3') = \delta(\boldsymbol{r}_2-\boldsymbol{r}_3)\delta(\boldsymbol{r}_2'-\boldsymbol{r}_3')B_f(\boldsymbol{r}_2,\boldsymbol{r}_2'), \tag{8.56}$$

where $B_f(\boldsymbol{r}_2,\boldsymbol{r}_2') = \langle f(\boldsymbol{r}_2)f(\boldsymbol{r}_2')\rangle$ is the correlation function of field $f(\boldsymbol{r})$.

8.3.2 Nonlinear Integral Equations

Consider now integral equation (8.26) extended to the case of an equation with quadratic nonlinearity. There are two possible cases. In the first — simplest — case, the solution can be expressed through the integral of the solution to the linear equation with respect to an auxiliary parameter; the second — more complex — case describes the space-time structure of hydrodynamic turbulence and is described by the integro-differential equation (1.172), page 70.

The Simplest Nonlinear Integral Equation

Consider the nonlinear equation

$$S(\boldsymbol{r},\boldsymbol{r}') = S_0(\boldsymbol{r},\boldsymbol{r}') + \int d\boldsymbol{r}_1 \int d\boldsymbol{r}_2 \int d\boldsymbol{r}_3 S(\boldsymbol{r},\boldsymbol{r}_1)\Lambda(\boldsymbol{r}_1,\boldsymbol{r}_2,\boldsymbol{r}_3)f(\boldsymbol{r}_2)S(\boldsymbol{r}_3,\boldsymbol{r}'). \tag{8.57}$$

Along with this equation we draw once again Eq. (8.26) and mark its solution by index Λ

$$S_\Lambda(\boldsymbol{r},\boldsymbol{r}')=S_0(\boldsymbol{r},\boldsymbol{r}') + \int d\boldsymbol{r}_1 \int d\boldsymbol{r}_2 \int d\boldsymbol{r}_3 S_0(\boldsymbol{r},\boldsymbol{r}_1)\Lambda(\boldsymbol{r}_1,\boldsymbol{r}_2,\boldsymbol{r}_3)f(\boldsymbol{r}_2)S_\Lambda(\boldsymbol{r}_3,\boldsymbol{r}'). \tag{8.58}$$

As was mentioned earlier, the solution to Eq. (8.58) can be represented as the iteration series

$$S_\Lambda = \sum_{n=0}^{\infty} (S_0 \Lambda f)^n S_0.$$

It is obvious that the solution to integral equation (8.57) has a similar iteration structure

$$S = \sum_{n=0}^{\infty} A_n (S_0 \Lambda f)^n S_0 \tag{8.59}$$

with an additional numeric parameter A_n. This parameter can be easily found from the quadratic equation

$$y = 1 + \lambda y^2$$

whose solution is

$$y = \frac{1-\sqrt{1-4\lambda}}{2\lambda} = \sum_{n=0}^{\infty} A_n \lambda^n.$$

Consequently,

$$A_n = 2^n \frac{(2n-1)!!}{(n+1)!} = \frac{2^{2n+1}}{\pi} \frac{\Gamma(n+1/2)\Gamma(3/2)}{\Gamma(n+2)} = \frac{2^{2n+1}}{\pi} B\left(n+\frac{1}{2},\frac{3}{2}\right),$$

where $B(\gamma,\delta)$ is the beta-function whose integral representation is

$$B(\gamma,\delta) = \int_0^1 dp p^{\gamma-1}(1-p)^{\delta-1}.$$

As a result, we have

$$A_n = \frac{2}{\pi} \int_0^1 dp\, (4p)^n \sqrt{\frac{1-p}{p}}.$$

Substituting this expression in Eq. (8.59), we obtain

$$S = \frac{2}{\pi}\int\limits_0^1 dp\sqrt{\frac{1-p}{p}}\sum_{n=0}^{\infty}(S_0 4p\Lambda f)^n S_0 = \frac{2}{\pi}\int\limits_0^1 dp\sqrt{\frac{1-p}{p}}S_{4p\Lambda}. \tag{8.60}$$

Thus, the solution to the nonlinear equation (8.57) is expressed through the solution to the linear equation (8.58) as the integral with respect to an auxiliary parameter [113].

Space-Time Description of Hydrodynamic Turbulence

Consider now the nonlinear integral equation (1.172), page 70 for the space-time harmonics of the turbulent velocity field

$$(i\omega+\nu\boldsymbol{k}^2)\widehat{u}_i(\boldsymbol{K})+\frac{i}{2}\int d^4\boldsymbol{K}_1\int d^4\boldsymbol{K}_2\Lambda_i^{\alpha\beta}(\boldsymbol{K}_1,\boldsymbol{K}_2,\boldsymbol{K})\,\widehat{u}_\alpha(\boldsymbol{K}_1)\,\widehat{u}_\beta(\boldsymbol{K}_2) = \widehat{f}_i(\boldsymbol{K}), \tag{8.61}$$

where

$$\Lambda_i^{\alpha\beta}(\boldsymbol{K}_1,\boldsymbol{K}_2,\boldsymbol{K}) = \frac{1}{(2\pi)^3}\left\{k_\alpha\Delta_{i\beta}(\boldsymbol{k})+k_\beta\Delta_{i\alpha}(\boldsymbol{k})\right\}\delta(\boldsymbol{k}_1+\boldsymbol{k}_2-\boldsymbol{k})\delta(\omega_1+\omega_2-\omega),$$

$$\Delta_{ij}(\boldsymbol{k}) = \delta_{ij}-\frac{k_ik_j}{\boldsymbol{k}^2}\quad (i,\,\alpha,\,\beta\,=\,1,\,2,\,3).$$

Here, $\boldsymbol{K}$ is the four-dimensional wave vector with components $\{\boldsymbol{k},\omega\}$ and $\widehat{f}_i(\boldsymbol{K})$ are the space-time Fourier harmonics of external forces; in view of the fact that $u_i(\boldsymbol{x},t)$ is the real-valued field, we have

$$\widehat{u}_i^*(\boldsymbol{K}) = \widehat{u}_i(-\boldsymbol{K}).$$

Equation (8.61) can be juxtaposed with the equivalent linear variational differential Hopf equation

$$(i\omega+\nu\boldsymbol{k}^2)\frac{\delta}{\delta z_i(\boldsymbol{K})}\varphi[\boldsymbol{z}] = i\widehat{f}_i(\boldsymbol{K})\,\varphi[\boldsymbol{z}]$$
$$-\frac{1}{2}\int d^4\boldsymbol{K}_1\int d^4\boldsymbol{K}_2\Lambda_i^{\alpha\beta}(\boldsymbol{K}_1,\boldsymbol{K}_2,\boldsymbol{K})\frac{\delta^2\varphi[\boldsymbol{z}]}{\delta z_\alpha(\boldsymbol{K}_1)\,\delta z_\beta(\boldsymbol{K}_2)} \tag{8.62}$$

for the functional

$$\varphi[\boldsymbol{z}] = \exp\left\{i\int d^4\boldsymbol{K}'\widehat{\boldsymbol{u}}(\boldsymbol{K}')\,\boldsymbol{z}(\boldsymbol{K}')\right\}. \tag{8.63}$$

Averaging Eq. (8.62) over an ensemble of realizations of the external force field $\widehat{\boldsymbol{f}}(\boldsymbol{K})$, we obtain the equation

$$(i\omega + \nu \boldsymbol{k}^2)\frac{\delta}{\delta z_i(\boldsymbol{K})}\Phi[\boldsymbol{z}] = i\left\langle \widehat{f}_i(\boldsymbol{K})\,\varphi[\boldsymbol{z}]\right\rangle$$
$$-\frac{1}{2}\int d^4\boldsymbol{K}_1\int d^4\boldsymbol{K}_2 \Lambda_i^{\alpha\beta}(\boldsymbol{K}_1,\boldsymbol{K}_2,\boldsymbol{K})\frac{\delta^2\Phi[\boldsymbol{z}]}{\delta z_\alpha(\boldsymbol{K}_1)\,\delta z_\beta(\boldsymbol{K}_2)}. \quad (8.64)$$

for the characteristic functional

$$\Phi[\boldsymbol{z}] = \langle\varphi[\boldsymbol{z}]\rangle .$$

We will now assume that the random field of external force $\widehat{\boldsymbol{f}}(\boldsymbol{K})$ is the Gaussian field homogeneous in space and stationary in time whose different statistical characteristics are determined by the space-time spectral function

$$\left\langle \widehat{f}_i(\boldsymbol{K}_1)\widehat{f}_j(\boldsymbol{K}_2)\right\rangle = \frac{1}{2}\delta^4(\boldsymbol{K}_1+\boldsymbol{K}_2)\,F_{ij}(\boldsymbol{K}_1).$$

Because $\widehat{\boldsymbol{f}}(\boldsymbol{K})$ is the nondivergent (solenoidal) field, we have

$$F_{ij}(\boldsymbol{K}) = \Delta_{ij}(\boldsymbol{k})F(\boldsymbol{K}),$$

where $F(\boldsymbol{K})$ is the space-time spectrum of external forces.

Splitting the correlator in the right-hand side of Eq. (8.64) by the Furutsu–Novikov formula, we can rewrite Eq. (8.64) in the form

$$(i\omega + \nu \boldsymbol{k}^2)\frac{\delta}{\delta z_i(\boldsymbol{K})}\Phi[\boldsymbol{z}] = -\frac{1}{2}F_{ij}(\boldsymbol{K})\int d^4\boldsymbol{K}_1 z_\alpha(\boldsymbol{K}_1)\,G_{\alpha j}[\boldsymbol{K}_1,-\boldsymbol{K};\boldsymbol{z}]$$
$$-\frac{1}{2}\int d^4\boldsymbol{K}_1\int d^4\boldsymbol{K}_2 \Lambda_i^{\alpha\beta}(\boldsymbol{K}_1,\boldsymbol{K}_2,\boldsymbol{K})\frac{\delta^2\Phi[\boldsymbol{z}]}{\delta z_\alpha(\boldsymbol{K}_1)\,\delta z_\beta(\boldsymbol{K}_2)}, \quad (8.65)$$

where we introduced the new functional

$$G_{ij}\left[\boldsymbol{K},\boldsymbol{K}';\boldsymbol{z}\right] = \left\langle \frac{\delta u_i(\boldsymbol{K})}{\delta \widehat{f}_j(\boldsymbol{K}')}\varphi[\boldsymbol{z}]\right\rangle .$$

We can obtain the equation for quantity $\delta u_i(\boldsymbol{K})/\delta\widehat{f}_j(\boldsymbol{K}')$ by varying Eq. (8.61),

$$(i\omega+\nu\boldsymbol{k}^2)\frac{\delta\widehat{u}_i(\boldsymbol{K})}{\delta\widehat{f}_j(\boldsymbol{K}')}+i\int d^4\boldsymbol{K}_1\int d^4\boldsymbol{K}_2\Lambda_i^{\alpha\beta}(\boldsymbol{K}_1,\boldsymbol{K}_2,\boldsymbol{K})\,\widehat{u}_\alpha(\boldsymbol{K}_1)\frac{\delta\widehat{u}_\beta(\boldsymbol{K}_2)}{\delta\widehat{f}_j(\boldsymbol{K}')}$$
$$=\delta_{ij}\delta^4\left(\boldsymbol{K}-\boldsymbol{K}'\right), \quad (8.66)$$

As a consequence, functional $G_{ij}\left[\boldsymbol{K},\boldsymbol{K}';\boldsymbol{z}\right]$ satisfies the equation

$$(i\omega+\nu\boldsymbol{k}^2)G_{ij}\left[\boldsymbol{K},\boldsymbol{K}';\boldsymbol{z}\right]=\delta_{ij}\delta^4\left(\boldsymbol{K}-\boldsymbol{K}'\right)\Phi[\boldsymbol{z}]$$
$$-\int d^4\boldsymbol{K}_1\int d^4\boldsymbol{K}_2\Lambda_i^{\alpha\beta}\left(\boldsymbol{K}_1,\boldsymbol{K}_2,\boldsymbol{K}\right)\frac{\delta}{\delta z_\alpha(\boldsymbol{K}_1)}G_{\beta j}\left[\boldsymbol{K}_2,\boldsymbol{K}';\boldsymbol{z}\right]. \quad (8.67)$$

The system of functional equations (8.65) and (8.67) for $\Phi[\boldsymbol{z}]$ and $G_{ij}\left[\boldsymbol{K},\boldsymbol{K}';\boldsymbol{z}\right]$ is closed and completely governs the statistical description of the velocity field [42] (see also [46, 49, 112]).

It is easy to show that average energy income of the velocity field harmonics due to work of the external force is given by the expression

$$\left\langle\widehat{u}_i\left(\boldsymbol{K}\right)\widehat{f}_j\left(\boldsymbol{K}'\right)\right\rangle=\frac{1}{2}F_{jl}\left(\boldsymbol{K}'\right)\left\langle\frac{\delta\widehat{u}_i\left(\boldsymbol{K}\right)}{\delta\widehat{f}_l\left(-\boldsymbol{K}'\right)}\right\rangle=\frac{1}{2}F_{j\alpha}\left(\boldsymbol{K}'\right)G_{i\alpha}\left[\boldsymbol{K},-\boldsymbol{K}';0\right], \quad (8.68)$$

which defines the physical meaning of quantity $G_{ij}\left[\boldsymbol{K},\boldsymbol{K}';\boldsymbol{z}\right]$ as the functional that describes correlations between the velocity field and the rate of energy income (force power). Here, quantity $\delta\widehat{u}_i\left(\boldsymbol{K}\right)/\delta\widehat{f}_l\left(\boldsymbol{K}'\right)$ can be considered as some sort of Green's function for Eq. (8.61). Indeed, if we add some deterministic force $\boldsymbol{\eta}\left(\boldsymbol{K}\right)$ to the right-hand side of Eq. (8.61), then the solution of the resulting equation will be a functional of argument $\widehat{\boldsymbol{f}}\left(\boldsymbol{K}\right)+\boldsymbol{\eta}\left(\boldsymbol{K}\right)$, i.e.,

$$\widehat{u}_i\left(\boldsymbol{K}\right)=\widehat{u}_i\left[\boldsymbol{K};\widehat{\boldsymbol{f}}\left(\boldsymbol{K}'\right)+\boldsymbol{\eta}\left(\boldsymbol{K}'\right)\right]. \quad (8.69)$$

Expand solution (8.69) in the functional series in $\boldsymbol{\eta}\left(\boldsymbol{K}\right)$

$$\widehat{u}_i\left(\boldsymbol{K}\right)=\widehat{u}_i\left[\boldsymbol{K};\widehat{\boldsymbol{f}}\left(\boldsymbol{K}'\right)\right]+\int d^4\boldsymbol{K}'\left.\frac{\delta\widehat{u}_i\left(\boldsymbol{K}\right)}{\delta\eta_j\left(\boldsymbol{K}'\right)}\right|_{\boldsymbol{\eta}=0}\eta_j\left(\boldsymbol{K}'\right)+\cdots$$
$$=\widehat{u}_i^0\left(\boldsymbol{K}\right)+\int d^4\boldsymbol{K}'\frac{\delta\widehat{u}_i\left(\boldsymbol{K}\right)}{\delta\widehat{f}_j\left(\boldsymbol{K}'\right)}\eta_j\left(\boldsymbol{K}'\right)+\cdots. \quad (8.70)$$

The first term of the expansion is simply the solution of problem (8.61). The second term describes the dynamic system response on the infinitely small deterministic force, and quantity $\delta\widehat{u}_i\left(\boldsymbol{K}\right)/\delta\widehat{f}_j\left(\boldsymbol{K}'\right)$ appears the analog of Green's function for linear systems. Averaging Eq. (8.70) over an ensemble of realizations of random forces and taking into account the equality $\left\langle\widehat{u}_i^0\left(\boldsymbol{K}\right)\right\rangle=0$, we obtain the expression for the system average response

$$\left\langle\widehat{u}_i\left(\boldsymbol{K}\right)\right\rangle=\int d^4\boldsymbol{K}'\left\langle\frac{\delta\widehat{u}_i\left(\boldsymbol{K}\right)}{\delta\widehat{f}_j\left(\boldsymbol{K}'\right)}\right\rangle\eta_j\left(\boldsymbol{K}'\right)+\cdots.$$

Turn back to the system of equations (8.65) and (8.67) and represent functionals $\Phi[\boldsymbol{z}]$ and $G_{ij}\left[\boldsymbol{K},\boldsymbol{K}';\boldsymbol{z}\right]$ in the form

$$\Phi[\boldsymbol{z}] = e^{\phi[\boldsymbol{z}]}, \quad G_{ij}\left[\boldsymbol{K},\boldsymbol{K}';\boldsymbol{z}\right] = S_{ij}\left[\boldsymbol{K},\boldsymbol{K}';\boldsymbol{z}\right]e^{\phi[\boldsymbol{z}]}.$$

Then, equations for functionals $\phi[\boldsymbol{z}]$ and $S_{ij}\left[\boldsymbol{K},\boldsymbol{K}';\boldsymbol{z}\right]$ assume the form

$$\begin{aligned}
&\frac{\delta}{\delta z_i(\boldsymbol{K})}\phi[\boldsymbol{z}] \\
&= -\frac{1}{2}\int d^4\boldsymbol{K}'\int d^4\boldsymbol{K}_1 S^0_{i\gamma}(\boldsymbol{K},\boldsymbol{K}')F_{\gamma j}(\boldsymbol{K}')z_\alpha(\boldsymbol{K}_1)G_{\alpha j}\left[\boldsymbol{K}_1,-\boldsymbol{K};\boldsymbol{z}\right] \\
&\quad -\frac{1}{2}\int d^4\boldsymbol{K}'\int d^4\boldsymbol{K}_1\int d^4\boldsymbol{K}_2 S^0_{i\gamma}(\boldsymbol{K},\boldsymbol{K}')\Lambda^{\alpha\beta}_\gamma(\boldsymbol{K}_1,\boldsymbol{K}_2,\boldsymbol{K}) \\
&\qquad\qquad \times\left\{\frac{\delta^2\phi[\boldsymbol{z}]}{\delta z_\alpha(\boldsymbol{K}_1)\,\delta z_\beta(\boldsymbol{K}_2)}+\frac{\delta^2\phi[\boldsymbol{z}]}{\delta z_\alpha(\boldsymbol{K}_1)}\frac{\delta\phi[\boldsymbol{z}]}{\delta z_\beta(\boldsymbol{K}_2)}\right\},
\end{aligned} \tag{8.71}$$

$$\begin{aligned}
S_{ij}\left[\boldsymbol{K},\boldsymbol{K}';\boldsymbol{z}\right] &= S^0_{ij}(\boldsymbol{K},\boldsymbol{K}') \\
&\quad -\int d^4\boldsymbol{K}''\int d^4\boldsymbol{K}_1\int d^4\boldsymbol{K}_2 S^0_{i\gamma}(\boldsymbol{K},\boldsymbol{K}'')\Lambda^{\alpha\beta}_\gamma(\boldsymbol{K}_1,\boldsymbol{K}_2,\boldsymbol{K}'') \\
&\quad \times\left\{\frac{\delta\phi[\boldsymbol{z}]}{\delta z_\alpha(\boldsymbol{K}_1)}G_{\beta j}\left[\boldsymbol{K}_2,\boldsymbol{K}';\boldsymbol{z}\right]+\frac{\delta}{\delta z_\alpha(\boldsymbol{K}_1)}G_{\beta j}\left[\boldsymbol{K}_2,\boldsymbol{K}';\boldsymbol{z}\right]\right\},
\end{aligned} \tag{8.72}$$

where

$$S^0_{ij}(\boldsymbol{K},\boldsymbol{K}') = \left(i\omega+\nu\boldsymbol{k}^2\right)^{-1}\delta_{ij}\delta^4(\boldsymbol{K}-\boldsymbol{K}').$$

The last equation is analogous to the *Schwinger equation* in the quantum field theory.

Note that expansion of functional $\phi[\boldsymbol{z}]$ in the functional Taylor series in $\boldsymbol{z}(\boldsymbol{K})$ determines the velocity field cumulants, and expansion of functional $S_{ij}\left[\boldsymbol{K},\boldsymbol{K}';\boldsymbol{z}\right]$ determines the correlators between Green's function analog $G_{ij}(\boldsymbol{K},\boldsymbol{K}') = \dfrac{\delta u_i(\boldsymbol{K})}{\delta f_j(\boldsymbol{K}')}$ and the velocity field.

If only the behavior of the velocity correlation function is of interest, the system of functional equations (8.71), (8.72) appears redundant, and we can filter out useless information by representing the spectral function of velocity in terms of a specific perturbation series. To construct such a series, we introduce, as in the linear case, quantity $S^{-1}_{ij}\left[\boldsymbol{K},\boldsymbol{K}';\boldsymbol{z}\right]$ by the formula

$$\int d^4\boldsymbol{K}' S_{ij}\left[\boldsymbol{K},\boldsymbol{K}';\boldsymbol{z}\right]S^{-1}_{j\delta}\left[\boldsymbol{K}',\boldsymbol{K}'';\boldsymbol{z}\right] = \delta_{i\delta}\delta^4(\boldsymbol{K}-\boldsymbol{K}''). \tag{8.73}$$

One can easily see that the relationship

$$\int d^4\boldsymbol{K}' S^{-1}_{ij}\left[\boldsymbol{K},\boldsymbol{K}';\boldsymbol{z}\right]S_{j\delta}\left[\boldsymbol{K}',\boldsymbol{K}'';\boldsymbol{z}\right] = \delta_{i\delta}\delta^4(\boldsymbol{K}-\boldsymbol{K}'') \tag{8.74}$$

will also hold.

Introduce additionally the three-index functional

$$\Gamma_{\gamma}^{j\delta}\left[\boldsymbol{P},\boldsymbol{K}',\boldsymbol{K}'';\boldsymbol{z}\right]=\frac{\delta}{\delta z_{\gamma}\left(\boldsymbol{P}\right)}S_{j\delta}^{-1}\left[\boldsymbol{K}',\boldsymbol{K}'';\boldsymbol{z}\right],\tag{8.75}$$

which is similar to the *mass operator vertex portion* in the quantum field theory. Varying Eq. (8.73) with respect to $z_{\gamma}(\boldsymbol{P})$, we can express $\delta S/\delta z$ in terms of S and Γ

$$\begin{aligned}&\frac{\delta}{\delta z_{\gamma}\left(\boldsymbol{P}\right)}S_{i\mu}\left[\boldsymbol{K},\boldsymbol{Q};\boldsymbol{z}\right]\\&=-\int d^{4}\boldsymbol{K}'\int d^{4}\boldsymbol{K}''S_{ij}\left[\boldsymbol{K},\boldsymbol{K}';\boldsymbol{z}\right]\Gamma_{\gamma}^{j\delta}\left[\boldsymbol{P},\boldsymbol{K}',\boldsymbol{K}'';\boldsymbol{z}\right]S_{\delta\mu}\left[\boldsymbol{K}'',\boldsymbol{Q};\boldsymbol{z}\right].\end{aligned}\tag{8.76}$$

Using Eq. (8.76), we can rewrite Eq. (8.72) in the form

$$\begin{aligned}S_{ij}\left[\boldsymbol{K},\boldsymbol{K}';\boldsymbol{z}\right]&=S_{ij}^{0}\left(\boldsymbol{K},\boldsymbol{K}'\right)-\int d^{4}\boldsymbol{P}\int d^{4}\boldsymbol{K}_{1}\int d^{4}\boldsymbol{K}_{2}S_{i\gamma}^{0}\left(\boldsymbol{K},\boldsymbol{P}\right)\Lambda_{\gamma}^{\alpha\beta}\left(\boldsymbol{K}_{1},\boldsymbol{K}_{2},\boldsymbol{P}\right)\\&\times\left\{\frac{\delta\phi[\boldsymbol{z}]}{\delta z_{\alpha}(\boldsymbol{K}_{1})}G_{\beta j}\left[\boldsymbol{K}_{2},\boldsymbol{K}';\boldsymbol{z}\right]-\int d^{4}\boldsymbol{K}'\int d^{4}\boldsymbol{K}''S_{\beta\sigma}\left[\boldsymbol{K}_{2},\boldsymbol{K}';\boldsymbol{z}\right]\right.\\&\left.\times\,\Gamma_{\alpha}^{\sigma\nu}\left[\boldsymbol{K}_{1},\boldsymbol{K}',\boldsymbol{K}'';\boldsymbol{z}\right]S_{\nu j}\left[\boldsymbol{K}'',\boldsymbol{K}';\boldsymbol{z}\right]\right\}.\end{aligned}\tag{8.77}$$

Setting $\boldsymbol{z}=0$ in Eq. (8.77), we obtain the equation that interconnects quantities $S|_{\boldsymbol{z}=0}$ and $\Gamma|_{\boldsymbol{z}=0}$ and is similar to the *Dyson equation* of the quantum field theory ($\delta\phi/\delta\boldsymbol{z}=0$ at $\boldsymbol{z}=0$).

Multiplying Eq. (8.77) by S^{-1} from the right and by S_0^{-1} from the left, integrating over the corresponding arguments, and varying the result with respect to $\boldsymbol{z}$, we obtain the following functional equation for Γ

$$\begin{aligned}\Gamma_{\gamma}^{\mu\rho}[\boldsymbol{P}_{3},\boldsymbol{P}_{2},\boldsymbol{P}_{1};\boldsymbol{z}]&=\int d^{4}\boldsymbol{K}_{1}\Lambda_{\mu}^{\alpha,\rho}(\boldsymbol{K}_{1},\boldsymbol{P}_{1},\boldsymbol{P}_{2})\frac{\delta^{2}\phi[\boldsymbol{z}]}{\delta z_{\alpha}\left(\boldsymbol{K}_{1}\right)\delta z_{\gamma}\left(\boldsymbol{P}_{3}\right)}\\&-\int d^{4}\boldsymbol{K}_{1}\int d^{4}\boldsymbol{K}_{2}\int d^{4}\boldsymbol{K}'\Lambda_{\mu}^{\alpha,\beta}(\boldsymbol{K}_{1},\boldsymbol{P}_{1},\boldsymbol{P}_{2})\\&\times\frac{\delta}{\delta z_{\gamma}(\boldsymbol{P}_{3})}\left\{S_{\beta\sigma}\left[\boldsymbol{K}_{2},\boldsymbol{K}';\boldsymbol{z}\right]\Gamma_{\alpha}^{\sigma\rho}\left[\boldsymbol{K},\boldsymbol{K}',\boldsymbol{P}_{2};\boldsymbol{z}\right]\right\}.\end{aligned}\tag{8.78}$$

The system of equations (8.71), (8.77), and (8.78) is closed; however, the solutions to this system are interconnected additionally by the relationship (8.76).

If we will now construct the perturbation series with absolute terms of Eqs. (8.71) and (8.78) as the zero-order approximations and will express appearing variations of S with respect to $\boldsymbol{z}$ using relationship (8.76), then we obtain the space-time velocity spectrum and function $\Gamma|_{\boldsymbol{z}=0}$ in the form of the infinite series, every term of which includes these very functions. These series will be integral equations with infinite number of terms and, being combined with Eq. (8.72) at $\boldsymbol{z} = 0$, they form the closed system of equations for quantities $\delta^2\phi/\delta\boldsymbol{z}\delta\boldsymbol{z}$, $S|_{\boldsymbol{z}=0}$, and $\Gamma|_{\boldsymbol{z}=0}$. However, explicit representation of even a few terms of these series is hardly possible because of cumbersome rearrangements and complicated structure of functional equations (8.71), (8.77), and (8.78). The reader can find an analysis of possible simplifications in Ref. [42] (see also [46, 49, 112]).

8.4 Completely Solvable Stochastic Dynamic Systems

Consider now several dynamic systems that allow sufficiently adequate statistical analysis for arbitrary random parameters.

8.4.1 *Ordinary Differential Equations*

Multiplicative Action

As the first example, we consider the vector stochastic equation with initial value

$$\frac{d}{dt}\boldsymbol{x}(t) = z(t)g(t)\boldsymbol{F}(\boldsymbol{x}), \quad \boldsymbol{x}(0) = \boldsymbol{x}_0, \tag{8.79}$$

where $g(t)$ and $F_i(\boldsymbol{x})$, $i = 1, \cdots, N$ are the deterministic functions and $z(t)$ is the random process whose statistical characteristics are described by the characteristic functional

$$\Phi[t; v(\tau)] = \left\langle \exp\left\{ i \int_0^t d\tau z(\tau)v(\tau) \right\} \right\rangle = e^{\Theta[t; v(\tau)]}.$$

Equation (8.79) has a feature that offers a possibility of determining statistical characteristics of its solutions in the general case of arbitrarily distributed process $z(t)$. The point is that introduction of new 'random' time

$$T = \int_0^t d\tau z(\tau)g(\tau)$$

reduces Eq. (8.79) to the equation looking formally deterministic

$$\frac{d}{dT}\boldsymbol{x}(T) = \boldsymbol{F}(\boldsymbol{x}), \quad \boldsymbol{x}(0) = \boldsymbol{x}_0,$$

so that the solution to Eq. (8.79) has the following structure

$$\boldsymbol{x}(t) = \boldsymbol{x}(T) = \boldsymbol{x}\left(\int_0^t d\tau z(\tau) g(\tau)\right). \tag{8.80}$$

Varying Eq. (8.80) with respect to $z(\tau)$ and using Eq. (8.79), we obtain the equality

$$\frac{\delta}{\delta z(\tau)}\boldsymbol{x}(t) = g(\tau)\frac{d}{dT}\boldsymbol{x}(T) = g(\tau)\boldsymbol{F}(\boldsymbol{x}(t)). \tag{8.81}$$

Thus, the variational derivative of solution $\boldsymbol{x}(t)$ is expressed in terms of the same solution at the same time. This fact makes it possible to immediately write the closed equations for statistical characteristics of problem (8.79).

Let us derive the equation for the one-time probability density

$$P(\boldsymbol{x}, t) = \langle \delta(\boldsymbol{x}(t) - \boldsymbol{x})\rangle .$$

It has the form

$$\frac{\partial}{\partial t}P(\boldsymbol{x}, t) = \left\langle \dot{\Theta}_t\left[t; \frac{\delta}{i\delta z(\tau)}\right] \delta\left(\boldsymbol{x}(t) - \boldsymbol{x}\right)\right\rangle . \tag{8.82}$$

Consider now the result of operator $\delta/\delta z(\tau)$ applied to the indicator function $\varphi(\boldsymbol{x}, t) = \delta(\boldsymbol{x}(t) - \boldsymbol{x})$. Using formula (8.81), we obtain the expression

$$\frac{\delta}{\delta z(\tau)}\delta\left(\boldsymbol{x}(t) - \boldsymbol{x}\right) = -g(\tau)\frac{\partial}{\partial \boldsymbol{x}}\left\{\boldsymbol{F}(\boldsymbol{x})\varphi(\boldsymbol{x}, t)\right\}.$$

Consequently, we can rewrite Eq. (8.82) in the form of the closed operator equation

$$\frac{\partial}{\partial t}P(\boldsymbol{x}, t) = \dot{\Theta}_t\left[t; ig(\tau)\frac{\partial}{\partial \boldsymbol{x}}\boldsymbol{F}(\boldsymbol{x})\right] P(\boldsymbol{x}, t), \quad P(\boldsymbol{x}, 0) = \delta(\boldsymbol{x} - \boldsymbol{x}_0), \tag{8.83}$$

whose particular form depends on the behavior of process $z(t)$.

For the two-time probability density

$$P(\boldsymbol{x}, t; \boldsymbol{x}_1, t_1) = \langle \delta(\boldsymbol{x}(t) - \boldsymbol{x})\delta(\boldsymbol{x}(t_1) - \boldsymbol{x}_1)\rangle$$

we obtain similarly the equation (for $t > t_1$)

$$\frac{\partial}{\partial t}P(\boldsymbol{x}, t; \boldsymbol{x}_1, t_1) = \dot{\Theta}_t\left[t; ig(\tau)\left\{\frac{\partial}{\partial \boldsymbol{x}}\boldsymbol{F}(\boldsymbol{x}) + \theta(t_1 - \tau)\frac{\partial}{\partial \boldsymbol{x}_1}\boldsymbol{F}(\boldsymbol{x}_1)\right\}\right] P(\boldsymbol{x}, t; \boldsymbol{x}_1, t_1) \tag{8.84}$$

with the initial value

$$P(\boldsymbol{x}, t_1; \boldsymbol{x}_1, t_1) = \delta(\boldsymbol{x} - \boldsymbol{x}_1) P(\boldsymbol{x}_1, t_1),$$

where function $P(\boldsymbol{x}_1, t_1)$ satisfies Eq. (8.83).

One can see from Eq. (8.84) that multidimensional probability density cannot be factorized in terms of the transition probability (see Sect. 3.3), so that process $\boldsymbol{x}(t)$ is not the Markovian process. The particular forms of Eqs. (8.83) and (8.84) are governed by the statistics of process $z(t)$.

If $z(t)$ is the Gaussian process whose mean value and correlation function are

$$\langle z(t) \rangle = 0, \quad B(t, t') = \langle z(t) z(t') \rangle ,$$

then functional $\Theta[t; v(\tau)]$ has the form

$$\Theta[t; v(\tau)] = -\frac{1}{2} \int\limits_0^t dt_1 \int\limits_0^t dt_2 B(t_1, t_2) v(t_1) v(t_2)$$

and Eq. (8.83) assumes the following form

$$\frac{\partial}{\partial t} P(\boldsymbol{x}, t) = g(t) \int\limits_0^t d\tau B(t, \tau) g(\tau) \frac{\partial}{\partial x_j} F_j(\boldsymbol{x}) \frac{\partial}{\partial x_k} F_k(\boldsymbol{x}) P(\boldsymbol{x}, t) \tag{8.85}$$

and can be considered as the *extended Fokker–Planck equation* .

The class of problems formulated in terms of the system of equations

$$\frac{d}{dt} \boldsymbol{x}(t) = z(t) \boldsymbol{F}(\boldsymbol{x}) - \lambda \boldsymbol{x}(t), \quad \boldsymbol{x}(0) = \boldsymbol{x}_0, \tag{8.86}$$

where $\boldsymbol{F}(\boldsymbol{x})$ are the homogeneous polynomials of power k, can be reduced to problem (8.79). Indeed, introducing new functions

$$\boldsymbol{x}(t) = \widetilde{\boldsymbol{x}}(t) e^{-\lambda t},$$

we arrive at problem (8.79) with function $g(t) = e^{-\lambda(k-1)t}$. In the important special case with $k = 2$ and functions $\boldsymbol{F}(\boldsymbol{x})$ such that $\boldsymbol{x}\boldsymbol{F}(\boldsymbol{x}) = 0$, the system of equations (8.79) describes hydrodynamic systems with the linear friction (see, e.g., [14, 23]). In this case, the interaction between the components appears random.

If $\lambda = 0$, energy conservation holds in hydrodynamic systems for any realization of process $z(t)$. For $t \to \infty$, there is the steady-state probability distribution $P(\boldsymbol{x})$, which is, under the assumption that no additional integrals of motion exist, the uniform distribution over sphere $x_i^2 = E_0$. If additional integrals of motion exist (as it is the case for finite-dimensional approximation of the two-dimensional motion of fluid), the domain of the steady-state probability distribution will coincide with the phase space region allowed by the integrals of motion.

Note that, in the special case of the Gaussian process $z(t)$ appeared in the one-dimensional linear equation of type Eq. (8.86)

$$\frac{d}{dt}x(t) = -\lambda x(t) + z(t)x(t), \quad x(0) = 1,$$

which determines the simplest logarithmic-normal random process, we obtain, instead of Eq. (8.85), the extended Fokker–Planck equation

$$\left(\frac{\partial}{\partial t} + \lambda\frac{\partial}{\partial x}x\right)P(x,t) = \int\limits_0^t d\tau B(t,\tau)\frac{\partial}{\partial x}x\frac{\partial}{\partial x}xP(x,t), \qquad (8.87)$$
$$P(x,0) = \delta(x-1).$$

Additive Action

Consider now the class of linear equations

$$\frac{d}{dt}\boldsymbol{x}(t) = A(t)\boldsymbol{x}(t) + \boldsymbol{f}(t), \quad \boldsymbol{x}(0) = \boldsymbol{x}_0, \qquad (8.88)$$

where $A(t)$ is the deterministic matrix and $\boldsymbol{f}(t)$ is the random vector function whose characteristic functional $\varPhi[t;\boldsymbol{v}(\tau)]$ is known.

For the probability density of the solution to Eq. (8.88), we have

$$\frac{\partial}{\partial t}P(\boldsymbol{x},t) = -\frac{\partial}{\partial x_i}\left(A_{ik}(t)x_kP(\boldsymbol{x},t)\right) + \left\langle \dot{\Theta}_t\left[t;\frac{\delta}{i\delta\boldsymbol{f}(\tau)}\right]\delta\left(\boldsymbol{x}(t)-\boldsymbol{x}\right)\right\rangle. \quad (8.89)$$

In the problem under consideration, the variational derivative $\dfrac{\delta\boldsymbol{x}(t)}{\delta\boldsymbol{f}(\tau)}$ also satisfies (for $\tau < t$) the linear equation with the initial value

$$\frac{d}{dt}\frac{\delta}{\delta\boldsymbol{f}(\tau)}x_i(t) = A_{ik}(t)\frac{\delta}{\delta\boldsymbol{f}(\tau)}x_k(t), \quad \left.\frac{\delta}{\delta f_l(\tau)}x_i(t)\right|_{t=\tau} = \delta_{il}. \qquad (8.90)$$

Equation (8.90) has no randomness and governs Green's function $G_{il}(t,\tau)$ of homogeneous system (8.88), which means that

$$\frac{\delta}{\delta f_l(\tau)}x_i(t) = G_{il}(t,\tau).$$

As a consequence, we have

$$\frac{\delta}{\delta f_l(\tau)}\delta(\boldsymbol{x}(t)-\boldsymbol{x}) = -\frac{\partial}{\partial x_k}G_{kl}(t,\tau)\delta(\boldsymbol{x}(t)-\boldsymbol{x}),$$

and Eq. (8.89) appears converted into the closed equation

$$\frac{\partial}{\partial t}P(\boldsymbol{x},t) = -\frac{\partial}{\partial \boldsymbol{x}_i}\left(A_{ik}(t)x_k P(\boldsymbol{x},t)\right) + \dot{\Theta}_t\left[t; iG_{kl}(t,\tau)\frac{\partial}{\partial x_k}\right]P(\boldsymbol{x},t). \quad (8.91)$$

From Eq. (8.91) follows that any moment of quantity $\boldsymbol{x}(t)$ will satisfy a closed linear equation that will include only a finite number of cumulant functions whose order will not exceed the order of the moment of interest.

For the two-time probability density

$$P(\boldsymbol{x},t;\boldsymbol{x}_1,t_1) = \left\langle \delta(\boldsymbol{x}(t)-\boldsymbol{x})\delta(\boldsymbol{x}(t_1)-\boldsymbol{x}_1)\right\rangle,$$

we quite similarly obtain the equation

$$\frac{\partial}{\partial t}P(\boldsymbol{x},t;\boldsymbol{x}_1,t_1) = -\frac{\partial}{\partial x_i}\left(A_{ik}(t)x_k P(\boldsymbol{x},t;\boldsymbol{x}_1,t_1)\right)$$
$$+\dot{\Theta}_t\left[t; i\left\{G_{kl}(t,\tau)+G_{kl}(t_1,\tau)\right\}\frac{\partial}{\partial x_l}\right]P(\boldsymbol{x},t;\boldsymbol{x}_1,t_1) \quad (t>t_1) \quad (8.92)$$

with the initial value

$$P(\boldsymbol{x},t_1;\boldsymbol{x}_1,t_1) = \delta(\boldsymbol{x}-\boldsymbol{x}_1)P(\boldsymbol{x}_1,t_1),$$

where $P(\boldsymbol{x}_1,t_1)$ is the one-time probability density satisfying Eq. (8.91). From Eq. (8.92) follows that $\boldsymbol{x}(t)$ is not the Markovian process. The particular form of Eq. (8.91) and (8.92) depends on the structure of functional $\Phi[t;v(\tau)]$, i.e., on the random behavior of function $\boldsymbol{f}(t)$.

For the Gaussian vector process $\boldsymbol{f}(t)$ whose mean value and correlation function are

$$\langle \boldsymbol{f}(t)\rangle = 0, \quad B_{ij}(t,t') = \langle f_i(t)f_j(t')\rangle$$

Eq. (8.91) assumes the form of the extended Fokker–Planck equation

$$\frac{\partial}{\partial t}P(\boldsymbol{x},t) = -\frac{\partial}{\partial x_i}\left(A_{ik}(t)x_k P(\boldsymbol{x},t)\right) + \int_0^t d\tau B_{jl}(t,\tau)G_{kj}(t,t)G_{ml}(t,\tau)\frac{\partial^2}{\partial x_k \partial x_m}P(\boldsymbol{x},t). \quad (8.93)$$

Consider the dynamics of a particle under random forces in the presence of friction [69] as an example of such a problem.

Inertial Particle under Random Forces

The simplest example of particle diffusion under the action of random external force $\boldsymbol{f}(t)$ and linear friction is described by the linear system of equations (1.94), page 47

$$\frac{d}{dt}\boldsymbol{r}(t) = \boldsymbol{v}(t), \quad \frac{d}{dt}\boldsymbol{v}(t) = -\lambda\left[\boldsymbol{v}(t) - \boldsymbol{f}(t)\right],$$
$$\boldsymbol{r}(0) = 0, \quad \boldsymbol{v}(0) = 0. \tag{8.94}$$

The stochastic solution to Eqs. (8.94) has the form

$$\boldsymbol{v}(t) = \lambda \int_0^t d\tau e^{-\lambda(t-\tau)} \boldsymbol{f}(\tau), \quad \boldsymbol{r}(t) = \int_0^t d\tau \left[1 - e^{-\lambda(t-\tau)}\right] \boldsymbol{f}(\tau). \tag{8.95}$$

In the case of stationary random process $\boldsymbol{f}(t)$ with the correlation tensor

$$\langle f_i(t) f_j(t')\rangle = B_{ij}(t - t')$$

and temporal correlation radius τ_0 determined from the relationship

$$\int_0^\infty d\tau B_{ii}(\tau) = \tau_0 B_{ii}(0),$$

Eq. (8.94) allows obtaining analytical expressions for correlators between particle velocity components and coordinates

$$\langle v_i(t) v_j(t)\rangle = \lambda \int_0^t d\tau B_{ij}(\tau) \left[e^{-\lambda\tau} - e^{-\lambda(2t-\tau)}\right],$$

$$\frac{1}{2}\frac{d}{dt}\langle r_i(t) r_j(t)\rangle = \langle r_i(t) v_j(t)\rangle = \int_0^t d\tau B_{ij}(\tau) \left[1 - e^{-\lambda t}\right] \left[1 - e^{-\lambda(t-\tau)}\right]. \tag{8.96}$$

In the steady-state regime, when $\lambda t \gg 1$ and $t/\tau_0 \gg 1$, but parameter $\lambda\tau_0$ can be arbitrary, particle velocity is the stationary process with the correlation tensor

$$\langle v_i(t) v_j(t)\rangle = \lambda \int_0^\infty d\tau B_{ij}(\tau) e^{-\lambda\tau}, \tag{8.97}$$

and correlations $\langle r_i(t) v_j(t)\rangle$ and $\langle r_i(t) r_j(t)\rangle$ are as follows

$$\langle r_i(t) v_j(t)\rangle = \int_0^\infty d\tau B_{ij}(\tau), \quad \langle r_i(t) r_j(t)\rangle = 2t \int_0^\infty d\tau B_{ij}(\tau). \tag{8.98}$$

If we additionally assume that $\lambda\tau_0 \gg 1$, the correlation tensor grades into

$$\langle v_i(t) v_j(t)\rangle = B_{ij}(0), \tag{8.99}$$

which is consistent with (8.94), because $\boldsymbol{v}(t) \equiv \boldsymbol{f}(t)$ in this limit.

If the opposite condition $\lambda\tau_0 \ll 1$ holds, then

$$\langle v_i(t)v_j(t)\rangle = \lambda\int\limits_0^\infty d\tau B_{ij}(\tau).$$

This result corresponds to the delta-correlated approximation of random process $\boldsymbol{f}(t)$.

Probability Distribution Function

Introduce now the indicator function of the solution to Eq. (8.94)

$$\varphi(\boldsymbol{r},\boldsymbol{v},t) = \delta\left(\boldsymbol{r}(t)-\boldsymbol{r}\right)\delta\left(\boldsymbol{v}(t)-\boldsymbol{v}\right),$$

which satisfies the Liouville equation

$$\begin{gathered}\left(\frac{\partial}{\partial t}+\boldsymbol{v}\frac{\partial}{\partial \boldsymbol{r}}-\lambda\frac{\partial}{\partial \boldsymbol{v}}\boldsymbol{v}\right)\varphi(\boldsymbol{r},\boldsymbol{v},t) = -\lambda\boldsymbol{f}(t)\frac{\partial}{\partial \boldsymbol{v}}\varphi(\boldsymbol{r},\boldsymbol{v},t),\\ \varphi(\boldsymbol{r},\boldsymbol{v};0) = \delta\left(\boldsymbol{r}\right)\delta\left(\boldsymbol{v}\right).\end{gathered} \tag{8.100}$$

The mean value of the indicator function $\varphi(\boldsymbol{r},\boldsymbol{v},t)$ over an ensemble of realizations of random process $\boldsymbol{f}(t)$ is the joint one-time probability density of particle position and velocity

$$P(\boldsymbol{r},\boldsymbol{v},t) = \langle\varphi(\boldsymbol{r},\boldsymbol{v};t)\rangle = \langle\delta\left(\boldsymbol{r}(t)-\boldsymbol{r}\right)\delta\left(\boldsymbol{v}(t)-\boldsymbol{v}\right)\rangle_{\boldsymbol{f}}.$$

Averaging Eq. (8.100) over an ensemble of realizations of random process $\boldsymbol{f}(t)$, we obtain the unclosed equation

$$\begin{gathered}\left(\frac{\partial}{\partial t}+\boldsymbol{v}\frac{\partial}{\partial \boldsymbol{r}}-\lambda\frac{\partial}{\partial \boldsymbol{v}}\boldsymbol{v}\right)P(\boldsymbol{r},\boldsymbol{v},t) = -\lambda\frac{\partial}{\partial \boldsymbol{v}}\left\langle\boldsymbol{f}(t)\varphi(\boldsymbol{r},\boldsymbol{v},t)\right\rangle,\\ P(\boldsymbol{r},\boldsymbol{v};0) = \delta\left(\boldsymbol{r}\right)\delta\left(\boldsymbol{v}\right).\end{gathered} \tag{8.101}$$

This equation contains correlation $\langle\boldsymbol{f}(t)\varphi(\boldsymbol{r},\boldsymbol{v},t)\rangle$ and is equivalent to the equality

$$\begin{gathered}\left(\frac{\partial}{\partial t}+\boldsymbol{v}\frac{\partial}{\partial \boldsymbol{r}}-\lambda\frac{\partial}{\partial \boldsymbol{v}}\boldsymbol{v}\right)P(\boldsymbol{r},\boldsymbol{v},t) = \left\langle\dot{\Theta}\left[t;\frac{\delta}{i\delta\boldsymbol{f}(\tau)}\right]\varphi(\boldsymbol{r},\boldsymbol{v},t)\right\rangle,\\ P(\boldsymbol{r},\boldsymbol{v};0) = \delta\left(\boldsymbol{r}\right)\delta\left(\boldsymbol{v}\right),\end{gathered} \tag{8.102}$$

where functional $\Theta\left[t;\boldsymbol{v}(\tau)\right]$ is related to the characteristic functional of random process $\boldsymbol{f}(t)$

$$\Phi\left[t;\boldsymbol{\psi}(\tau)\right]=\left\langle\exp\left\{i\int\limits_0^t d\tau\,\boldsymbol{\psi}(\tau)\boldsymbol{f}(\tau)\right\}\right\rangle=e^{\Theta[t;\boldsymbol{\psi}(\tau)]}$$

by the formula

$$\dot{\Theta}\left[t;\boldsymbol{\psi}(\tau)\right]=\frac{d}{dt}\ln\Phi\left[t;\boldsymbol{\psi}(\tau)\right]=\frac{d}{dt}\Theta\left[t;\boldsymbol{\psi}(\tau)\right].$$

Functional $\Theta[t;\boldsymbol{\psi}(\tau)]$ can be expanded in the functional power series

$$\Theta[t;\boldsymbol{\psi}(\tau)]=\sum_{n=1}^{\infty}\frac{i^n}{n!}\int\limits_0^t dt_1\cdots\int\limits_0^t dt_n K^{(n)}_{i_1,\cdots,i_n}(t_1,\cdots,t_n)\psi_{i_1}(t_1)\cdots\psi_{i_n}(t_n),$$

where functions

$$K^{(n)}_{i_1,\cdots,i_n}(t_1,\cdots,t_n)=\frac{1}{i^n}\frac{\delta^n}{\delta\psi_{i_1}(t_1)\cdots\delta\psi_{i_n}(t_n)}\Theta[t;\boldsymbol{\psi}(\tau)]\bigg|_{\boldsymbol{\psi}=0}$$

are the n-th order cumulant functions of random process $\boldsymbol{f}(t)$.

Consider the variational derivative

$$\frac{\delta}{\delta f_j(t')}\varphi(\boldsymbol{r},\boldsymbol{v},t)=-\left[\frac{\partial}{\partial r_k}\frac{\delta r_k(t)}{\delta f_j(t')}+\frac{\partial}{\partial v_k}\frac{\delta v_k(t)}{\delta f_j(t')}\right]\varphi(\boldsymbol{r},\boldsymbol{v},t).\tag{8.103}$$

In the context of dynamic problem (8.94), the variational derivatives of functions $\boldsymbol{r}(t)$ and $\boldsymbol{v}(t)$ in Eq. (8.103) can be calculated from Eqs. (8.95) and have the forms

$$\frac{\delta v_k(t)}{\delta f_j(t')}=\lambda\delta_{kj}e^{-\lambda(t-t')},\quad\frac{\delta r_k(t)}{\delta f_j(t')}=\delta_{kj}\left[1-e^{-\lambda(t-t')}\right].\tag{8.104}$$

Using Eq. (8.104), we can now rewrite Eq. (8.103) in the form

$$\frac{\delta}{\delta\boldsymbol{f}(t')}\varphi(\boldsymbol{r},\boldsymbol{v};t)=-\left\{\left[1-e^{-\lambda(t-t')}\right]\frac{\partial}{\partial\boldsymbol{r}}+\lambda e^{-\lambda(t-t')}\frac{\partial}{\partial\boldsymbol{v}}\right\}\varphi(\boldsymbol{r},\boldsymbol{v},t),$$

after which Eq. (8.102) assumes the closed form

$$\begin{aligned}&\left(\frac{\partial}{\partial t}+\boldsymbol{v}\frac{\partial}{\partial\boldsymbol{r}}-\lambda\frac{\partial}{\partial\boldsymbol{v}}\boldsymbol{v}\right)P(\boldsymbol{r},\boldsymbol{v},t)\\&\qquad=\dot{\Theta}\left[t;i\left\{\left[1-e^{-\lambda(t-t')}\right]\frac{\partial}{\partial\boldsymbol{r}}+\lambda e^{-\lambda(t-t')}\frac{\partial}{\partial\boldsymbol{v}}\right\}\right]P(\boldsymbol{r},\boldsymbol{v},t),\end{aligned}\tag{8.105}$$

$$P(\boldsymbol{r},\boldsymbol{v};0)=\delta(\boldsymbol{r})\,\delta(\boldsymbol{v})\,.$$

Note that from Eq. (8.105) follows that equations for the n-th order moment functions include cumulant functions of orders not higher than n.

The Gaussian Process $\boldsymbol{f}(t)$

Assume now that $\boldsymbol{f}(t)$ is the Gaussian stationary process with the zero mean value and correlation tensor

$$B_{ij}(t-t')=\langle f_i(t)f_j(t')\rangle\,.$$

In this case, the characteristic functional of process $\boldsymbol{f}(t)$ is

$$\Phi\left[t;\boldsymbol{\psi}(\tau)\right]=\exp\left\{-\frac{1}{2}\int\limits_0^t\int\limits_0^t dt_1dt_2B_{ij}(t_1-t_2)\psi_i(t_1)\psi_j(t_2)\right\},$$

functional $\dot{\Theta}[t;\boldsymbol{\psi}(\tau)]$ is given by the formula

$$\dot{\Theta}\left[t;\boldsymbol{\psi}(\tau)\right]=-\psi_i(t)\int\limits_0^t dt'B_{ij}(t-t')\psi_j(t'),$$

and Eq. (8.105) appears an extension of the Fokker–Planck equation

$$\left(\frac{\partial}{\partial t}+\boldsymbol{v}\frac{\partial}{\partial \boldsymbol{r}}-\lambda\frac{\partial}{\partial \boldsymbol{v}}\boldsymbol{v}\right)P(\boldsymbol{r},\boldsymbol{v},t)=\lambda^2\int\limits_0^t d\tau B_{ij}(\tau)e^{-\lambda\tau}\frac{\partial^2}{\partial v_i\partial v_j}P(\boldsymbol{r},\boldsymbol{v},t)$$

$$+\lambda\int\limits_0^t d\tau B_{ij}(\tau)\left[1-e^{-\lambda\tau}\right]\frac{\partial^2}{\partial v_i\partial r_j}P(\boldsymbol{r},\boldsymbol{v},t),\quad(8.106)$$

$$P(\boldsymbol{r},\boldsymbol{v};0)=\delta(\boldsymbol{r})\,\delta(\boldsymbol{v})\,.$$

Equation (8.106) is the exact equation and remains valid for arbitrary times t. From this equation follows that $\boldsymbol{r}(t)$ and $\boldsymbol{v}(t)$ are the Gaussian functions. For moment functions of processes $\boldsymbol{r}(t)$ and $\boldsymbol{v}(t)$, we obtain in the ordinary way the system of equations

$$\frac{d}{dt}\left\langle r_i(t)r_j(t)\right\rangle = 2\left\langle r_i(t)v_j(t)\right\rangle,$$

$$\left(\frac{d}{dt}+\lambda\right)\left\langle r_i(t)v_j(t)\right\rangle = \left\langle v_i(t)v_j(t)\right\rangle + \lambda\int_0^t d\tau B_{ij}(\tau)\left[1-e^{-\lambda\tau}\right], \quad (8.107)$$

$$\left(\frac{d}{dt}+2\lambda\right)\left\langle v_i(t)v_j(t)\right\rangle = 2\lambda^2\int_0^t d\tau B_{ij}(\tau)e^{-\lambda\tau}.$$

From system (8.107) follows that steady-state values of all one-time correlators for $\lambda t \gg 1$ and $t/\tau_0 \gg 1$ are given by the expressions

$$\left\langle v_i(t)v_j(t)\right\rangle = \lambda\int_0^\infty d\tau B_{ij}(\tau)e^{-\lambda\tau}, \quad \left\langle r_i(t)v_j(t)\right\rangle = D_{ij}, \qquad (8.108)$$
$$\left\langle r_i(t)r_j(t)\right\rangle = 2tD_{ij},$$

where

$$D_{ij} = \int_0^\infty d\tau B_{ij}(\tau) \qquad (8.109)$$

is the spatial diffusion tensor, which agrees with expressions (8.97) and (8.98).

Remark 8.1. ***Temporal Correlation Tensor and Temporal Correlation Radius of Process $v(t)$.***

We can additionally calculate the temporal correlation radius of velocity $v(t)$, i.e., the scale of correlator $\left\langle v_i(t)v_j(t_1)\right\rangle$. Using equalities (8.104), we obtain for $t_1 < t$ the equation

$$\left(\frac{d}{dt}+\lambda\right)\left\langle v_i(t)v_j(t_1)\right\rangle = \lambda^2\int_0^{t_1} dt' B_{ij}(t-t')e^{-\lambda(t_1-t')} = \lambda^2 e^{\lambda(t-t_1)}\int_{t-t_1}^{t} d\tau B_{ij}(\tau)e^{-\lambda\tau} \qquad (8.110)$$

with the initial value

$$\left\langle v_i(t)v_j(t_1)\right\rangle|_{t=t_1} = \left\langle v_i(t_1)v_j(t_1)\right\rangle. \qquad (8.111)$$

In the steady-state regime, i.e., for $\lambda t \gg 1$ and $\lambda t_1 \gg 1$, but at fixed difference $(t-t_1)$, we obtain the equation with initial value $(\tau = t - t_1)$

$$\left(\frac{d}{d\tau}+\lambda\right)\left\langle v_i(t+\tau)v_j(t)\right\rangle = \lambda^2 e^{\lambda\tau}\int_\tau^\infty d\tau_1 B_{ij}(\tau_1)e^{-\lambda\tau_1}, \qquad (8.112)$$
$$\left\langle v_i(t+\tau)v_j(t)\right\rangle_{\tau=0} = \left\langle v_i(t)v_j(t)\right\rangle.$$

One can easily write the solution to Eq. (8.112); however, our interest here concerns only the temporal correlation radius $\tau_{\boldsymbol{v}}$ of random process $\boldsymbol{v}(t)$. To obtain this quantity, we integrate Eq. (8.112) with respect to parameter τ over the interval $(0, \infty)$. The result is

$$\lambda \int_0^\infty d\tau \left\langle v_i(t+\tau) v_j(t) \right\rangle = \left\langle v_i(t) v_j(t) \right\rangle + \lambda \int_0^\infty d\tau_1 B_{ij}(\tau_1) \left[1 - e^{-\lambda \tau_1}\right],$$

and we, using Eq. (8.108), arrive at the expression

$$\tau_{\boldsymbol{v}} \left\langle \boldsymbol{v}^2(t) \right\rangle = D_{ii} = \tau_0 B_{ii}(0), \tag{8.113}$$

i.e.,

$$\tau_{\boldsymbol{v}} = \frac{\tau_0 B_{ii}(0)}{\left\langle \boldsymbol{v}^2(t) \right\rangle} = \frac{\tau_0 B_{ii}(0)}{\lambda \int_0^\infty d\tau B_{ii}(\tau) e^{-\lambda\tau}} = \begin{cases} \tau_0 & \text{for} \quad \lambda\tau_0 \gg 1, \\ 1/\lambda & \text{for} \quad \lambda\tau_0 \ll 1. \end{cases} \tag{8.114}$$

♦

Integrating Eq.(8.106) over $\boldsymbol{r}$, we obtain the closed equation for the probability density of particle velocity

$$\left(\frac{\partial}{\partial t} - \lambda \frac{\partial}{\partial \boldsymbol{v}} \boldsymbol{v} \right) P(\boldsymbol{v}, t) = \lambda^2 \int_0^t d\tau B_{ij}(\tau) e^{-\lambda\tau} \frac{\partial^2}{\partial v_i \partial v_j} P(\boldsymbol{r}, \boldsymbol{v}, t),$$

$$P(\boldsymbol{r}, \boldsymbol{v}; 0) = \delta\left(\boldsymbol{v}\right).$$

The solution to this equation corresponds to the Gaussian process $\boldsymbol{v}(t)$ with correlation tensor (8.96), which follows from the fact that the second equation of system (8.94) is closed. It can be shown that, if the steady-state probability density exists under the condition $\lambda t \gg 1$, then this probability density satisfies the equation

$$-\frac{\partial}{\partial \boldsymbol{v}} \boldsymbol{v} P(\boldsymbol{v}, t) = \lambda \int_0^\infty d\tau B_{ij}(\tau) e^{-\lambda\tau} \frac{\partial^2}{\partial v_i \partial v_j} P(\boldsymbol{v}, t),$$

and the rate of establishing this distribution depends on parameter λ.

The equation for the probability density of particle coordinate $P(\boldsymbol{r}, t)$ cannot be derived immediately from Eq. (8.106). Indeed, integrating Eq. (8.106) over $\boldsymbol{v}$, we obtain the equality

$$\frac{\partial}{\partial t} P(\boldsymbol{r}, t) = -\frac{\partial}{\partial \boldsymbol{r}} \int \boldsymbol{v} P(\boldsymbol{r}, \boldsymbol{v}, t) d\boldsymbol{v}, \quad P(\boldsymbol{r}, 0) = \delta\left(\boldsymbol{r}\right). \tag{8.115}$$

For function $\int v_k P(\boldsymbol{r}, \boldsymbol{v}, t) d\boldsymbol{v}$, we have the equality

$$\left(\frac{\partial}{\partial t}+\lambda\right)\int v_k P(\boldsymbol{r},\boldsymbol{v},t)d\boldsymbol{v} = -\frac{\partial}{\partial \boldsymbol{r}}\int v_k \boldsymbol{v} P(\boldsymbol{r},\boldsymbol{v},t)d\boldsymbol{v}$$

$$-\lambda\int\limits_0^t d\tau B_{kj}(\tau)\left[1-e^{-\lambda\tau}\right]\frac{\partial}{\partial r_j}P(\boldsymbol{r},t), \quad (8.116)$$

and so on, i.e., this approach results in an infinite system of equations.

Random function $\boldsymbol{r}(t)$ satisfies the first equation of system (8.94) and, if we would know the complete statistics of function $\boldsymbol{v}(t)$ (i.e., the multi-time statistics), we could calculate all statistical characteristics of function $\boldsymbol{r}(t)$. Unfortunately, Eq. (8.106) describes only one-time statistical quantities, and only the infinite system of equations similar to Eqs. (8.115), (8.116), and so on appears equivalent to the multi-time statistics of function $\boldsymbol{v}(t)$. Indeed, function $\boldsymbol{r}(t)$ can be represented in the form

$$\boldsymbol{r}(t) = \int\limits_0^t dt_1 \boldsymbol{v}(t_1),$$

so that the spatial diffusion coefficient in the steady-state regime assumes, in view of Eq. (8.113), the form

$$\frac{1}{2}\frac{d}{dt}\left\langle \boldsymbol{r}^2(t)\right\rangle = \int\limits_0^\infty d\tau \left\langle \boldsymbol{v}(t+\tau)\boldsymbol{v}(t)\right\rangle = \tau_{\boldsymbol{v}}\left\langle \boldsymbol{v}^2(t)\right\rangle = D_{ii} = \tau_0 B_{ii}(0), \quad (8.117)$$

from which follows that it depends on the temporal correlation radius $\tau_{\boldsymbol{v}}$ and the variance of random function $\boldsymbol{v}(t)$.

However, in the case of this simplest problem, we know both variances and all correlations of functions $\boldsymbol{v}(t)$ and $\boldsymbol{r}(t)$ (see Eqs. (8.96)) and, consequently, can draw the equation for the probability density of particle coordinate $P(\boldsymbol{r};t)$. This equation is the diffusion equation

$$\frac{\partial}{\partial t}P(\boldsymbol{r};t) = D_{ij}(t)\frac{\partial^2}{\partial r_i \partial r_j}P(\boldsymbol{r},t), \quad P(\boldsymbol{r},0) = \delta(\boldsymbol{r}),$$

where

$$D_{ij}(t) = \frac{1}{2}\frac{d}{dt}\left\langle r_i(t) r_j(t)\right\rangle = \frac{1}{2}\left\{\left\langle r_i(t) v_j(t)\right\rangle + \left\langle r_j(t) v_i(t)\right\rangle\right\}$$

$$= \int\limits_0^t d\tau B_{ij}(\tau)\left[1-e^{-\lambda t}\right]\left[1-e^{-\lambda(t-\tau)}\right]$$

is the diffusion tensor (8.96). Under the condition $\lambda t \gg 1$, we obtain the equation

$$\frac{\partial}{\partial t}P(\boldsymbol{r};t) = D_{ij}\frac{\partial^2}{\partial r_i \partial r_j}P(\boldsymbol{r},t), \quad P(\boldsymbol{r},0) = \delta(\boldsymbol{r}) \tag{8.118}$$

with the diffusion tensor (8.109).

Note that conversion from Eq. (8.106) to the equation for the probability density of particle coordinate (8.118) with the diffusion coefficient (8.109) corresponds to the so-called *Kramers problem* (see, e.g., [131]).

Delta-Correlated Approximation ($\lambda\tau_0 \ll 1$)

Under the assumption that $\lambda\tau_0 \ll 1$, where τ_0 is the temporal correlation radius of process $\boldsymbol{f}(t)$, Eq. (8.106) becomes simpler

$$\left(\frac{\partial}{\partial t} + \boldsymbol{v}\frac{\partial}{\partial \boldsymbol{r}} - \lambda\frac{\partial}{\partial \boldsymbol{v}}\boldsymbol{v}\right)P(\boldsymbol{r},\boldsymbol{v},t) = \lambda^2\int_0^t d\tau B_{ij}(\tau)\frac{\partial^2}{\partial v_i \partial v_j}P(\boldsymbol{r},\boldsymbol{v},t),$$

$$P(\boldsymbol{r},\boldsymbol{v};0) = \delta(\boldsymbol{r})\,\delta(\boldsymbol{v})$$

and corresponds to the approximation of random function $\boldsymbol{f}(t)$ by the delta-correlated process. If $t \gg \tau_0$, we can replace the upper limit of the integral with the infinity and proceed to the standard diffusion Fokker–Planck equation

$$\left(\frac{\partial}{\partial t} + \boldsymbol{v}\frac{\partial}{\partial \boldsymbol{r}} - \lambda\frac{\partial}{\partial \boldsymbol{v}}\boldsymbol{v}\right)P(\boldsymbol{r},\boldsymbol{v},t) = \lambda^2 D_{ij}\frac{\partial^2}{\partial v_i \partial v_j}P(\boldsymbol{r},\boldsymbol{v},t),$$
$$P(\boldsymbol{r},\boldsymbol{v};0) = \delta(\boldsymbol{r})\,\delta(\boldsymbol{v}) \tag{8.119}$$

with the diffusion tensor (8.109). In this approximation, the combined random process $\{\boldsymbol{r}(t),\ \boldsymbol{v}(t)\}$ is the Markovian process.

Under the condition $\lambda t \gg 1$, there are the steady-state equation for the probability density of particle velocity

$$-\lambda\frac{\partial}{\partial \boldsymbol{v}}\boldsymbol{v}P(\boldsymbol{v}) = \lambda^2 D_{ij}\frac{\partial^2}{\partial v_i \partial v_j}P(\boldsymbol{v})$$

and the nonsteady-state equation for the probability density of particle coordinate

$$\frac{\partial}{\partial t}P(\boldsymbol{r};t) = D_{ij}\frac{\partial^2}{\partial r_i \partial r_j}P(\boldsymbol{r},t), \quad P(\boldsymbol{r},0) = \delta(\boldsymbol{r})\,.$$

Another Asymptotic Limit ($\lambda\tau_0 \gg 1$)

Consider now the limit $\lambda\tau_0 \gg 1$. In this case, we can rewrite Eq. (8.106) in the form

$$\left(\frac{\partial}{\partial t} + \boldsymbol{v}\frac{\partial}{\partial \boldsymbol{r}} - \lambda\frac{\partial}{\partial \boldsymbol{v}}\boldsymbol{v}\right) P(\boldsymbol{r},\boldsymbol{v},t) = \lambda B_{ij}(0)\left[1 - e^{-\lambda t}\right]\frac{\partial^2}{\partial v_i \partial v_j} P(\boldsymbol{r},\boldsymbol{v},t)$$

$$- B_{ij}(0)\left[1 - e^{-\lambda t}\right]\frac{\partial^2}{\partial v_i \partial r_j} P(\boldsymbol{r},\boldsymbol{v},t) + \lambda\int\limits_0^t d\tau B_{ij}(\tau)\frac{\partial^2}{\partial v_i \partial r_j} P(\boldsymbol{r},\boldsymbol{v},t),$$

$$P(\boldsymbol{r},\boldsymbol{v};0) = \delta(\boldsymbol{r})\,\delta(\boldsymbol{v})\,.$$

Integrating this equation over $\boldsymbol{r}$, we obtain the equation for the probability density of particle velocity

$$\left(\frac{\partial}{\partial t} - \lambda\frac{\partial}{\partial \boldsymbol{v}}\boldsymbol{v}\right) P(\boldsymbol{v},t) = \lambda B_{ij}(0)\left[1 - e^{-\lambda t}\right]\frac{\partial^2}{\partial v_i \partial v_j} P(\boldsymbol{v},t),$$

$$P(\boldsymbol{v};0) = \delta(\boldsymbol{v})$$

and, under the condition $\lambda t \gg 1$, we arrive at the steady-state Gaussian probability density with variance

$$\langle v_i(t) v_j(t)\rangle = B_{ij}(0).$$

As regards the probability density of particle position, it satisfies, under the condition $\lambda t \gg 1$, the equation

$$\frac{\partial}{\partial t} P(\boldsymbol{r};t) = D_{ij}\frac{\partial^2}{\partial r_i \partial r_j} P(\boldsymbol{r},t), \quad P(\boldsymbol{r},0) = \delta(\boldsymbol{r}) \tag{8.120}$$

with the same diffusion coefficient as previously. This is a consequence of the fact that Eq. (8.113) is independent of parameter λ. Note that this equation corresponds to the limit process $\lambda \to \infty$ in Eq. (8.94)

$$\frac{d}{dt}\boldsymbol{r}(t) = \boldsymbol{v}(t), \quad \boldsymbol{v}(t) = \boldsymbol{f}(t), \quad \boldsymbol{r}(0) = 0.$$

In the limit $\lambda \to \infty$ (or $\lambda\tau_0 \gg 1$), we have the equality

$$\boldsymbol{v}(t) \approx \boldsymbol{f}(t), \tag{8.121}$$

and all multi-time statistics of random functions $\boldsymbol{v}(t)$ and $\boldsymbol{r}(t)$ will be described in terms of statistical characteristics of process $\boldsymbol{f}(t)$. In particular, the one-time probability density of particle velocity $\boldsymbol{v}(t)$ is the Gaussian probability density with variance $\langle v_i(t)v_j(t)\rangle = B_{ij}(0)$, and the spatial diffusion coefficient is

$$D = \frac{1}{2}\frac{d}{dt}\left\langle \boldsymbol{r}^2(t)\right\rangle = \int\limits_0^\infty d\tau B_{ii}(\tau) = \tau_0 B_{ii}(0).$$

As we have seen earlier, in the case of process $\boldsymbol{f}(t)$ such that it can be correctly described in the delta-correlated approximation (i.e., if $\lambda\tau_0 \ll 1$), the approximate equality (8.121) appears inappropriate to determine statistical characteristics of process $\boldsymbol{v}(t)$. Nevertheless, Eq. (8.120) with the same diffusion tensor remains as before valid for the one-time statistical characteristics of process $\boldsymbol{r}(t)$, which follows from the fact that Eq. (8.117) is valid for any parameter λ and arbitrary probability density of random process $\boldsymbol{f}(t)$.

Above, we considered several types of stochastic ordinary differential equations that allow obtaining closed statistical description in the general form. It is clear that similar situations can appear in dynamic systems formulated in terms of partial differential equations.

8.4.2 Partial Differential Equations

First of all, we note that the first-order partial differential equation

$$\left(\frac{\partial}{\partial t} + z(t)g(t)\frac{\partial}{\partial \boldsymbol{x}}\boldsymbol{F}(\boldsymbol{x})\right)\rho(\boldsymbol{r},t) = 0, \tag{8.122}$$

is equivalent to the system of ordinary differential equations (8.79), page 218 and, consequently, also allows the complete statistical description for arbitrary given random process $z(t)$. Solution $\rho(\boldsymbol{r},t)$ to Eq. (8.122) is a functional of random process $z(t)$ of the form

$$\rho(\boldsymbol{x},t) = \rho[\boldsymbol{x},t;\, z(\tau)] = \rho(\boldsymbol{x},T[t;\, z(\tau)])$$

with functional

$$T[t;\, z(\tau)] = \int\limits_0^t d\tau\, z(\tau)g(\tau), \tag{8.123}$$

so that Eq. (8.122) can be rewritten in the form

$$\left(\frac{\partial}{\partial T} + \frac{\partial}{\partial \boldsymbol{x}}\boldsymbol{F}(\boldsymbol{x})\right)\rho(\boldsymbol{x},T) = 0.$$

The variational derivative is expressed in this case as follows

$$\frac{\delta}{\delta z(\tau)}\rho(\boldsymbol{x},t) = \frac{\delta}{\delta z(\tau)}\rho(\boldsymbol{x},T[t;\, z(\tau)]) = \frac{d\rho(\boldsymbol{x},T)}{dT}\frac{\delta T[t;\, z(\tau)]}{\delta z(\tau)}$$

$$= \theta(t-\tau)g(\tau)\frac{d\rho(\boldsymbol{x},T)}{dT} = -\theta(t-\tau)g(\tau)\frac{\partial}{\partial \boldsymbol{x}}\boldsymbol{F}(\boldsymbol{x})\rho(\boldsymbol{x},t).$$

Consider now the class of nonlinear partial differential equations whose parameters are independent of spatial variable $\boldsymbol{x}$,

$$\left(\frac{\partial}{\partial t} + \boldsymbol{z}(t)\frac{\partial}{\partial \boldsymbol{x}}\right) q(\boldsymbol{x},t) = F\left(t, q, \frac{\partial q}{\partial \boldsymbol{x}}, \frac{\partial}{\partial \boldsymbol{x}} \otimes \frac{\partial q}{\partial \boldsymbol{x}}, \dots\right), \tag{8.124}$$

where $\boldsymbol{z}(t)$ is the vector random process and F is the deterministic function. Solution to this equation is representable in the form

$$q(\boldsymbol{x},t) = \exp\left\{-\int_0^t d\tau\, \boldsymbol{z}(\tau)\frac{\partial}{\partial \boldsymbol{x}}\right\} Q\left(\boldsymbol{x},t\right) = Q\left(\boldsymbol{x} - \int_0^t d\tau\, \boldsymbol{z}(\tau), t\right), \tag{8.125}$$

where we introduced the shift operator. Function $Q(x,t)$ satisfies the equation

$$\exp\left\{-\int_0^t d\tau\, \boldsymbol{z}(\tau)\frac{\partial}{\partial \boldsymbol{x}}\right\}\frac{\partial Q(\boldsymbol{x},t)}{\partial t}$$

$$= F\left(t, \left[\exp\left\{-\int_0^t d\tau\, \boldsymbol{z}(\tau)\frac{\partial}{\partial \boldsymbol{x}}\right\}Q\right], \left[\exp\left\{-\int_0^t d\tau\, \boldsymbol{z}(\tau)\frac{\partial}{\partial \boldsymbol{x}}\right\}\frac{\partial Q}{\partial \boldsymbol{x}}\right],\right.$$

$$\left.\frac{\partial}{\partial \boldsymbol{x}} \otimes \left[\exp\left\{-\int_0^t d\tau\, \boldsymbol{z}(\tau)\frac{\partial}{\partial \boldsymbol{x}}\right\}\frac{\partial Q}{\partial \boldsymbol{x}}\right], \dots,\right). \tag{8.126}$$

The shift operator can be factored out from arguments of function

$$F\left(t, q, \frac{\partial q}{\partial \boldsymbol{x}}, \frac{\partial}{\partial \boldsymbol{x}} \otimes \frac{\partial q}{\partial \boldsymbol{x}}, \dots\right),$$

and Eq. (8.126) assumes the form of the deterministic equation

$$\frac{\partial Q(\boldsymbol{x},t)}{\partial t} = F\left(t, Q, \frac{\partial Q}{\partial \boldsymbol{x}}, \frac{\partial}{\partial \boldsymbol{x}} \otimes \exp\frac{\partial Q}{\partial \boldsymbol{x}}, \dots\right). \tag{8.127}$$

Thus, the variational derivative can be expressed in the form

$$\frac{\delta q(\boldsymbol{x},t)}{\delta \boldsymbol{z}(\tau)} = \frac{\delta Q\left(\boldsymbol{x} - \int_0^t d\tau\, \boldsymbol{z}(\tau), t\right)}{\delta \boldsymbol{z}(\tau)} = -\theta(t-\tau)\frac{\partial}{\partial \boldsymbol{x}}q(\boldsymbol{x},t), \tag{8.128}$$

i.e., the variational derivatives of the solution to problem (8.124) are expressed in terms of the spatial derivatives.

Consider the *Burgers equation* with random drift $\boldsymbol{z}(t)$ as a specific example:

$$\frac{\partial}{\partial t}\boldsymbol{v}(\boldsymbol{x},t)+\left[(\boldsymbol{v}+\boldsymbol{z}(t))\frac{\partial}{\partial \boldsymbol{x}}\right]\boldsymbol{v}(\boldsymbol{x},t)=\nu\frac{\partial^2}{\partial \boldsymbol{x}^2}\boldsymbol{v}(\boldsymbol{x},t). \tag{8.129}$$

In this case, the solution has the form

$$\boldsymbol{v}(\boldsymbol{x},t)=\exp\left\{-\int\limits_0^t d\tau\,\boldsymbol{z}(\tau)\frac{\partial}{\partial \boldsymbol{x}}\right\}\boldsymbol{V}(\boldsymbol{x},t)=\boldsymbol{V}\left(\boldsymbol{x}-\int\limits_0^t d\tau\,z(\tau),t\right),$$

where function $\boldsymbol{V}(x,t)$ satisfies the standard Burgers equation

$$\frac{\partial}{\partial t}\boldsymbol{V}(\boldsymbol{x},t)+\left(\boldsymbol{V}(\boldsymbol{x},t)\frac{\partial}{\partial \boldsymbol{x}}\right)\boldsymbol{V}(\boldsymbol{x},t)=\nu\frac{\partial^2}{\partial \boldsymbol{x}^2}\boldsymbol{V}(\boldsymbol{x},t). \tag{8.130}$$

In this example, the variational derivative is given by the expression

$$\frac{\delta}{\delta z_i(\tau)}\boldsymbol{v}(\boldsymbol{x},t)=\frac{\delta}{\delta z_i(\tau)}\boldsymbol{V}\left(\boldsymbol{x}-\int\limits_0^t d\tau\,z(\tau),t\right)=-\theta(t-\tau)\frac{\partial}{\partial x_i}\boldsymbol{v}(\boldsymbol{x},t). \tag{8.131}$$

In the case of such problems, statistical characteristics of the solution can be determined immediately by averaging the corresponding expressions constructed from the solution to the last equation. Proceeding in this way, one obtains that the desired function, say, function $\langle q(\boldsymbol{x},t)\rangle$, satisfies a closed equation containing derivatives of all orders with respect to $\boldsymbol{x}$.

Consider two examples.

One-Dimensional Diffusion of Passive Tracers

Consider the one-dimensional diffusion of passive tracers in random velocity field. This problem is formulated in terms of the equation

$$\frac{\partial}{\partial t}\rho(x,t)+v(t)\frac{\partial}{\partial x}f(x)\rho(x,t)=0, \tag{8.132}$$

where we will assume that $v(t)$ is the stationary random Gaussian process with parameters

$$\langle v(t)\rangle=0,\quad B(t-t')=\langle v(t)v(t')\rangle\quad \left(B_v(0)=\langle v^2(t)\rangle\right)$$

and $f(x)$ is the deterministic function. The indicator function $\varphi(x,t;\rho)=\delta\left(\rho(x,t)-\rho\right)$ for Eq. (8.132) satisfies the equation

$$\left(\frac{\partial}{\partial t}+v(t)f(x)\frac{\partial}{\partial x}\right)\varphi(x,t;\rho)=v(t)\frac{df(x)}{dx}\frac{\partial}{\partial\rho}\rho\varphi(x,t;\rho),$$

$$\varphi(x,0;\rho)=\delta(\rho_0(x)-\rho).$$

We rewrite this equation in the form

$$\frac{\partial}{\partial t}\varphi(x,t;\rho)=-v(t)\left\{\frac{\partial}{\partial x}f(x)-\frac{df(x)}{dx}\left(1+\frac{\partial}{\partial\rho}\rho\right)\right\}\varphi(x,t;\rho),$$
$$\varphi(x,0;\rho)=\delta(\rho_0(x)-\rho). \tag{8.133}$$

Averaging Eq. (8.133) over an ensemble of realizations of random process $v(t)$, we obtain the expression

$$\frac{\partial}{\partial t}P(x,t;\rho)=-\int\limits_0^t dt'B(t-t')\left\{\frac{\partial}{\partial x}f(x)-\frac{df(x)}{dx}\left(1+\frac{\partial}{\partial\rho}\rho\right)\right\}\left\langle\frac{\delta}{\delta v(t')}\varphi(t,x;\rho)\right\rangle,$$
$$P(x,0;\rho)=\delta(\rho_0(x)-\rho). \tag{8.134}$$

The solution to Eq. (8.133) has the form

$$\varphi(x,t;\rho)=\varphi(x,T(t);\rho),$$

where $T(t)=\int_0^t d\tau v(\tau)$ is the new (random) time and function $\varphi(x,T;\rho)$ as a function of its arguments satisfies the deterministic equation

$$\frac{\partial}{\partial T}\varphi(x,T;\rho)=-\left\{\frac{\partial}{\partial x}f(x)-\frac{df(x)}{dx}\left(1+\frac{\partial}{\partial\rho}\rho\right)\right\}\varphi(x,T;\rho),$$
$$\varphi(x,0;\rho)=\delta(\rho_0(x)-\rho). \tag{8.135}$$

Consequently,

$$\frac{\delta}{\delta v(t')}\varphi(x,t;\rho)=\frac{\partial\varphi(x,T;\rho)}{\partial T}\frac{\delta}{\delta v(t')}T(t)=\frac{\partial\varphi(x,T;\rho)}{\partial T}\theta(t-t')$$
$$=-\theta(t-t')\left\{\frac{\partial}{\partial x}f(x)-\frac{df(x)}{dx}\left(1+\frac{\partial}{\partial\rho}\rho\right)\right\}\varphi(x,t;\rho),$$

where $\theta(t)$ is the Heaviside step function (1 for $t>0$ and 0 for $t<0$), and Eq. (8.134) assumes the form of a closed equation

$$\frac{\partial}{\partial t}P(x,t;\rho) = \int\limits_0^t dt' B(t-t') \left\{ \frac{\partial}{\partial x} f(x) - \frac{df(x)}{dx} \left(1 + \frac{\partial}{\partial \rho}\rho\right) \right\}$$
$$\times \left\{ \frac{\partial}{\partial x} f(x) - \frac{df(x)}{dx} \left(1 + \frac{\partial}{\partial \rho}\rho\right) \right\} P(x,t;\rho),$$
$$P(x,0;\rho) = \delta(\rho_0(x) - \rho).$$

Burgers Equation with Random Drift

Consider the one-dimensional Burgers equation with random drift

$$\frac{\partial}{\partial t}q(x,t) + (q + z(t))\frac{\partial}{\partial x}q(x,t) = \nu\frac{\partial^2}{\partial x^2}q(x,t). \tag{8.136}$$

In this case, we have for the variational derivative of $q(x,t)$ with respect to $z(\tau)$

$$\frac{\delta}{\delta z(\tau)}q(x,t) = \frac{\delta}{\delta z(\tau)}Q\left(x - \int\limits_0^t d\tau z(\tau), t\right) = -\theta(t-\tau)\frac{\partial}{\partial x}q(x,t). \tag{8.137}$$

Assume now that random process $z(t)$ is the Gaussian process stationary in time and described by correlation function $B(t-t')$. Let us average Eq. (8.136) over an ensemble of realizations of process $z(t)$ to obtain

$$\frac{\partial}{\partial t}\langle q(x,t)\rangle + \frac{1}{2}\frac{\partial}{\partial x}\langle q^2(x,t)\rangle + \frac{\partial}{\partial x}\langle z(t)q(x,t)\rangle = \nu\frac{\partial^2}{\partial x^2}\langle q(x,t)\rangle. \tag{8.138}$$

We split the correlators in the left-hand side of this equation using formulas (see Sect. 7.2, page 168)

$$\langle q[z(\tau)+\eta_1(\tau)]q[z(\tau)+\eta_2(\tau)]\rangle$$
$$= \exp\left\{\int\limits_0^t d\tau_1 \int\limits_0^t d\tau_2 B(\tau_1-\tau_2)\frac{\delta^2}{\delta\eta_1(\tau_1)\delta\eta_2(\tau_2)}\right\}\langle q[z(\tau)+\eta_1(\tau)]q[z(\tau)+\eta_2(\tau)]\rangle,$$

$$\langle z(t)q(x,t)\rangle = \int\limits_0^t d\tau B(t-\tau)\left\langle \frac{\delta}{\delta z(\tau)}q(x,t)\right\rangle.$$

In view of Eq. (8.137) we can represent these formulas in the form

$$\left\langle q^2(x,t)\right\rangle = \exp\left\{2\int\limits_0^t d\tau(t-\tau)B(\tau)\frac{\partial^2}{\partial\eta_1\partial\eta_2}\right\}\left\langle q(x+\eta_1,t)\right\rangle\left\langle q(x+\eta_2,t)\right\rangle|_{\eta=0}$$

$$= \sum_{n=0}^{\infty}\frac{2^n}{n!}\left[\int\limits_0^t d\tau(t-\tau)B(\tau)\right]^n\left[\frac{\partial^n}{\partial x^n}\left\langle q(x,t)\right\rangle\right]^2,$$

$$\left\langle z(t)q(x,t)\right\rangle = -\int\limits_0^t d\tau B(\tau)\frac{\partial}{\partial x}\left\langle q(x,t)\right\rangle.$$

As a result, Eq. (8.138) becomes the closed equation

$$\frac{\partial}{\partial t}\left\langle q(x,t)\right\rangle + \frac{1}{2}\frac{\partial}{\partial x}\sum_{n=0}^{\infty}\frac{2^n}{n!}\left[\int\limits_0^t d\tau(t-\tau)B(\tau)\right]^n\left[\frac{\partial^n}{\partial x^n}\left\langle q(x,t)\right\rangle\right]^2$$
$$= \left(\nu + \int\limits_0^t d\tau B(\tau)\right)\frac{\partial^2}{\partial x^2}\left\langle q(x,t)\right\rangle, \quad (8.139)$$

However, in contrast to (8.136), this equation depends on all derivatives with respect to spatial variable x [48, 49].

Unfortunately, there is only limited number of equations that allow sufficiently complete analysis. In the general case, the analysis of dynamic systems appears possible only on the basis of various asymptotic and approximate techniques. In physics, techniques based on approximating actual random processes and fields by the fields delta-correlated in time are often and successfully used.

8.5 Delta-Correlated Fields and Processes

8.5.1 General Remarks

In the case of random field $\boldsymbol{f}(\boldsymbol{x},t)$ delta-correlated in time, the following equality holds (see Sect. 7.8, page 181)

$$\dot{\Theta}_t[t,t_0;\boldsymbol{v}(\boldsymbol{y},\tau)] \equiv \dot{\Theta}_t[t,t_0;\boldsymbol{v}(\boldsymbol{y},t)]$$

and situation becomes significantly simpler. The fact that field $\boldsymbol{f}(\boldsymbol{x},t)$ is delta-correlated means that

$$\Theta[t,t_0;\boldsymbol{v}(\boldsymbol{y},\tau)] = \sum_{i=1}^{\infty} \frac{i^n}{n!} \int\limits_{t_0}^{t} d\tau \int d\boldsymbol{y}_1 \cdots \int d\boldsymbol{y}_n K_n^{i_1,\cdots,i_n}(\boldsymbol{y}_1,\cdots,\boldsymbol{y}_n;\tau)$$

$$\times\, v_{i_1}(\boldsymbol{y}_1,\tau)\cdots v_{i_n}(\boldsymbol{y}_n,\tau),$$

which, in turn, means that field $\boldsymbol{f}(\boldsymbol{x},t)$ is characterized by cumulant functions of the form

$$K_n^{i_1,\cdots,i_n}(\boldsymbol{y}_1,t_1;\cdots;\boldsymbol{y}_n,t_n) = K_n^{i_1,\cdots,i_n}(\boldsymbol{y}_1,\cdots,\boldsymbol{y}_n;t_1)\delta(t_1-t_2)\cdots\delta(t_{n-1}-t_n).$$

In view of Eq. (8.3), page 194 Eqs. (8.7), (8.9), and (8.12) appear to be in this case the closed operator equations in functions $P(\boldsymbol{x},t)$, $p(\boldsymbol{x},t|\boldsymbol{x}_0,t_0)$, and $P_m(\boldsymbol{x}_1,t_1;\cdots;\boldsymbol{x}_m,t_m)$. Indeed, Eq. (8.7), page 195 is reduced to the equation

$$\left(\frac{\partial}{\partial t} + \frac{\partial}{\partial \boldsymbol{x}}\boldsymbol{v}(\boldsymbol{x},t)\right) P(\boldsymbol{x},t) = \dot{\Theta}_t\left[t,t_0;i\frac{\partial}{\partial \boldsymbol{x}}\delta(\boldsymbol{y}-\boldsymbol{x})\right] P(\boldsymbol{x},t),$$
$$P(\boldsymbol{x},0) = \delta(\boldsymbol{x}-\boldsymbol{x}_0), \tag{8.140}$$

whose concrete form is governed by functional $\Theta[t,t_0;\boldsymbol{v}(\boldsymbol{y},\tau)]$, i.e., by statistical behavior of random field $\boldsymbol{f}(\boldsymbol{x},t)$. Correspondingly, Eq. (8.12) for the m-time probability density is reduced to the operator equation ($t_1 \le t_2 \le \cdots \le t_m$)

$$\left(\frac{\partial}{\partial t} + \frac{\partial}{\partial \boldsymbol{x}_m}\boldsymbol{v}(\boldsymbol{x}_m,t_m)\right) P_m(\boldsymbol{x}_1,t_1;\cdots;\boldsymbol{x}_m,t_m)$$

$$= \dot{\Theta}_{t_m}\left[t_m,t_0;i\frac{\partial}{\partial \boldsymbol{x}_m}\delta(\boldsymbol{y}-\boldsymbol{x}_m)\right] P_m(\boldsymbol{x}_1,t_1;\cdots;\boldsymbol{x}_m,t_m), \tag{8.141}$$

$$P_m(\boldsymbol{x}_1,t_1;\cdots;\boldsymbol{x}_m,t_{m-1}) = \delta(\boldsymbol{x}_m-\boldsymbol{x}_{m-1})P_{m-1}(\boldsymbol{x}_1,t_1;\cdots;\boldsymbol{x}_{m-1},t_{m-1}).$$

We can seek the solution to Eq. (8.141) in the form

$$P_m(\boldsymbol{x}_1,t_1;\cdots;\boldsymbol{x}_m,t_m) = p(\boldsymbol{x}_m,t_m|\boldsymbol{x}_{m-1},t_{m-1})P_{m-1}(\boldsymbol{x}_1,t_1;\cdots;\boldsymbol{x}_{m-1},t_{m-1}). \tag{8.142}$$

Because all differential operations in Eq. (8.141) concern only t_m and $\boldsymbol{x}_m$, we can substitute Eq. (8.142) in Eq. (8.141) to obtain the following equation for the transition probability density

$$\left(\frac{\partial}{\partial t} + \frac{\partial}{\partial \boldsymbol{x}}\boldsymbol{v}(\boldsymbol{x},t)\right) p(\boldsymbol{x},t|\boldsymbol{x}_0,t_0) = \dot{\Theta}_t\left[t,t_0;i\frac{\partial}{\partial \boldsymbol{x}}\delta(\boldsymbol{y}-\boldsymbol{x})\right] p(\boldsymbol{x},t|\boldsymbol{x}_0,t_0),$$
$$p(\boldsymbol{x},t|\boldsymbol{x}_0,t_0)|_{t\to t_0} = \delta(\boldsymbol{x}-\boldsymbol{x}_0). \tag{8.143}$$

Here, we denoted variables $\boldsymbol{x}_m$ and t_m as $\boldsymbol{x}$ and t and variables $\boldsymbol{x}_{m-1}$ and t_{m-1} as $\boldsymbol{x}_0$ and t_0.

Using formula (8.142) $(m-1)$ times, we obtain the relationship

$$P_m(\boldsymbol{x}_1,t_1;\cdots;\boldsymbol{x}_m,t_m) = p(\boldsymbol{x}_m,t_m|\boldsymbol{x}_{m-1},t_{m-1})\cdots p(\boldsymbol{x}_2,t_2|\boldsymbol{x}_1,t_1)P(\boldsymbol{x}_1,t_1), \tag{8.144}$$

where $P(\boldsymbol{x}_1,t_1)$ is the one-time probability density governed by Eq. (8.140). Equality (8.144) expresses the many-time probability density in terms of the product of transition probability densities, which means that random process $\boldsymbol{x}(t)$ is the Markovian process. The transition probability density is defined in this case as follows:

$$p(\boldsymbol{x},t|\boldsymbol{x}_0,t_0) = \langle\delta(\boldsymbol{x}(t)-\boldsymbol{x})|\boldsymbol{x}_0,t_0\rangle\,.$$

Special models of parameter fluctuations can significantly simplify the obtained equations.

For example, in the case of the Gaussian delta-correlated field $\boldsymbol{f}(\boldsymbol{x},t)$, the correlation tensor has the form $(\langle\boldsymbol{f}(\boldsymbol{x},t)\rangle = 0)$

$$B_{ij}(\boldsymbol{x},t;\boldsymbol{x}',t') = 2\delta(t-t')F_{ij}(\boldsymbol{x},\boldsymbol{x}';t).$$

Then, functional $\Theta[t,t_0;\boldsymbol{v}(\boldsymbol{y},\tau)]$ assumes the form

$$\Theta[t,t_0;\boldsymbol{v}(\boldsymbol{y},\tau)] = -\int\limits_{t_0}^{t} d\tau\int d\boldsymbol{y}_1\int d\boldsymbol{y}_2 F_{ij}(\boldsymbol{y}_1,\boldsymbol{y}_2;\tau)v_i(\boldsymbol{y}_1,\tau)v_j(\boldsymbol{y}_2,\tau),$$

and Eq. (8.140) reduces to the Fokker–Planck equation

$$\left(\frac{\partial}{\partial t}+\frac{\partial}{\partial x_k}\left[v_k(\boldsymbol{x},t)+A_k(\boldsymbol{x},t)\right]\right)P(\boldsymbol{x},t) = \frac{\partial^2}{\partial x_k\partial x_l}\left[F_{kl}(\boldsymbol{x},\boldsymbol{x},t)P(\boldsymbol{x},t)\right], \tag{8.145}$$

where

$$A_k(\boldsymbol{x},t) = \left.\frac{\partial}{\partial x'_l}F_{kl}(\boldsymbol{x},\boldsymbol{x}';t)\right|_{\boldsymbol{x}'=\boldsymbol{x}}.$$

Note that Eq. (8.9), page 197 in this case assumes the form of the backward Fokker–Planck equation (see, e.g., [21])

$$\left(\frac{\partial}{\partial t_0}+\left[v_k(\boldsymbol{x}_0,t_0)+A_k(\boldsymbol{x}_0,t_0)\right]\frac{\partial}{\partial x_{0k}}\right)P(\boldsymbol{x},t|\boldsymbol{x}_0,t_0)$$
$$= -F_{kl}(\boldsymbol{x}_0,\boldsymbol{x}_0;t_0)\frac{\partial^2}{\partial x_{0k}\partial x_{0l}}P(\boldsymbol{x},t|\boldsymbol{x}_0,t_0) \tag{8.146}$$

with initial condition

$$P(\boldsymbol{x},t|\boldsymbol{x}_0,t)=\delta(\boldsymbol{x}-\boldsymbol{x}_0).$$

In view of the special role that the Gaussian delta-correlated field $\boldsymbol{f}(\boldsymbol{x},t)$ plays in physics, we give an alternative and more detailed discussion of this approximation commonly called the *approximation of the Gaussian delta-correlated field* in Chapter 10, page 305.

For random field $\boldsymbol{f}(\boldsymbol{x},t)$ related to the delta-correlated Poisson process (see Sect. 7.8 page 181) one can obtain the forward and backward equations of type of the Kolmogorov–Feller equation.

We illustrate the above general theory by the examples of several equations.

8.5.2 One-Dimensional Nonlinear Differential Equation

Consider the one-dimensional stochastic equation

$$\frac{d}{dt}x(t)=f(x,t)+z(t)g(x,t),\quad x(0)=x_0, \tag{8.147}$$

where $f(x,t)$ and $g(x,t)$ are the deterministic functions and $z(t)$ is the random function of time. For indicator function $\varphi(x,t)=\delta(x(t)-x)$, we have the Liouville equation

$$\left(\frac{\partial}{\partial t}+\frac{\partial}{\partial x}f(x,t)\right)\varphi(x,t)=-z(t)\frac{\partial}{\partial x}\left\{g(x,t)\varphi(x,t)\right\},$$

so that the equation for the one-time probability density $P(x,t)$ has the form

$$\left(\frac{\partial}{\partial t}+\frac{\partial}{\partial x}f(x,t)\right)P(x,t)=\left\langle\dot{\Theta}_t\left[t,\frac{\delta}{i\delta z(\tau)}\right]\varphi(x,t)\right\rangle.$$

In the case of the delta-correlated random process $z(t)$, the equality

$$\dot{\Theta}_t[t,v(\tau)]=\dot{\Theta}_t[t,v(t)]$$

holds. Taking into account the equality

$$\frac{\delta}{\delta z(t-0)}\varphi(x,t)=-\frac{\partial}{\partial x}\left\{g(x,t)\varphi(x,t)\right\},$$

we obtain the closed operator equation

$$\left(\frac{\partial}{\partial t}+\frac{\partial}{\partial x}f(x,t)\right)P(x,t)=\dot{\Theta}_t\left[t,i\frac{\partial}{\partial x}g(x,t)\right]P(x,t). \tag{8.148}$$

For the Gaussian delta-correlated process , we have

$$\Theta[t, v(\tau)] = -\frac{1}{2}\int\limits_0^t d\tau B(\tau) v^2(\tau), \tag{8.149}$$

and Eq. (8.148) assumes the form of the Fokker–Planck equation

$$\left(\frac{\partial}{\partial t} + \frac{\partial}{\partial x} f(x,t)\right) P(x,t) = \frac{1}{2} B(t) \frac{\partial}{\partial x} g(x,t) \frac{\partial}{\partial x} g(x,t) P(x,t). \tag{8.150}$$

For the Poisson delta-correlated process $z(t)$, we have

$$\Theta[t, v(\tau)] = \nu \int\limits_0^t d\tau \left\{ \int\limits_{-\infty}^{\infty} d\xi p(\xi) e^{i\xi v(\tau)} - 1 \right\}, \tag{8.151}$$

and Eq. (8.148) reduces to the form

$$\left(\frac{\partial}{\partial t} + \frac{\partial}{\partial x} f(x,t)\right) P(x,t) = \nu \left\{ \int\limits_{-\infty}^{\infty} d\xi p(\xi) e^{-\xi \frac{\partial}{\partial x} g(x,t)} - 1 \right\} P(x,t). \tag{8.152}$$

If we set $g(x,t) = 1$, Eq. (8.147) assumes the form

$$\frac{d}{dt} x(t) = f(x,t) + z(t), \quad x(0) = x_0.$$

In this case, the operator in the right-hand side of Eq. (8.152) is the shift operator, and Eq. (8.152) assumes the form of the Kolmogorov–Feller equation

$$\left(\frac{\partial}{\partial t} + \frac{\partial}{\partial x} f(x,t)\right) P(x,t) = \nu \int\limits_{-\infty}^{\infty} d\xi p(\xi) P(x-\xi, t) - \nu P(x,t).$$

Define now $g(x,t) = x$, so that Eq. (8.147) reduces to the form

$$\frac{d}{dt} x(t) = f(x,t) + z(t) x(t), \quad x(0) = x_0.$$

In this case, Eq. (8.152) assumes the form

$$\left(\frac{\partial}{\partial t} + \frac{\partial}{\partial x} f(x,t)\right) P(x,t) = \nu \left\{ \int\limits_{-\infty}^{\infty} d\xi p(\xi) e^{-\xi \frac{\partial}{\partial x} x} - 1 \right\} P(x,t). \tag{8.153}$$

To determine the action of the operator in the right-hand side of Eq. (8.153), we expand it in series in ξ

$$\left\{e^{-\xi\frac{\partial}{\partial x}x}-1\right\}P(x,t)=\sum_{n=1}^{\infty}\frac{(-\xi)^n}{n!}\left(\frac{\partial}{\partial x}x\right)^n P(x,t)$$

and consider the action of every term.

Representing x in the form $x = e^{\varphi}$, we can transform this formula as follows (the fact that x is the alternating quantity is insignificant here)

$$\sum_{n=1}^{\infty}\frac{(-\xi)^n}{n!}e^{-\varphi}\frac{\partial^n}{\partial\varphi^n}e^{\varphi}P(e^{\varphi},t)=e^{-\varphi}\left\{e^{-\xi\frac{\partial}{\partial\varphi}}-1\right\}e^{\varphi}P(e^{\varphi},t)=$$
$$e^{-\xi}P(e^{\varphi-\xi},t)-P(e^{\varphi},t).$$

Reverting to variable x, we can represent Eq. (8.153) in the final form of the integro-differential equation similar to the Kolmogorov–Feller equation

$$\left(\frac{\partial}{\partial t}+\frac{\partial}{\partial x}f(x,t)\right)P(x,t)=\nu\int\limits_{-\infty}^{\infty}d\xi p(\xi)e^{-\xi}P(xe^{-\xi},t)-\nu P(x,t).$$

In Chapter 3, page 137, we mentioned that formula

$$x(t)=\int\limits_0^t d\tau g(t-\tau)z(\tau)$$

relates the Poisson process $x(t)$ with arbitrary impulse function $g(t)$ to the Poisson delta-correlated random process $z(t)$. Let $g(t) = e^{-\lambda t}$. In this case, process $x(t)$ satisfies the stochastic differential equation

$$\frac{d}{dt}x(t)=-\lambda x(t)+z(t)$$

and, consequently, both transition probability density and one-time probability density of this process satisfy, according to Eq. (8.152), the equations

$$\frac{\partial}{\partial t}p(x,t|x_0,t_0)=\widehat{L}(x)p(x,t|x_0,t_0),\quad \frac{\partial}{\partial t}P(x,t)=\widehat{L}(x)P(x,t),$$

where operator

$$\widehat{L}(x)=\lambda\frac{\partial}{\partial x}x+\nu\left\{\int\limits_{-\infty}^{\infty}d\xi p(\xi)e^{-\xi\frac{\partial}{\partial x}}-1\right\}. \tag{8.154}$$

8.5.3 Linear Operator Equation

Consider now the linear operator equation

$$\frac{d}{dt}\boldsymbol{x}(t) = \widehat{A}(t)\boldsymbol{x}(t) + z(t)\widehat{B}(t)\boldsymbol{x}(t), \quad \boldsymbol{x}(0) = \boldsymbol{x}_0, \tag{8.155}$$

where $\widehat{A}(t)$ and $\widehat{B}(t)$ are the deterministic operators (e.g., differential operators with respect to auxiliary variables or regular matrixes). We will assume that function $z(t)$ is the random delta-correlated function.

Averaging system (8.155), we obtain, according to general formulas,

$$\frac{d}{dt}\left\langle \boldsymbol{x}(t)\right\rangle = \widehat{A}(t)\left\langle \boldsymbol{x}(t)\right\rangle + \left\langle \dot{\Theta}_t\left[t, \frac{\delta}{i\delta z(t)}\right]\boldsymbol{x}(t)\right\rangle. \tag{8.156}$$

Then, taking into account the equality

$$\frac{\delta}{\delta z(t-0)}\boldsymbol{x}(t) = \widehat{B}(t)\boldsymbol{x}(t)$$

that follows immediately from Eq. (8.155), we can rewrite Eq. (8.156) in the form

$$\frac{d}{dt}\left\langle \boldsymbol{x}(t)\right\rangle = \widehat{A}(t)\left\langle \boldsymbol{x}(t)\right\rangle + \dot{\Theta}_t\left[t, -i\widehat{B}\right]\left\langle \boldsymbol{x}(t)\right\rangle. \tag{8.157}$$

Thus, in the case of linear system (8.155), equations for average values also are the linear equations.

We can expand the logarithm of the characteristic functional $\Theta[t; v(\tau)]$ of delta-correlated processes in the functional Fourier series

$$\Theta[t; v(\tau)] = \sum_{n=1}^{\infty}\frac{i^n}{n!}\int\limits_0^t d\tau K_n(\tau)v^n(\tau), \tag{8.158}$$

where $K_n(t)$ determine the cumulant functions of process $z(t)$. Substituting Eq. (8.158) in Eq. (8.157), we obtain the equation

$$\frac{d}{dt}\left\langle \boldsymbol{x}(t)\right\rangle = \widehat{A}(t)\left\langle \boldsymbol{x}(t)\right\rangle + \sum_{n=1}^{\infty}\frac{1}{n!}K_n(t)\left[\widehat{B}(t)\right]^n\left\langle \boldsymbol{x}(t)\right\rangle. \tag{8.159}$$

If there exists power l such that $\widehat{B}^l(t) = 0$, then Eq. (8.159) assumes the form

$$\frac{d}{dt}\left\langle \boldsymbol{x}(t)\right\rangle = \widehat{A}(t)\left\langle \boldsymbol{x}(t)\right\rangle + \sum_{n=1}^{l-1}\frac{1}{n!}K_n(t)\left[\widehat{B}(t)\right]^n\left\langle \boldsymbol{x}(t)\right\rangle. \tag{8.160}$$

In this case, the equation for average value depends only on a finite number of cumulants of process $z(t)$. This means that there is no necessity in knowledge of probability distribution of function $z(t)$ in the context of the equation for average value; sufficient information includes only certain cumulants of process and knowledge of the fact that process $z(t)$ can be considered as the delta-correlated random process. Statistical description of an oscillator with fluctuating frequency is a good example of such system in physics.

Stochastic Parametric Resonance

Consider statistical description of an oscillator with fluctuating frequency (1.32), page 22 as an example of simple linear dynamic system that allows a sufficiently complete analysis. The problem on such an oscillator is formulated in terms of the initial-value problem for the second-order differential equation

$$\frac{d^2}{dt^2}x(t) + \omega_0^2[1 + z(t)]x(t) = 0,$$
$$x(0) = x_0, \quad \frac{d}{dt}x(0) = y_0, \tag{8.161}$$

which is equivalent to the system of equations

$$\frac{d}{dt}x(t) = y(t), \quad \frac{d}{dt}y(t) = -\omega_0^2[1 + z(t)]x(t),$$
$$x(0) = x_0, \quad y(0) = y_0. \tag{8.162}$$

For system (8.162), indicator function

$$\varphi(x, y, t) = \delta\left(x(t) - x\right)\delta\left(y(t) - y\right)$$

satisfies the Liouville equation

$$\left(\frac{\partial}{\partial t} + y\frac{\partial}{\partial x} - \omega_0^2 x\frac{\partial}{\partial y}\right)\varphi(x, y, t) = \omega_0^2 z(t) x \frac{\partial}{\partial y}\varphi(x, y, t).$$

The joint one-time probability density of solutions to system (8.162) is defined by the equality $P(x, y, t) = \langle\varphi(x, y, t)\rangle$ and satisfies the operator equation

$$\left(\frac{\partial}{\partial t} + y\frac{\partial}{\partial x} - \omega_0^2 x\frac{\partial}{\partial y}\right)P(x, y, t) = \left\langle \dot{\Theta}_t\left[t; \frac{\delta}{i\delta z(\tau)}\right]\varphi(x, y, t)\right\rangle, \tag{8.163}$$

where $\dot{\Theta}_t\left[t; v(\tau)\right] = \dfrac{d}{dt}\Theta\left[t; v(\tau)\right]$, and $\Theta\left[t; v(\tau)\right]$ is the logarithm of the characteristic functional of process $z(t)$,

$$\Theta\left[t; v(\tau)\right] = \ln\left\langle \exp\left\{ i \int\limits_0^t d\tau z(\tau) v(\tau) \right\} \right\rangle.$$

In the case of delta-correlated process $z(t)$, the joint one-time probability density of solutions to system (8.162) satisfies the simplified operator equation

$$\left(\frac{\partial}{\partial t} + y\frac{\partial}{\partial x} - \omega_0^2 x \frac{\partial}{\partial y}\right) P(x,y,t) = \left\langle \dot{\Theta}_t\left[t; \frac{\delta}{i\delta z(t)}\right] \varphi(x,y,t) \right\rangle, \qquad (8.164)$$

which, in view of the equality

$$\frac{\delta}{\delta z(t-0)}\varphi(x,y,t) = \omega_0^2 x \frac{\partial}{\partial y}\varphi(x,y,t)$$

immediately following from the Liouville equation, can be represented as the closed operator equation

$$\left(\frac{\partial}{\partial t} + y\frac{\partial}{\partial x} - \omega_0^2 x \frac{\partial}{\partial y}\right) P(x,y,t) = \dot{\Theta}_t\left[t; -i\omega_0^2 x \frac{\partial}{\partial y}\right] P(x,y,t). \qquad (8.165)$$

Equation (8.165) offers a possibility of deriving closed systems of equations for moments of arbitrary orders. This possibility follows from the fact that the operator in the right-hand side of Eq. (8.165) depends only on homogeneous combination $x\dfrac{\partial}{\partial y}$ whose action cannot increase the order of the moment under consideration, which is, of cause, a consequence of linearity of the initial system of equations (8.162). Hence, the equations for moments will depend only on the process $z(t)$ cumulants whose orders are smaller or equal to the order of the moment of interest.

Indeed, consider the vector quantity

$$A_k(t) = x^k(t) y^{N-k}(t) \quad (k = 0, \ldots, N).$$

One can derive from system (8.162) that this quantity satisfies the stochastic equation

$$\frac{d}{dt}A_k(t) = kA_{k-1}(t) - \omega_0^2(N-k)[1+z(t)]A_{k+1}(t) \quad (k = 0, \ldots, N),$$

which corresponds to the linear operator equation (8.155), page 243 with constant matrixes

$$\widehat{A}_{ij} = i\delta_{i,j+1} - \omega_0^2(N-i)\delta_{i,j-1}, \quad \widehat{B}_{ij} = -\omega_0^2(N-i)\delta_{i,j-1}.$$

It is obvious that the square of matrix $\widehat{B}_{ij}$ is

$$\widehat{B}^2_{ij} = -\omega_0^4(N-i)(N-j+i)\delta_{i,j-2}$$

and so on for higher powers; consequently, for power $N+1$ we have

$$\widehat{B}^{N+1} \equiv 0.$$

According to (8.159), page 243, averages $\langle A_k(t)\rangle$ ($k = 0, \cdots, N$) satisfy the equation

$$\frac{d}{dt}\langle A_k(t)\rangle = k\langle A_{k-1}(t)\rangle - \omega_0^2(N-k)\langle A_{k+1}(t)\rangle + \sum_{n=1}^{N}\frac{1}{n!}K_n\left[\widehat{B}^n\right]_{kl}\langle A_l(t)\rangle, \tag{8.166}$$

where K_n are the cumulants of random process $z(t)$ and the summation is truncated at $n = N$ because, as was mentioned, the equation for average value can depend only on the cumulants of process $z(t)$ whose orders are smaller or equal to N. In particular, the first moments of the solution to the system of stochastic equations (8.162) for the delta-correlated process $z(t)$ are independent of fluctuations of system parameters in view of the equality $K_1 = 0$, and the second moments satisfy the system of equations that coincides with the system derived for the Gaussian fluctuations of system parameters.

In the case of the delta-correlated process $z(t)$, we can additionally obtain the correlation functions of solutions to system of equations (8.162). Indeed, multiplying system (8.162) by $x(t')$, where $t' < t$, and averaging the result over an ensemble of realizations of process $z(t)$, we obtain the closed system

$$\frac{d}{dt}\langle x(t)x(t')\rangle = \langle y(t)x(t')\rangle, \quad \frac{d}{dt}\langle y(t)x(t')\rangle = -\omega_0^2\langle x(t)x(t')\rangle, \tag{8.167}$$

because

$$\frac{\delta}{\delta z(t-0)}x(t)x(t') = 0.$$

The initial values of this system are as follows

$$\langle x(t)x(t')\rangle_{t=t'} = \left\langle x^2(t')\right\rangle, \quad \langle y(t)x(t')\rangle_{t=t'} = \langle x(t')y(t')\rangle. \tag{8.168}$$

The system of equations for the other pair of correlation functions for $t > t'$ is derived similarly

$$\frac{d}{dt}\langle x(t)y(t')\rangle = \langle y(t)y(t')\rangle, \quad \frac{d}{dt}\langle y(t)y(t')\rangle = -\omega_0^2\langle x(t)y(t')\rangle. \tag{8.169}$$

The corresponding initial conditions are

$$\langle x(t)y(t')\rangle_{t=t'} = \langle x(t')y(t')\rangle, \quad \langle y(t)y(t')\rangle_{t=t'} = \left\langle y^2(t')\right\rangle. \tag{8.170}$$

Solutions to systems of equations (8.167) and (8.169) with the respective initial values (8.168) and (8.170) have the form

$$
\begin{aligned}
\langle x(t)x(t')\rangle &= \left\langle x^2(t')\right\rangle \cos\omega_0(t-t') + \frac{1}{\omega_0}\left\langle x(t')y(t')\right\rangle \sin\omega_0(t-t'),\\
\langle y(t)x(t')\rangle &= -\omega_0\left\langle x^2(t')\right\rangle \sin\omega_0(t-t') + \left\langle x(t')y(t')\right\rangle \cos\omega_0(t-t'),\\
\langle x(t)y(t')\rangle &= \left\langle x(t')y(t')\right\rangle \cos\omega_0(t-t') + \frac{1}{\omega_0}\left\langle y^2(t')\right\rangle \sin\omega_0(t-t'),\\
\langle y(t)y(t')\rangle &= -\omega_0\left\langle x(t')y(t')\right\rangle \sin\omega_0(t-t') + \left\langle y^2(t')\right\rangle \cos\omega_0(t-t').
\end{aligned}
\tag{8.171}
$$

Gaussian Delta-Correlated Fluctuations of Parameters

For the Gaussian stationary delta-correlated process $z(t)$, functional $\Theta[t;v(\tau)]$ is given by the formula $\left(\langle z(t)\rangle = 0,\ \langle z(t)z(t')\rangle = 2\sigma^2\tau_0\delta(t-t')\right)$

$$
\Theta[t;v(\tau)] = -\sigma^2\tau_0\int_0^t d\tau v^2(\tau),
$$

where σ^2 is the variance and τ_0 is the temporal correlation radius of process $z(t)$, so that Eq. (8.165) assumes the form of the Fokker–Planck equation

$$
\begin{aligned}
&\left(\frac{\partial}{\partial t} + y\frac{\partial}{\partial x} - \omega_0^2 x\frac{\partial}{\partial y}\right) P(x,y,t) = D\omega_0^2 x^2 \frac{\partial^2}{\partial y^2} P(x,y,t),\\
&P(x,y,0) = \delta(x-x_0)\,\delta(y-y_0),
\end{aligned}
\tag{8.172}
$$

where $D = \sigma^2\tau_0\omega_0^2$ is the diffusion coefficient in space $\{x, y/\omega_0\}$.

Let us derive the equations for the two first moments of solutions to system (8.162).

For average values of $x(t)$ and $y(t)$, we obtain the system of equations

$$
\begin{aligned}
&\frac{d}{dt}\langle x(t)\rangle = \langle y(t)\rangle, \quad \frac{d}{dt}\langle y(t)\rangle = -\omega_0^2\langle x(t)\rangle,\\
&x(0) = x_0, \quad y(0) = y_0
\end{aligned}
\tag{8.173}
$$

that coincides with system (8.162) without fluctuations, which agrees with the above discussion. Consequently, we have

$$
\begin{aligned}
\langle x(t)\rangle &= x_0\cos\omega_0(t-t') + \frac{1}{\omega_0}y_0\sin\omega_0(t-t'),\\
\langle y(t)\rangle &= -\omega_0 x_0\sin\omega_0(t-t') + y_0\cos\omega_0(t-t').
\end{aligned}
\tag{8.174}
$$

The second moments of quantities $x(t)$ and $y(t)$ satisfy the system of equations

$$\frac{d}{dt}\left\langle x^2(t)\right\rangle = 2\left\langle x(t)y(t)\right\rangle,$$

$$\frac{d}{dt}\left\langle x(t)y(t)\right\rangle = \left\langle y^2(t)\right\rangle - \omega_0^2\left\langle x^2(t)\right\rangle, \tag{8.175}$$

$$\frac{d}{dt}\left\langle y^2(t)\right\rangle = -2\omega_0^2\left\langle x(t)y(t)\right\rangle + D\omega_0^2\left\langle x^2(t)\right\rangle.$$

From this system, we can derive the closed third-order equation for any particular moment. For example, for quantity $\langle U(t)\rangle = \left\langle x^2(t)\right\rangle$ that describes the average potential energy of an oscillator, we obtain the equation

$$\frac{d^3}{dt^3}\left\langle U(t)\right\rangle + 4\omega_0^2\frac{d}{dt}\left\langle U(t)\right\rangle - 4D\omega_0^2\left\langle U(t)\right\rangle = 0, \tag{8.176}$$

which corresponds to the following stochastic initial-value problem for $U(t) = x^2(t)$

$$\frac{d^3}{dt^3}U(t) + 4\omega_0^2\frac{d}{dt}U(t) + 2\omega_0^2\left(z(t)\frac{d}{dt}U(t) + \frac{d}{dt}z(t)U(t)\right) = 0,$$

$$U(0) = x_0^2, \quad \left.\frac{d}{dt}U(t)\right|_{t=0} = 2x_0y_0, \tag{8.177}$$

$$\left.\frac{d^2}{dt^2}U(t)\right|_{t=0} = 2y_0^2 - \omega_0^2[1+z(0)]x_0^2$$

that can also be obtained immediately from system (8.162).

To simplify the calculations, we will assume that the initial values of system (8.162) have the form

$$x(0) = 0, \quad y(0) = \omega_0. \tag{8.178}$$

Assuming that the problem has a small parameter related to the intensity of process $z(t)$ fluctuations, we can approximately (to terms of order of D/ω_0) represent the solution to system (8.175) in the form

$$\left\langle x^2(t)\right\rangle = \frac{1}{2}\left\{e^{Dt} - e^{-\frac{Dt}{2}}\left[\cos(2\omega_0 t) + \frac{3D}{4\omega_0}\sin(2\omega_0 t)\right]\right\},$$

$$\left\langle x(t)y(t)\right\rangle = \frac{\omega_0}{4}\left\{2e^{-\frac{Dt}{2}}\sin(2\omega_0 t) + \frac{D}{\omega_0}\left[e^{Dt} - e^{-\frac{Dt}{2}}\cos(2\omega_0 t)\right]\right\}, \tag{8.179}$$

$$\left\langle y^2(t)\right\rangle = \frac{\omega_0^2}{2}\left\{e^{Dt} + e^{-\frac{Dt}{2}}\left[\cos(2\omega_0 t) - \frac{D}{4\omega_0}\sin(2\omega_0 t)\right]\right\}.$$

Thus, solution (8.179) of system of equations (8.175) has terms increasing with time, which corresponds to statistical parametric build-up of fluctuations in dynamic system (8.162) at the expense of frequency fluctuations. In the case of weak fluctuations, the increment of fluctuations is

$$\mu = D \quad (D/\omega_0 \ll 1).$$

From Eqs. (8.179) follows that solutions to statistical problem (8.162) have two characteristic temporal scales $t_1 \sim 1/\omega_0$ and $t_2 \sim 1/D$. The first temporal scale corresponds to the period of oscillations in system (8.162) without fluctuations (fast processes), and the second scale characterizes slow variations of statistical characteristics, which appear due to fluctuations (slow processes). The ratio of these scales is small:

$$t_1/t_2 = D/\omega_0 \ll 1.$$

We can explicitly obtain slow variations of statistical characteristics of processes $x(t)$ and $y(t)$ by excluding fast motions by means of averaging the corresponding quantities over the period $T = 2\pi/\omega_0$. Denoting such averaging with the overbar, we have

$$\overline{\langle x^2(t)\rangle} = \frac{1}{2}e^{Dt}, \quad \overline{\langle x(t)y(t)\rangle} = 0, \quad \overline{\langle y^2(t)\rangle} = \frac{\omega_0^2}{2}e^{Dt}.$$

Stochastic Problem with Linear Friction

If we add the linear friction to the system of equations (8.162), i.e., if we consider the dynamic system

$$\frac{d}{dt}x(t) = y(t), \quad \frac{d}{dt}y(t) = -2\gamma y(t) - \omega_0^2[1 + z(t)]x(t), \tag{8.180}$$

then the corresponding Fokker–Planck equation will have the form

$$\left(\frac{\partial}{\partial t} - 2\gamma\frac{\partial}{\partial y}y + y\frac{\partial}{\partial x} - \omega_0^2 x\frac{\partial}{\partial y}\right)P(x,y,t) = D\omega_0^2 x^2\frac{\partial^2}{\partial y^2}P(x,y,t)$$

and the system of equations for the second moments assumes, instead of (8.175), the form

$$\frac{d}{dt}\left\langle x^2(t)\right\rangle = 2\left\langle x(t)y(t)\right\rangle,$$

$$\frac{d}{dt}\left\langle x(t)y(t)\right\rangle = \left\langle y^2(t)\right\rangle - 2\gamma\left\langle x(t)y(t)\right\rangle - \omega_0^2\left\langle x^2(t)\right\rangle,$$

$$\frac{d}{dt}\left\langle y^2(t)\right\rangle = -4\gamma\left\langle y^2(t)\right\rangle - 2\omega_0^2\left\langle x(t)y(t)\right\rangle + D\omega_0^2\left\langle x^2(t)\right\rangle.$$

For this system, we will seek the solution proportional to $e^{\lambda t}$. The corresponding characteristic equation for λ assumes then the form

$$\lambda^3 + 6\gamma\lambda^2 + 4(\omega_0^2 + 2\gamma^2)\lambda + 4\omega_0^2(2\gamma - D) = 0.$$

As is known, the necessary and sufficient conditions of solution stability (which means the absence of roots λ_k with positive real parts) is formulated as the Routh–Hurwitz condition, which is equivalent in our case to the inequality $D < 2\gamma$. Thus, if this condition is not satisfied, i.e., if

$$2\gamma < D, \tag{8.181}$$

the second moments grow in time exponentially, which means the occurrence of the statistical parametric excitation of second moments. Note that conditions of the statistical parametric excitation differ for different moments. For example, the condition of exciting the fourth moments appears weaker than condition (8.181) and has the form [122]

$$D > \frac{2\gamma}{3}\frac{\omega_0^2 + 3\gamma^2}{\omega_0^2 + 6\gamma^2}.$$

For the stochastic parametric oscillator with friction, we can consider the problem on the steady-state regime that steadies under the action of random forces statistically independent of frequency fluctuations. This problem is formulated as the stochastic system of equations

$$\frac{d}{dt}x(t) = y(t), \quad \frac{d}{dt}y(t) = -2\gamma y(t) - \omega_0^2[1 + z(t)]x(t) + f(t), \tag{8.182}$$

where $f(t)$ is the Gaussian process statistically independent of process $z(t)$; it is assumed that $f(t)$ is the delta-correlated process with the following parameters

$$\langle f(t)\rangle = 0, \quad \langle f(t)f(t')\rangle = 2\sigma_f^2\tau_f\delta(t - t'),$$

where σ_f^2 is the variance and τ_f is the temporal correlation radius of process $f(t)$.

The one-time probability density of the solutions to stochastic system (8.182) satisfies the Fokker–Planck equation

$$\left(\frac{\partial}{\partial t} - 2\gamma\frac{\partial}{\partial y}y + y\frac{\partial}{\partial x} - \omega_0^2 x\frac{\partial}{\partial y}\right)P(x,y,t)$$
$$= D\omega_0^2 x^2\frac{\partial^2}{\partial y^2}P(x,y,t) + \sigma_f^2\tau_f\frac{\partial^2}{\partial y^2}P(x,y,t), \tag{8.183}$$

and, consequently, we have

$$\langle x(t)\rangle = 0, \quad \langle y(t)\rangle = 0.$$

Equations for the second moments form in this case the system

$$\frac{d}{dt}\left\langle x^2(t)\right\rangle = 2\left\langle x(t)y(t)\right\rangle,$$

$$\frac{d}{dt}\left\langle x(t)y(t)\right\rangle = \left\langle y^2(t)\right\rangle - 2\gamma\left\langle x(t)y(t)\right\rangle - \omega_0^2\left\langle x^2(t)\right\rangle, \tag{8.184}$$

$$\frac{d}{dt}\left\langle y^2(t)\right\rangle = -4\gamma\left\langle y^2(t)\right\rangle - 2\omega_0^2\left\langle x(t)y(t)\right\rangle + D\omega_0^2\left\langle x^2(t)\right\rangle + 2\sigma_f^2\tau_f,$$

whose steady-state solution exists for $t \to \infty$ if the condition (8.181) is satisfied. This solution behaves as follows

$$\left\langle x(t)y(t)\right\rangle = 0, \quad \left\langle x^2(t)\right\rangle = \frac{\sigma_f^2\tau_f}{\omega_0^2(D-2\gamma)}, \quad \left\langle y^2(t)\right\rangle = \frac{\sigma_f^2\tau_f}{D-2\gamma}.$$

Poisson Delta-Correlated Fluctuations of Parameters

Functional $\Theta[t;v(\tau)]$ of the Poisson delta-correlated random process $z(t)$ is given by Eq. (5.33)

$$\Theta[t;v(\tau)] = \nu\int_0^t d\tau\int_{-\infty}^{\infty} d\xi p(\xi)\left[e^{i\xi v(\tau)} - 1\right],$$

so that Eq. (8.164) assumes the form of the Kolmogorov–Feller equation

$$\left(\frac{\partial}{\partial t} + y\frac{\partial}{\partial x} - \omega_0^2 x\frac{\partial}{\partial y}\right)P(x,y,t) = \nu\int_{-\infty}^{\infty} d\xi p(\xi)P(x,y+\xi\omega_0^2 x,t) - \nu P(x,y,t). \tag{8.185}$$

For sufficiently small parameter ξ, the logarithm of the characteristic functional grades into the expression

$$\Theta[t;v(\tau)] = -\nu\left\langle\xi^2\right\rangle\int_0^t d\tau v^2(\tau),$$

and Eq. (8.185) grades into the Fokker–Planck equation (8.172) with the diffusion coefficient

$$D = \frac{1}{2}\nu\left\langle\xi^2\right\rangle\omega_0^2 x.$$

8.5.4 Partial Differential Equations

Statistical Interpretation of Solutions to Deterministic Equations

In a number of cases, solutions to many deterministic problems can be treated as a result of averaging certain functionals over random trajectories. Such interpretation can be useful in the context of various applications.

Let us derive the conditions under which such interpretation is applicable to some simple equations.

Consider the problem formulated as the initial-value problem for the partial differential equation

$$\frac{\partial}{\partial t}u(\boldsymbol{r},t) = -q(\boldsymbol{r},t)u(t,\boldsymbol{r}) + Q(\boldsymbol{\nabla},t)u(\boldsymbol{r},t), \quad u(\boldsymbol{r},0) = u_0(\boldsymbol{r}). \tag{8.186}$$

Along with Eq. (8.186), we consider the first-order partial differential equation

$$\frac{\partial}{\partial t}\phi(\boldsymbol{r},t) = -q(\boldsymbol{r},t)\phi(t,\boldsymbol{r}) + \boldsymbol{z}(t)\boldsymbol{\nabla}\phi(\boldsymbol{r},t), \quad \phi(\boldsymbol{r},0) = u_0(\boldsymbol{r}) \tag{8.187}$$

whose solution has the form

$$\phi[\boldsymbol{r},t;\boldsymbol{z}(\tau)] = u_0\left(\boldsymbol{r} + \int\limits_0^t d\tau \boldsymbol{z}(\tau)\right) \exp\left\{-\int\limits_0^t d\tau q\left(\tau,\boldsymbol{r} + \int\limits_\tau^t d\tau' \boldsymbol{z}(\tau')\right)\right\}. \tag{8.188}$$

We will assume that $z(t)$ is the random function delta-correlated in time t with characteristic functional $\varPhi[t;v(\tau)]$. Averaging Eq. (8.187) over an ensemble of realizations $z(t)$, we obtain the equation

$$\frac{\partial}{\partial t}\left\langle\phi(\boldsymbol{r},t)\right\rangle = -q(\boldsymbol{r},t)\left\langle\phi(\boldsymbol{r},t)\right\rangle + \left\langle \dot{\Theta}_t\left[t,\frac{\delta}{i\delta\boldsymbol{z}(t)}\right]\phi(\boldsymbol{r},t)\right\rangle. \tag{8.189}$$

Taking into account the equality

$$\frac{\delta}{\delta\boldsymbol{z}(t-0)}\phi(\boldsymbol{r},t) = \boldsymbol{\nabla}\phi(\boldsymbol{r},t),$$

which is a consequence of the initial dynamic equation (8.187), we can rewrite Eq. (8.189) in the form

$$\frac{\partial}{\partial t}\left\langle\phi(\boldsymbol{r},t)\right\rangle = -q(\boldsymbol{r},t)\left\langle\phi(\boldsymbol{r},t)\right\rangle + \dot{\Theta}_t\left[t,-i\boldsymbol{\nabla}\right]\left\langle\phi(\boldsymbol{r},t)\right\rangle. \tag{8.190}$$

Comparing now Eq. (8.190) with Eq. (8.186), we can see that

$$u(\boldsymbol{r},t) = \left\langle\phi[\boldsymbol{r},t;\boldsymbol{z}(\tau)]\right\rangle_{\boldsymbol{z}} \tag{8.191}$$

if

$$Q(\boldsymbol{\nabla}, t) = \dot{\Theta}_t[t, -i\boldsymbol{\nabla}]. \tag{8.192}$$

In this case, we can treat Eq. (8.191) as the solution to Eq. (8.186) written in the form of the continual integral.

In addition, we can give the operator form of Eq. (8.191) by introducing the functional shift operator

$$u(\boldsymbol{r}, t) = \langle \phi[\boldsymbol{r}, t; \boldsymbol{z}(\tau) + \boldsymbol{v}(\tau)] \rangle_{\boldsymbol{z}} |_{\boldsymbol{v}=0} = \Phi\left[t, \frac{\delta}{i\delta \boldsymbol{v}(\tau)}\right] \phi[\boldsymbol{r}, t; \boldsymbol{v}(\tau)]\Big|_{\boldsymbol{v}=0}, \tag{8.193}$$

where $\Phi[t; \boldsymbol{v}(\tau)]$ is the characteristic functional of process $\boldsymbol{z}(t)$.

For the Gaussian process $z(t)$, we have

$$\Theta[t, \boldsymbol{v}(\tau)] = -\frac{1}{2} \int\limits_0^t d\tau B(\tau) \boldsymbol{v}^2(\tau) \quad \left(Q(\boldsymbol{\nabla}, t) = \frac{1}{2} B(t) \Delta, \quad B(t) > 0 \right).$$

As a consequence, we obtain the well-known result that the solution to the diffusion equation

$$\frac{\partial}{\partial t} u(\boldsymbol{r}, t) = -q(\boldsymbol{r}, t) u(t, \boldsymbol{r}) + \frac{1}{2} B(t) \Delta u(\boldsymbol{r}, t), \quad u(0, \boldsymbol{r}) = u_0(\boldsymbol{r}) \tag{8.194}$$

can be treated as the result of averaging the functional $\phi[\boldsymbol{r}, t; \boldsymbol{z}(\tau)]$ over the Gaussian delta-correlated process $z(t)$, i.e.

$$\begin{aligned} u(\boldsymbol{r}, t) &= \left\langle u_0 \left(\boldsymbol{r} + \int\limits_0^t d\tau \boldsymbol{z}(\tau) \right) \exp\left\{ -\int\limits_0^t d\tau q \left(\boldsymbol{r} + \int\limits_\tau^t d\tau' \boldsymbol{z}(\tau'), \tau \right) \right\} \right\rangle_{\boldsymbol{z}} \\ &= \exp\left\{ -\frac{1}{2} \int\limits_0^t d\tau B(\tau) \frac{\delta^2}{\delta \boldsymbol{v}^2(\tau)} \right\} \\ &\times \left\{ u_0 \left(\boldsymbol{r} + \int\limits_0^t d\tau \boldsymbol{v}(\tau) \right) \exp\left[-\int\limits_0^t d\tau q \left(\boldsymbol{r} + \int\limits_\tau^t d\tau' \boldsymbol{v}(\tau'), \tau \right) \right] \right\}_{\boldsymbol{v}=0}. \end{aligned}$$

For the Poisson random process of the type of *shot noise*

$$\boldsymbol{z}(t) = \sum_{i=1}^{\infty} \boldsymbol{\xi}_i \delta(t - t_i),$$

we have the functional

$$\dot{\Theta}_t[t, \boldsymbol{v}(t)] = \nu \left\{ \int d\boldsymbol{\xi} p(\boldsymbol{\xi}) e^{i\boldsymbol{\xi}\boldsymbol{v}(t)} - 1 \right\},$$

and, consequently,

$$Q(\boldsymbol{\nabla}, t) = \nu \left\{ \int d\boldsymbol{\xi}\, p(\boldsymbol{\xi}) e^{\boldsymbol{\xi}\boldsymbol{\nabla}} - 1 \right\}.$$

Consequently, the solution to the integro-differential equation

$$\frac{\partial}{\partial t} u(\boldsymbol{r}, t) = -q(\boldsymbol{r}, t) u(\boldsymbol{r}, t) + \nu \int d\boldsymbol{\xi} p(\boldsymbol{\xi}) u(\boldsymbol{r}, t + \boldsymbol{\xi}) - \nu u(\boldsymbol{r}, t),$$
$$u(\boldsymbol{r}, 0) = u_0(\boldsymbol{r}) \tag{8.195}$$

can be represented as the result of averaging the functional $\phi[\, \boldsymbol{r}, t; \boldsymbol{z}(\tau)]$, i.e., in the form

$$u(\boldsymbol{r}, t) = \left\langle u_0 \left(\boldsymbol{r} + \int\limits_0^t d\tau \boldsymbol{z}(\tau) \right) \exp\left\{ - \int\limits_0^t d\tau q \left(\boldsymbol{r} + \int\limits_\tau^t d\tau' \boldsymbol{z}(\tau'), \tau \right) \right\} \right\rangle_{\boldsymbol{z}},$$

if function $p(\boldsymbol{\xi})$ in Eq. (8.195) can be interpreted as the probability density of random quantity $\boldsymbol{\xi}$.

If $p(\boldsymbol{\xi}) = \delta(\boldsymbol{\xi} - \boldsymbol{r}_0)$, then Eq. (8.195) assumes simpler form

$$\frac{\partial}{\partial t} u(\boldsymbol{r}, t) = -q(\boldsymbol{r}, t) u(\boldsymbol{r}, t) + \nu \left[u(\boldsymbol{r} + \boldsymbol{r}_0, t) - u(\boldsymbol{r}, t) \right],$$
$$u(\boldsymbol{r}, 0) = u_0(\boldsymbol{r}). \tag{8.196}$$

In this case, Eqs. (8.195) and (8.196) have the form of the transfer equation.

If a dynamic system is formulated in terms of partial differential equations of orders higher than the first and fluctuations of system parameters are delta-correlated, we can obtain closed equations, but for characteristic functionals in functional spaces. Consider two examples of such equations.

Parabolic Equation of Quasi-Optics

If we assume that field $\varepsilon(x, \boldsymbol{R})$ in the linear parabolic equation (1.153), page 63

$$\frac{\partial}{\partial x} u(x, \boldsymbol{R}) = \frac{i}{2k} \Delta_{\boldsymbol{R}} u(x, \boldsymbol{R}) + i \frac{k}{2} \varepsilon(x, \boldsymbol{R}) u(x, \boldsymbol{R}), \quad u(0, \boldsymbol{R}) = u_0(\boldsymbol{R})$$

is the homogeneous delta-correlated random field, then Eq. (8.21) for the characteristic functional $\Phi[x; v, v^*]$ of the solution to this equation assumes the form

$$\frac{\partial}{\partial x}\Phi[x; v, v^*] = \left\langle \dot{\Theta}_x \left[x; \frac{\delta}{i\delta\varepsilon(x, \boldsymbol{R}')}\right] \varphi[x; v,v^*] \right\rangle$$
$$+ \frac{i}{2k}\left\{\int d\boldsymbol{R}' \left[v(\boldsymbol{R}')\Delta_{\boldsymbol{R}'}\frac{\delta}{\delta v(\boldsymbol{R}')} - v^*(\boldsymbol{R}')\Delta_{\boldsymbol{R}'}\frac{\delta}{\delta v^*(\boldsymbol{R}')}\right]\right\}\Phi[x; v, v^*].$$

Taking into account Eq. (8.20), we can represent the last equation in the closed operator form [46, 48, 49]

$$\frac{\partial}{\partial x}\Phi[x; v, v^*] = \dot{\Theta}_x \left[x, \frac{k}{2}\widehat{M}(\boldsymbol{R}')\right] \Phi[x; v, v^*)]$$
$$+ \frac{i}{2k}\left\{\int d\boldsymbol{R}' \left[v(\boldsymbol{R}')\Delta_{\boldsymbol{R}'}\frac{\delta}{\delta v(\boldsymbol{R}')} - v^*(\boldsymbol{R}')\Delta_{\boldsymbol{R}'}\frac{\delta}{\delta v^*(\boldsymbol{R}')}\right]\right\}\Phi[x; v, v^*], \tag{8.197}$$

where $\widehat{M}(\boldsymbol{R}')$ is given by the formula

$$\widehat{M}(\boldsymbol{R}') = v(\boldsymbol{R}')\frac{\delta}{\delta v(\boldsymbol{R}')} - v^*(\boldsymbol{R}')\frac{\delta}{\delta v^*(\boldsymbol{R}')}$$

and functional

$$\dot{\Theta}_x \left[x; \psi(\xi, \boldsymbol{R}')\right] = \frac{d}{dx}\ln\left\langle \exp\left\{ i\int\limits_0^x d\xi \int d\boldsymbol{R}'\varepsilon(\xi, \boldsymbol{R}')\psi(\xi, \boldsymbol{R}')\right\}\right\rangle$$

is the derivative of the logarithm of the characteristic functional of filed $\varepsilon(x, \boldsymbol{R})$.

Equation (8.197) yields the equations for the moment functions of filed $u(x, \boldsymbol{R})$,

$$M_{m,n}(x; \boldsymbol{R}_1, \cdots, \boldsymbol{R}_m; \boldsymbol{R}'_1, \cdots, \boldsymbol{R}'_n) = \langle u(x, \boldsymbol{R}_1)\cdots u(x, \boldsymbol{R}_m)u^*(x, \boldsymbol{R}'_1)\cdots u^*(x, \boldsymbol{R}'_n)\rangle$$

(for $m = n$, these functions are usually called the *coherence functions* of order $2n$), which are a consequence of the linearity of the initial dynamic equation (8.13). These equations have the form

$$\frac{\partial}{\partial x}M_{m,n} = \frac{i}{2k}\left(\sum_{p=1}^{m}\Delta_{\boldsymbol{R}_p} - \sum_{q=1}^{n}\Delta_{\boldsymbol{R}'_q}\right)M_{m,n}$$

$$+\dot{\Theta}_x\left[x, \frac{1}{k}\left(\sum_{p=1}^{m}\delta(\boldsymbol{R}'-\boldsymbol{R}_p) - \sum_{q=1}^{n}\delta(\boldsymbol{R}'-\boldsymbol{R}'_q)\right)\right]M_{m,n}. \quad (8.198)$$

If we assume now that $\varepsilon(x,\boldsymbol{R})$ is the homogeneous Gaussian delta-correlated field with the correlation function

$$B_\varepsilon(x,\boldsymbol{R}) = A(\boldsymbol{R})\delta(x), \quad A(\boldsymbol{R}) = \int_{-\infty}^{\infty} dx B_\varepsilon(x,\boldsymbol{R}),$$

so that functional $\Theta\left[x;\psi(\xi,\boldsymbol{R}')\right]$ has the form

$$\Theta\left[x;\psi(\xi,\boldsymbol{R}')\right] = -\frac{1}{2}\int_0^x d\xi\int d\boldsymbol{R}'\int d\boldsymbol{R}A(\boldsymbol{R}'-\boldsymbol{R})\psi(\xi,\boldsymbol{R}')\psi(\xi,\boldsymbol{R}),$$

then Eq. (8.197) assumes the closed operator form [46, 48, 49, 122, 129]

$$\frac{\partial}{\partial x}\Phi[x;v,v^*] = -\frac{k^2}{8}\int d\boldsymbol{R}'\int d\boldsymbol{R}\,A(\boldsymbol{R}'-\boldsymbol{R})\widehat{M}(\boldsymbol{R}')\widehat{M}(\boldsymbol{R})\Phi[x;v,v^*]$$

$$+\frac{i}{2k}\left\{\int d\boldsymbol{R}'\left[v(\boldsymbol{R}')\Delta_{\boldsymbol{R}'}\frac{\delta}{\delta v(\boldsymbol{R}')} - v^*(\boldsymbol{R}')\Delta_{\boldsymbol{R}'}\frac{\delta}{\delta v^*(\boldsymbol{R}')}\right]\right\}\Phi[x;v,v^*], \quad (8.199)$$

and Eqs. (8.198) for the moment functions of wavefield $u(x,\boldsymbol{R})$ assume the form

$$\frac{\partial}{\partial x}M_{m,n} = \frac{i}{2k}\left(\sum_{p=1}^{m}\Delta_{\boldsymbol{R}_p} - \sum_{q=1}^{n}\Delta_{\boldsymbol{R}'_q}\right)M_{m,n} - \frac{k^2}{8}Q(\boldsymbol{R}_1,\ldots,\boldsymbol{R}_m;\boldsymbol{R}'_1,\ldots,\boldsymbol{R}'_m)M_{m,n}, \quad (8.200)$$

where

$$Q(\boldsymbol{R}_1,\cdots,\boldsymbol{R}_m;\boldsymbol{R}'_1,\cdots,\boldsymbol{R}'_m)$$

$$=\sum_{i=1}^{m}\sum_{j=1}^{m}A(\boldsymbol{R}_i-\boldsymbol{R}_j) - 2\sum_{i=1}^{m}\sum_{j=1}^{n}A(\boldsymbol{R}_i-\boldsymbol{R}'_j) + \sum_{i=1}^{n}\sum_{j=1}^{n}A(\boldsymbol{R}'_i-\boldsymbol{R}'_j). \quad (8.201)$$

Remark 8.2. ***Another Derivation of Eqs. (8.198) and (8.200).***

In the case of the delta-correlated fluctuations of medium parameters, there is another, physically clearer way of deriving Eqs. (8.198) and (8.200) for the moment functions of wavefield $u(x, \boldsymbol{R})$ [45, 46].

As was mentioned earlier, field $u(x, \boldsymbol{R})$ depends functionally only on the preceding values of field $\varepsilon(\xi, \boldsymbol{R}')$, i.e., for $\xi \leqslant x$. However, in the general case, there is statistical relationship between field $u(x, \boldsymbol{R})$ and subsequent values of field $\varepsilon(\xi, \boldsymbol{R}')$ for $\xi \geqslant x$. In the approximation of the delta-correlated fluctuations of medium parameters, this statistical relationship disappears, and fields $u(\xi_i, \boldsymbol{R})$ for $\xi_i < x$ are independent of $\varepsilon(\eta_j, \boldsymbol{R}')$ for $\eta_j > x$ not only functionally, but also statistically; i.e., for $\xi_i < x; \eta_j > x$, the following equality holds:

$$\left\langle \prod_{i,j} u(\xi_i, \boldsymbol{R}_i)\varepsilon(\eta_j, \boldsymbol{R}_j) \right\rangle = \left\langle \prod_{i} u(\xi_i, \boldsymbol{R}_i) \right\rangle \left\langle \prod_{j} \varepsilon(\eta_j, \boldsymbol{R}_j) \right\rangle. \tag{8.202}$$

Using Eq. (8.202), we can easily obtain the equations for statistical moments of field $u(x, \boldsymbol{R})$. We derive the equation for average field $\langle u(x, \boldsymbol{R})\rangle$ as an example. With this goal in view, we rewrite the initial stochastic equation (8.13) in the form of the integral equation

$$u(x, \boldsymbol{R}) = u_0(\boldsymbol{R}) \exp\left\{ i\frac{k}{2} \int_0^x d\xi \varepsilon(\xi, \boldsymbol{R}) \right\} + \frac{i}{2k} \int_0^x d\xi \exp\left\{ i\frac{k}{2} \int_\xi^x d\eta \varepsilon(\eta, \boldsymbol{R}) \right\} \Delta_{\boldsymbol{R}} u(\xi, \boldsymbol{R}). \tag{8.203}$$

Averaging Eq. (8.203) over an ensemble of realizations of random field $\varepsilon(\xi, \boldsymbol{R})$, we take into account Eq. (8.202) to obtain the closed integral equation

$$\langle u(x, \boldsymbol{R})\rangle = u_0(\boldsymbol{R}) \left\langle \exp\left\{ i\frac{k}{2} \int_0^x d\xi \varepsilon(\xi, \boldsymbol{R}) \right\} \right\rangle + \frac{i}{2k} \int_0^x d\xi \left\langle \exp\left\{ i\frac{k}{2} \int_\xi^x d\eta \varepsilon(\eta, \boldsymbol{R}) \right\} \right\rangle \Delta_{\boldsymbol{R}} \langle u(\xi, \boldsymbol{R})\rangle. \tag{8.204}$$

To transform the integral equation into the differential equation, we use the fact that the equality

$$\left\langle \exp\left\{ i\frac{k}{2} \int_0^x d\xi \varepsilon(\xi, \boldsymbol{R}) \right\} \right\rangle = \left\langle \exp\left\{ i\frac{k}{2} \int_0^\xi d\eta \varepsilon(\eta, \boldsymbol{R}) \right\} \right\rangle \left\langle \exp\left\{ i\frac{k}{2} \int_\xi^x d\eta \varepsilon(\eta, \boldsymbol{R}) \right\} \right\rangle,$$

holds in the case of the delta-correlated fluctuations of medium parameter for any point $0 \leqslant \xi \leqslant x$. Thus, introducing function

$$\Phi(x, \boldsymbol{R}) = \left\langle \exp\left\{ i\frac{k}{2}\int\limits_0^x d\eta\varepsilon(\eta, \boldsymbol{R}) \right\} \right\rangle,$$

we can rewrite Eq. (8.204) in the form

$$\langle u(x, \boldsymbol{R})\rangle = u_0(\boldsymbol{R})\Phi(x, \boldsymbol{R}) + \frac{i}{2k}\int\limits_0^x d\xi \frac{\Phi(x, \boldsymbol{R})}{\Phi(\xi, \boldsymbol{R})}\Delta_{\boldsymbol{R}} \langle u(\xi, \boldsymbol{R})\rangle, \tag{8.205}$$

from which easily follows the differential equation for $\langle u(x, \boldsymbol{R})\rangle$

$$\frac{\partial}{\partial x}\langle u(x, \boldsymbol{R})\rangle = \frac{i}{2k}\Delta_{\boldsymbol{R}}\langle u(x, \boldsymbol{R})\rangle + \langle u(x, \boldsymbol{R})\rangle \frac{\partial}{\partial x}\ln \Phi(x, \boldsymbol{R})$$

coinciding with Eq. (8.198) for $m = 1,\ n = 0$. Equations for the higher-order moments of field $u(x, \boldsymbol{R})$ can be derived similarly. ♦

Random Forces in Hydrodynamic Turbulence

In the case of the hydrodynamic equation (8.22), page 202 under the assumption that random field $\boldsymbol{f}(\boldsymbol{x}, t)$ is homogeneous in space and stationary and delta-correlated in time, Eq. (8.25), page 202 for the characteristic functional of the Fourier transform of the velocity field

$$\Phi[t; \boldsymbol{z}(\boldsymbol{k}')] = \Phi[t; \boldsymbol{z}] = \langle \varphi[t; \boldsymbol{z}(\boldsymbol{k}')]\rangle = \left\langle \exp\left\{ i\int d\boldsymbol{k}'\boldsymbol{z}(\boldsymbol{k}')\widehat{\boldsymbol{u}}(\boldsymbol{k}', t)\right\}\right\rangle$$

assumes the form

$$\frac{\partial}{\partial t}\Phi[t; \boldsymbol{z}] = \left\langle \dot{\Theta}_t\left[t; \frac{\delta}{i\delta \boldsymbol{f}(\boldsymbol{\kappa}, t)}\right]\varphi[t; \boldsymbol{z}]\right\rangle$$
$$-\int d\boldsymbol{k} z_i(\boldsymbol{k})\left\{\frac{1}{2}\int d\boldsymbol{k}_1\int d\boldsymbol{k}_2 \Lambda_{i,\alpha\beta}(\boldsymbol{k}_1, \boldsymbol{k}_2, \boldsymbol{k})\frac{\delta^2}{\delta z_\alpha(\boldsymbol{k}_1)\delta z_\beta(\boldsymbol{k}_2)} + \nu k^2 \frac{\delta}{\delta z_i(\boldsymbol{k})}\Phi[t; \boldsymbol{z}]\right\}, \tag{8.206}$$

where

$$\dot{\Theta}_t\left[t; \boldsymbol{\psi}(\boldsymbol{\kappa}, \tau)\right] = \frac{d}{dt}\ln\left\langle \exp\left\{ i\int\limits_0^t d\tau\int d\boldsymbol{\kappa}\, \widehat{\boldsymbol{f}}(\boldsymbol{\kappa}, \tau)\boldsymbol{\psi}(\boldsymbol{\kappa}, \tau)\right\}\right\rangle$$

is the derivative of the logarithm of the characteristic functional of external forces $\widehat{\boldsymbol{f}}(\boldsymbol{k},t)$. By virtue of equality $\dfrac{\delta}{\delta\widehat{\boldsymbol{f}}(\boldsymbol{k},t-0)}\varphi[t;\boldsymbol{z}] = iz(k)\varphi[t;\boldsymbol{z}]$ (see Eq. (8.24), page 202), we can rewrite Eq. (8.206) in the form of the closed equation

$$\frac{\partial}{\partial t}\Phi[t;\boldsymbol{z}] = \dot{\Theta}_t\left[t;\boldsymbol{z}(\boldsymbol{k})\right]\Phi[t;\boldsymbol{z}]$$
$$-\int d\boldsymbol{k} z_i(\boldsymbol{k})\left\{\frac{1}{2}\int d\boldsymbol{k}_1\int d\boldsymbol{k}_2\Lambda_{i,\alpha\beta}(\boldsymbol{k}_1,\boldsymbol{k}_2,\boldsymbol{k})\frac{\delta^2}{\delta z_\alpha(\boldsymbol{k}_1)\delta z_\beta(\boldsymbol{k}_2)} + \nu k^2\frac{\delta}{\delta z_i(\boldsymbol{k})}\right\}\Phi[t;\boldsymbol{z}]. \tag{8.207}$$

If we assume now that $\boldsymbol{f}(\boldsymbol{x},t)$ is the Gaussian random field homogeneous and isotropic in space and stationary in time with the correlation tensor

$$B_{ij}(\boldsymbol{x}_1-\boldsymbol{x}_2,t_1-t_2) = \langle f_i(\boldsymbol{x}_1,t_1)f_j(\boldsymbol{x}_2,t_2)\rangle,$$

then the field $\widehat{\boldsymbol{f}}(\boldsymbol{k},t)$ will also be the Gaussian stationary random field with the correlation tensor

$$\left\langle \widehat{f}_i(\boldsymbol{k},t+\tau)\widehat{f}_j(\boldsymbol{k}',t)\right\rangle = \frac{1}{2}F_{ij}(\boldsymbol{k},\tau)\delta(\boldsymbol{k}+\boldsymbol{k}),$$

where $F_{ij}(\boldsymbol{k},\tau)$ is the spatial spectrum of the external force given by the formula

$$F_{ij}(\boldsymbol{k},\tau) = 2(2\pi)^3\int d\boldsymbol{x}B_{ij}(\boldsymbol{x},\tau)e^{-i\boldsymbol{k}\boldsymbol{x}}.$$

In view of the fact that forces are spatially isotropic, we have $F_{ij}(\boldsymbol{k},\tau) = F(k,\tau)\Delta_{ij}(\boldsymbol{k})$. As long as field $\widehat{\boldsymbol{f}}(\boldsymbol{k},t)$ is delta-correlated in time, we have $F(k,\tau) = F(k)\delta(\tau)$, so that functional $\Theta\left[t;\boldsymbol{\psi}(\boldsymbol{\kappa},\tau)\right]$ is given by the formula

$$\Theta\left[t;\boldsymbol{\psi}(\boldsymbol{\kappa},\tau)\right] = -\frac{1}{4}\int\limits_0^t d\tau\int d\boldsymbol{\kappa}\, F(\kappa)\Delta_{ij}(\boldsymbol{\kappa})\psi_i(\boldsymbol{\kappa},\tau)\psi_j(-\boldsymbol{\kappa},\tau),$$

and Eq. (8.207) assumes the closed form [114]

$$\frac{\partial}{\partial t}\Phi[t;\boldsymbol{z}] = -\frac{1}{4}\int d\boldsymbol{k}F(k)\Delta_{ij}(\boldsymbol{k})z_i(\boldsymbol{k})z_j(-\boldsymbol{k})\Phi[t;\boldsymbol{z}]$$
$$-\int d\boldsymbol{k} z_i(\boldsymbol{k})\left\{\frac{1}{2}\int d\boldsymbol{k}_1\int d\boldsymbol{k}_2\Lambda_{i,\alpha\beta}(\boldsymbol{k}_1,\boldsymbol{k}_2,\boldsymbol{k})\frac{\delta^2}{\delta z_\alpha(\boldsymbol{k}_1)\delta z_\beta(\boldsymbol{k}_2)} + \nu k^2\frac{\delta}{\delta z_i(\boldsymbol{k})}\right\}\Phi[t;\boldsymbol{z}]. \tag{8.208}$$

Equation (8.208) plays the role of the Fokker–Planck equation of the problem under consideration. The unknown in this equation is the characteristic functional, and this fact distinguishes this equation from the standard equation of this type, where the unknown is the probability density expressed as the Fourier transform of this functional.

Another distinction consists in the fact that Eq. (8.208) is the diffusion equation in the infinite-dimensional space, because of which it is the variational differential equation. The diffusion coefficient can be different for different wave components; it is given by the spectral tensor of external forces $F(k)\Delta_{ij}(\boldsymbol{k})$.

Chapter 9
Stochastic Equations with the Markovian Fluctuations of Parameters

In the preceding chapter, we dealt with the statistical description of dynamic systems in terms of the general methods that assumed the knowledge of the characteristic functional of fluctuating parameters. However, this functional is unknown in most cases, and we are forced to resort either to assumptions on the model of parameter fluctuations, or to asymptotic approximations.

The methods based on approximating the fluctuating parameters with the Markovian random processes and fields with a finite temporal correlation radius are widely used. Such approximations can be found, for example, as solutions to the dynamic equations with delta-correlated fluctuations of parameters. Consider such methods in greater detail by the examples of the Markovian random processes [47–49].

9.1 General Remarks

Consider stochastic equations of the form

$$\frac{d}{dt}\boldsymbol{x}(t) = f\left(t, \boldsymbol{x}, \boldsymbol{z}(t)\right), \quad \boldsymbol{x}(0) = \boldsymbol{x}_0, \tag{9.1}$$

where $f\left(t, \boldsymbol{x}, \boldsymbol{z}(t)\right)$ is the deterministic function. The vector process

$$\boldsymbol{z}(t) = \{z_1(t), \cdots, z_n(t)\}$$

is the Markovian vector process whose transition probability density satisfies the equation (see Chapter 3, page 150)

$$\frac{\partial}{\partial t} p(\boldsymbol{z}, t|\boldsymbol{z}_0, t_0) = \widehat{L}(\boldsymbol{z}) p(\boldsymbol{z}, t|\boldsymbol{z}_0, t_0).$$

In this equation, operator $\widehat{L}(\boldsymbol{z})$ is called the *kinetic operator*.

V.I. Klyatskin, *Stochastic Equations: Theory and Applications in Acoustics, Hydrodyn., Magnetohydrodyn., and Radiophys., Vol. 1*, Understanding Complex Systems,
DOI: 10.1007/978-3-319-07587-7_9,

Our task consists in the determination of statistical characteristics of the solution to Eq. (9.1) from known statistical characteristics of process $\boldsymbol{z}(t)$, for example, from the kinetic operator $\widehat{L}(\boldsymbol{z})$.

In the general case of arbitrary Markovian process $\boldsymbol{z}(t)$, we cannot judge about process $\boldsymbol{x}(t)$. We can only assert that the joint process $\{\boldsymbol{x}(t), \boldsymbol{z}(t)\}$ is the Markovian process. Indeed, as we showed in Chapter 4, page 179 the following differentiation formula

$$\frac{d}{dt}\left\langle \delta\left(\boldsymbol{z}(t)-\boldsymbol{z}\right) R[t;\boldsymbol{z}(\boldsymbol{\tau})]\right\rangle = \left\langle \delta\left(\boldsymbol{z}(t)-\boldsymbol{z}\right)\frac{d}{dt}R[t;\boldsymbol{z}(\boldsymbol{\tau})]\right\rangle$$
$$+\widehat{L}(\boldsymbol{z})\left\langle \delta\left(\boldsymbol{z}(t)-\boldsymbol{z}\right) R[t;\boldsymbol{z}(\boldsymbol{\tau})]\right\rangle, \qquad (9.2)$$

holds for arbitrary functional $R[t;\boldsymbol{z}(\tau)]$, $\tau \le t$ if $\boldsymbol{z}(t)$ is the Markovian process. Multiplying Eq. (9.2) by arbitrary function $F(\boldsymbol{z})$ and integrating the result over $\boldsymbol{z}$, we obtain another representation of the differentiation formula

$$\frac{d}{dt}\left\langle F\left(\boldsymbol{z}(t)\right) R[t;\boldsymbol{z}(\boldsymbol{\tau})]\right\rangle = \left\langle F\left(\boldsymbol{z}(t)\right)\frac{d}{dt}R[t;\boldsymbol{z}(\boldsymbol{\tau})]\right\rangle$$
$$+\left\langle R[t;\boldsymbol{z}(\boldsymbol{\tau})]\left[\widehat{L}^{+}(\boldsymbol{z})F\left(\boldsymbol{z}(t)\right)\right]\right\rangle, \qquad (9.3)$$

where $\widehat{L}^{+}(\boldsymbol{z})$ is the operator conjugated to operator $\widehat{L}(\boldsymbol{z})$.

Now, we specify functional $R[t;\boldsymbol{z}(\tau)]$ in the form of the indicator function

$$R[\boldsymbol{x},t;\boldsymbol{z}(\tau)] = \delta(\boldsymbol{x}(t)-\boldsymbol{x}),$$

where $\boldsymbol{x}(t)$ is the solution to Eq. (9.1). In this case, function $R[\boldsymbol{x},t;\boldsymbol{z}(\tau)]$ satisfies the equation

$$\frac{\partial}{\partial t}R[\boldsymbol{x},t;\boldsymbol{z}(\tau)] = -\frac{\partial}{\partial x_i}f_i(t,\boldsymbol{x},\boldsymbol{z})R[\boldsymbol{x},t;\boldsymbol{z}(\tau)],$$

which is the stochastic Liouville equation for our problem. Note that the correlator

$$\langle\delta(\boldsymbol{z}(t)-\boldsymbol{z})R[\boldsymbol{x},t;\boldsymbol{z}(\tau)]\rangle = P(\boldsymbol{x},\boldsymbol{z},t)$$

appears in this case to be the one-time joint probability density of processes $\boldsymbol{x}(t)$ and $\boldsymbol{z}(t)$. Consequently, the differentiation formula (9.2) assumes the form of the closed equation for the one-time probability density

$$\frac{\partial}{\partial t}P(\boldsymbol{x},\boldsymbol{z},t) = -\frac{\partial}{\partial x_i}f_i(t,\boldsymbol{x},\boldsymbol{z})P(\boldsymbol{x},\boldsymbol{z},t) + \widehat{L}(\boldsymbol{z})P(\boldsymbol{x},\boldsymbol{z},t). \qquad (9.4)$$

It is obvious that the transition probability density of the joint process $\{\boldsymbol{x}(t), \boldsymbol{z}(t)\}$ also satisfies Eq. (9.4), which means that process $\{\boldsymbol{x}(t), \boldsymbol{z}(t)\}$ is the Markovian process. If we would able to solve Eq. (9.4), then we could integrate the solution over $\boldsymbol{z}$ to obtain the probability density of the solution to Eq. (9.1), i.e., function $P(\boldsymbol{x}, t)$. In this case, process $\boldsymbol{x}(t)$ would not be the Markovian process.

There are several types of processes $\boldsymbol{z}(t)$ that allow obtaining equations for density $P(\boldsymbol{x}, t)$ without solving Eq. (9.4) for $P(\boldsymbol{x}, \boldsymbol{z}, t)$. Among these processes, we mention first of all the telegrapher's and generalized telegrapher's processes, Markovian processes with finite number of states, and Gaussian Markovian processes. Below, we discuss these processes in more detail as examples of processes widely used in different applications.

9.2 Telegrapher's Processes

Recall that telegrapher's random process $z(t)$ (the two-state, or binary process) is defined by the equality

$$z(t) = a(-1)^{n(0,t)},$$

where random quantity a assumes values $a = \pm a_0$ with probabilities 1/2 and $n(t_1, t_2)$, $t_1 < t_2$ is the Poisson integer-valued process with mean value $\overline{n(t_1, t_2)} = \nu|t_1 - t_2|$.

Telegrapher's process $z(t)$ is stationary in time and its correlation function

$$\langle z(t)z(t')\rangle = a_0^2 e^{-2\nu|t-t'|}$$

has the temporal correlation radius $\tau_0 = 1/(2\nu)$.

For splitting the correlation between telegrapher's process $z(t)$ and arbitrary functional $R[t; z(\tau)]$, where $\tau \leq t$, we obtained the relationship (7.36), page 175

$$\langle z(t)R[t; z(\tau)]\rangle = a_0^2 \int_0^t dt_1 e^{-2\nu(t-t_1)} \left\langle \frac{\delta}{\delta z(t_1)} \widetilde{R}[t; z(\tau)] \right\rangle, \tag{9.5}$$

where functional $\widetilde{R}[t; z(\tau)]$ is given by the formula

$$\widetilde{R}[t, t_1; z(\tau)] = R[t; z(\tau)\theta(t_1 - \tau + 0)] \quad (t_1 < t). \tag{9.6}$$

Formula (9.5) is appropriate for analyzing stochastic equations linear in process $z(t)$. Let functional $R[t; z(\tau)]$ is the solution to a system of differential equations of the first order in time with initial values at $t = 0$. Functional $\widetilde{R}[t, t_1; z(\tau)]$ will also satisfy the same system of equations with product $z(t)\theta(t_1 - t)$ instead of $z(t)$. Consequently, we obtain that functional $\widetilde{R}[t, t_1; z(\tau)] = R[t; 0]$ for all times $t > t_1$; moreover, it satisfies the same system of equations for absent fluctuations (i.e., at $z(t) = 0$) with the initial value $\widetilde{R}[t_1, t_1; z(\tau)] = R[t_1; z(\tau)]$.

Another formula convenient in the context of stochastic equations linear in random telegrapher's process $z(t)$ concerns the differentiation of the correlation of this process with arbitrary functional $R[t; z(\tau)]$ $(\tau \leq t)$ (7.42), page 176

$$\frac{d}{dt}\left\langle z(t)R[t;z(\tau)]\right\rangle = -2\nu\left\langle z(t)R[t;z(\tau)]\right\rangle + \left\langle z(t)\frac{d}{dt}R[t;z(\tau)]\right\rangle. \tag{9.7}$$

In addition, we have the equality

$$\left\langle z(t')R[t;z(\tau)]\right\rangle = e^{-2\nu|t-t'|}\left\langle z(t)R[t;z(\tau)]\right\rangle, \quad t' \geqslant t, \quad \tau \leqslant t. \tag{9.8}$$

Formula (9.7) determines the rule of factoring the differentiation operation out of averaging brackets

$$\left\langle z(t)\frac{d^n}{dt^n}R[t;z(\tau)]\right\rangle = \left(\frac{d}{dt} + 2\nu\right)^n \left\langle z(t)R[t;z(\tau)]\right\rangle. \tag{9.9}$$

We consider some special examples to show the usability of these formulas. It is evident that both methods give the same result. However, the method based on the differentiation formula appears to be more practicable.

9.2.1 System of Linear Operator Equations

The first example concerns the system of linear operator equations

$$\frac{d}{dt}\boldsymbol{x}(t) = \widehat{A}(t)\boldsymbol{x}(t) + z(t)\widehat{B}(t)\boldsymbol{x}(t), \quad \boldsymbol{x}(0) = \boldsymbol{x}_0, \tag{9.10}$$

where $\widehat{A}(t)$ and $\widehat{B}(t)$ are certain differential operators (they may include differential operators with respect to auxiliary variables). If operators $\widehat{A}(t)$ and $\widehat{B}(t)$ are matrixes, then Eqs. (9.10) describe the linear dynamic system.

Average Eqs. (9.10) over an ensemble of random functions $z(t)$. The result will be the vector equation

$$\frac{d}{dt}\left\langle \boldsymbol{x}(t)\right\rangle = \widehat{A}(t)\left\langle \boldsymbol{x}(t)\right\rangle + \widehat{B}(t)\boldsymbol{\psi}(t), \quad \left\langle \boldsymbol{x}(0)\right\rangle = \boldsymbol{x}_0, \tag{9.11}$$

where we introduced new functions

$$\boldsymbol{\psi}(t) = \left\langle z(t)\boldsymbol{x}(t)\right\rangle.$$

We can use formula (9.7) for these functions; as a result, we obtain the equality

$$\frac{d}{dt}\boldsymbol{\psi}(t) = -2\nu\boldsymbol{\psi}(t) + \left\langle z(t)\frac{d}{dt}\boldsymbol{x}(t)\right\rangle. \tag{9.12}$$

Substituting now derivative $d\boldsymbol{x}/dt$ (9.10) in Eq. (9.12), we obtain the equation for the vector function $\boldsymbol{\psi}(t)$

$$\left(\frac{d}{dt} + 2\nu\right)\boldsymbol{\psi}(t) = \widehat{A}(t)\boldsymbol{\psi}(t) + \widehat{B}(t)\left\langle z^2(t)\boldsymbol{x}(t)\right\rangle. \tag{9.13}$$

Because $z^2(t) \equiv a_0^2$ for telegrapher's process, we obtain finally the closed system of linear equations for vectors $\langle \boldsymbol{x}(t)\rangle$ and $\boldsymbol{\psi}(t)$

$$\begin{aligned} &\frac{d}{dt}\langle \boldsymbol{x}(t)\rangle = \widehat{A}(t)\langle \boldsymbol{x}(t)\rangle + \widehat{B}(t)\boldsymbol{\psi}(t), \quad \langle \boldsymbol{x}(0)\rangle = \boldsymbol{x}_0, \\ &\left(\frac{d}{dt} + 2\nu\right)\boldsymbol{\psi}(t) = \widehat{A}(t)\boldsymbol{\psi}(t) + a_0^2\widehat{B}(t)\langle \boldsymbol{x}(t)\rangle, \quad \boldsymbol{\psi}(0) = (0). \end{aligned} \tag{9.14}$$

If operators $\widehat{A}(t)$ and $\widehat{B}(t)$ are the time-independent matrixes A and B, we can solve system (9.14) using the Laplace transform. After the Laplace transform, system (9.14) becomes the algebraic system of equations

$$(pE - A)\langle \boldsymbol{x}\rangle_p - B\boldsymbol{\psi}_p = \boldsymbol{x}_0, \quad [(p + 2\nu)E - A]\boldsymbol{\psi}_p - a_0^2 B\langle \boldsymbol{x}\rangle_p = 0, \tag{9.15}$$

where E is the unit matrix. From this system, we obtain solution $\langle \boldsymbol{x}\rangle_p$ in the form

$$\langle \boldsymbol{x}\rangle_p = \left[(pE - A) - a_0^2 B\frac{1}{(p + 2\nu)E - A}B\right]^{-1}\boldsymbol{x}_0. \tag{9.16}$$

Stochastic Parametric Resonance

Consider the problem of statistical description of an oscillator with fluctuating frequency as a simple example of the linear dynamic system (9.10). This problem is formulated as the second-order equation (1.32), page 22 with initial values

$$\frac{d^2}{dt^2}x(t) + \omega_0^2[1 + z(t)]x(t) = 0, \quad x(0) = x_0, \quad \left.\frac{d}{dt}x(t)\right|_{t=0} = y_0, \tag{9.17}$$

which is equivalent to the system of equations

$$\frac{d}{dt}x(t) = y(t), \quad \frac{d}{dt}y(t) = -\omega_0^2[1 + z(t)]x(t),$$
$$x(0) = x_0, \quad y(0) = y_0. \tag{9.18}$$

If our interest concerns only the average value of the solution to statistical problem (9.17), we can deal without rewriting it in the form of the system of equations (9.18). Averaging Eq. (9.17) over an ensemble of realizations $z(t)$, we obtain the unclosed equation

$$\left(\frac{d^2}{dt^2} + \omega_0^2\right)\langle x(t)\rangle + \omega_0^2\langle z(t)x(t)\rangle = 0. \tag{9.19}$$

To split the correlator in the right-hand side of Eq. (9.19), we multiply Eq. (9.17) by function $z(t)$ and average the result to obtain the equation

$$\left\langle z(t)\left(\frac{d^2}{dt^2} + \omega_0^2\right)x(t)\right\rangle + \omega_0^2 a_0^2\langle x(t)\rangle = 0. \tag{9.20}$$

Deriving Eq. (9.20), we took into account that quantity $z^2(t) \equiv a_0^2$ is not random in the case of telegrapher's process.

Then, we use the rule of factoring the derivative out of averaging brackets (9.9), page 264 to rewrite Eq. (9.20) in the form

$$\left[\left(\frac{d}{dt} + 2\nu\right)^2 + \omega_0^2\right]\langle z(t)x(t)\rangle + \omega_0^2 a_0^2\langle x(t)\rangle = 0. \tag{9.21}$$

Now, Eqs. (9.19) and (9.21) form the closed system of equations.

From Eq. (9.21), we obtain

$$\langle z(t)x(t)\rangle = \omega_0 a_0^2 \int_0^t dt' e^{-2\nu(t-t')}\sin\omega_0(t-t')\langle x(t')\rangle.$$

Consequently, Eq. (9.19) can be represented in the form of the integro-differential equation

$$\left(\frac{d^2}{dt^2} + \omega_0^2\right)\langle x(t)\rangle + \omega_0^3 a_0^2 \int_0^t dt' e^{-2\nu(t-t')}\sin\omega_0(t-t')\langle x(t')\rangle = 0. \tag{9.22}$$

We can again use the Laplace transform to solve either the system of equations (9.19) and (9.21) or Eq. (9.22); in both cases, the solution has the form

$$\langle x \rangle_p = F(p) \frac{L(p+2\nu)}{L(p)L(p+2\nu) - \omega_0^4 a_0^2} = F(p) \frac{1}{p^2 + \omega_0^2 - \dfrac{\omega_0^4 a_0^2}{(p+2\nu)^2 + \omega_0^2}}, \tag{9.23}$$

where

$$F(p) = p x_0 + y_0, \quad L(p) = p^2 + \omega_0^2.$$

Under the conditions

$$\omega_0 \ll 2\nu, \quad \frac{\omega_0^2 a_0^2}{4\nu^2} \ll 1,$$

solution (9.23) grades into the Laplace transform of Eq. (8.174), i.e., corresponds to the Gaussian random process $z(t)$ delta-correlated in time.

Consider now the problem on the second moments of the solution to Eq. (9.17). Here, the use of system of equations (9.18) appears necessary. In a way similar to the above derivation of the system of equations (9.19) and (9.21), we obtain the system of six equations for second moments

$$\begin{aligned}
&\frac{d}{dt} \left\langle x^2(t) \right\rangle = 2 \left\langle x(t)y(t) \right\rangle, \\
&\frac{d}{dt} \left\langle x(t)y(t) \right\rangle = \left\langle y^2(t) \right\rangle - \omega_0^2 \left\langle x^2(t) \right\rangle - \omega_0^2 \left\langle z(t)x^2(t) \right\rangle, \\
&\frac{d}{dt} \left\langle y^2(t) \right\rangle = -2\omega_0^2 \left\langle x(t)y(t) \right\rangle - 2\omega_0^2 \left\langle z(t)x(t)y(t) \right\rangle, \\
&\left(\frac{d}{dt} + 2\nu\right) \left\langle z(t)x^2(t) \right\rangle = 2 \left\langle z(t)x(t)y(t) \right\rangle, \\
&\left(\frac{d}{dt} + 2\nu\right) \left\langle z(t)x(t)y(t) \right\rangle = \left\langle z(t)y^2(t) \right\rangle - \omega_0^2 \left[\left\langle z(t)x^2(t) \right\rangle - a_0^2 \left\langle x^2(t) \right\rangle \right], \\
&\left(\frac{d}{dt} + 2\nu\right) \left\langle z(t)y^2(t) \right\rangle = -2\omega_0^2 \left[\left\langle z(t)x(t)y(t) \right\rangle - a_0^2 \left\langle x(t)y(t) \right\rangle \right].
\end{aligned} \tag{9.24}$$

System of equations (9.24) allows one to obtain closed systems for every unknown function $\left\langle x^2(t) \right\rangle$, $\left\langle x(t)y(t) \right\rangle$, and $\left\langle y^2(t) \right\rangle$. For example, the average value of the potential energy $\langle U(t) \rangle$, where $U(t) = x^2(t)$, satisfies the closed system of two equations (every of which is the third-order equation)

$$
\begin{gathered}
\frac{d^3}{dt^3}\langle U(t)\rangle + 4\omega_0^2\frac{d}{dt}\langle U(t)\rangle + 4\omega_0^2\left(\frac{d}{dt}+\nu\right)\langle z(t)U(t)\rangle = 0,\\
\left(\frac{d}{dt}+2\nu\right)\left\{\left(\frac{d}{dt}+2\nu\right)^2 + 4\omega_0^2\right\}\langle z(t)U(t)\rangle + 4\omega_0^2 a_0^2\left(\frac{d}{dt}+\nu\right)\langle U(t)\rangle = 0.
\end{gathered}
\tag{9.25}
$$

It is clear that we could obtain system (9.25) without deriving the complete system of equations (9.24). Indeed, random quantity $U(t)$ satisfies the stochastic third-order equation (8.177)

$$
\frac{d^3}{dt^3}U(t) + 4\omega_0^2\frac{d}{dt}U(t) + 2\omega_0^2\left(z(t)\frac{d}{dt}U(t) + \frac{d}{dt}z(t)U(t)\right) = 0 \tag{9.26}
$$

with the initial value that can generally depend on process $z(t)$ and its derivatives. Averaging Eq. (9.26) over an ensemble of random process realizations and using rule (9.9), page 264 to factor the derivative out of averaging brackets, we obtain the first equation of system (9.25). Then, multiplying Eq. (9.26) by $z(t)$ and using again the rule (9.9), we obtain the second equation of system (9.25).

Systems of equations (9.24) and (9.25) can be solved using the Laplace transform. For example, in the case of the conditions $x(0) = x_0$, $y(0) = y_0$, we have

$$
U(0) = x_0^2, \quad \left.\frac{d}{dt}U(t)\right|_{t=0} = 0, \quad \left.\frac{d^2}{dt^2}U(t)\right|_{t=0} = 2y_0^2, \tag{9.27}
$$

and we obtain the solution of Eqs. (9.25) in the form

$$
\begin{gathered}
\langle U\rangle_p = 2y_0^2\frac{L(p+2\nu)}{L(p)L(p+2\nu) - a_0^2M^2(p)},\\
L(p) = p\left(p^2+4\omega_0^2\right), \quad M(p) = 4\omega_0^2\left(p^2+\nu\right).
\end{gathered}
\tag{9.28}
$$

In the limiting case of great parameters ν and a_0^2, but finite ratio $a_0^2/2\nu = \sigma^2\tau_0$, we obtain from the second equation of system (9.25)

$$
\langle z(t)U(t)\rangle = -\frac{\omega_0^2\sigma^2\tau_0}{\nu}\langle U(t)\rangle.
$$

Consequently, average potential energy $\langle U(t)\rangle$ satisfies in this limiting case the closed third-order equation

$$
\frac{d^3}{dt^3}\langle U(t)\rangle + 4\omega_0^2\frac{d}{dt}\langle U(t)\rangle - 4\omega_0^4\sigma^2\tau_0\langle U(t)\rangle = 0,
$$

which coincides with Eq. (8.176) and corresponds to the Gaussian delta-correlated process $z(t)$.

The system of equations for correlation functions $\langle x(t)x(t')\rangle$ and $\langle y(t)x(t')\rangle$ for $t > t'$ can be obtained in a way similar to the derivation of Eqs. (9.24); it has, obviously, the form

$$\frac{d}{dt}\langle x(t)x(t')\rangle = \langle y(t)x(t')\rangle,$$

$$\frac{d}{dt}\langle y(t)x(t')\rangle = -\omega_0^2\langle x(t)x(t')\rangle - \omega_0^2\langle z(t)x(t)x(t')\rangle,$$

$$\left(\frac{d}{dt}+2\nu\right)\langle z(t)x(t)x(t')\rangle = \langle z(t)y(t)x(t')\rangle,$$

$$\left(\frac{d}{dt}+2\nu\right)\langle z(t)y(t)x(t')\rangle = -\omega_0^2\langle z(t)x(t)x(t')\rangle - \omega_0^2 a_0^2\langle x(t)x(t')\rangle.$$

The initial values for this system are obtained as the solution to system (9.24) at $t = t'$. In a similar way, one can derive the second pair of equations for correlation functions $\langle x(t)y(t')\rangle$, $\langle y(t)y(t')\rangle$ for $t > t'$. In the limit $\nu \to \infty$, $a_0^2 \to \infty$, but finite ratio $a_0^2/2\nu = \sigma^2\tau_0$, we revert to systems of equations (8.167) and (8.169), which correspond to the Gaussian delta-correlated process $z(t)$.

9.2.2 One-Dimensional Nonlinear Differential Equation

Consider now the nonlinear one-dimensional equation

$$\frac{d}{dt}x(t) = f(x,t) + z(t)g(x,t), \quad x(0) = x_0. \tag{9.29}$$

In this case, the indicator function $\varphi(x,t) = \delta(x(t)-x)$ satisfies the stochastic Liouville equation

$$\frac{\partial}{\partial t}\varphi(x,t) = -\frac{\partial}{\partial x}f(x,t)\varphi(x,t) - z(t)\frac{\partial}{\partial x}g(x,t)\varphi(x,t). \tag{9.30}$$

Averaging Eq. (9.30) over an ensemble of realizations of functions $z(t)$ yields the equation for the probability density of solutions to Eq. (9.29) $P(x,t) = \langle\varphi(x,t)\rangle$ in the form

$$\frac{\partial}{\partial t}P(x,t) = -\frac{\partial}{\partial x}f(x,t)P(x,t) - \frac{\partial}{\partial x}g(x,t)\Psi(x,t), \tag{9.31}$$

where we introduced new function

$$\Psi(x,t) = \langle z(t)\varphi(x,t)\rangle \,.$$

Since solution to Eq. (9.30) is a functional of process $z(t)$, we can apply formula (9.7), page 264 to $\Psi(x,t)$ to obtain the equality

$$\left(\frac{d}{dt}+2\nu\right)\Psi(x,t) = \left\langle z(t)\frac{\partial}{\partial t}\varphi(x,t)\right\rangle . \tag{9.32}$$

Substitution of the right-hand side of Eq. (9.30) in Eq (9.32) yields the equation

$$\left(\frac{d}{dt}+2\nu\right)\Psi(x,t) = -\frac{\partial}{\partial x}f(x,t)\Psi(x,t) - \frac{\partial}{\partial x}g(x,t)\left\langle z^2(t)\varphi(x,t)\right\rangle , \tag{9.33}$$

and we obtain the closed system of equations

$$\begin{aligned}\frac{\partial}{\partial t}P(x,t) &= -\frac{\partial}{\partial x}f(x,t)P(x,t) - \frac{\partial}{\partial x}g(x,t)\Psi(x,t),\\ \left(\frac{d}{dt}+2\nu\right)\Psi(x,t) &= -\frac{\partial}{\partial x}f(x,t)\Psi(x,t) - a_0^2\frac{\partial}{\partial x}g(x,t)P(x,t).\end{aligned} \tag{9.34}$$

If functions $f(x,t)$ and $g(x,t)$ are independent of time, the steady-state probability distribution satisfies (if it exists) the equations

$$\begin{aligned} f(x)P(x) &= -g(x)\Psi(x),\\ \left(2\nu+\frac{d}{dx}f(x)\right)\Psi(x) &= -a_0^2\frac{d}{dx}g(x)P(x).\end{aligned} \tag{9.35}$$

Eliminating function $\Psi(x)$, we obtain the first-order differential equation [47–49]

$$\left(2\nu+\frac{d}{dx}f(x)\right)\frac{f(x)}{g(x)}P(x) = a_0^2\frac{d}{dx}g(x)P(x),$$

whose solution can be represented in the form of the quadrature ($|f(x)| < a_0|g(x)|$)

$$P(x) = \frac{C|g(x)|}{a_0^2g^2(x)-f^2(x)}\exp\left\{\frac{2\nu}{a_0^2}\int dx\frac{f(x)}{a_0^2g^2(x)-f^2(x)}\right\}, \tag{9.36}$$

where the positive constant C is determined from the normalization condition.

Note that, in the limit $\nu\to\infty$ and $a_0^2\to\infty$ under the condition that $a_0^2\tau_0 = \text{const}$ ($\tau_0 = 1/(2\nu)$), probability distribution (9.36) grades into the expression

$$P(x) = \frac{C}{|g(x)|}\exp\left\{\frac{2\nu}{a_0^2}\int dx\frac{f(x)}{g^2(x)}\right\}$$

corresponding to the Gaussian delta-correlated process $z(t)$, i.e., the Gaussian process with the correlation function

$$\langle z(t)z(t')\rangle = 2a_0^2\tau_0\delta(t-t').$$

To obtain an idea of system dynamics under the condition of finite correlation radius of process $z(t)$, we consider the simple example with $g(x)=1$, $f(x)=-x$ and $a_0=1$. In this case, we obtain from Eq. (9.36) the probability distribution

$$P(x) = \frac{1}{B(\nu,1/2)}(1-x^2)^{\nu-1} \quad (|x|<1), \tag{9.37}$$

where $B(\nu,1/2)$ is the beta-function. This distribution has essentially different behaviors for $\nu>1$, $\nu=1$ and $\nu<1$, which is schematically shown in Fig. 9.1.

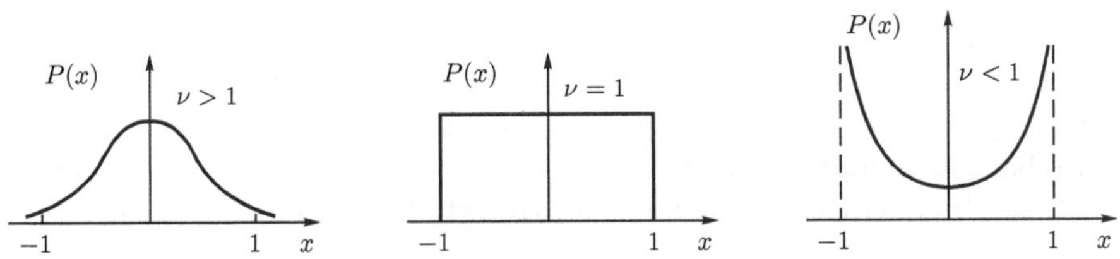

Fig. 9.1 Steady-state probability distribution (9.37) of the solution to the equation $\frac{d}{dt}x(t) = -x + z(t)$ versus parameter ν

One can easily see that this system resides mainly near state $x=0$ if $\nu>1$, and near states $x=\pm 1$ if $\nu<1$. In the case $\nu=1$, we obtain the uniform probability distribution on segment $[-1,1]$.

9.2.3 Particle in the One-Dimensional Potential Field

Another example of nonlinear system concerns the one-dimensional motion of a particle in the filed $U(x)$ under the condition that random forces have a finite temporal correlation radius. We will describe the motion of the particle by the stochastic system of equations

$$\frac{d}{dt}x(t) = y(t), \quad \frac{d}{dt}y(t) = -\frac{dU(x)}{dx} - \lambda y(t) + \mu z(t), \tag{9.38}$$

where function $z(t)$ is assumed to be telegrapher's process $(z^2(t)=1)$. Similarly to the derivation of Eq. (9.34), we obtain the operator equation for the steady-state joint probability density of particle coordinate x and velocity y

$$\widehat{L}(x,y)P(x,y)=\mu^2\frac{\partial}{\partial y}\frac{1}{2\nu+\widehat{L}(x,y)}\frac{\partial}{\partial y}P(x,y), \tag{9.39}$$

where $\widehat{L}(x,y)$ is the Liouville operator,

$$\widehat{L}(x,y)=y\frac{\partial}{\partial x}-\frac{dU(x)}{dx}\frac{\partial}{\partial y}-\lambda\frac{\partial}{\partial y}y.$$

For $\nu\to\infty$, Eq. (9.39) grades into the steady-state Fokker–Planck equation

$$\widehat{L}(x,y)P(x,y)=\frac{\mu^2}{2\nu}\frac{\partial^2}{\partial y^2}P(x,y),$$

whose solution is the Gibbs distribution

$$P(x,y)=C\exp\left\{-\beta\left(\frac{y^2}{2}+U(x)\right)\right\}\quad\left(\beta=\frac{2\nu\lambda}{\mu^2}\right). \tag{9.40}$$

But in the general case, Eq. (9.39) describes the deformation of distribution (9.40) because of finite correlation time $\tau_0=1/(2\nu)$ of process $z(t)$. Equation (9.39) can be rewritten in the form of the partial differential equation

$$\left\{\left(2\nu+\widehat{L}\right)^2\widehat{L}-\lambda\left(2\nu+\widehat{L}\right)\widehat{L}\right\}P(x,y)$$
$$+\left\{-\mu^2\left(2\nu+\widehat{L}\right)\frac{\partial^2}{\partial y^2}+\mu^2\frac{\partial^2}{\partial x\partial y}+\frac{\partial^2U(x)}{\partial x^2}\widehat{L}\right\}P(x,y)=0. \tag{9.41}$$

Deriving Eq. (9.41), we used the differentiation formula for the inverse operator $\widehat{L}^{-1}(\alpha)$

$$\frac{\partial}{\partial\alpha}\widehat{L}^{-1}(\alpha)=-\widehat{L}^{-1}(\alpha)\frac{\partial\widehat{L}(\alpha)}{\partial\alpha}\widehat{L}^{-1}(\alpha).$$

Equation (9.41) is rather complicated, and it hardly can be solved for arbitrary field $U(x)$. However, one can easily see that solution to Eq. (9.41) will not be a function of sole particle energy as it is the case in Eq. (9.40); in addition, particle coordinate and velocity will be statistically dependent quantities.

9.2.4 Ordinary Differential Equation of the n-th Order

Let now operators $\widehat{A}(t)$ and $\widehat{B}(t)$ in Eq. (9.10) be the matrices, i.e.,

$$\frac{d}{dt}\boldsymbol{x}(t) = A(t)\boldsymbol{x}(t) + z(t)B(t)\boldsymbol{x}(t), \quad \boldsymbol{x}(0) = \boldsymbol{x}_0.$$

If only one component is of our interest, we can obtain for it the operator equation

$$\widehat{L}\left(\frac{d}{dt}\right)x(t) + \sum_{i+j=0}^{n-1} b_{ij}(t)\frac{d^i}{dt^i}z(t)\frac{d^j}{dt^j}x(t) = f(t), \tag{9.42}$$

where operator

$$\widehat{L}\left(\frac{d}{dt}\right) = \sum_{i=0}^{n} a_i(t)\frac{d^i}{dt^i}$$

and n is the order of matrices A and B in Eq. (9.10).

The initial values for x are included in function $f(t)$ through the corresponding derivatives of the delta function. Note that function $f(t)$ can depend additionally on derivatives of random process $z(t)$ at $t = 0$, i.e., $f(t)$ is also the random function statistically related to process $z(t)$.

Averaging Eq. (9.42) over an ensemble of realizations of process $z(t)$ with the use of formula (9.9), we obtain the equation

$$\widehat{L}\left(\frac{d}{dt}\right)\langle x(t)\rangle + M\left[\frac{d}{dt}, \frac{d}{dt} + 2\nu\right]\langle z(t)x(t)\rangle = \langle f(t)\rangle, \tag{9.43}$$

where

$$M[p,q] = \sum_{i+j=0}^{n-1} b_{ij}(t)p^i q^j.$$

However, Eq. (9.43) is unclosed because of the presence of correlator $\langle z(t)x(t)\rangle$. Multiplying Eq. (9.42) by $z(t)$ and averaging the result, we obtain the equation for correlator $\langle z(t)x(t)\rangle$

$$\widehat{L}\left(\frac{d}{dt} + 2\nu\right)\langle z(t)x(t)\rangle + a_0^2 M\left[\frac{d}{dt} + 2\nu, \frac{d}{dt}\right]\langle x(t)\rangle = \langle z(t)f(t)\rangle. \tag{9.44}$$

System of equations (9.43) and (9.44) is the closed system. If functions $a_i(t)$ and $b_{ij}(t)$ are independent of time, this system can be solved using the Laplace transform. This solution is as follows:

$$\langle x\rangle_p = \frac{\widehat{L}(p+2\nu)\langle f\rangle_p - M[p,p+2\nu]\langle zf\rangle_p}{\widehat{L}(p)\widehat{L}(p+2\nu) - a_0^2 M[p+2\nu,p]M[p,p+2\nu]}. \tag{9.45}$$

Note that Eq. (9.26) considered earlier is a special case of Eq. (9.42) and, consequently, Eq. (9.28) is a special case of Eq. (9.45).

9.2.5 Statistical Interpretation of Telegrapher's Equation

In the preceding chapter, page 252, we showed that solutions to certain class of partial differential equations can be treated as the result of averaging certain functional over the random process delta-correlated in time. A similar situation occurs for telegrapher's random process.

Consider the initial-value problem for the wave equation with linear friction

$$\left(\frac{\partial^2}{\partial t^2} + 2\nu\frac{\partial}{\partial t} - v^2\frac{\partial^2}{\partial x^2}\right) F(x,t) = 0,$$
$$F(x,0) = \varphi(x), \quad \frac{\partial}{\partial t}F(x,t)\bigg|_{t=0} = \psi(x). \tag{9.46}$$

We can rewrite Eq. (9.46) as the integro-differential equation

$$\frac{\partial}{\partial t}F(x,t) = \psi(x)e^{-2\nu t} + v^2\int_0^t dt_1 e^{-2\nu(t-t_1)}\frac{\partial^2}{\partial x^2}F(x,t). \tag{9.47}$$

Introduce now the auxiliary stochastic equation

$$\frac{\partial}{\partial t}f(x,t) = \psi(x)e^{-2\nu t} + vz(t)\frac{\partial}{\partial x}f(x,t), \quad f(x,0) = \varphi(x), \tag{9.48}$$

where $z(t)$ is telegrapher's process ($z^2 = 1$). From the above material obviously follows that

$$F(x,t) = \langle f(x,t)\rangle_z.$$

The solution to Eq. (9.48) has the form

$$f(x,t) = \varphi\left(x + v\int_0^t d\tau z(\tau)\right) + \int_0^t dt_1 e^{-2\nu t_1}\psi\left(x + v\int_{t_1}^t d\tau z(\tau)\right).$$

Consequently, the solution to Eq. (9.46) can be represented as the statistical average over random process $z(t)$:

$$F(x,t) = \left\langle\varphi\left(x + v\int_0^t d\tau z(\tau)\right)\right\rangle_z + \int_0^t dt_1 e^{-2\nu t_1}\left\langle\psi\left(x + v\int_{t_1}^t d\tau z(\tau)\right)\right\rangle_z.$$

9.3 Generalized Telegrapher's Process

Generalized telegrapher's process is defined by the formula

$$z(t) = a_{n(0,t)}, \tag{9.49}$$

where $n(t_1, t_2)$, $t_1 < t_2$ is the integer-valued Poisson random process statistically independent of random quantities a_i, which are also statistically independent and have probability density $p(a)$.

Generalized telegrapher's process $z(t)$ is stationary in time and its correlation function

$$\langle z(t)z(t')\rangle = \langle a^2\rangle e^{-\nu|t-t'|}$$

is characterized by the temporal correlation radius $\tau_0 = 1/\nu$.

As in the case of telegrapher's process, two alternative methods are appropriate for analyzing stochastic equations whose parameter fluctuations can be described by generalized telegrapher's process.

The first method immediately deals with the formula (7.45), page 178 for splitting the correlation of process $z(t)$ with arbitrary functional $R[t; z(\tau)]$ of this process,

$$\langle z(t)R[t; z(\tau)]\rangle = \langle \widetilde{a}R[t; \widetilde{a}]\rangle e^{-vt} + a_0^2 \int\limits_0^t dt_1 e^{-v(t-t_1)} \left\langle \widetilde{a}\widetilde{R}[t, t_1; \widetilde{a}, z(\tau)]\right\rangle, \tag{9.50}$$

where functional $\widetilde{R}[t, t_1; \widetilde{a}, z(\tau)]$ is given by the formula

$$\widetilde{R}[t, t_1; \widetilde{a}, z(\tau)] = R[t; \widetilde{a}\theta(\tau - t_1 + 0) + z(\tau)\theta(t_1 - \tau + 0)], \tag{9.51}$$

and random quantity $\widetilde{a}$ is statistically independent of process $z(t)$.

In contrast to telegrapher's process, the second method appears here more formal and deals with the differentiation formula (7.52), (7.54), page 180 that has the form

$$\left(\frac{d}{dt} + \nu\right)\langle F(z(t))\, R[t; z(\tau)]\rangle = \left\langle F(z(t))\, \frac{d}{dt} R[t; z(\tau)]\right\rangle, \quad (\tau \le t) \tag{9.52}$$

under the condition that $\langle F(z(t))\rangle = \langle F(a)\rangle_a = 0$. In particular, we have the formula

$$\left(\frac{d}{dt} + \nu\right)^n \langle F(z(t))\, R[t; z(\tau)]\rangle = \left\langle F(z(t))\, \frac{d^n}{dt^n} R[t; z(\tau)]\right\rangle \tag{9.53}$$

defining the rule of factoring the differential operator out of averaging brackets.

The further analysis becomes simpler if we define function $F(z(t))$ as follows:

$$F(t) = F(z(t)) = \frac{1}{1+\lambda z(t)} - C_0(\lambda), \tag{9.54}$$

where

$$C_k(\lambda) = \left\langle \frac{a^k}{1+\lambda a} \right\rangle_a \tag{9.55}$$

and λ is arbitrary parameter. This function $F(t)$ satisfies the identity

$$z(t)F(t) = -\frac{1}{\lambda}F(t) + C_1(t) - z(t)C_0(\lambda). \tag{9.56}$$

Now, we consider several examples of working according to the above formalisms.

9.3.1 Stochastic Linear Equation

First of all, we consider, as in the previous section, the stochastic linear equation (9.10), page 264 assuming that linear operators $\widehat{A}(t)$ and $\widehat{B}(t)$ in this equation are constant matrices A and B. In this case, the equation for the mean value $\langle \boldsymbol{x}(t) \rangle$ is

$$\frac{d}{dt} \langle \boldsymbol{x}(t) \rangle = A \langle \boldsymbol{x}(t) \rangle + B \langle z(t) \boldsymbol{x}(t) \rangle . \tag{9.57}$$

Using Eq. (9.50), we can rewrite this equation in the form

$$\frac{d}{dt} \langle \boldsymbol{x}(t) \rangle = A \langle \boldsymbol{x}(t) \rangle + B \langle \widetilde{a} \boldsymbol{x}[t; \widetilde{a}] \rangle e^{-\nu t} + \nu B \int\limits_0^t dt_1 e^{-\nu(t-t_1)} \langle \widetilde{a} \boldsymbol{x}[t, t_1; \widetilde{a}, z(\tau)] \rangle . \tag{9.58}$$

According to Eq. (9.51), functional $\boldsymbol{x}[t, t_1; \widetilde{a}, z(\tau)]$ satisfies the equation

$$\frac{d}{dt} \boldsymbol{x}(t) = A\boldsymbol{x}(t) + \widetilde{a} B \boldsymbol{x}(t) \quad (t > t_1) \tag{9.59}$$

with the initial value

$$\boldsymbol{x}[t_1, t_1; \widetilde{a}, z(\tau)] = \boldsymbol{x}[t_1; z(\tau)]. \tag{9.60}$$

Hence, functional $\boldsymbol{x}[t, t_1; \widetilde{a}, z(\tau)]$ has the form

$$\boldsymbol{x}[t, t_1; \widetilde{a}, z(\tau)] = e^{(A+\widetilde{a}B)(t-t_1)} \boldsymbol{x}(t_1),$$

and Eq. (9.58) turns into the integro-differential equation

$$\frac{d}{dt}\langle \boldsymbol{x}(t)\rangle = A\langle \boldsymbol{x}(t)\rangle + e^{-\nu t}B\left\langle \widetilde{a}e^{(A+\widetilde{a}B)t}\right\rangle \boldsymbol{x}_0$$

$$+\nu B\int\limits_0^t dt_1 e^{-\nu(t-t_1)}\left\langle \widetilde{a}e^{(A+\widetilde{a}B)(t-t_1)}\right\rangle \langle \boldsymbol{x}(t_1)\rangle \quad (9.61)$$

with the initial value $\langle \boldsymbol{x}(0)\rangle = \boldsymbol{x}_0$.

We can easily solve Eq. (9.61) using the Laplace transform. The solution has the form

$$\langle \boldsymbol{x}\rangle_p = (E-\nu C)^{-1}C\boldsymbol{x}_0, \quad (9.62)$$

where

$$C = \left\langle \{(p+\nu)E - A - \widetilde{a}B\}^{-1}\right\rangle_a$$

and E is the unit matrix.

Use now the alternative method for splitting the correlator in Eq. (9.57). According to the differentiation formula (9.52), we have

$$\left(\frac{d}{dt}+\nu\right)\langle F(t)\,\boldsymbol{x}(t)\rangle = \left\langle F(t)\,\frac{d}{dt}\boldsymbol{x}(t)\right\rangle = A\langle F(t)\boldsymbol{x}(t)\rangle + B\langle z(t)F(t)\boldsymbol{x}(t)\rangle\,. \quad (9.63)$$

Using then Eq. (9.56), we can rewrite Eq. (9.63) as the identity

$$\left[\left(\frac{d}{dt}+\nu\right)E - A + \frac{1}{\lambda}B\right]\langle F(t)\,\boldsymbol{x}(t)\rangle = BC_1(\lambda)\langle \boldsymbol{x}(t)\rangle - BC_0(\lambda)\langle z(t)\boldsymbol{x}(t)\rangle\,. \quad (9.64)$$

Performing the Laplace transform of Eqs. (9.57) and (9.64), we obtain the unclosed system of equations

$$(pE-A)\langle \boldsymbol{x}\rangle_p - B\langle z\boldsymbol{x}\rangle_p = \boldsymbol{x}_0,$$

$$\left[(p+\nu)E - A + \frac{1}{\lambda}B\right]\langle F(t)\,\boldsymbol{x}(t)\rangle_p = BC_1(\lambda)\langle \boldsymbol{x}\rangle_p - BC_0(\lambda)\langle z\boldsymbol{x}\rangle_p\,, \quad (9.65)$$

which is valid for arbitrary λ. For $\frac{1}{\lambda} = [A-(p+\nu)E]B^{-1}$, the left-hand side of the second equation vanishes, and we obtain the algebraic relationship between $\langle \boldsymbol{x}\rangle_p$ and $\langle z\boldsymbol{x}\rangle_p$; together with the first equation of system (9.65), this relationship yields the solution that coincides with Eq. (9.62).

Stochastic Parametric Resonance

We consider the statistical description of solution to problem (9.17), page 265 as a specific example. Averaging Eq. (9.17) over an ensemble of realizations of generalized telegrapher's process $z(t)$, we obtain the unclosed equation

$$\left(\frac{d^2}{dt^2}+\omega_0^2\right)\langle x(t)\rangle+\omega_0^2\langle z(t)x(t)\rangle=0, \tag{9.66}$$

with the initial values $\langle x(0)\rangle = x_0$, $\left\langle \frac{dx(t)}{dt}\right\rangle_{t=0} = y_0$.

To split the correlator appeared in (9.66), we multiply Eq. (9.17), page 265 by function $F(z(t))$ and average the result. Using then formula (9.52), page 275 that defines the rule of factoring the derivative out of averaging brackets, we obtain the equation

$$\left[\left(\frac{d}{dt}+\nu\right)^2+\omega_0^2\right]\langle F(z(t))\,x(t)\rangle+\omega_0^2\langle z(t)F(z(t))\,x(t)\rangle=0 \tag{9.67}$$

with zero-valued initial values $\langle F(z(t))\,x(0)\rangle = 0$, $\left\langle F(z(t))\frac{dx(t)}{dt}\right\rangle_{t=0} = 0$.

The further analysis becomes simpler if we use function $F(z(t))$ in form (9.54), page 276 and rewrite Eq. (9.67) as follows:

$$\left[\left(\frac{d}{dt}+\nu\right)^2+\omega_0^2\left(1-\frac{1}{\lambda}\right)\right]\langle F(z(t))\,x(t)\rangle+\omega_0^2C_1(\lambda)\langle x(t)\rangle-\omega_0^2C_0(\lambda)\langle z(t)x(t)\rangle=0. \tag{9.68}$$

Performing the Laplace transform of Eqs. (9.66) and (9.68), we obtain

$$(p^2+\omega_0^2)\langle x\rangle_p+\omega_0^2\langle zx\rangle_p=y_0+px_0,$$

$$\left[(p+\nu)^2+\omega_0^2\left(1-\frac{1}{\lambda}\right)\right]\langle Fx\rangle_p++\omega_0^2C_1(\lambda)\langle x\rangle_p-\omega_0^2C_0(\lambda)\langle zx\rangle_p=0. \tag{9.69}$$

In Eqs. (9.68) and (9.69), parameter λ is arbitrary parameter. Now, we specify it as follows:

$$\lambda=\lambda_p=\frac{\omega_0^2}{L(p+\nu)},\quad L(p)=p^2+\omega_0^2. \tag{9.70}$$

In this case, the first term in the second equation of system (9.69) vanishes, and we obtain the relationship between correlator $\langle zx\rangle_p$ and average solution to Eq. (8.161) $\langle x\rangle_p$ in the form

$$\langle zx\rangle_p=\frac{C_1(p)}{C_0(p)}\langle x\rangle_p, \tag{9.71}$$

where

$$C_k(p)=\left\langle\frac{a^k}{L(p+\nu)+a\omega_0^2}\right\rangle_a.$$

Substituting Eq. (9.71) in the first equation of system (9.69), we obtain the solution in the form

$$\langle x\rangle_p = \frac{y_0 + p x_0}{L(p) + \omega_0^2 \dfrac{C_1(p)}{C_0(p)}}. \tag{9.72}$$

As was noted earlier, the mean value of the solution to problem (9.17), page 265 can be obtained with the use of the other — alternative and more intuitive — method. Using Eq. (9.50), page 275, we can rewrite Eq. (9.66) in the form

$$\left(\frac{d^2}{dt^2} + \omega_0^2\right)\langle x(t)\rangle = -\omega_0^2 \left\langle a\widetilde{x}[t;a]\right\rangle_a e^{-\nu t} - \nu\omega_0^2 \int\limits_0^t dt_1 e^{-\nu(t-t_1)} \left\langle a\widetilde{x}[t,t_1;a,z(\tau)]\right\rangle_a, \tag{9.73}$$

where functional $\widetilde{x}[t,t_1;a,z(\tau)]$ satisfies the equation

$$\left(\frac{d^2}{dt^2} + \omega_0^2\right)\widetilde{x}(t) + \omega_0^2 a\widetilde{x}(t) = 0$$

with the initial values

$$\widetilde{x}[t,t_1;a,z(\tau)]_{t=t_1} = x(t_1), \qquad \frac{d}{dt}\widetilde{x}[t,t_1;a,z(\tau)]\bigg|_{t=t_1} = \frac{d}{dt_1}x(t_1)$$

and $\widetilde{x}[t;a] = \widetilde{x}[t,0;a,z(\tau)]$.

The solution to this equation is as follows

$$\widetilde{x}[t,t_1;a,z(\tau)] = x(t_1)\cos\left[\omega_0\sqrt{1+a}(t-t_1)\right] + \frac{dx(t_1)}{dt_1}\frac{\sin\left[\omega_0\sqrt{1+a}(t-t_1)\right]}{\omega_0\sqrt{1+a}},$$

and, consequently, Eq. (9.28) can be rewritten in the closed form

$$\begin{aligned}
\left(\frac{d^2}{dt^2}+\omega_0^2\right)\langle x(t)\rangle &= -x_0\omega_0^2 e^{-\nu t}\left\langle a\cos\left(\omega_0\sqrt{1+a}t\right)\right\rangle_a - y_0\omega_0^2 e^{-\nu t}\left\langle a\frac{\sin\left(\omega_0\sqrt{1+a}t\right)}{\omega_0\sqrt{1+a}}\right\rangle_a \\
&\quad - \nu\omega_0^2\int\limits_0^t dt_1 e^{-\nu(t-t_1)}\langle x(t_1)\rangle\left\langle a\cos\left[\omega_0\sqrt{1+a}(t-t_1)\right]\right\rangle_a \\
&\quad - \nu\omega_0^2\int\limits_0^t dt_1 e^{-\nu(t-t_1)}\frac{d\langle x(t_1)\rangle}{dt_1}\left\langle a\frac{\sin\left[\omega_0\sqrt{1+a}(t-t_1)\right]}{\omega_0\sqrt{1+a}}\right\rangle_a.
\end{aligned} \tag{9.74}$$

Equation (9.74) can be easily solved using the Laplace transform; the result coincides with Eq. (9.72).

9.3.2 One-Dimensional Nonlinear Differential Equation

Consider now the one-dimensional equation (9.29), page 269

$$\frac{d}{dt}x(t) = f(x) + z(t)g(x), \quad x(0) = x_0$$

assuming that $z(t)$ is generalized telegrapher's process and functions $f(x)$ and $g(x)$ are independent of time. In this case, the indicator function satisfies the Liouville equation (9.30), page 269 that assumes here the form

$$\frac{\partial}{\partial t}\varphi(x,t) = -\frac{\partial}{\partial x}f(x)\varphi(x,t) - z(t)\frac{\partial}{\partial x}g(x)\varphi(x,t). \tag{9.75}$$

Averaging Eq. (9.75), we obtain the equation for the one-time probability density

$$\frac{\partial}{\partial t}P(x,t) + \frac{\partial}{\partial x}f(x)P(x,t) = -\frac{\partial}{\partial x}g(x)\left\langle z(t)\varphi(x,t)\right\rangle$$

$$= -e^{-\nu t}\frac{\partial}{\partial x}g(x)\left\langle \widetilde{a}\varphi[x,t;\widetilde{a}]\right\rangle - \nu\frac{\partial}{\partial x}g(x)\int\limits_0^t dt_1 e^{-\nu(t-t_1)}\left\langle \widetilde{a}\widetilde{\varphi}[x,t,t_1;\widetilde{a},z(\tau)]\right\rangle. \tag{9.76}$$

Functional $\widetilde{\varphi}[x,t,t_1;\widetilde{a},z(\tau)]$ will satisfy now the equation

$$\frac{\partial}{\partial t}\widetilde{\varphi}(x,t) = -\frac{\partial}{\partial x}f(x)\widetilde{\varphi}(x,t) - \widetilde{a}\frac{\partial}{\partial x}g(x)\widetilde{\varphi}(x,t)$$

with the initial value $\widetilde{\varphi}(x,t_1) = \varphi(x,t_1)$. In the operator form, the solution to this equation will be

$$\widetilde{\varphi}(x,t) = e^{-(t-t_1)\frac{\partial}{\partial x}[f(x)+\widetilde{a}g(x)]}\varphi(x,t_1).$$

Hence, we can rewrite the equation for the probability density (9.76) in the closed operator form

$$\left(\frac{\partial}{\partial t} + \frac{\partial}{\partial x}f(x)\right)P(x,t) = -e^{-\nu t}\frac{\partial}{\partial x}g(x)\left\langle \widetilde{a}e^{-t\frac{\partial}{\partial x}[f(x)+\widetilde{a}g(x)]}\right\rangle \varphi(x,0)$$

$$-\nu\frac{\partial}{\partial x}g(x)\int\limits_0^t dt_1 e^{-\nu(t-t_1)}\left\langle \widetilde{a}e^{-(t-t_1)\frac{\partial}{\partial x}[f(x)+\widetilde{a}g(x)]}\right\rangle P(x,t_1). \tag{9.77}$$

The steady-state probability distribution (if it exists) satisfies the operator equation

$$f(x)P(x) = -\nu g(x) \int\limits_0^\infty d\tau e^{-\nu\tau} \left\langle \widetilde{a} e^{-\tau \frac{d}{dx}[f(x)+\widetilde{a}g(x)]} \right\rangle P(x)$$

that can be rewritten as follows:

$$f(x)P(x) = -\nu g(x) \left\langle \frac{\widetilde{a}}{\nu + \dfrac{d}{dx}[f(x) + \widetilde{a}g(x)]} \right\rangle P(x). \tag{9.78}$$

To convert Eq. (9.78) to the differential equation, we must specify the probability distribution of random quantity a.

Assume for example that quantity a is characterized by sufficiently small intensity of fluctuations and $\langle a \rangle = 0$. Then, expanding the operator in the right-hand side of Eq. (9.78) in powers of a and neglecting all terms higher than $\langle a^2 \rangle$, we obtain the operator equation

$$f(x)P(x) = -\nu \left\langle a^2 \right\rangle g(x) \frac{1}{\nu + \dfrac{d}{dx}f(x)} \frac{d}{dx} \frac{1}{\nu + \dfrac{d}{dx}f(x)} P(x). \tag{9.79}$$

If we represent now function $P(x)$ in the form

$$P(x) = \left[\nu + \frac{d}{dx}f(x)\right]\psi(x),$$

then we obtain the second-order differential equation for function $\psi(x)$

$$\left[\nu + \frac{d}{dx}f(x)\right] \frac{f(x)}{g(x)} \left[\nu + \frac{d}{dx}f(x)\right]\psi(x) = -\nu \left\langle a^2 \right\rangle \frac{d}{dx} g(x)\psi(x). \tag{9.80}$$

For $\nu \to \infty$, we can expand the mean value in the right-hand side of Eq. (9.78) in powers of $1/\nu$ and obtain the steady-state Fokker–Planck equation

$$f(x)P(x) = g(x) \frac{\nu}{\langle a^2 \rangle} \frac{d}{dx} g(x)P(x)$$

corresponding to the Gaussian delta-correlated process $z(t)$.

9.3.3 Ordinal Differential Equation of the n-th Order

Consider now Eq. (9.42), page 273

$$\widehat{L}\left(\frac{d}{dt}\right) x(t) + \sum_{i+j=0}^{n-1} b_{ij}(t) \frac{d^i}{dt^i} z(t) \frac{d^j}{dt^j} x(t) = f(t)$$

with generalized telegrapher's process $z(t)$. For simplicity, we will assume that initial values for Eq. (9.42) are independent of $z(t)$ and coefficients a_i of operator $\widehat{L}$ (9.151) and b_{ij} are constants.

Averaging Eq. (9.42) with the use of formula (9.53), page 275, we obtain the equation

$$\widehat{L}\left(\frac{d}{dt}\right) \langle x(t)\rangle + M\left[\frac{d}{dt}, \frac{d}{dt} + \nu\right] \langle z(t)x(t)\rangle = f(t), \tag{9.81}$$

where $M[p,q] = \sum\limits_{i+j=0}^{n-1} b_{ij}(t) p^i q^j$, as before.

Consider now correlator $\langle F(t)x(t)\rangle$, where $F(t)$ is given by Eq. (9.54). In accordance with the differentiation formula (9.53), this function satisfies the equation

$$\widehat{L}\left(\frac{d}{dt} + \nu\right) \langle F(t)x(t)\rangle = \left\langle F(t)\widehat{L}\left(\frac{d}{dt}\right) x(t)\right\rangle$$
$$= -\sum_{i+j=0}^{n-1} b_{ij}(t) \left\langle F(t) \frac{d^i}{dt^i} z(t) \frac{d^j}{dt^j} x(t)\right\rangle - \sum_{i+j=0}^{n-1} b_{ij}(t) \left(\frac{d}{dt} + \nu\right)^i \left\langle z(t)F(t) \frac{d^j}{dt^j} x(t)\right\rangle. \tag{9.82}$$

Using now Eq. (9.56), we can rewrite the right-hand side of Eq. (9.82) in the form

$$-\sum_{i+j=0}^{n-1} b_{ij}(t) \left(\frac{d}{dt} + \nu\right)^i - \frac{1}{\lambda}\left\langle F(t) \frac{d^j}{dt^j} x(t)\right\rangle$$
$$+ C_1(\lambda) \frac{d^j}{dt^j} \langle x(t)\rangle - C_0(\lambda) \left\langle z(t) \frac{d^j}{dt^j} x(t)\right\rangle = \frac{1}{\lambda} M\left[\frac{d}{dt} + \nu, \frac{d}{dt} + \nu\right] \langle F(t)x(t)\rangle$$
$$- C_1(\lambda) M\left[\frac{d}{dt} + \nu, \nu\right] \langle x(t)\rangle + C_0(\lambda) M\left[\frac{d}{dt} + \nu, \frac{d}{dt} + \nu\right] \langle z(t)x(t)\rangle. \tag{9.83}$$

Consequently, Eq. (9.82) assumes the form

$$\left\{\widehat{L}\left(\frac{d}{dt} + \nu\right) - \frac{1}{\lambda} M\left[\frac{d}{dt} + \nu, \frac{d}{dt} + \nu\right]\right\} \langle F(t)x(t)\rangle$$
$$= -C_1(\lambda) M\left[\frac{d}{dt} + \nu, \frac{d}{dt}\right] \langle x(t)\rangle + C_0(\lambda) M\left[\frac{d}{dt} + \nu, \frac{d}{dt} + \nu\right] \langle z(t)x(t)\rangle \tag{9.84}$$

with the initial value $\langle F(t)x(t)\rangle|_{t=0} = 0$.

Perform now the Laplace transform of Eqs. (9.81) and (9.84). As a result, we obtain the algebraic system of equations

$$\widehat{L}(p)\langle x\rangle_p + M[p, p+\nu]\langle zx\rangle_p = f(p),$$

$$\left\{\widehat{L}(p+\nu) - \frac{1}{\lambda}M[p+\nu, p+\nu]\right\}\langle Fx\rangle_p$$
$$= -C_1(\lambda)M[p+\nu, p]\langle x\rangle_p + C_0(\lambda)M[p+\nu, p+\nu]\langle zx\rangle_p. \quad (9.85)$$

Equations (9.85) are valid for arbitrary λ. If we set

$$\lambda = \lambda_p = M[p+\nu, p+\upsilon]/L(p+\nu), \quad (9.86)$$

the second equation of system (9.85) becomes the algebraic relationship between $\langle zx\rangle_p$ and $\langle x\rangle_p$:

$$\langle z(t)x(t)\rangle_p = \frac{\widetilde{C}_1(p)}{\widetilde{C}_0(p)}\frac{M[p+\nu, p]}{M[p+\nu, p+\nu]}\langle x\rangle_p, \quad (9.87)$$

where

$$\widetilde{C}_k(p) = \left\langle \frac{a^k}{\widehat{L}(p+\nu) + aM[p+\nu, p+\nu]}\right\rangle_a.$$

Substituting (9.87) in the first equation of system (9.85), we obtain the algebraic equation for $\langle x\rangle_p$, whose solution has the form [78]

$$\langle x\rangle_p = f(p)\left[\widehat{L}(p) + \frac{M[p+\nu, p]\,M[p, p+\nu]}{M[p+\nu, p+\nu]}\frac{\widetilde{C}_1(p)}{\widetilde{C}_0(p)}\right]^{-1}. \quad (9.88)$$

9.4 Gaussian Markovian Processes

Here, we consider several examples associated with the Gaussian Markovian processes.

9.4.1 General Remarks

Define random process $z(t)$ by the formula

$$z(t) = z_1(t) + \cdots + z_N(t), \quad (9.89)$$

where $z_i(t)$ are statistically independent telegrapher's processes with correlation functions

$$\langle z_i(t) z_j(t')\rangle = \delta_{ij} \left\langle z^2 \right\rangle e^{-\alpha|t-t'|} \quad (\alpha = 2\nu).$$

If we set $\left\langle z^2 \right\rangle = \sigma^2/N$, then this process grades for $N \to \infty$ into the Gaussian Markovian process with correlation function (see Remark 3.1, page 139)

$$\langle z_i(t) z_j(t')\rangle = \left\langle z^2 \right\rangle e^{-\alpha|t-t'|}.$$

Thus, process $z(t)$ (9.89) approximates the Gaussian Markovian process in terms of the Markovian process with a finite number of states.

It is evident that the differentiation formula and the rule of factoring the derivative out of averaging brackets assume for process $z(t)$ the forms

$$\left(\frac{d}{dt} + \alpha k\right) \langle z_1(t) \cdots z_k(t) R[t; z(\tau)]\rangle = \left\langle z_1(t) \cdots z_k(t) \frac{d}{dt} R[t; z(\tau)]\right\rangle,$$

$$\left\langle z_1(t) \cdots z_k(t) \frac{d^n}{dt^n} R[t; z(\tau)]\right\rangle = \left(\frac{d}{dt} + \alpha k\right)^n \langle z_1(t) \cdots z_k(t) R[t; z(\tau)]\rangle, \tag{9.90}$$

where $R[t; z(\tau)]$ is arbitrary functional of process $z(t)$ $(\tau \leq t)$.

9.4.2 Stochastic Linear Equation

Consider again Eq. (9.10), which we represent here in the form

$$\frac{d}{dt}\boldsymbol{x}(t) = \widehat{A}(t)\boldsymbol{x}(t) + [z_1(t) + \cdots + z_N(t)]\widehat{B}(t)\boldsymbol{x}(t), \quad \boldsymbol{x}(0) = \boldsymbol{x}_0, \tag{9.91}$$

and introduce vector-function

$$\boldsymbol{X}_k(t) = \langle z_1(t) \cdots z_k(t) \boldsymbol{x}(t)\rangle, \quad k = 1, \cdots, N; \quad \boldsymbol{X}_0(t) = \langle \boldsymbol{x}(t)\rangle. \tag{9.92}$$

Using formula (9.90) for differentiating correlations (9.92) and Eq. (9.91), we obtain the recurrence equation for $\boldsymbol{x}_k(t)$ $(k = 0, 1, \cdots, N)$

$$\begin{aligned}\left(\frac{d}{dt} + \alpha k\right) \boldsymbol{X}_k(t) &= \widehat{A}(t)\boldsymbol{X}_k(t) + \left\langle z_1(t) \cdots z_k(t)[z_1(t) + \cdots + z_N(t)]\widehat{B}(t)\boldsymbol{x}(t)\right\rangle \\ &= \widehat{A}(t)\boldsymbol{X}_k(t) + k\left\langle z^2 \right\rangle \widehat{B}(t)\boldsymbol{X}_{k-1}(t) + (N-k)\widehat{B}(t)\boldsymbol{X}_{k+1}(t),\end{aligned} \tag{9.93}$$

with the initial value

$$\boldsymbol{X}_k(0) = \boldsymbol{x}_0 \delta_{k,0}.$$

Thus, the mean value of the solution to system (9.91) satisfies the closed system of $(N+1)$ vector equations. If operators $\widehat{A}(t)$, $\widehat{B}(t)$ are time-independent matrices, system (9.93) can be easily solved using the Laplace transform. It is clear that such a solution will have the form of a finite segment of continued fraction. If we set $\langle z^2 \rangle = \sigma^2/N$ and proceed to the limit $N \to \infty$, then random process (9.89) will grade, as was mentioned earlier, into the Gaussian Markovian process and solution to system (9.93) will assume the form of the infinite continued fraction.

9.4.3 Ordinal Differential Equation of the n-th Order

Consider stochastic equation (9.42), page 273

$$\widehat{L}\left(\frac{d}{dt}\right)x(t) + \sum_{i+j=0}^{n-1} b_{ij}(t)\frac{d^i}{dt^i}z(t)\frac{d^j}{dt^j}x(t) = f(t)$$

with random process $z(t)$ given by Eq. (9.89) and introduce, as in the previous example, functions

$$X_k(t) = \langle z_1(t)\cdots z_k(t)x(t)\rangle\,, \quad k = 1,\cdots,N; \quad X_0(t) = \langle x(t)\rangle\,.$$

To obtain equations for these functions, we multiply Eq. (9.42) by $z_1(t)\cdots z_k(t)$ and average the result over an ensemble of realizations of $z_i(t)$. Using Eqs. (9.90), we obtain that function $X_k(t)$ satisfies the closed system of recurrence equations

$$\widehat{L}\left(\frac{d}{dt}+\alpha k\right)X_k(t) = F_k(t) - k\langle z^2\rangle M\left[\frac{d}{dt}+\alpha k, \frac{d}{dt}+\alpha(k-1)\right]X_{k-1}(t)$$
$$-(N-k)\langle z^2\rangle M\left[\frac{d}{dt}+\alpha k, \frac{d}{dt}+\alpha(k+1)\right]X_{k+1}(t), \quad (9.94)$$

where

$$F_k(t) = \langle z_1(t)\cdots z_k(t)f(t)\rangle\,.$$

If operator $\widehat{L}$ and functions b_{ij} are independent of time t, the Laplace transform reduces system (9.94) to the algebraic system

$$\widehat{L}(p+\alpha k)X_k(p) = F_k(p) - k\langle z^2\rangle M[p+\alpha k, p+\alpha(k-1)]X_{k-1}(p)$$
$$-(N-k)\langle z^2\rangle M[p+\alpha k, p+\alpha(k+1)]X_{k+1}(p). \quad (9.95)$$

In the special case of function $f(t)$ independent of $z_k(t)$, when $F_k(p) = f(p)\delta_{k,0}$, Eq. (9.95) can be easily solved, and the solution has the form of the finite segment of continued fraction

$$X_0(p) = f(p)K_0(p), \quad K_l(p) = \frac{1}{A_l(p) - B_l(p)K_{l+1}(p)}, \quad (9.96)$$

where

$$A_l(p) = \widehat{L}\left(p + \alpha l\right),$$

$$B_l(p) = \left\langle z^2 \right\rangle (N-l)(l+1) M\left[p + \alpha l, p + \alpha(l+1)\right] M\left[p + \alpha(l+1), p + \alpha l\right]. \tag{9.97}$$

If $N = 1$, i.e., if we deal with only one telegrapher's process, the solution (9.96), (9.97) assumes the form of Eq. (9.45), page 273, which corresponds to the two-level continued fraction.

If we set $\left\langle z^2 \right\rangle = \sigma^2/N$ and proceed to the limit $N \to \infty$, we obtain solution $\langle x(p) \rangle$ for the Gaussian Markovian process $z(t)$ in the form of the infinite continued fraction (9.96) with parameters [78]

$$\begin{aligned} A_l(p) &= \widehat{L}\left(p + \alpha l\right), \\ B_l(p) &= \sigma^2 (l+1) M\left[p + \alpha l, p + \alpha(l+1)\right] M\left[p + \alpha(l+1), p + \alpha l\right]. \end{aligned} \tag{9.98}$$

Stochastic Parametric Resonance

We illustrate the above material by the example of statistical description of solution to problem (9.17), page 265 for the Gaussian Markovian process $z(t)$.

We introduce function

$$X_l(t) = \langle z_1(t) \cdots z_l(t) x(t) \rangle, \tag{9.99}$$

where $x(t)$ is the solution to problem (9.17). Multiplying then Eq. (9.17) by product $z_1(t) \cdots z_l(t)$, averaging the result over an ensemble of realizations of all processes $z_i(t)$, and using Eq. (9.90), we obtain the recurrent equality

$$\widehat{L}\left(\frac{d}{dt} + \alpha l\right) X_l(t) + \omega_0^2 \left\langle z^2 \right\rangle l X_{l-1} + \omega_0^2 (N-l) X_{l+1} = 0, \ (l = 0, \cdots, N), \tag{9.100}$$

where

$$\widehat{L}\left(\frac{d}{dt} + \alpha l\right) = \left(\frac{d}{dt} + \alpha l\right)^2 + \omega_0^2.$$

Equality (9.100) can be considered as the closed system of N equations with the initial values

$$X_0(0) = 0, \quad \left.\frac{d}{dt} X_0(t)\right|_{t=0} = y_0.$$

Performing the Laplace transform, we obtain recurrent algebraic system of equations

$$L(p + \alpha l) X_l(p) + \omega_0^2 \left\langle z^2 \right\rangle X_{l-1}(p) + \omega_0^2 (N-l) X_{l+1}(p) = F(p) \delta_{l,0}, \tag{9.101}$$

where $F(p) = y_0 + px_0$. Now, we set

$$X_l(p) = -\omega_0^2 \left\langle z^2 \right\rangle l K_l(p) X_{l-1} \tag{9.102}$$

for $l \neq 0$. Substituting Eq. (9.102) in Eq. (9.101), we obtain the finite segment of continued fraction

$$K_l(p) = \frac{1}{L(p+\alpha l) - \omega_0^4 \left\langle z^2 \right\rangle (N-l)(l+1) K_{l+1}(p)}, \tag{9.103}$$

and the solution to problem (8.161), page 244 is

$$\langle x \rangle_p = X_0(p) = F(p) K_0(p). \tag{9.104}$$

At $N = 1$, equality (9.104) grades into Eq. (9.23) for single telegrapher's process and corresponds to the two-level continued fraction (9.103).

Setting $\left\langle z^2 \right\rangle = \sigma^2/N$ and proceeding to the limit $N \to \infty$, we obtain the solution for the Gaussian Markovian process in the form of the infinite continued fraction (9.104), where

$$K_l(p) = \frac{1}{L(p+\alpha l) - \omega_0^4 \sigma^2 (l+1) K_{l+1}(p)}. \tag{9.105}$$

The second moments of the solution to problem (8.161), page 244 can be considered similarly. For example, considering potential energy $U(t) = x^2(t)$ that satisfies dynamic equation (9.26) with initial values (9.27), page 268, we obtain the mean value $\langle U(t) \rangle$ in the form of the finite segment of continued fraction (in the case of N telegrapher's processes)

$$\langle U \rangle_p = 2y_0^2 K_0(p), \quad K_l(p) = \frac{1}{A_l(p) - B_l(p) K_{l+1}(p)}, \tag{9.106}$$

where

$$A_l(p) = (p+\alpha l)\left[(p+\alpha l)^2 + 4\omega_0^2\right],$$
$$B_l(p) = 4\left\langle z^2 \right\rangle \omega_0^4 (l+1)(N-l)\left[2p + \alpha(2l+1)^2\right].$$

At $N = 1$, we obtain the solution (9.28), page 268 corresponding to single telegrapher's process. Setting $\left\langle z^2 \right\rangle = \sigma^2/N$ and proceeding to the limit $N \to \infty$, we obtain the solution for the Gaussian Markovian process in the form of the infinite continued fraction (9.106), where

$$A_l(p) = (p+\alpha l)\left[(p+\alpha l)^2 + 4\omega_0^2\right], \quad B_l(p) = 4\sigma^2\omega_0^4(l+1)\left[2p + \alpha(2l+1)^2\right].$$

9.4.4 The Square of the Gaussian Markovian Process

The finite-dimensional approximation of the Gaussian Markovian process (9.89), page 283 is practicable for describing fluctuations of dynamic systems of the form $F(z(t))$, where $z(t)$ is the Gaussian Markovian process, too.

For example, for system $F(z(t)) = z^2(t) - \langle z^2(t) \rangle$, the finite-dimensional approximation has the form (see Remark 3.1, page 139)

$$F(z(t)) = \sum_{i \neq k}^{N} z_i(t) z_k(t).$$

In this case, the mean value of the solution to system of equations (9.10), page 264 (we assume that operators $\widehat{A}(t)$ and $\widehat{B}(t)$ are matrices)

$$\frac{d}{dt} \boldsymbol{x}(t) = A\boldsymbol{x}(t) + \sum_{i \neq k}^{N} z_i(t) z_k(t) B\boldsymbol{x}(t) \tag{9.107}$$

will satisfy the closed system of $([N/2] - 1)$ vector equations in functions

$$\boldsymbol{X}_n(t) = \langle z_1(t) \cdots z_{2n}(t) \boldsymbol{x}(t) \rangle, \quad n = 1, \cdots, [N/2]; \quad \boldsymbol{X}_0(t) = \langle \boldsymbol{x}(t) \rangle .$$

Here, $[N/2]$ is the integer part of $N/2$.

It is obvious that functions $\boldsymbol{X}_n(t)$ satisfy the equations

$$\left(\frac{d}{dt} + 2\alpha n \right) \boldsymbol{X}_n(t) - A\boldsymbol{X}_n(t) = B \left\langle z_1(t) \cdots z_{2n}(t) \sum_{i \neq k}^{N} z_i(t) z_k(t) \boldsymbol{x}(t) \right\rangle . \tag{9.108}$$

The further analysis is similar to the derivation of system of equations (9.94). Divide the sum over i and k in the right-hand side of Eq. (9.108) into four regions (Fig. 9.2).

In region (1), both functions $z_i(t)$ and $z_k(t)$ will be extinguished by the corresponding functions of product

$$z_1(t) \cdots z_{2n}(t).$$

The number of such terms is $2n(2n-1)$; consequently, in region (1), the right-hand side of Eq. (9.108) assumes the form

$$2n(2n-1) \langle z^2 \rangle^2 B\boldsymbol{X}_{n-1}(t)$$

In region (2), none of functions $z_i(t)$ and $z_k(t)$ is extinguished, and we obtain that the corresponding term in the right-hand side of Eq. (9.108) has the form

$$(N-2n)(N-2n-1) B\boldsymbol{X}_{n+1}(t)$$

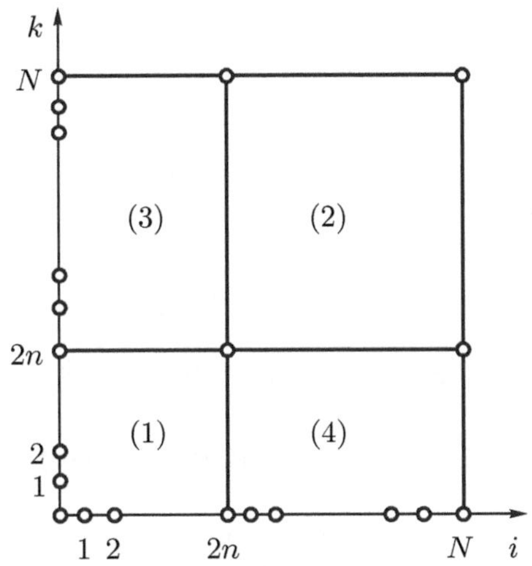

Fig. 9.2 Schematic of index division in sum (9.108)

In regions (3) and (4), only one of functions $z_i(t)$ and $z_k(t)$ is extinguished. The number of such terms is $4n(N-2n)$, so that the corresponding term in the right-hand side of Eq. (9.108) has the form

$$4n(N-2n)\left\langle z^2\right\rangle B\boldsymbol{X}_n(t)$$

As a result, Eq. (9.108) assumes the form of the closed system of recurrence equations

$$\left[\left(\frac{d}{dt}+2\alpha n\right)E-A-4n(N-2n)\left\langle z^2\right\rangle B\right]\boldsymbol{X}_n(t)$$
$$=2n(2n-1)\left\langle z^2\right\rangle^2 B\boldsymbol{X}_{n-1}(t)+(N-2n)(N-2n-1)B\boldsymbol{X}_{n+1}(t),$$

where $n=0,\,1,\cdots,\,[N/2]$. It is obvious that, for constant matrices A and B, the solution to this system again has the form of a finite segment of continued fraction. The simplest approximations with $N=2$ and $N=3$ give the closed systems of only two vector equations.

9.5 Markovian Processes with Finite-Dimensional Phase Space

All considered processes — telegrapher's and generalized telegrapher's processes and process $z(t)=z_1(t)+\cdots+z_N(t)$, where $z_i(t)$ are statistically independent telegrapher's processes, — are the special cases of the Markovian processes $z(t)$ with a finite number of states (or with the finite-dimensional phase space). We assume that possible values of process $z(t)$ are in the general case $z_1, \cdots, z_n$. As a result, all realizations of process $z(t)$ satisfy the identity

$$(z(t) - z_1)(z(t) - z_2) \cdots (z(t) - z_n) \equiv 0,$$

and, consequently,

$$z^n(t) = (z_1 + \cdots + z_n) z^{n-1}(t) + (-1)^{n-1} z_1 \cdots z_n. \tag{9.109}$$

In this case, the mean value of the solution to the system of equations

$$\frac{d}{dt}\boldsymbol{x}(t) = \widehat{A}(t)\boldsymbol{x}(t) + z(t)\widehat{B}(t)\boldsymbol{x}(t), \quad \boldsymbol{x}(0) = \boldsymbol{x}_0 \tag{9.110}$$

will again satisfy a closed system of equations. Indeed, averaging Eq. (9.110) and repeatedly using the differentiation formula (9.3), page 262 for correlators

$$\left\langle z^k(t)\boldsymbol{x}(t)\right\rangle \quad (k = 1, \cdots, n-1),$$

we reach function $\langle z^n(t)\boldsymbol{x}(t)\rangle$ at the last step. Because this function is expressed in terms of the functions of preceding steps (see Eq. (9.109)), we obtain the closed system of vector equations of the n-th order.

9.5.1 Two-State Process

Consider the process with two states z_1, z_2 and respective transition probabilities ν and μ as an example. In this case, Eq. (9.109) assumes the form

$$z^2(t) = (z_1 + z_2) z(t) - z_1 z_2. \tag{9.111}$$

Averaging Eq. (9.110), we obtain

$$\left(E\frac{d}{dt} - \widehat{A}(t)\right)\langle \boldsymbol{x}(t)\rangle = \widehat{B}(t)\langle z(t)\boldsymbol{x}(t)\rangle, \quad \langle \boldsymbol{x}(0)\rangle = \boldsymbol{x}_0. \tag{9.112}$$

According to Eq. (9.3), page 262, correlation $\langle z(t)\boldsymbol{x}(t)\rangle$ is given by the formula

$$\frac{d}{dt}\langle z(t)\boldsymbol{x}(t)\rangle = \left\langle z(t)\frac{d}{dt}\boldsymbol{x}(t)\right\rangle + \left\langle \boldsymbol{x}(t)\left[\widehat{L}^{+}(z) z(t)\right]\right\rangle,$$

where the kinetic and conjugated operators are the following matrices (see page 145)

$$\widehat{L}(z) = \begin{pmatrix} -\nu & \mu \\ \nu & -\mu \end{pmatrix}, \quad \widehat{L}^{+}(z) = \begin{pmatrix} -\nu & \nu \\ \mu & -\mu \end{pmatrix}. \tag{9.113}$$

Because the action of operator $\widehat{L}^{+}(z)$ on $z(t)$ is representable in the form

$$\widehat{L}^{+}(z)z(t) = \begin{pmatrix} -\nu & \nu \\ \mu & -\mu \end{pmatrix} \begin{pmatrix} z_1 \\ z_2 \end{pmatrix} = \begin{pmatrix} -\nu\,(z_1 - z_2) \\ \mu\,(z_1 - z_2) \end{pmatrix} = (\nu z_2 + \mu z_1) - (\nu + \mu) z(t),$$

we can rewrite the equation for correlation $\langle z(t)\boldsymbol{x}(t)\rangle$ as follows

$$\left\{E\left(\frac{d}{dt} + \mu + \nu\right) - \widehat{A}(t) - (z_1 + z_2)\widehat{B}(t)\right\}\langle z(t)\boldsymbol{x}(t)\rangle$$
$$= \left\{(\nu z_2 + \mu z_1)E - z_1 z_2 \widehat{B}(t)\right\}\langle \boldsymbol{x}(t)\rangle . \quad (9.114)$$

Equations (9.112) and (9.114) form the closed system of two vector equations.

Note that, in the special case of the scalar equation with parameters $A = 0$ and $B(t) = iv(t)$, the solution to Eq. (9.110) is

$$x(t) = \exp\left\{ i \int_0^t d\tau z(\tau) v(\tau) \right\},$$

so that the mean value of this solution coincides with the characteristic functional of random process $z(t)$

$$\varPhi[t; v(\tau)] = \langle x(t)\rangle = \left\langle \exp\left\{ i \int_0^t d\tau z(\tau) v(\tau) \right\}\right\rangle .$$

In this case, we can obtain the differential equation for functional $\varPhi[t; v(\tau)]$ by eliminating function $\langle z(t)x(t)\rangle$ from Eqs. (9.112) and (9.114):

$$\frac{d^2}{dt^2}\varPhi[t; v(\tau)] + \left[\mu + \nu - \frac{1}{v(t)}\frac{dv(t)}{dt} - iv(t)(z_1 + z_2)\right]\frac{d}{dt}\varPhi[t; v(\tau)]$$
$$- \left[iv(t)(\nu z_2 + \mu z_1) + z_1 z_2 v^2(t)\right]\varPhi[t; v(\tau)] = 0. \quad (9.115)$$

9.6 Causal Stochastic Integral Equations

9.6.1 General Remarks

Consider the simplest one-dimensional integral equation

$$S(t, t') = g(t - t')\theta(t - t') + \Lambda \int_0^t d\tau g(t - \tau) z(\tau) S(\tau, t'), \quad (9.116)$$

where $z(t)$ is the random function of time, $g(t-t')$ is the deterministic function, Λ is a constant parameter, and $\theta(t)$ is the Heaviside step function . Iterating this equation, we can see that its solution $S(t,t')$ depends on random function $z(\tau)$ only for $t' \le \tau \le t$, which means that the causality condition

$$\frac{\delta}{\delta z(\tau)} S(t,t') = 0 \quad \text{for} \quad \tau < t', \quad \tau > t$$

holds. In addition, $S(t,t') \sim \theta(t-t')$.

Average Eq. (9.116) over an ensemble of realizations of function $z(t)$. In the case of stationary process $z(t)$, function

$$\langle S(t,t')\rangle = \langle S(t-t')\rangle ,$$

and the result of averaging assumes the form

$$\langle S(t-t')\rangle = g(t-t')\theta(t-t') + \Lambda \int_0^t d\tau g(t-\tau) \int_0^\tau d\tau' Q(\tau-\tau') \langle S(\tau'-t')\rangle , \tag{9.117}$$

where $Q(t) \sim \theta(t)$ is the mass function defined by the equality

$$\Lambda \langle z(t)S(t,t')\rangle = \int_0^t d\tau Q(t-\tau) \langle S(\tau-t')\rangle .$$

Performing the Laplace transform in Eq. (9.117) with respect to $t-t'$, we obtain

$$\langle S\rangle_p = g(p) + g(p)Q(p) \langle S\rangle_p , \tag{9.118}$$

where

$$\Lambda \langle z(t)S(t,t')\rangle_p = Q(p) \langle S\rangle_p . \tag{9.119}$$

Note that if the integral equation (9.116) can be reduced to the differential equation

$$L\left(\frac{d}{dt}\right) S(t,t') = \Lambda z(t)S(t,t') + \delta(t-t'),$$

then $g(p) = L^{-1}(p)$.

According to the material of Sect. 5.3, the structure of function $Q(p)$ can be determined from the auxiliary equation

$$\widetilde{S}(t,t') = g(t-t')\theta(t-t') + \Lambda \int_0^t d\tau g(t-\tau) \left[z(\tau)+\eta(\tau)\right] \widetilde{S}(\tau,t'), \tag{9.120}$$

where $\eta(\tau)$ is arbitrary deterministic function. If we average Eq. (9.120) and denote the solution to averaged equation $G[t,t';\eta(\tau)]$, then vertex function

$\varGamma(t,t_1,t') = \varGamma(t-t_1,t_1-t')$ will be given by the equality

$$\varGamma(t,t_1,t') = -\left.\frac{\delta}{\delta\eta(t_1)} G^{-1}[t,t';\eta]\right|_{\eta=0}.$$

The variational derivative $\delta G/\delta\eta$ at $\eta(\tau)=0$ can be expressed in terms of vertex function $\varGamma(t,t_1,t')$ and average Green's function by the relationship (8.37), page 207

$$\left.\frac{\delta}{\delta\eta(t_1)} G[t,t';\eta]\right|_{\eta=0} = \int d\tau_1 \int d\tau_2 \left\langle S(t-\tau_1)\right\rangle \varGamma(\tau_1-t_1,t_1-\tau_2) \left\langle S(\tau_2-t')\right\rangle, \tag{9.121}$$

where the domain of integration is defined by the condition of positiveness of all arguments. Performing the Laplace transform in (9.121) with respect to $(t-t_1)$ and (t_1-t'), we obtain the equality

$$\frac{\delta G}{\delta\eta}(p,q) = \left\langle S\right\rangle_p \varGamma(p,q) \left\langle S\right\rangle_q \tag{9.122}$$

that makes it possible to determine the Laplace transform of the vertex function. In this case, the mass function is expressed through the characteristic functional of process $z(t)$.

Variational derivative $\delta G/\delta\eta$ in the right-hand side of Eq. (9.122) can be obtained by varying Eq. (9.120) with respect to $\eta(t_1)$ followed by setting $\eta(t)=0$ and averaging the obtained equation. If Eq. (9.120) can be averaged analytically, variational derivative $\delta G/\delta\eta$ can be obtained by varying the averaged equation with respect to $\eta(t)$.

Consider the realization of the above scheme for different processes $z(t)$.

9.6.2 Telegrapher's Random Process

Let $z(t)$ is telegrapher's process with correlation function

$$\left\langle z(t)z(t')\right\rangle = \left\langle z^2\right\rangle e^{-\alpha|t-t'|}.$$

Averaging Eq. (9.120), we obtain

$$G(t,t') = g(t-t')\theta(t-t') + \Lambda\int\limits_0^t d\tau g(t-\tau)\eta(\tau)G(\tau,t')$$
$$+\Lambda\int\limits_0^t d\tau g(t-\tau)\left\langle z(\tau)\widetilde{S}(\tau,t')\right\rangle. \tag{9.123}$$

Equation (9.123) is unclosed because it contains new unknown function $\left\langle z(\tau)\widetilde{S}(\tau,t')\right\rangle$. To obtain the equation for this function, we multiply Eq. (9.120) by $z(t)$ and average the result

$$\left\langle z(t)\widetilde{S}(t,t')\right\rangle = \Lambda\int_0^t d\tau g(t-\tau)\eta(\tau)\left\langle z(t)\widetilde{S}(\tau,t')\right\rangle$$

$$+\Lambda\int_0^t d\tau g(t-\tau)\left\langle z(t)z(\tau)\widetilde{S}(\tau,t')\right\rangle. \quad (9.124)$$

Taking into account formula (7.37), page 175

$$\langle z(t)R[t';z(\tau)]\rangle = e^{-\alpha(t-t')}\langle z(t')R[t';z(\tau)]\rangle, \quad (9.125)$$

which is valid for arbitrary functional $R[t';z(\tau)]$ for $t'\leq t$, we obtain the equation ($z^2(t)\equiv\langle z^2\rangle$)

$$\left\langle z(t)\widetilde{S}(t,t')\right\rangle = \Lambda\int_0^t d\tau g(t-\tau)\eta(\tau)e^{-\alpha(t-\tau)}\left\langle z(\tau)\widetilde{S}(\tau,t')\right\rangle$$

$$+\Lambda\left\langle z^2\right\rangle\int_0^t d\tau g(t-\tau)e^{-\alpha(t-\tau)}G(\tau,t'). \quad (9.126)$$

System of equations (9.123) and (9.126) is the closed system. Setting $\eta=0$ in this system and performing then the Laplace transform with respect to $(t-t')$, we obtain the algebraic system

$$\langle S\rangle_p = g(p)+\Lambda g(p)\langle zS\rangle_p, \quad \langle zS\rangle_p = \Lambda\left\langle z^2\right\rangle g(p)\langle S\rangle_p, \quad (9.127)$$

whose solution is as follows:

$$\langle S\rangle_p = \frac{g(p)}{1-\Lambda^2\left\langle z^2\right\rangle g(p)g(p+\alpha)}, \quad \langle zS\rangle_p = \frac{\Lambda\left\langle z^2\right\rangle g(p)g(p+\alpha)}{1-\Lambda^2\left\langle z^2\right\rangle g(p)g(p+\alpha)}. \quad (9.128)$$

According to Eq. (9.119), the mass function $Q(p)$ is

$$Q(p) = \Lambda^2\left\langle z^2\right\rangle g(p+\alpha). \quad (9.129)$$

In order to determine the vertex function , we vary Eqs. (9.123) and (9.126) with respect to $\eta(t_1)$, set $\eta(t)=0$, and perform the Laplace transform with respect to $(t-t_1)$ and (t_1-t'). As a result, we obtain the algebraic system

$$\frac{\delta G}{\delta \eta}(p,q) = \Lambda g(p)\langle S\rangle_p + \Lambda g(p)\left\langle z\frac{\delta \widetilde{S}}{\delta \eta}\right\rangle_{p,q},$$

$$\left\langle z\frac{\delta \widetilde{S}}{\delta \eta}\right\rangle_{p,q} = \Lambda g(p+\alpha)\langle zS\rangle_q + \Lambda\langle z^2\rangle g(p+\alpha)\frac{\delta G}{\delta \eta}(p,q), \tag{9.130}$$

whose solution is

$$\frac{\delta G}{\delta \eta}(p,q) = \Lambda\langle S\rangle_p\left\{1+\Lambda^2\langle z^2\rangle g(p+\alpha)g(q+\alpha)\right\}\langle S\rangle_q, \tag{9.131}$$

where we used Eq. (9.127). Comparing Eq. (9.131) with Eq. (9.122), we obtain the vertex function in the form

$$\Gamma(p,q) = \Lambda\left\{1+\Lambda^2\langle z^2\rangle g(p+\alpha)g(q+\alpha)\right\}. \tag{9.132}$$

9.6.3 Generalized Telegrapher's Random Process

Let $z(t)$ is the generalized telegrapher's process. Averaging Eq. (9.120), we obtain Eq. (9.123). Then we should derive the equation for function $\left\langle F_\lambda(t)\widetilde{S}(t,t')\right\rangle$, where

$$F_\lambda(t) = \frac{1}{1+\lambda z(t)} - C_0(\lambda), \quad C_k(\lambda) = \left\langle \frac{a^k}{1+\lambda a}\right\rangle_a \quad (\langle F_\lambda(t)\rangle_a = 0),$$

and λ is arbitrary parameter. Multiplying Eq. (9.120) by $F_\lambda(t)$ and averaging the result, we obtain

$$\left\langle F_\lambda(t)\widetilde{S}(t,t')\right\rangle = \Lambda\int\limits_0^t d\tau g(t-\tau)e^{-\alpha(t-\tau)}\eta(\tau)\left\langle F_\lambda(t)\widetilde{S}(\tau,t')\right\rangle$$

$$-\Lambda\int\limits_0^t d\tau g(t-\tau)e^{-\alpha(t-\tau)}\left\{\frac{1}{\lambda}\left\langle F_\lambda(\tau)\widetilde{S}(\tau,t')\right\rangle - C_1(\lambda)G(\tau,t') + C_0(\lambda)\left\langle z(\tau)\widetilde{S}(\tau,t')\right\rangle\right\}. \tag{9.133}$$

Deriving Eq. (9.133), we used the equality

$$\langle F_\lambda(t)R[t';z(\tau)]\rangle = e^{-\alpha(t-t')}\langle F_\lambda(t')R[t';z(\tau)]\rangle,$$

which is valid for arbitrary functional $R[t;z(\tau)]$ of random process $z(t)$ for $t' \geqslant t$, and the identity (9.128)

$$z(t)F(t) = -\frac{1}{\lambda}F(t) + C_1(t) - z(t)C_0(\lambda).$$

To determine the mass function, we set $\eta(t) = 0$ in Eqs. (9.123) and (9.133) and perform the Laplace transform. As a result, we obtain the system of equations

$$\langle S \rangle_p = g(p) + \Lambda g(p) \langle zS \rangle_p ,$$

$$\langle F_\lambda(t)S(t,t') \rangle_p \left\{ 1 + \frac{\Lambda}{\lambda} g(p+\alpha) \right\} = \Lambda g(p+\alpha) \left\{ C_1(\lambda) \langle S \rangle_p - C_0(\lambda) \langle F_\lambda(t)S(t,t') \rangle_p \right\}$$

valid for arbitrary λ. Setting

$$\lambda = \lambda_p = -\Lambda g(p+\alpha),$$

we obtain the algebraic relationship between $\langle zS \rangle_p$ and $\langle S \rangle_p$

$$\langle z(t)S(t,t') \rangle_p = \langle S \rangle_p \frac{C_1(\lambda_p)}{C_0(\lambda_p)} \tag{9.134}$$

and, consequently,

$$\langle S \rangle_p = \frac{g(p)}{1 - \Lambda g(p) \dfrac{C_1(\lambda_p)}{C_0(\lambda_p)}}. \tag{9.135}$$

Using (9.134), function $\langle F_\lambda(t)S(t,t') \rangle_p$ for arbitrary λ can be represented in the form

$$\langle F_\lambda(t)S(t,t') \rangle_p = -\frac{\lambda\lambda_p \langle S \rangle_p \left[C_1(\lambda)C_0(\lambda_p) - C_0(\lambda)C_1(\lambda_p) \right]}{(\lambda - \lambda_p)\, C_0(\lambda_p)}. \tag{9.136}$$

In this case, the mass function , as follows from Eq. (9.134), is

$$Q(p) = \Lambda \frac{C_1(\lambda_p)}{C_0(\lambda_p)}. \tag{9.137}$$

To determine the vertex function , we vary Eqs. (9.123) and (9.133) with respect to $\eta(t_1)$, set $\eta(t) = 0$, and perform the Laplace transform. As a result, we obtain the system

$$\frac{\delta G}{\delta \eta}(p,q) = \Lambda g(p) \langle S \rangle_q + \Lambda g(p) \left\langle z \frac{\delta \widetilde{S}}{\delta \eta} \right\rangle_{p,q},$$

$$\left\langle F_\lambda \frac{\delta \widetilde{S}}{\delta \eta} \right\rangle_p = \left\{ 1 + \frac{\Lambda}{\lambda} g(p+\alpha) \right\} \Lambda g(p+\alpha) F_\lambda S_q$$

$$+ \Lambda g(p+\alpha) \left\{ C_1(\lambda) \frac{\delta G}{\delta \eta}(p,q) - C_0(\lambda) \left\langle z \frac{\delta \widetilde{S}}{\delta \eta} \right\rangle_{p,q} \right\}. \tag{9.138}$$

Then, setting $\lambda = \lambda_p$ in Eq. (9.138), we obtain the algebraic system in $\frac{\delta G}{\delta\eta}(p,q)$ and $\left\langle z\frac{\delta\widetilde{S}}{\delta\eta}\right\rangle_{p,q}$, whose solution can be represented as follows

$$\frac{\delta G}{\delta\eta}(p,q) = \Lambda \langle S\rangle_p \left\{1 + \Lambda\frac{g(p+\alpha)g(q+\alpha)}{g(p+\alpha)-g(q+\alpha)}\left[\frac{C_1(\lambda_p)}{C_0(\lambda_p)} - \frac{C_1(\lambda_q)}{C_0(\lambda_q)}\right]\right\}\langle S\rangle_q, \tag{9.139}$$

where we used Eqs. (9.135), (9.136). Consequently, the vertex function is

$$\Gamma(p,q) = \Lambda\left\{1 + \Lambda\frac{g(p+\alpha)g(q+\alpha)}{g(p+\alpha)-g(q+\alpha)}\left[\frac{C_1(\lambda_p)}{C_0(\lambda_p)} - \frac{C_1(\lambda_q)}{C_0(\lambda_q)}\right]\right\}. \tag{9.140}$$

If the probability distribution of quantity a has the form

$$p(a) = \frac{1}{2}\left[\delta(a-a_0)+\delta(a+a_0)\right],$$

then $C_1(\lambda)/C_0(\lambda) = -\lambda a_0^2$, and we turn back to telegrapher's process with parameter $\langle z^2\rangle = a_0^2$.

If a is the continuous random quantity with zero-valued mean and sufficiently small variance, then

$$C_0(\lambda) \approx 1, \quad C_1(\lambda) \approx -\lambda\langle a^2\rangle,$$

and the vertex function assumes the form

$$\Gamma(p,q) = \Lambda\left\{1 + \Lambda^2\langle a^2\rangle g(p+\alpha)g(q+\alpha)\right\}. \tag{9.141}$$

However, Eq. (9.141) is valid only if obvious inequalities

$$|\lambda^2|\langle a^2\rangle \ll 1 \quad (\lambda = \lambda_p, \lambda_q) \tag{9.142}$$

are satisfied.

9.6.4 Gaussian Markovian Process

Let $z(t)$ is the Gaussian Markovian process with correlation function

$$\langle z(t)z(t)\rangle = \langle z^2\rangle e^{-\alpha|t-t'|}.$$

This process can be obtained from the process with a finite number of states (see Remark 3.1, page 139)

$$\xi_N = z_1(t) + \cdots + z_N(t),$$

where $z_i(t)$ are the statistically independent telegrapher's processes with $\langle z^2 \rangle = \sigma^2/N$, using limit process $N \to \infty$.

So, we consider Eq. (9.120) with $z(t) = \xi_N(t)$ and introduce functions

$$G_l(t,t') = \left\langle z_1(t) \cdots z_l(t) \widetilde{S}(t,t') \right\rangle \quad (G_0(t,t') = G(t,t')) \,. \tag{9.143}$$

Multiplying Eq. (9.120) by product $z_1(t) \cdots z_N(t)$, averaging the result, and using
Eq. (9.125), we can obtain the recurrence equation for $l = 0, 1, \cdots, N$

$$G_l(t,t') = g(t-t')\theta(t-t')\delta_{l,0} + \Lambda \int\limits_0^t d\tau g(t-\tau)\eta(\tau) e^{-\alpha l(t-\tau)} G_l(\tau,t')$$
$$+ \Lambda \int\limits_0^t d\tau g(t-\tau) e^{-\alpha l(t-\tau)} \left\{ l \left\langle z^2 \right\rangle G_{l-1}(\tau,t') + (N-l) G_{l+1}(\tau,t') \right\} . \tag{9.144}$$

Setting $\eta(t) = 0$ and performing the Laplace transform with respect to $(t-t')$, we obtain the algebraic recurrence equation

$$G_l(p) = g_0(p)\delta_{l,0} + \Lambda g_l(p) \left\{ l \left\langle z^2 \right\rangle G_{l-1}(p) + (N-l) G_{l+1}(p) \right\} , \tag{9.145}$$

where $g_l(p) = g(l + \alpha l)$.

The solution to Eq. (9.145) has the form of a finite segment of the continued fraction

$$G_l(p) = \Lambda g_l(p) l \left\langle z^2 \right\rangle K_l(p) G_{l-1}(p), \quad l = 1, \cdots, N, \tag{9.146}$$

where

$$K_l(p) = \frac{1}{1 - \gamma_l(p) K_{l+1}(p)}, \quad \gamma_l(p) = \Lambda^2 \left\langle z^2 \right\rangle (l+1)(N-l) g_l(p) g_{l+1}(p). \tag{9.147}$$

Consequently,

$$G_l(p) = \Lambda^l \left\langle z^2 \right\rangle^l l! \left\{ g_l(p) K_l(p) \right\}! g_0(p) K_0(p), \tag{9.148}$$

where $f_l!$ stands for the product $f_1 \cdots f_l$. Taking into account the fact that

$$\left\langle \xi_N(t) S(t,t') \right\rangle_p = N G_1(p),$$

we obtain the expression for the mass function :

$$Q_N(p) = \Lambda^2 N \left\langle z^2 \right\rangle g_1(p) K_1(p). \tag{9.149}$$

Setting now $\left\langle z^2 \right\rangle = \sigma^2/N$ and proceeding to the limit $N \to \infty$, we obtain the mass function for the Gaussian Markovian process:

$$Q(p) = \Lambda^2 \sigma^2 g_1(p) K_1(p), \tag{9.150}$$

where $K_1(p)$ is the infinite continued fraction (9.147) with parameter

$$\widetilde{\gamma}_l(p) = \Lambda^2 \sigma^2 (l+1) g_l(p) g_{l+1}(p). \tag{9.151}$$

Calculating the vertex function in the cases of telegrapher's and generalized telegrapher's processes, we straightforwardly followed the procedure valid for arbitrary integral equations. The goal of that consideration was to illustrate the general procedure. In the case of Eq. (9.116), we can immediately obtain the expression for the vertex function if only the solution to the Dyson equation is known. Indeed, according to Eqs. (9.121) and (8.30), page 205, we have the following relationship for Eq. (9.116)

$$\Lambda \left\langle S(t,t_0) S(t_0,t') \right\rangle = \int d\tau_1 \int d\tau_2 \left\langle S(t-\tau_1) \right\rangle \Gamma(\tau_1 - t_1, t_1 - \tau_2) \left\langle S(\tau_2 - t') \right\rangle. \tag{9.152}$$

Let now random process $z(t)$ is a function of process $\xi_N(t)$. Then, we can split the correlator in the left-hand side of Eq. (9.152) using formula (7.32), which assumes in this case the form

$$\left\langle S(t,t_0) S(t_0,t') \right\rangle = \sum_{k=0}^{N} C_N^k \frac{1}{\left\langle z^2 \right\rangle^k} \left\langle z_1(t_0) \cdots z_k(t_0) S(t,t_0) \right\rangle \left\langle z_1(t_0) \cdots z_k(t_0) S(t_0,t') \right\rangle. \tag{9.153}$$

Performing the Laplace transform with respect to $(t - t_0)$ and $(t_0 - t')$, we obtain the equality

$$\left\langle SS \right\rangle_{p,q} = \sum_{k=0}^{N} C_N^k \frac{1}{\left\langle z^2 \right\rangle^k} G_k(p) G_k(q), \tag{9.154}$$

where function $G_k(p)$ is given by Eq. (9.143). Consequently, we obtain the following expression for the vertex function $\Gamma(p,q)$

$$\Gamma(p,q) = \frac{1}{\Lambda} \sum_{k=0}^{N} C_N^k \frac{1}{\left\langle z^2 \right\rangle^k} \frac{G_k(p) G_k(q)}{G_0(p) G_0(q)}. \tag{9.155}$$

For $z(t) = \xi_N(t)$, functions $G_k(p)$ are given by Eq. (9.148), and we obtain

$$\Gamma_N(p,q) = \Lambda \left[1 + \sum_{k=0}^{N} \Lambda^{2k} \left\langle z^2 \right\rangle^k \frac{N!k!}{(N-k)!} \left\{ g_k(p) g_k(q) K_k(p) K_k(q) \right\}! \right]. \tag{9.156}$$

At $N = 1$, we turn back to the case of single telegrapher's process, and Eq. (9.156) grades into (9.132).

Setting $\langle z^2 \rangle = \sigma^2/N$ and proceeding to the limit $N \to \infty$, we obtain the vertex function for the Gaussian Markovian process in the form of the infinite series

$$\Gamma(p,q) = \Lambda \left[1 + \sum_{k=0}^{\infty} k! \Lambda^{2k} \sigma^{2k} \left\{ g_k(p) g_k(q) K_k(p) K_k(q) \right\}! \right], \tag{9.157}$$

whose terms include infinite continued fractions (9.147) with parameter (9.151). Two first terms of series (9.157) are as follows

$$\Gamma(p,q) = \Lambda \left[1 + \Lambda^2 \sigma^2 g_1(p) g_1(q) K_1(p) K_1(q) + \cdots \right].$$

Now, we dwell on approximations commonly used in analyzing stochastic integral equations.

First of all, we consider the Gaussian Markovian process. In this case, the mass function is related to the vertex function by the formula

$$Q(t-t') = \Lambda \sigma^2 \int \int d\tau_1 d\tau_2 e^{-\alpha(t-\tau)} \langle S(t-\tau_1) \rangle \Gamma(\tau_1 - \tau, \tau - t'), \tag{9.158}$$

where the domain of integration is defined by the condition of positiveness of all arguments. Performing the Laplace transform in Eq. (9.158) with respect to $(t-t')$, we obtain the equality

$$Q(p) = \Lambda \sigma^2 \langle S \rangle_{p+\alpha} \Gamma(p+\alpha, p). \tag{9.159}$$

The Kraichnan approximation corresponds to the replacement of vertex function $\Gamma(p+\alpha,p)$ in Eq. (9.158) by Λ, and the Bourret approximation assumes additionally substitution of $\langle S \rangle_{p+\alpha}$ with $g_1(p)$.

The solution to the Dyson equation (9.118) depends primarily on the poles and other significant singularities of function $g_1(p)$. Denote p_0 the singular point of this function. Then, if the condition

$$\Lambda^2 \sigma^2 |g_1(p_0)|^2 |K_1(p_0)|^2 \ll 1 \tag{9.160}$$

holds, we can neglect all terms of series (9.157) excluding the first one. Functions $K_1(p_0)$ themselves depend on parameter $\beta^2 = \Lambda^2 |g_1(p_0)|^2$, and $|K_1(p_0)| \sim 1$ for $\beta^2 \ll 1$.

Thus, we can replace vertex function $\Gamma(p,q)$ by Λ under the condition that

$$\beta^2 = \Lambda^2 |g_1(p_0)|^2 \ll 1. \tag{9.161}$$

Earlier, we showed that function $\langle S \rangle_{p_0+\alpha}$ also has a small parameter. In the first approximation with respect to this small parameter, the mass function is

$$Q(p) = \Lambda^2 g_1(p), \tag{9.162}$$

which corresponds to the Bourret approximation. Thus, the Kraichnan approximation fails in the context of this problem, whereas the Bourret approximation represents the first term of the asymptotic expansion of the solution in the above small parameter.

Note that the mass function in the Bourret approximation (9.162) coincides with the mass function for telegrapher's process (9.129).

The limit process $\alpha \to \infty$ in the solutions obtained for all above processes results in the Gaussian delta-correlated process with correlation function

$$\langle z(t)z(t')\rangle = 2\sigma^2\tau_0\delta(t-t'), \quad \tau_0 = 1/\alpha.$$

It is clear that this solution can be obtained immediately from Eq. (9.116).

Part IV
Asymptotic and Approximate Methods for Analyzing Stochastic Equations

Chapter 10
Gaussian Random Field Delta-Correlated in Time (Ordinary Differential Equations)

In the foregoing chapters, we considered in detail the general methods of analyzing stochastic equations. Here, we give an alternative and more detailed consideration of the approximation of the Gaussian random delta-correlated (in time) field in the context of stochastic equations and discuss the physical meaning of this widely used approximation.

10.1 The Fokker–Planck Equation

Let vector function $\boldsymbol{x}(t) = \{x_1(t), x_2(t), \cdots, x_n(t)\}$ satisfies the dynamic equation

$$\frac{d}{dt}\boldsymbol{x}(t) = \boldsymbol{v}(\boldsymbol{x}, t) + \boldsymbol{f}(\boldsymbol{x}, t), \quad \boldsymbol{x}(t_0) = \boldsymbol{x}_0, \tag{10.1}$$

where $v_i(\boldsymbol{x}, t)$ $(i = 1, \cdots, n)$ are the deterministic functions and $f_i(\boldsymbol{x}, t)$ are the random functions of $(n+1)$ variable that have the following properties:

(a) $f_i(\boldsymbol{x}, t)$ is the Gaussian random field in the $(n+1)$ -dimensional space $(\boldsymbol{x}, t)$;
(b) $\langle f_i(\boldsymbol{x}, t)\rangle = 0$.

For definiteness, we assume that t is the temporal variable and $\boldsymbol{x}$ is the spatial variable.

Statistical characteristics of field $f_i(\boldsymbol{x}, t)$ are completely described by correlation tensor

$$B_{ij}(\boldsymbol{x}, t; \boldsymbol{x}', t') = \langle f_i(\boldsymbol{x}, t) f_j(\boldsymbol{x}', t')\rangle .$$

Because Eq. (10.1) is the first-order equation with the initial value, its solution satisfies the dynamic causality condition

$$\frac{\delta}{\delta f_j(\boldsymbol{x}', t')} x_i(t) = 0 \quad \text{for} \quad t' < t_0 \quad \text{and} \quad t' > t, \tag{10.2}$$

V.I. Klyatskin, *Stochastic Equations: Theory and Applications in Acoustics, Hydrodyn., Magnetohydrodyn., and Radiophys., Vol. 1*, Understanding Complex Systems,
DOI: 10.1007/978-3-319-07587-7_10,

which means that solution $\boldsymbol{x}(t)$ depends only on values of functions $f_j(\boldsymbol{x}, t')$ for times t' preceding time t, i.e., $t_0 \leq t' \leq t$. In addition, we have the following equality for the variational derivative

$$\frac{\delta}{\delta f_j(\boldsymbol{x}', t-0)} x_i(t) = \delta_{ij} \delta(\boldsymbol{x}(t) - \boldsymbol{x}'). \tag{10.3}$$

Nevertheless, the statistical relationship between $\boldsymbol{x}(t)$ and function values $f_j(\boldsymbol{x}, t'')$ for consequent times $t'' > t$ can exist, because such function values $f_j(\boldsymbol{x}, t'')$ are correlated with values $f_j(\boldsymbol{x}, t')$ for $t' \leq t$. It is obvious that the correlation between function $\boldsymbol{x}(t)$ and consequent values $f_j(\boldsymbol{x}, t'')$ is appreciable only for $t'' - t \leq \tau_0$, where τ_0 is the correlation radius of field $\boldsymbol{f}(\boldsymbol{x}, t)$ with respect to variable t.

For many actual physical processes, characteristic temporal scale T of function $\boldsymbol{x}(t)$ significantly exceeds correlation radius τ_0 ($T \gg \tau_0$); in this case, the problem has the small parameter τ_0/T that can be used to construct an approximate solution.

In the first approximation with respect to this small parameter, one can consider the asymptotic solution for $\tau_0 \to 0$. In this case values of function $\boldsymbol{x}(t')$ for $t' < t$ will be independent of values $\boldsymbol{f}(\boldsymbol{x}, t'')$ for $t'' > t$ not only functionally, but also statistically. This approximation is equivalent to the replacement of correlation tensor B_{ij} with the effective tensor

$$B_{ij}^{\text{eff}}(\boldsymbol{x}, t; \boldsymbol{x}', t') = 2\delta(t - t') F_{ij}(\boldsymbol{x}, \boldsymbol{x}'; t). \tag{10.4}$$

Here, quantity $F_{ij}(\boldsymbol{x}, \boldsymbol{x}'; t)$ is determined from the condition that integrals of $B_{ij}(\boldsymbol{x}, t; \boldsymbol{x}', t')$ and $B_{ij}^{\text{eff}}(\boldsymbol{x}, t; \boldsymbol{x}', t')$ over t' coincide

$$F_{ij}(\boldsymbol{x}, \boldsymbol{x}'; t) = \frac{1}{2} \int\limits_{-\infty}^{\infty} dt' B_{ij}(\boldsymbol{x}, t; \boldsymbol{x}', t'),$$

which just corresponds to the passage to the Gaussian random field delta-correlated in time t.

Introduce the indicator function

$$\varphi(\boldsymbol{x}, t) = \delta(\boldsymbol{x}(t) - \boldsymbol{x}), \tag{10.5}$$

where $\boldsymbol{x}(t)$ is the solution to Eq. (10.1), which satisfies the Liouville equation

$$\left(\frac{\partial}{\partial t} + \frac{\partial}{\partial \boldsymbol{x}} \boldsymbol{v}(\boldsymbol{x}, t) \right) \varphi(\boldsymbol{x}, t) = -\frac{\partial}{\partial \boldsymbol{x}} \boldsymbol{f}(\boldsymbol{x}, t) \varphi(\boldsymbol{x}, t) \tag{10.6}$$

and the equality

$$\frac{\delta}{\delta f_j(\boldsymbol{x}', t-0)} \varphi(\boldsymbol{x}, t) = -\frac{\partial}{\partial x_j} \left\{ \delta(\boldsymbol{x} - \boldsymbol{x}') \varphi(\boldsymbol{x}, t) \right\}. \tag{10.7}$$

The equation for the probability density of the solution to Eq. (10.1)

$$P(\boldsymbol{x},t) = \langle \varphi(\boldsymbol{x},t) \rangle = \langle \delta(\boldsymbol{x}(t) - \boldsymbol{x}) \rangle$$

can be obtained by averaging Eq. (10.6) over an ensemble of realizations of field $\boldsymbol{f}(\boldsymbol{x},t)$,

$$\left(\frac{\partial}{\partial t} + \frac{\partial}{\partial \boldsymbol{x}} \boldsymbol{v}(\boldsymbol{x},t) \right) P(\boldsymbol{x},t) = -\frac{\partial}{\partial \boldsymbol{x}} \langle \boldsymbol{f}(\boldsymbol{x},t) \varphi(\boldsymbol{x},t) \rangle . \tag{10.8}$$

We rewrite Eq. (10.8) in the form

$$\left(\frac{\partial}{\partial t} + \frac{\partial}{\partial \boldsymbol{x}} \boldsymbol{v}(\boldsymbol{x},t) \right) P(\boldsymbol{x},t) = -\frac{\partial}{\partial x_i} \int d\boldsymbol{x}' \int\limits_{t_0}^{t} dt' B_{ij}(\boldsymbol{x},t;\boldsymbol{x}',t') \left\langle \frac{\delta \varphi(\boldsymbol{x},t)}{\delta f_j(\boldsymbol{x}',t')} \right\rangle , \tag{10.9}$$

where we used the Furutsu–Novikov formula (7.12), page 168

$$\langle f_k(\boldsymbol{x},t) R[t;\boldsymbol{f}(\boldsymbol{y},\tau)] \rangle = \int d\boldsymbol{x}' \int dt' B_{kl}(\boldsymbol{x},t;\boldsymbol{x}',t') \left\langle \frac{\delta R[t;\boldsymbol{f}(\boldsymbol{y},\tau)]}{\delta f_l(\boldsymbol{x}',t')} \right\rangle \tag{10.10}$$

for the correlator of the Gaussian random field $\boldsymbol{f}(\boldsymbol{x},t)$ with arbitrary functional $R[t;\boldsymbol{f}(\boldsymbol{y},\tau)]$ of it and the dynamic causality condition (10.2).

Equation (10.9) shows that the one-time probability density of solution $\boldsymbol{x}(t)$ at instant t is governed by functional dependence of solution $\boldsymbol{x}(t)$ on field $\boldsymbol{f}(\boldsymbol{x}',t)$ for all times in the interval (t_0,t).

In the general case, there is no closed equation for the probability density $P(\boldsymbol{x},t)$. However, if we use approximation (10.4) for the correlation function of field $\boldsymbol{f}(\boldsymbol{x},t)$, there appear terms related to variational derivatives $\delta\varphi[\boldsymbol{x},t;\boldsymbol{f}(\boldsymbol{y},\tau)]/\delta f_j(\boldsymbol{x}',t')$ at coincident temporal arguments $t' = t - 0$,

$$\left(\frac{\partial}{\partial t} + \frac{\partial}{\partial \boldsymbol{x}} \boldsymbol{v}(\boldsymbol{x},t) \right) P(\boldsymbol{x},t) = -\frac{\partial}{\partial x_i} \int d\boldsymbol{x}' F_{ij}(\boldsymbol{x},\boldsymbol{x}';t) \left\langle \frac{\delta \varphi(\boldsymbol{x},t)}{\delta f_j(\boldsymbol{x}',t-0)} \right\rangle .$$

According to Eq. (10.7), these variational derivatives can be expressed immediately in terms of quantity $\varphi[\boldsymbol{x},t;\boldsymbol{f}(\boldsymbol{y},\tau)]$. Thus, we obtain the closed Fokker-Planck equation

$$\left(\frac{\partial}{\partial t} + \frac{\partial}{\partial x_k} \left[v_k(\boldsymbol{x},t) + A_k(\boldsymbol{x},t) \right] \right) P(\boldsymbol{x},t) = \frac{\partial^2}{\partial x_k \partial x_l} \left[F_{kl}(\boldsymbol{x},\boldsymbol{x};t) P(\boldsymbol{x},t) \right] , \tag{10.11}$$

where

$$A_k(\boldsymbol{x},t) = \left. \frac{\partial}{\partial x'_l} F_{kl}(\boldsymbol{x},\boldsymbol{x}';t) \right|_{\boldsymbol{x}'=\boldsymbol{x}} .$$

Equation (10.11) should be solved with the initial condition $P(\boldsymbol{x}, t_0) = \delta(\boldsymbol{x} - \boldsymbol{x}_0)$, or with a more general initial condition $P(\boldsymbol{x}, t_0) = W_0(\boldsymbol{x})$ if the initial conditions are also random, but statistically independent of field $\boldsymbol{f}(\boldsymbol{x}, t)$.

The Fokker-Planck equation (10.11) is a partial differential equation and its further analysis essentially depends on boundary conditions with respect to $\boldsymbol{x}$ whose form can vary depending on the problem at hand.

Consider the quantities appeared in Eq. (10.11). In this equation, the terms containing $A_k(\boldsymbol{x}, t)$ and $F_{kl}(\boldsymbol{x}, \boldsymbol{x}'; t)$ are stipulated by fluctuations of field $\boldsymbol{f}(\boldsymbol{x}, t)$. If field $\boldsymbol{f}(\boldsymbol{x}, t)$ is stationary in time, quantities $A_k(\boldsymbol{x})$ and $F_{kl}(\boldsymbol{x}, \boldsymbol{x}')$ are independent of time. If field $\boldsymbol{f}(\boldsymbol{x}, t)$ is additionally homogeneous and isotropic in all spatial coordinates, then $F_{kl}(\boldsymbol{x}, \boldsymbol{x}, t) = \text{const}$, which corresponds to the constant tensor of diffusion coefficients, and $A_k(\boldsymbol{x}, t) = 0$ (note however that quantities $F_{kl}(\boldsymbol{x}, \boldsymbol{x}'; t)$ and $A_k(\boldsymbol{x}, t)$ can depend on $\boldsymbol{x}$ because of the use of a curvilinear coordinate system).

10.2 Transitional Probability Distributions

Turn back to dynamic system (10.1) and consider the m-time probability density

$$P_m(\boldsymbol{x}_1, t_1; \cdots; \boldsymbol{x}_m, t_m) = \langle \delta(\boldsymbol{x}(t_1) - \boldsymbol{x}_1) \cdots \delta(\boldsymbol{x}(t_m) - \boldsymbol{x}_m) \rangle \tag{10.12}$$

for m different instants $t_1 < t_2 < \cdots < t_m$. Differentiating Eq. (10.12) with respect to time t_m and using then dynamic equation (10.1), dynamic causality condition (10.2), definition of function $F_{kl}(\boldsymbol{x}, \boldsymbol{x}'; t)$, and the Furutsu–Novikov formula (10.10), one can obtain the equation similar to the Fokker–Planck equation (10.11),

$$\frac{\partial}{\partial t_m} P_m(\boldsymbol{x}_1, t_1; \cdots; \boldsymbol{x}_m, t_m) + \sum_{k=1}^{n} \frac{\partial}{\partial x_{mk}} \left[v_k(\boldsymbol{x}_m, t_m) + A_k(\boldsymbol{x}_m, t_m) \right] P_m(\boldsymbol{x}_1, t_1; \cdots; \boldsymbol{x}_m, t_m)$$

$$= \sum_{k=1}^{n} \sum_{l=1}^{n} \frac{\partial^2}{\partial x_{mk} \partial x_{ml}} \left[F_{kl}(\boldsymbol{x}_m, \boldsymbol{x}_m; t_m) P_m(\boldsymbol{x}_1, t_1; \cdots; \boldsymbol{x}_m, t_m)) \right]. \tag{10.13}$$

No summation over index m is performed here. The initial value to Eq. (10.13) can be determined from Eq. (10.12). Setting $t_m = t_{m-1}$ in (10.12), we obtain

$$P_m(\boldsymbol{x}_1, t_1; \cdots; \boldsymbol{x}_m, t_{m-1}) = \delta(\boldsymbol{x}_m - \boldsymbol{x}_{m-1}) P_{m-1}(\boldsymbol{x}_1, t_1; \cdots; \boldsymbol{x}_{m-1}, t_{m-1}). \tag{10.14}$$

Equation (10.13) assumes the solution in the form

$$P_m(\boldsymbol{x}_1, t_1; \cdots; \boldsymbol{x}_m, t_m) = p(\boldsymbol{x}_m, t_m | \boldsymbol{x}_{m-1}, t_{m-1}) P_{m-1}(\boldsymbol{x}_1, t_1; \cdots; \boldsymbol{x}_{m-1}, t_{m-1}). \tag{10.15}$$

Because all differential operations in Eq. (10.13) concern only t_m and $\boldsymbol{x}_m$, we can find the equation for the *transitional probability density* by substituting Eq. (10.15) in Eqs. (10.13) and (10.14):

$$\left(\frac{\partial}{\partial t}+\frac{\partial}{\partial x_k}\left[v_k(\boldsymbol{x},t)+A_k(\boldsymbol{x},t)\right]\right)p(\boldsymbol{x},t|\boldsymbol{x}_0,t_0)=\frac{\partial^2}{\partial x_k\partial x_l}\left[F_{kl}(\boldsymbol{x},\boldsymbol{x};t)p(\boldsymbol{x},t|\boldsymbol{x}_0,t_0)\right] \tag{10.16}$$

with the initial condition $p(\boldsymbol{x},t|\boldsymbol{x}_0,t_0)|_{t\to t_0}=\delta(\boldsymbol{x}-\boldsymbol{x}_0)$, where

$$p(\boldsymbol{x},t|\boldsymbol{x}_0,t_0)=\left\langle\delta(\boldsymbol{x}(t)-\boldsymbol{x})|\boldsymbol{x}(t_0)=\boldsymbol{x}_0\right\rangle.$$

In Eq. (10.16) we denoted variables $\boldsymbol{x}_m$ and t_m as $\boldsymbol{x}$ and t, and variables $\boldsymbol{x}_{m-1}$ and t_{m-1} as $\boldsymbol{x}_0$ and t_0.

Using formula (10.15) $(m-1)$ times, we obtain the relationship

$$P_m(\boldsymbol{x}_1,t_1;\cdots;\boldsymbol{x}_m,t_m)=p(\boldsymbol{x}_m,t_m|\boldsymbol{x}_{m-1},t_{m-1})\cdots p(\boldsymbol{x}_2,t_2|\boldsymbol{x}_1,t_1)P(\boldsymbol{x}_1,t_1), \tag{10.17}$$

where $P(\boldsymbol{x}_1,t_1)$ is the one-time probability density governed by Eq. (10.11). Equality (10.17) expresses the multi-time probability density as the product of transitional probability densities, which means that random process $\boldsymbol{x}(t)$ is the Markovian process.

Equation (10.11) is usually called the *forward Fokker–Planck equation* . The *backward Fokker–Planck equation* (it describes the transitional probability density as a function of the initial parameters t_0 and $\boldsymbol{x}_0$) can also be easily derived.

Indeed, we obtained the backward Liouville equation (3.9), page 97 for indicator function

$$\left(\frac{\partial}{\partial t_0}+\boldsymbol{v}(\boldsymbol{x}_0,t_0)\frac{\partial}{\partial\boldsymbol{x}_0}\right)\varphi(\boldsymbol{x},t|\boldsymbol{x}_0,t_0)=-\boldsymbol{f}(\boldsymbol{x}_0,t_0)\frac{\partial}{\partial\boldsymbol{x}_0}\varphi(\boldsymbol{x},t|\boldsymbol{x}_0,t_0), \tag{10.18}$$

with the initial value $\varphi(\boldsymbol{x},t|\boldsymbol{x}_0,t)=\delta(\boldsymbol{x}-\boldsymbol{x}_0)$.

This equation describes the dynamic system evolution in terms of initial parameters t_0 and $\boldsymbol{x}_0$. From Eq. (10.18) follows the equality similar to Eq. (10.7),

$$\frac{\delta}{\delta f_j(\boldsymbol{x}',t_0+0)}\varphi(\boldsymbol{x},t|\boldsymbol{x}_0,t_0)=\delta(\boldsymbol{x}-\boldsymbol{x}')\frac{\partial}{\partial x_{0j}}\varphi(\boldsymbol{x},t|\boldsymbol{x}_0,t_0). \tag{10.19}$$

Averaging now the backward Liouville equation (10.18) over an ensemble of realizations of random field $\boldsymbol{f}(\boldsymbol{x},t)$ with the effective correlation tensor (10.4), using the Furutsu–Novikov formula (10.10), and relationship (10.19) for the variational derivative, we obtain the backward Fokker–Planck equation (see also [21])

$$\left(\frac{\partial}{\partial t_0} + [v_k(\boldsymbol{x}_0, t_0) + A_k(\boldsymbol{x}_0, t_0)] \frac{\partial}{\partial x_{0k}}\right) p(\boldsymbol{x}, t|\boldsymbol{x}_0, t_0)$$

$$= -F_{kl}(\boldsymbol{x}_0, \boldsymbol{x}_0; t_0) \frac{\partial^2}{\partial x_{0k} \partial x_{0l}} p(\boldsymbol{x}, t|\boldsymbol{x}_0, t_0), \quad (10.20)$$

$$p(\boldsymbol{x}, t|\boldsymbol{x}_0, t) = \delta(\boldsymbol{x} - \boldsymbol{x}_0).$$

The forward and backward Fokker–Planck equations are equivalent. The forward equation is more convenient for analyzing the temporal behavior of statistical characteristics of the solution to Eq. (10.1). The backward equation appears to be more convenient for studying statistical characteristics related to initial values, such as the time during which process $\boldsymbol{x}(t)$ resides in certain spatial region and the time at which the process arrives at region's boundary. In this case the probability of the fact that random process $\boldsymbol{x}(t)$ rests in spatial region V is given by the integral

$$G(t; \boldsymbol{x}_0, t_0) = \int\limits_V d\boldsymbol{x} p(\boldsymbol{x}, t|\boldsymbol{x}_0, t_0),$$

which, according to Eq. (10.20), satisfies the closed equation

$$\left(\frac{\partial}{\partial t_0} + [v_k(\boldsymbol{x}_0, t_0) + A_k(\boldsymbol{x}_0, t_0)] \frac{\partial}{\partial x_{0k}}\right) G(t; \boldsymbol{x}_0, t_0)$$

$$= -F_{kl}(\boldsymbol{x}_0, \boldsymbol{x}_0; t_0) \frac{\partial^2}{\partial x_{0k} \partial x_{0l}} G(t; \boldsymbol{x}_0, t_0), \quad (10.21)$$

$$G(t; \boldsymbol{x}_0, t_0) = \begin{cases} 1 \ (\boldsymbol{x}_0 \in V), \\ 0 \ (\boldsymbol{x}_0 \notin V). \end{cases}$$

For Eq. (10.21) we must formulate additional boundary conditions, which depend on characteristics of both region V and its boundaries.

10.3 Applicability Range of the Fokker–Planck Equation

To estimate the applicability range of the Fokker–Planck equation, we must include into consideration the finite-valued correlation radius τ_0 of field $\boldsymbol{f}(\boldsymbol{x}, t)$ with respect to time. In this case, the equation for the probability density (10.11) is replaced with the equation

$$\widehat{E} P(\boldsymbol{x}, t) = -\frac{\partial}{\partial x_k} S'(\boldsymbol{x}, t),$$

where $\widehat{E}$ — is the operator appeared in the left-hand side of Eq. (10.11) in which quantity $F_{kl}(\boldsymbol{x}, \boldsymbol{x}', t)$ is replaced with

$$\widetilde{F}_{kl}(\boldsymbol{x}, \boldsymbol{x}', t) = \int_0^t dt' B_{kl}(\boldsymbol{x}, \boldsymbol{x}', t)$$

and term $S'(\boldsymbol{x}, t)$ includes corrections to the factor of the probability flux density because of finiteness of τ_0. For $\tau_0 \to 0$, we turn back to Eq. (10.11). Thus, smallness of parameter τ_0/T is the necessary, but generally not sufficient condition in order that one could describe statistical characteristics of the solution to Eq. (10.1) using the approximation of the delta-correlated random field of which a consequence is the Fokker–Planck equation. Every particular problem requires more detailed investigations. Below, we give a more physical method called the *diffusion approximation* . This method also leads to the Markovian property of the solution to Eq. (10.1); however, it considers to some extent the finite value of the temporal correlation radius.

Here, we emphasize that the approximation of the delta-correlated random field does not reduce to the formal replacement of random field $\boldsymbol{f}(\boldsymbol{x}, t)$ in Eq. (10.1) with the random field with correlation function (10.4). This approximation corresponds to the construction of an asymptotic expansion for temporal correlation radius τ_0 of filed $\boldsymbol{f}(\boldsymbol{x}, t)$ approaching to zero. It is in such limit process that exact average quantities like $\langle \boldsymbol{f}(\boldsymbol{x}, t) R[t; \boldsymbol{f}(\boldsymbol{x}', \tau)] \rangle$ grade into the expressions obtained by the formal replacement of the correlation tensor of field $\boldsymbol{f}(\boldsymbol{x}, t)$ with the effective tensor (10.4).

10.3.1 Langevin Equation

We illustrate the above material by the example of the *Langevin equation* that allows an exhaustive statistical analysis [51]. This equation has the form

$$\frac{d}{dt} x(t) = -\lambda x(t) + f(t), \quad x(t_0) = 0 \tag{10.22}$$

and assumes that the sufficiently fine smooth function $f(t)$ is the stationary Gaussian process with zero-valued average and correlation function

$$\langle f(t) f(t') \rangle = B_f(t - t').$$

For any particular realization of random force $f(t)$, the solution to Eq. (10.22) has the form

$$x(t) = \int_{t_0}^t d\tau f(\tau) e^{-\lambda(t-\tau)}.$$

Consequently, this solution $x(t)$ is also the Gaussian process with the parameters

$$\langle x(t)\rangle = 0, \quad \langle x(t)x(t')\rangle = \int_{t_0}^{t} d\tau_1 \int_{t_0}^{t'} d\tau_2 B_f(\tau_1 - \tau_2) e^{-\lambda(t+t'-\tau_1-\tau_2)}.$$

In addition, we have, for example,

$$\langle f(t)x(t)\rangle = \int_{0}^{t-t_0} d\tau B_f(\tau) e^{-\lambda\tau}.$$

Note that the one-time probability density $P(x,t) = \langle\delta(x(t)-x)\rangle$ for Eq. (10.22) satisfies the equation

$$\left(\frac{\partial}{\partial t} - \lambda\frac{\partial}{\partial x}x\right) P(x,t) = \int_{0}^{t-t_0} d\tau B_f(\tau) e^{-\lambda\tau} \frac{\partial^2}{\partial x^2} P(x,t), \quad P(x,t_0) = \delta(x),$$

which rigorously follows from Eq. (8.93), page 222. As a consequence, we obtain

$$\frac{d}{dt}\left\langle x^2(t)\right\rangle = -2\lambda\left\langle x^2(t)\right\rangle + 2\int_{0}^{t-t_0} d\tau B_f(\tau) e^{-\lambda\tau}.$$

For $t_0 \to -\infty$, process $x(t)$ grades into the stationary Gaussian process with the following one-time statistical parameters ($\langle x(t)\rangle = 0$)

$$\sigma_x^2 = \left\langle x^2(t)\right\rangle = \frac{1}{\lambda}\int_{0}^{\infty} d\tau B_f(\tau) e^{-\lambda\tau}, \quad \langle f(t)x(t)\rangle = \int_{0}^{\infty} d\tau B_f(\tau) e^{-\lambda\tau}.$$

In particular, for exponential correlation function $B_f(t)$,

$$B_f(t) = \sigma_f^2 e^{-|\tau|/\tau_0},$$

we obtain the expressions

$$\langle x(t)\rangle = 0, \quad \left\langle x^2(t)\right\rangle = \frac{\sigma_f^2\tau_0}{\lambda(1+\lambda\tau_0)}, \quad \langle f(t)x(t)\rangle = \frac{\sigma_f^2\tau_0}{1+\lambda\tau_0}, \tag{10.23}$$

which grade into the asymptotic expressions

$$\left\langle x^2(t)\right\rangle = \frac{\sigma_f^2\tau_0}{\lambda}, \quad \langle f(t)x(t)\rangle = \sigma_f^2\tau_0 \tag{10.24}$$

for $\tau_0 \to 0$.

Multiply now Eq. (10.22) by $x(t)$. Assuming that function $x(t)$ is sufficiently fine function, we obtain the equality

$$x(t)\frac{d}{dt}x(t) = \frac{1}{2}\frac{d}{dt}x^2(t) = -\lambda x^2(t) + f(t)x(t).$$

Averaging this equation over an ensemble of realizations of function $f(t)$, we obtain the equation

$$\frac{1}{2}\frac{d}{dt}\left\langle x^2(t)\right\rangle = -\lambda\left\langle x^2(t)\right\rangle + \left\langle f(t)x(t)\right\rangle, \tag{10.25}$$

whose steady-state solution (it corresponds to the limit process $t_0 \to -\infty$ and $\tau_0 \to 0$)

$$\left\langle x^2(t)\right\rangle = \frac{1}{\lambda}\left\langle f(t)x(t)\right\rangle$$

coincides with Eq. (10.23) and (10.24).

Taking into account the fact that $\delta x(t)/\delta f(t-0) = 1$, we obtain the same result for correlation $\langle f(t)x(t)\rangle$ by using the formula

$$\left\langle f(t)x(t)\right\rangle = \int\limits_{-\infty}^{t} d\tau B_f(t-\tau)\left\langle \frac{\delta}{\delta f(\tau)}x(t)\right\rangle \tag{10.26}$$

with the effective correlation function

$$B_f^{\text{eff}}(t) = 2\sigma_f^2\tau_0\delta(t).$$

Earlier, we mentioned that statistical characteristics of solutions to dynamic problems in the approximation of the delta-correlated random process (field) coincide with the statistical characteristics of the Markovian processes. However, one should clearly understand that this is the case only for statistical averages and equations for these averages. In particular, realizations of process $x(t)$ satisfying the Langevin equation (10.22) drastically differ from realizations of the corresponding Markovian process. The latter satisfies Eq. (10.22) in which function $f(t)$ in the right-hand side is the ideal white noise with the correlation function $B_f(t) = 2\sigma_f^2\tau_0\delta(t)$; moreover, this equation must be treated in the sense of generalized functions, because the Markovian processes are not differentiable in the ordinary sense. At the same time, process $x(t)$ – whose statistical characteristics coincide with the characteristics of the Markovian process – behaves as sufficiently fine function and is differentiable in the ordinary sense. For example,

$$x(t)\frac{d}{dt}x(t) = \frac{1}{2}\frac{d}{dt}x^2(t),$$

and we have for $t_0 \to -\infty$ in particular

$$\left\langle x(t)\frac{d}{dt}x(t)\right\rangle = 0. \tag{10.27}$$

On the other hand, in the case of the ideal Markovian process $x(t)$ satisfying (in the sense of generalized functions) the Langevin equation (10.22) with the white noise in the right-hand side, Eq. (10.27) makes no sense at all, and the meaning of the relationship

$$\left\langle x(t)\frac{d}{dt}x(t)\right\rangle = -\lambda\left\langle x^2(t)\right\rangle + \langle f(t)x(t)\rangle \tag{10.28}$$

depends on the definition of averages. Indeed, if we will mean Eq. (10.28) as the limit of the equality

$$\left\langle x(t+\Delta)\frac{d}{dt}x(t)\right\rangle = -\lambda\left\langle x(t)x(t+\Delta)\right\rangle + \langle f(t)x(t+\Delta)\rangle \tag{10.29}$$

for $\Delta \to 0$, the result will be essentially different depending on whether we use limit processes $\Delta \to +0$, or $\Delta \to -0$. For limit process $\Delta \to +0$, we have

$$\lim_{\Delta\to+0}\langle f(t)x(t+\Delta)\rangle = 2\sigma_f^2\tau_0,$$

and, taking into account Eq. (10.26), we can rewrite Eq. (10.29) in the form

$$\left\langle x(t+0)\frac{d}{dt}x(t)\right\rangle = \sigma_f^2\tau_0. \tag{10.30}$$

On the contrary, for limit process $\Delta \to -0$, we have

$$\langle f(t)x(t-0)\rangle = 0$$

because of the dynamic causality condition, and Eq. (10.29) assumes the form

$$\left\langle x(t-0)\frac{d}{dt}x(t)\right\rangle = -\sigma_f^2\tau_0. \tag{10.31}$$

Comparing Eq. (10.27) with Eqs. (10.30) and (10.31), we see that, for the ideal Markovian process described by the solution to the Langevin equation with the white noise in the right-hand side and commonly called the *Ohrnstein–Ulenbeck process* , we have

$$\left\langle x(t+0)\frac{d}{dt}x(t)\right\rangle \neq \left\langle x(t-0)\frac{d}{dt}x(t)\right\rangle \neq \frac{1}{2}\frac{d}{dt}\left\langle x^2(t)\right\rangle.$$

Note that equalities (10.30) and (10.31) can also be obtained from the correlation function

$$\langle x(t)x(t+\tau)\rangle = \frac{\sigma_f^2\tau_0}{\lambda}e^{-\lambda|\tau|}$$

of process $x(t)$.

To conclude with the discussion of the approximation of the delta-correlated random process (field), we emphasize that, in all further examples, we will treat the sentence like 'dynamic system (equation) with the delta-correlated parameter fluctuations' as the asymptotic limit in which these parameters have temporal correlation radii small in comparison with all characteristic temporal scales of the problem under consideration.

10.4 The Simplest Markovian Random Processes

There are only few Fokker–Planck equations that allow an exact solution. First of all, among them are the Fokker–Planck equations corresponding to the stochastic equations that are themselves solvable in the analytic form. Such problems often allow determination of not only the one-time and transitional probability densities, but also the characteristic functional and other statistical characteristics important for practice.

10.4.1 System of Linear Equations

Consider the system of linear equations for the components of vector function $\boldsymbol{x}(t)$

$$\frac{d}{dt}\boldsymbol{x}(t) = A\boldsymbol{x}(t) + \boldsymbol{f}(t), \quad \boldsymbol{x}(t_0) = \boldsymbol{x}_0 \tag{10.32}$$

with constant matrix A. In Sect. 10.3.1, we considered in detail a special one-dimensional case of this equation — the Langevin equation . We will assume functions $f_i(t)$ the Gaussian functions delta-correlated in time, i.e., we set

$$\langle f_i(t) f_j(t') \rangle = 2B_{ij}\delta(t - t').$$

The solution to system of equations (10.32) has the form

$$\boldsymbol{x}(t) = e^{(t-t_0)A}\boldsymbol{x}_0 + \int\limits_{t_0}^{t} d\tau e^{(t-t_0)A}\boldsymbol{f}(\tau),$$

so that quantity $\boldsymbol{x}(t)$ is the Gaussian vector function with the parameters

$$\langle \boldsymbol{x}(t) \rangle = e^{(t-t_0)A}\boldsymbol{x}_0,$$

$$\sigma_{ij}^2(t,t') = \langle[x_i(t) - \langle x_i(t)\rangle][x_j(t') - \langle x_j(t')\rangle]\rangle = \int\limits_{t_0}^{t} d\tau \left\{e^{(t-t_0)A} B e^{(t-t_0)A^T}\right\}_{ij}. \tag{10.33}$$

where A^T is the transposed matrix of A.

We can easily see in this case that the Gaussian probability distribution with parameters (10.33) satisfies the Fokker–Planck equation for the transitional probability density $p(\boldsymbol{x}, t|\boldsymbol{x}_0, t_0)$

$$\left(\frac{\partial}{\partial t} + \frac{\partial}{\partial x_i} A_{ik} x_k\right) p(\boldsymbol{x}, t|\boldsymbol{x}_0, t_0) = \frac{\partial}{\partial x_i} B_{ik} \frac{\partial}{\partial x_k} p(\boldsymbol{x}, t|\boldsymbol{x}_0, t_0) \tag{10.34}$$

corresponding to stochastic system (10.32).

We note that Eq. (10.34) by itself also can be easily solved by the Fourier transform with respect to spatial coordinates.

The simplest special case of Eq. (10.32) is the equation that defines the *Wiener random process* . In view of the significant role that such processes plays in physics (for example, they describe the *Brownian motion of particles*) , we consider the Wiener process in detail.

10.4.2 Wiener Random Process

The Wiener random process is defined as the solution to the stochastic equation

$$\frac{d}{dt} w(t) = z(t), \quad w(0) = 0,$$

where $z(t)$ is the Gaussian process delta-correlated in time and described by the parameters

$$\langle z(t)\rangle = 0, \quad \langle z(t)z(t')\rangle = 2\sigma^2\tau_0\delta(t-t').$$

The solution to this equation

$$w(t) = \int\limits_0^t d\tau z(\tau) \tag{10.35}$$

is the continuous Gaussian nonstationary random process with the parameters

$$\langle w(t)\rangle = 0, \quad \langle w(t)w(t')\rangle = 2\sigma^2\tau_0 \min(t,t').$$

As a consequence, its characteristic functional has the form

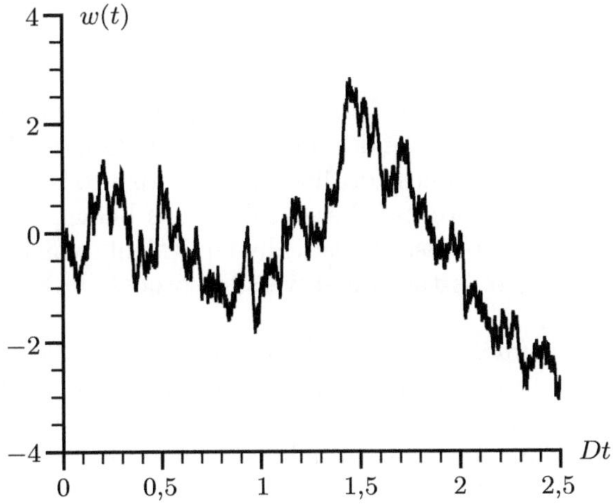

Fig. 10.1 Realization of the Wiener process (10.35)

$$\Phi[t; v(\tau)] = \left\langle \exp\left\{ i \int_0^t d\tau w(\tau) v(\tau) \right\} \right\rangle$$
$$= \exp\left\{ -\sigma^2 \tau_0 \int_0^t d\tau_1 \int_0^t d\tau_2 v(\tau_1) v(\tau_2) \min(\tau_1, \tau_2) \right\}. \quad (10.36)$$

Figure 10.1 shows a simulated realization of the Wiener process (10.35). Note that the increment of process $w(t)$ on the temporal interval (t_1, t_2)

$$w(t_1; t_2) = w(t_2) - w(t_1) = \int_{t_1}^{t_2} d\tau z(\tau)$$

has, like process $w(t)$ itself, the Gaussian statistics with the parameters

$$\langle w(t_1; t_2) \rangle = 0, \quad \left\langle [w(t_1; t_2)]^2 \right\rangle = 2\sigma^2 \tau_0 |t_2 - t_1|.$$

The Wiener random process $w(t)$ is the Gaussian continuous process with independent increments. This means that increments of process $w(t)$ on the nonoverlapping intervals $(t_1; t_2)$ and $(t_3; t_4)$ are statistically independent.

The characteristic functional of process

$$w(t_0; t_0 + t) = \int_{t_0}^{t_0+t} d\tau z(\tau)$$

coincides with the characteristic functional of process $w(t)$. This means that realizations of processes $w(t)$ and $w(t_0; t_0 + t)$ are *statistically equivalent* for any given parameter t_0. Thus, dealing solely with process realizations, we cannot decide to which process these realizations belong. In addition, processes $w(t)$ and $w(-t)$ are also statistically equivalent, which means that the Wiener random process is the time-reversible process in the sense specified above.

An additional — *fractal* — property inheres in realizations of the Wiener process. According to this property, realizations of the Wiener process $w(at)$ (compressed in time for $a > 1$) are statistically equivalent to realizations of process $a^{1/2}w(t)$ (elongated in amplitude). The fractal property of the Wiener process can be treated also as statistical equivalence of realizations of process $w(t)$ and realizations of process $w(at)/a^{1/2}$, which is compressed both in time t and amplitude, because their characteristic functionals coincide.

10.4.3 Wiener Random Process with Shear

Consider a more general process that includes additionally the drift dependent on parameter α

$$w(t; \alpha) = -\alpha t + w(t), \quad \alpha > 0.$$

Process $w(t; \alpha)$ is the Markovian process, and its probability density

$$P(w, t; \alpha) = \langle \delta(w(t; \alpha) - w) \rangle$$

satisfies the Fokker–Planck equation

$$\left(\frac{\partial}{\partial t} - \alpha \frac{\partial}{\partial w} \right) P(w, t; \alpha) = D \frac{\partial^2}{\partial w^2} P(w, t; \alpha), \quad P(w, 0; \alpha) = \delta(w), \quad (10.37)$$

where $D = \sigma^2 \tau_0$ is the diffusion coefficient. The solution to this equation has the form of the Gaussian distribution

$$P(w, t; \alpha) = \frac{1}{2\sqrt{\pi D t}} \exp\left\{ -\frac{(w + \alpha t)^2}{4Dt} \right\}. \quad (10.38)$$

The corresponding integral distribution function defined as the probability of the event that $w(t; \alpha) < w$ is given by the formula

$$F(w,t;\alpha) = \int\limits_{-\infty}^{w} dw P(w,t;\alpha) = \Pr\left(\frac{w}{\sqrt{2Dt}} + \alpha\sqrt{\frac{t}{2D}}\right), \tag{10.39}$$

where function $\Pr(z)$ is the error function (4.20), page 122. In this case, the curve of typical realization of the Wiener random process with shear is the linear function of time in accordance with Eqs. (5.22), page 132

$$w^*(t;\alpha) = -\alpha t.$$

In addition to the initial value, supplement Eq. (10.37) with the boundary condition

$$P(w,t;\alpha)|_{w=h} = 0, \quad (t > 0). \tag{10.40}$$

This condition breaks down realizations of process $w(t;\alpha)$ at the instant they reach boundary h. For $w < h$, the solution to the boundary-value problem (10.37), (10.40) (we denote it as $P(w,t;\alpha,h)$) describes the probability distribution of those realizations of process $w(t;\alpha)$ that survived instant t, i.e., never reached boundary h during the whole temporal interval. Correspondingly, the norm of the probability density appears not unity, but the probability of the event that $t < t^*$, where t^* is the instant at which process $w(t;\alpha)$ reaches boundary h for the first time

$$\int\limits_{-\infty}^{h} dw P(w,t;\alpha,h) = \mathrm{P}(t < t^*). \tag{10.41}$$

Introduce the integral distribution function and probability density of random instant at which the process reaches boundary h

$$\begin{aligned} F(t;\alpha,h) &= \mathrm{P}(t^* < t) = 1 - \mathrm{P}(t < t^*) = 1 - \int\limits_{-\infty}^{h} dw P(w,t;\alpha,h), \\ P(t;\alpha,h) &= \frac{\partial}{\partial t} F(t;\alpha,h) = -\frac{\partial}{\partial w} P(w,t;\alpha,h)|_{w=h}. \end{aligned} \tag{10.42}$$

If $\alpha > 0$, process $w(t;\alpha)$ moves on average out of boundary h; as a result, probability $\mathrm{P}(t < t^*)$ (10.41) tends for $t \to \infty$ to the probability of the event that process $w(t;\alpha)$ never reaches boundary h. In other words, limit

$$\lim_{t\to\infty} \int\limits_{-\infty}^{h} dw P(w,t;\alpha,h) = \mathrm{P}\left(w_{\max}(\alpha) < h\right) \tag{10.43}$$

is equal to the probability of the event that the process absolute maximum

$$w_{\max}(\alpha) = \max_{t \in (0,\infty)} w(t;\alpha)$$

is less than h. Thus, from Eq. (10.43) follows that the integral distribution function of the absolute maximum $w_{\max}(\alpha)$ is given by the formula

$$F(h;\alpha) = \mathrm{P}\left(w_{\max}(\alpha) < h\right) = \lim_{t\to\infty} \int\limits_{-\infty}^{h} dw P(w,t;\alpha,h). \tag{10.44}$$

After we solve boundary-value problem (10.37), (10.40) by using, for example, the reflection method, we obtain

$$P(w,t;\alpha,h) = \frac{1}{2\sqrt{\pi D t}} \left\{ \exp\left[-\frac{(w+\alpha t)^2}{4Dt} \right] - \exp\left[-\frac{h\alpha}{D} - \frac{(w-2h+\alpha t)^2}{4Dt} \right] \right\}. \tag{10.45}$$

Substituting this expression in Eq. (10.42), we obtain the probability density of instant t^* at which process $w(t;\alpha)$ reaches boundary h for the first time

$$P(t;\alpha,h) = \frac{1}{2Dt\sqrt{\pi D t}} \exp\left\{ -\frac{(h+\alpha t)^2}{4Dt} \right\}.$$

Finally, integrating Eq. (10.45) over w and setting $t \to \infty$, we obtain, in accordance with Eq. (10.44), the integral distribution function of absolute maximum $w_{\max}(\alpha)$ of process $w(t;\alpha)$ in the form [52, 71]

$$F(h;\alpha) = \mathrm{P}\left(w_{\max}(\alpha) < h\right) = 1 - \exp\left\{ -\frac{h\alpha}{D} \right\}. \tag{10.46}$$

Consequently, the absolute maximum of the Wiener process has the exponential probability density

$$P(h;\alpha) = \langle \delta\left(w_{\max}(\alpha) - h\right) \rangle = \frac{\alpha}{D} \exp\left\{ -\frac{h\alpha}{D} \right\}.$$

The Wiener random process offers a possibility of constructing other processes convenient for modeling different physical phenomena. In the case of positive quantities, the simplest approximation of such kind is the logarithmic-normal (lognormal) process. Consider this process in greater detail.

10.4.4 Lognormal Random Process, Intermittency and Dynamic Localization

We define the lognormal process (logarithmic-normal process) by the formula

$$y(t;\alpha) = e^{w(t;\alpha)} = \exp\left\{-\alpha t + \int_0^t d\tau z(\tau)\right\}, \tag{10.47}$$

where $z(t)$ is the Gaussian white noise process with the parameters

$$\langle z(t)\rangle = 0, \quad \langle z(t)z(t')\rangle = 2\sigma^2\tau_0\delta(t-t').$$

The lognormal process satisfies the stochastic equation

$$\frac{d}{dt}y(t;\alpha) = \{-\alpha + z(t)\}\, y(t;\alpha), \qquad y(0;\alpha) = 1.$$

We note that changing the sign of parameter α in (10.47) is statistically equivalent to switch to the process $1/y(t;\alpha)$ [61, 62].

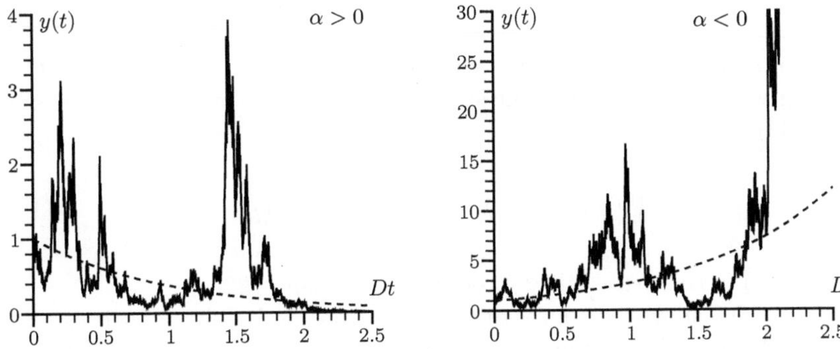

Fig. 10.2 Realizations of lognormal processes $y(t)$ for parameters $\alpha > 0$ and $\alpha < 0$ for the parameter ratio $|\alpha|/D = 1$

Figure 10.2 displays realizations of lognormal random processes $y(t)$ in Eq. (10.47) for positive and negative parameters α for the parameter ratio $|\alpha|/D = 1$ (the dashed curves show the functions $\exp\{-Dt\}$ and $\exp\{Dt\}$, correspondingly). The figure shows the presence of rare but strong *fluctuations* relative to the dashed curves towards both large values and zero. Such a property of random processes is called *intermittency.* The curve with respect to which the fluctuations are observed is referred to as the *typical realization curve* (see Sect. 5.1.1, page 127).

The one-time probability density of the lognormal process is given by the formula

$$P(y,t;\alpha) = \langle \delta\left(y(t;\alpha) - y\right)\rangle = \left\langle \delta\left(e^{w(t;\alpha)} - y\right)\right\rangle$$
$$= \frac{1}{y}\langle \delta\left(w(t;\alpha) - \ln y\right)\rangle = \frac{1}{y}\, P(w,t;\alpha)\big|_{w=\ln y},$$

where $P(w,t;\alpha)$ is the one-time probability density of the Wiener process with a drift, which is given by Eq. (10.38), so that

$$P(y,t;\alpha) = \frac{1}{2y\sqrt{\pi Dt}}\exp\left\{-\frac{(\ln y + \alpha t)^2}{4Dt}\right\} = \frac{1}{2y\sqrt{\pi Dt}}\exp\left\{-\frac{\ln^2\left(ye^{\alpha t}\right)}{4Dt}\right\}, \tag{10.48}$$

where $D = \sigma^2\tau_0$.

Note that the one-time probability density of random process $\widetilde{y}(t;\alpha) = 1/y(t;\alpha)$ is also lognormal and is given by the formula

$$P(\widetilde{y},t;\alpha) = \frac{1}{2\widetilde{y}\sqrt{\pi Dt}}\exp\left\{-\frac{\ln^2\left(\widetilde{y}e^{-\alpha t}\right)}{4Dt}\right\}, \tag{10.49}$$

which coincides with Eq. (10.48) with parameter α of opposite sign. Correspondingly, the integral distribution functions are given, in accordance with Eq. (10.39), by the expressions

$$F(y,t;\alpha) = \mathrm{P}\left(y(t;\alpha) < y\right) = \Pr\left(\frac{1}{\sqrt{2Dt}}\ln\left(ye^{\pm\alpha t}\right)\right), \tag{10.50}$$

where $\Pr(z)$ is the probability integral (4.20), page 122.

Figures 10.3 show the curves of the lognormal probability densities (10.48) and (10.49) for $\alpha/D = 1$ and dimensionless times $\tau = Dt = 0.1$ and 1. Figure 10.4 shows these probability densities at $\tau = 1$ in logarithmic scale.

Structurally, these probability distributions are absolutely different. The only common feature of these distributions consists in the existence of long

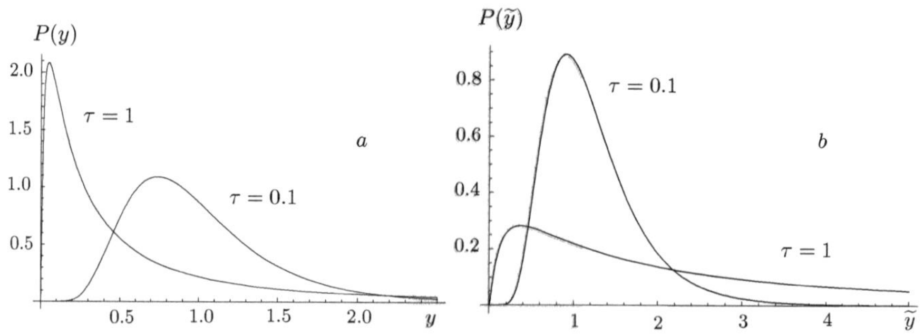

Fig. 10.3 Logarithmic-normal probability densities (10.48) (a) and (10.49) (b) for $\alpha/D = 1$ and dimensionless times $\tau = 0.1$ and 1

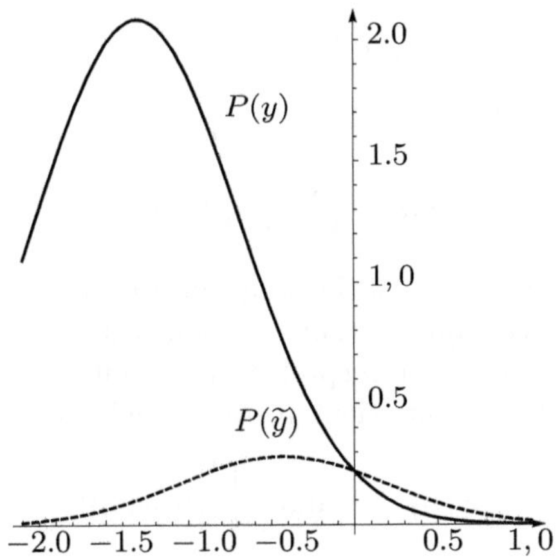

Fig. 10.4 Probability densities of processes $y(t;\alpha)$ (solid line) and $\widetilde{y}(t;\alpha)$ (dashed line) at $\tau = 1$ in common logarithmic scale

flat *tails* that appear in distributions at $\tau = 1$; these tails increase the role of high peaks of processes $y(t;\alpha)$ and $\widetilde{y}(t;\alpha)$ in the formation of the one-time statistics.

Having only the one-time statistical characteristics of process $y(t;\alpha)$, one can obtain a deeper insight into the behavior of realizations of process $y(t;\alpha)$ on the whole interval of times $(0,\infty)$ [52, 71]. In particular,

(**1**) The lognormal process $y(t;\alpha)$ is the Markovian process and its one-time probability density (10.48) satisfies the Fokker–Planck equation

$$\left(\frac{\partial}{\partial t} - \alpha\frac{\partial}{\partial y}y\right) P(y,t;\alpha) = D\frac{\partial}{\partial y}y\frac{\partial}{\partial y}yP(y,t;\alpha), \quad P(y,0;\alpha) = \delta(y-1). \tag{10.51}$$

From Eq. (10.51), one can easily derive the equations for the moment functions of process $y(t;\alpha)$; solutions to these equations are given by the formulas

$$\langle y^n(t;\alpha)\rangle = e^{n(n-\alpha/D)Dt}, \quad \left\langle \frac{1}{y^n(t;\alpha)}\right\rangle = e^{n(n+\alpha/D)Dt}, \quad n = 1,\, 2,\, \cdots \tag{10.52}$$

from which follows that moments exponentially grow with time.

From Eq. (10.51), one can easily obtain the equality

$$\alpha = -\lim_{t\to+\infty}\frac{\partial\,\langle\ln y(t)\rangle}{\partial t}. \tag{10.53}$$

Remark 10.1. ***Lyapunov Characteristic Index***

Note that many investigators give great attention to the approach based on the Lyapunov analysis of stability of solutions to deterministic ordinary differential equations

$$\frac{d}{dt}\boldsymbol{x}(t) = A(t)\boldsymbol{x}(t).$$

This approach deals with the upper limit of problem solution

$$\lambda_{\boldsymbol{x}(t)} = \overline{\lim_{t\to+\infty}} \frac{1}{t} \ln|\boldsymbol{x}(t)|$$

called the characteristic index of the solution. In the context of this approach applied to stochastic dynamic systems, these investigators often use statistical analysis at the last stage to interpret and simplify the obtained results; in particular, they calculate statistical averages such as

$$\left\langle \lambda_{\boldsymbol{x}(t)} \right\rangle = \overline{\lim_{t\to+\infty}} \frac{1}{t} \left\langle \ln|\boldsymbol{x}(t)| \right\rangle. \tag{10.54}$$

♦

Consequently, parameter α (10.53) is the *Lyapunov characteristic index* of the lognormal random process $y(t;\alpha)$ (see, e.g., [58–60]).

(**2**) From the integral distribution function, one can calculate the typical realization curve of lognormal process $y(t;\alpha)$; this curve appears the exponentially decaying curve

$$y^*(t;\alpha) = e^{-\alpha t}, \tag{10.55}$$

which agrees with Eq. (5.22), page 132.

Consequently, the exponential increase of moments is caused by deviations of process $y(t;\alpha)$ from the typical realization curve $y^*(t;\alpha)$ towards both large and small values of y.

At $\alpha/D = 1$, the average value of process $y(t;D)$ is independent of time and is equal to unity. Despite this fact, according to Eq. (10.50), the probability of the event that $y < 1$ for $Dt \gg 1$ rapidly approaches the unity by the law

$$\mathrm{P}\left(y(t;D) < 1\right) = \Pr\left(\sqrt{\frac{Dt}{2}}\right) = 1 - \frac{1}{\sqrt{\pi Dt}} e^{-Dt/4},$$

i.e., the curves of process realizations run mainly below the level of the process average $\langle y(t;D)\rangle = 1$, though namely large peaks of the process govern the behavior of statistical moments of process $y(t;D)$.

Here, we have a clear contradiction between the behavior of statistical characteristics of process $y(t;\alpha)$ and the behavior of process realizations.

(**3**) The behavior of realizations of process $y(t;\alpha)$ on the whole temporal interval can also be evaluated with the use of the p -majorant curves $M_p(t,\alpha)$ whose definition is as follows [52, 71]. We call the majorant curve the curve $M_p(t,\alpha)$ for which inequality $y(t;\alpha) < M_p(t,\alpha)$ is satisfied for all times t with probability p, i.e.,

$$\mathrm{P}\left\{y(t;\alpha) < M_p(t,\alpha) \text{ for all } t \in (0,\infty)\right\} = p.$$

The above statistics (10.46) of the absolute maximum of the Wiener process with a drift $w(t;\alpha)$ makes it possible to outline a wide enough class of the majorant curves. Indeed, let p be the probability of the event that the absolute maximum $w_{\max}(\beta)$ of the auxiliary process $w(t;\beta)$ with arbitrary parameter β in the interval $0 < \beta < \alpha$ satisfies inequality $w(t;\beta) < h = \ln A$. It is clear that the whole realization of process $y(t;\alpha)$ will run in this case below the majorant curve

$$M_p(t,\alpha,\beta) = Ae^{(\beta-\alpha)t}. \tag{10.56}$$

with the same probability p. As may be seen from Eq. (10.46), the probability of the event that process $y(t;\alpha)$ never exceeds majorant curve (10.56) depends on this curve parameters according to the formula

$$p = 1 - A^{-\beta/D}.$$

This means that we derived the one-parameter class of exponentially decaying majorant curves

$$M_p(t,\alpha,\beta) = \frac{1}{(1-p)^{D/\beta}}e^{(\beta-\alpha)t}. \tag{10.57}$$

Notice the remarkable fact that, despite statistical average $\langle y(t;D)\rangle$ remains constant ($\langle y(t;D)\rangle = 1$) and higher-order moments of process $y(t;D)$ are exponentially increasing functions, one can always select an exponentially decreasing majorant curve (10.57) such that realizations of process $y(t;D)$ will run below it with arbitrary predetermined probability $p < 1$. In particular, inequality ($\tau = Dt$)

$$y(t;D) < M_{1/2}(t,D,D/2) = M(\tau) = 4e^{-\tau/2} \tag{10.58}$$

is satisfied with probability $p = 1/2$ for any instant t from interval $(0,\infty)$.

Figure 10.5 schematically shows the behaviors of a realization of process $y(t;D)$ and the majorant curve (10.58). This schematic is an additional fact in favor of our conclusion that the exponential growth of moments of process $y(t;D)$ with time is the purely statistical effect caused by averaging over the whole ensemble of realizations.

Note that the area below the exponentially decaying majorant curves has a finite value. Consequently, high peaks of process $y(t;\alpha)$, which are the reason of the exponential growth of higher moments, only insignificantly contribute to the area below realizations; this area appears finite for almost all realizations, which means that the peaks of the lognormal process $y(t;\alpha)$ are sufficiently narrow .

(4) In this connection, it is of interest to investigate immediately the statistics of random area below realizations of process $y(t;\alpha)$

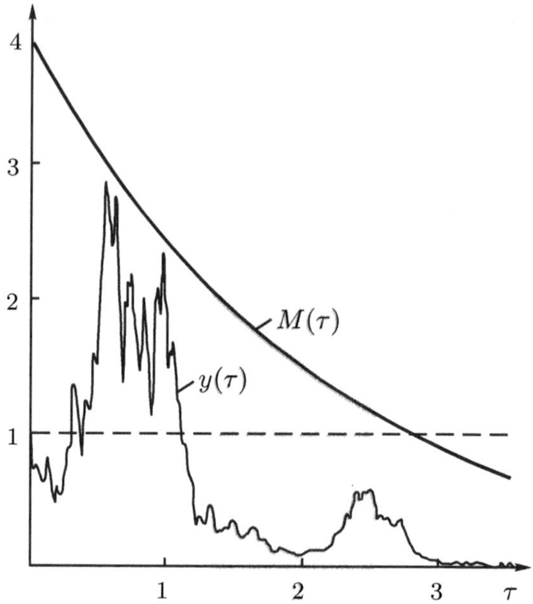

Fig. 10.5 Schematic behaviors of a realization of process $y(t; D)$ and majorant curve $M(\tau)$ (10.58)

$$S_n(t;\alpha) = \int_0^t d\tau y^n(\tau;\alpha). \tag{10.59}$$

This function satisfies the system of stochastic equations

$$\begin{aligned} &\frac{d}{dt} S_n(t;\alpha) = y^n(t;\alpha), \quad S_n(0;\alpha) = 0, \\ &\frac{d}{dt} y(t;\alpha) = \{-\alpha + z(t)\} y(t;\alpha), \quad y(0;\alpha) = 1, \end{aligned} \tag{10.60}$$

so that the two-component process $\{y(t;\alpha), S_n(t;\alpha)\}$ is the Markovian process whose one-time probability density

$$P(S_n, y, t;\alpha) = \langle \delta\left(S_n(t;\alpha) - S_n\right) \delta\left(y(t;\alpha) - y\right) \rangle$$

and transition probability density satisfy the Fokker–Planck equation

$$\begin{aligned} &\left(\frac{\partial}{\partial t} + y^n \frac{\partial}{\partial S_n} - \alpha \frac{\partial}{\partial y} y\right) P(S_n, y, t;\alpha) = D \frac{\partial}{\partial y} y \frac{\partial}{\partial y} y P(S_n, y, t;\alpha), \\ &P(S_n, y, 0;\alpha) = \delta(S_n)\delta(y-1). \end{aligned} \tag{10.61}$$

Unfortunately, Eq. (10.61) cannot be solved analytically, which prevents us from studying the statistics of process $S_n(t;\alpha)$ exhaustively. However, for the one-time statistical averages of process $S_n(t;\alpha)$, i.e., averages at a fixed instant, the corresponding statistics can be studied in sufficient detail.

With this goal in view, we rewrite Eq. (10.59) in the form

$$S_n(t;\alpha) = \int\limits_0^t d\tau \exp\left\{-n\alpha\tau + n\int\limits_0^\tau d\tau_1 z(\tau_1)\right\}$$
$$= \int\limits_0^t d\tau \exp\left\{-n\alpha(t-\tau) + n\int\limits_0^{t-\tau} d\tau_1 z(t-\tau-\tau_1)\right\},$$

from which follows that quantity $S_n(t;\alpha)$ in the context of the one-time statistics is statistically equivalent to the quantity

$$S_n(t;\alpha) = \int\limits_0^t d\tau e^{-n\alpha(t-\tau)+n\int\limits_0^{t-\tau} d\tau_1 z(\tau+\tau_1)}. \tag{10.62}$$

Differentiating now Eq. (10.62) with respect to time, we obtain the statistically equivalent stochastic equation

$$\frac{d}{dt}S_n(t;\alpha) = 1 - n\{\alpha - z(t)\}S_n(t;\alpha), \quad S_n(0;\alpha) = 0$$

whose one-time statistical characteristics are described by the one-time probability density $P(S_n,t;\alpha) = \langle\delta\left(S_n(t;\alpha) - S_n\right)\rangle$ that satisfies the Fokker–Planck equation

$$\left(\frac{\partial}{\partial t} + \frac{\partial}{\partial S_n} - n\alpha\frac{\partial}{\partial S_n}S_n\right)P(S_n,t;\alpha) = n^2 D\frac{\partial}{\partial S_n}S_n\frac{\partial}{\partial S_n}S_n P(S_n,y,t;\alpha). \tag{10.63}$$

As may be seen from Eq. (10.63), random integrals

$$S_n(\alpha) = \int\limits_0^\infty d\tau y^n(\tau;\alpha)$$

are distributed according to the steady-state probability density

$$P(S_n;\alpha) = \frac{1}{\left(n^2 D\right)^{\alpha/nD}\Gamma\left(\frac{\alpha}{D}\right)S_n^{1+\alpha/D}}\exp\left\{-\frac{1}{n^2 D S_n}\right\},$$

where $\Gamma(z)$ is the *gamma function* . In the special case $n = 1$, quantity

$$S(\alpha) = S_1(\alpha) = \int_0^\infty d\tau y(\tau;\alpha)$$

has the following probability density

$$P(S;\alpha) = \frac{1}{D^{\alpha/D}\Gamma\left(\frac{\alpha}{D}\right) S^{1+\alpha/D}} \exp\left\{-\frac{1}{DS}\right\}. \qquad (10.64)$$

If we set now $\alpha = D$, then the steady-state probability density and the corresponding integral distribution function will have the form

$$P(S;D) = \frac{1}{DS^2} \exp\left\{-\frac{1}{DS}\right\}, \quad F(S;D) = \exp\left\{-\frac{1}{DS}\right\}. \qquad (10.65)$$

The time-dependent behavior of the probability density of random process

$$\widetilde{S}(t,\alpha) = \int_t^\infty d\tau y(\tau;\alpha) \qquad (10.66)$$

gives an additional information about the behavior of realizations of process $y(t;\alpha)$ with time t. The integral in the right-hand side of Eq. (10.66) can be represented in the form

$$\widetilde{S}(t,\alpha) = y(t;\alpha) \int_0^\infty d\tau \exp\left\{-\alpha\tau + \int_0^\tau d\tau_1 z(\tau_1 + t)\right\}. \qquad (10.67)$$

In Eq. (10.67), random process $y(t;\alpha)$ is statistically independent of the integral factor, because they depend functionally on process $z(\tau)$ for nonoverlapping intervals of times τ; in addition, the integral factor by itself appears statistically equivalent to random quantity $S(\alpha)$. Consequently, the one-time probability density $P(\widetilde{S},t;\alpha) = \left\langle \delta\left(\widetilde{S}(t;\alpha) - \widetilde{S}\right)\right\rangle$ of random process $\widetilde{S}(t,\alpha)$ is described by the expression

$$P(\widetilde{S},t;\alpha) = \int_0^\infty\int_0^\infty dydS\delta(yS - \widetilde{S})P(y,t;\alpha) = \int_0^\infty \frac{dy}{y} P(y,t;\alpha) P\left(\frac{\widetilde{S}}{y};\alpha\right), \qquad (10.68)$$

where $P(y,t;\alpha)$ is the one-time probability density of lognormal process $y(t;\alpha)$ (10.47) and $P\left(\frac{\widetilde{S}}{y};\alpha\right)$ is the probability density (10.64) of random area.

The corresponding integral distribution function

$$F(\widetilde{S},t;\alpha) = \mathrm{P}\left(\widetilde{S}(t;\alpha) < \widetilde{S}\right) = \int\limits_0^{\widetilde{S}} d\widetilde{S} P\left(\widetilde{S},t;\alpha\right)$$

is given by the integral

$$F(\widetilde{S},t;\alpha) = \int\limits_0^\infty dy P(y,t;\alpha) F\left(\frac{\widetilde{S}}{y};\alpha\right),$$

where $F(S;\alpha)$ is the integral distribution function of random area $S(t;\alpha)$. In the special case $\alpha = D$, we obtain, according to Eq. (10.48) and (10.65), the expression

$$F(\widetilde{S},t;D) = \frac{1}{2\sqrt{\pi Dt}} \int\limits_0^\infty \frac{dy}{y} \exp\left\{-\frac{\ln^2\left(ye^{Dt}\right)}{4Dt} - \frac{y}{D\widetilde{S}}\right\}$$

from which follows that the probability of the event that inequality $\widetilde{S}(t;D) < \widetilde{S}$ is satisfied monotonously tends to unity with increasing Dt for any predetermined value of $D\widetilde{S}$. This is an additional evidence in favor of the fact that every separate realization of the lognormal process tends to zero with increasing Dt, though moment functions of process $y(t;\alpha)$ show the exponential growth caused by large spikes.

10.5 Logarithmic-Normal Random Fields, Intermittency and Clustering

10.5.1 Logarithmic-Normal Random Fields

Consider now a positive lognormal random field $f(\boldsymbol{r},t)$ whose one-point probability density

$$P(\boldsymbol{r},t;f) = \langle \delta\left(f(\boldsymbol{r},t) - f\right)\rangle$$

is determined by the equation

$$\frac{\partial}{\partial t} P(\boldsymbol{r},t;f) = \left\{D_0 \frac{\partial^2}{\partial \boldsymbol{r}^2} + \alpha \frac{\partial}{\partial f} f + D_f \frac{\partial}{\partial f} f \frac{\partial}{\partial f} f\right\} P(\boldsymbol{r},t;f) \qquad (10.69)$$

with initial condition $P(\boldsymbol{r},0;f) = \delta\left(f - f_0(\boldsymbol{r})\right)$, where D_0 is the diffusion coefficient in $\boldsymbol{r}$-space, and coefficients α and D_f characterize the diffusion in the f-space. The parameter α can be either positive, negative, or equal to zero (critical case). In the context of one-point characteristics, changing the sign of α means switch from the field $f(\boldsymbol{r},t)$ to the field $\widetilde{f}(\boldsymbol{r},t) = 1/f(\boldsymbol{r},t)$.

The solution to this equation has the form

$$P(\boldsymbol{r},t;f) = \frac{1}{2f\sqrt{\pi D_f t}} \exp\left\{D_0 t \frac{\partial^2}{\partial \boldsymbol{r}^2}\right\} \exp\left\{-\frac{\ln^2\left[fe^{\alpha t}/f_0(\boldsymbol{r})\right]}{4D_f t}\right\}. \quad (10.70)$$

For a positive conservative random field $f(\boldsymbol{r},t)$ with $\int d\boldsymbol{r}\, f(\boldsymbol{r},t) = \int d\boldsymbol{r}\, f_0(\boldsymbol{r})$, we have $\alpha = D_f$, and Eq. (10.69) assumes the form

$$\frac{\partial}{\partial t} P(\boldsymbol{r},t;f) = \left(D_0 \frac{\partial^2}{\partial \boldsymbol{r}^2} + \alpha \frac{\partial^2}{\partial f^2} f^2\right) P(\boldsymbol{r},t;f). \quad (10.71)$$

Of course, any random field always has the intermittency property. Indeed, time evolution of $f(\boldsymbol{r},t)$ at any fixed spatial point $\boldsymbol{r}$ is a random process to which all the above applies.

In the case of the problem statistically homogeneous in space and corresponding to the initial field distribution $f_0(\boldsymbol{r}) = f_0$, all the one-point statistical characteristics of the field $f(\boldsymbol{r},t)$ are independent of point $\boldsymbol{r}$, and positivity of parameter $\alpha = -\lim_{t\to\infty} \dfrac{\partial \langle \ln f(\boldsymbol{r},t)\rangle}{\partial t}$ implies that, at any point of space, realizations of a lognormal field $f(\boldsymbol{r},t)$ decrease with time despite the presence of rare large outbursts characteristic of the lognormal process. The characteristic decay time of the field is $t \sim 1/\alpha$. *But if this field decreases almost everywhere, it must somewhere concentrate, which means that clustering must occur.* For a negative value of α, the field grows at each fixed point of space.

In this case, probability density (10.70) is independent of $\boldsymbol{r}$ and is described by the equation

$$\frac{\partial}{\partial t} P(t;f) = \left\{\alpha \frac{\partial}{\partial f} f + D_f \frac{\partial}{\partial f} f \frac{\partial}{\partial f} f\right\} P(t;f), \qquad P(0;f) = \delta\left(f - f_0\right), \quad (10.72)$$

whose solution is given by the formula

$$P(t;f) = \frac{1}{2f\sqrt{\pi D_f t}} \exp\left\{-\frac{\ln^2\left[fe^{\alpha t}/f_0\right]}{4D_f t}\right\}. \quad (10.73)$$

Hence, the one-point statistical characteristics of random field $f(\boldsymbol{r},t)$ in a spatially homogeneous problem are statistically equivalent to the statistical characteristics of the lognormal process $f(t;\alpha)$ with probability density (10.73). A characteristic feature of this distribution is the appearance of a long gently sloping " *tail*" for $D_f t \gg 1$, which means the enhanced role of large outbursts of the process $f(t;\alpha)$ in forming the one-time statistics. For this distribution, all moment functions of any order n increase exponentially in time. In particular, for $n = 1$ and $D_f > \alpha$, the mean is given by

$$\langle f(\boldsymbol{r},t)\rangle = f_0 e^{(D_f - \alpha)t},$$

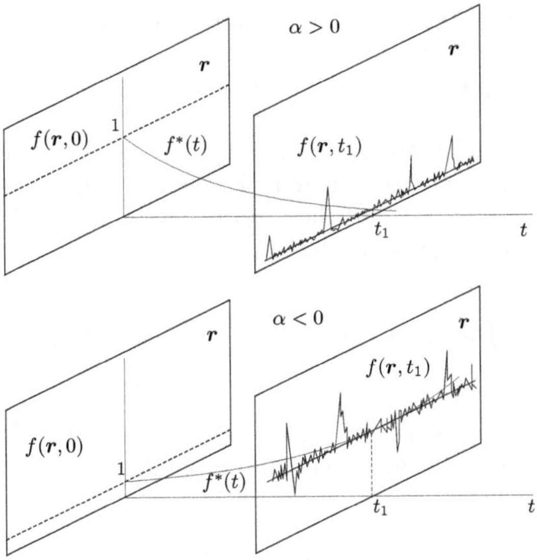

Fig. 10.6 Schematic behavior of realizations of random field $f(\boldsymbol{r}, t)$ for $\alpha > 0$ and $\alpha < 0$

and α is the Lyapunov characteristic index.

The papers [139, 140] mentioned in Sect. 5.1.1, 5.1.1 page 127 treat the intermittency of parametrically excited random fields (namely, the presence of great but rare outbursts) as the structure formation in the space of this field.

However, the above material clearly shows that this inference is contrary to fact because intermittency takes place always, while the structure formation (in the form of clustering) occurs only under certain conditions. Moreover, if statistically homogeneous field $f(\boldsymbol{r}, t)$ is a lognormal filed, field $1/f(\boldsymbol{r}, t)$ will be also the lognormal field and will have great rare outbursts. But if one of these fields shows clustering, the other does not show it though intermittency is characteristic of both these fields.

Figure 10.6 schematically shows realizations of random field $f(\boldsymbol{r}, t)$ with different signs of parameter α.

10.5.2 Statistical Topography of the Lognormal Random Fields

Analyzing one-point statistical characteristics of a spatially homogeneous problem, it is generally convenient to take into account that random field

$f(\boldsymbol{r},t)$ is statistically equivalent to certain random process $f(t)$ with the same statistical characteristics.

Knowing the one-point probability density of random field $f(\boldsymbol{r},t)$ (10.70), we can also obtain some general information about the spatial structure of $f(\boldsymbol{r},t)$. In particular, functionals of $f(\boldsymbol{r},t)$ such as the total mean volume (in the three-dimensional case) or area (in the two-dimensional case) of the domain in which $f(\boldsymbol{r},t) > f$ and the total mean "mass" of the field confined within that domain are given by the formulas

$$\langle V(t,f)\rangle = \int d\boldsymbol{r} \int\limits_f^\infty df'\, P(\boldsymbol{r},t;f'), \quad \langle M(t,f)\rangle = \int d\boldsymbol{r} \int\limits_f^\infty df'\, f' P(\boldsymbol{r},t;f').$$

The values of these functionals are independent of the diffusion in the $\boldsymbol{r}$ space (of the coefficient D_0), and in the case of probability distribution (10.70), we obtain the expressions

$$\begin{aligned} \langle V(t,f)\rangle &= \int d\boldsymbol{r}\, \mathrm{Pr}\left(\frac{1}{\sqrt{2Dt}} \ln\left(\frac{f_0(\boldsymbol{r})}{f} e^{-\alpha t}\right)\right), \\ \langle M(t,f)\rangle &= e^{(D-\alpha)t} \int d\boldsymbol{r}\, f_0(\boldsymbol{r})\, \mathrm{Pr}\left(\frac{1}{\sqrt{2Dt}} \ln\left(\frac{f_0(\boldsymbol{r})}{f} e^{(2D-\alpha)t}\right)\right), \end{aligned} \tag{10.74}$$

where the *probability integral* $\mathrm{Pr}(z)$ is defined in Eq. (4.20), page 122.

Using asymptotic forms (4.23), page 123 of function $\mathrm{Pr}(z)$, we can study the time evolution of these functionals. Namely, the asymptotic form of the mean volume as $t \to \infty$ decreases in time for $\alpha > 0$

$$\langle V(t,f)\rangle \approx \frac{1}{\alpha}\sqrt{\frac{D}{\pi f^{\alpha/D} t}} e^{-\alpha^2 t/4D} \int d\boldsymbol{r}\, \sqrt{f_0^{\alpha/D}(\boldsymbol{r})}.$$

For $\alpha < 0$, the mean volume occupies all the space as $t \to \infty$.

The asymptotic form of the total mean "mass" as $t \to \infty$ is (in the most interesting case $\alpha < 2D$)

$$\langle M(t,f)\rangle \approx e^{(D-\alpha)t} \int d\boldsymbol{r}\, f_0(\boldsymbol{r}) \left[1 - \frac{1}{(2D-\alpha)}\sqrt{\frac{D}{\pi t}\left(\frac{f}{f_0(\boldsymbol{r})}\right)^{(2D-\alpha)/D}} e^{-(2D-\alpha)^2 t/4}\right].$$

Therefore, for $\alpha > 0$, all the mean "mass" is collected in clusters in the limit $t \to \infty$.

For homogeneous initial conditions, the integrands in (10.74) represent specific values of the volume of large outbursts and their total "mass", i.e., the corresponding values per unit volume,

$$\langle v_{\text{hom}}(t,f)\rangle = \langle \theta\left(f(\boldsymbol{r},t)-f\right)\rangle = \mathrm{P}\{f(\boldsymbol{r},t)>f\} = \Pr\left(\frac{1}{\sqrt{2Dt}}\ln\left(\frac{f_0}{f}e^{-\alpha t}\right)\right),$$

$$\langle m_{\text{hom}}(t,f)\rangle = f_0 e^{(D-\alpha)t}\Pr\left(\frac{1}{\sqrt{2Dt}}\ln\left(\frac{f_0}{f}e^{(2D-\alpha)t}\right)\right). \tag{10.75}$$

If we choose the section level $f > f_0$, then $\langle v_{\text{hom}}(0,f)\rangle = 0$ and $\langle m_{\text{hom}}(0,f)\rangle = 0$ at the initial instant. After that, spatial perturbations of random field $f(\boldsymbol{r},t)$ occur, and we have the asymptotic expressions as $t\to\infty$ (for $2D>\alpha$)

$$\langle v_{\text{hom}}(t,f)\rangle = \mathrm{P}\{f(\boldsymbol{r},t)>f\} \approx \begin{cases} \dfrac{1}{\alpha}\sqrt{\dfrac{D}{\pi t}\left(\dfrac{f_0}{f}\right)^{\alpha/D}}\, e^{-\alpha^2 t/4D} & (\alpha>0),\\[2ex] 1-\dfrac{1}{|\alpha|}\sqrt{\dfrac{D}{\pi t}\left(\dfrac{f}{f_0}\right)^{|\alpha|/D}}\, e^{-\alpha^2 t/4D} & (\alpha<0),\end{cases} \tag{10.76}$$

$$\langle m_{\text{hom}}(t,f)\rangle \approx f_0 e^{(D-\alpha)t}\left[1-\frac{1}{(2D-\alpha)}\sqrt{\frac{D}{\pi t}\left(\frac{f}{f_0}\right)^{(2D-\alpha)/D}}\, e^{-(2D-\alpha)^2 t/4D}\right]. \tag{10.77}$$

Hence, for $\alpha>0$, the specific total volume tends to zero, and the specific total "mass" contained in that volume tends to the mean "mass" of the entire space, which corresponds to the criterion for structure formation with probability one for "ideal clustering" of the field $f(\boldsymbol{r},t)$ under consideration. *In these conditions, random field $f(r,t)$ is practically absent in the most part of space.* In addition, at each fixed point of space, the characteristic decay time of the field is $\alpha t\sim 1$, and the characteristic time of cluster structure formation of the field is $\alpha t \sim \max\left\{4\xi,\ 4\xi/(2\xi-1)^2\right\}$, where $\xi = D/\alpha$.

For $\alpha<0$, clustering is absent, and only the general increase of random field $f(\boldsymbol{r},t)$ occurs everywhere in the space. In this case, therefore, *chaos remains chaos*! Only clustering of the zeros of the field $f(\boldsymbol{r},t)$ occurs.

Remark 10.2. ***We Have the Following Theorem:***

In a statistically homogeneous problem, a conservative positive parametrically excited random log-normal field always experiences clustering with probability one, i.e., for almost all realizations of this field.

Indeed, we then have $f(\boldsymbol{r},t) = e^{\ln f(\boldsymbol{r},t)}$, and therefore

$$\langle f(\boldsymbol{r},t)\rangle = \left\langle e^{\ln f(\boldsymbol{r},t)}\right\rangle = \exp\left\{\langle \ln f(\boldsymbol{r},t)\rangle + \frac{1}{2}\sigma^2_{\ln f(\boldsymbol{r},t)}\right\},$$

where $\sigma^2_{\ln f(\boldsymbol{r},t)}$ is the variance of the random field $\ln f(\boldsymbol{r},t)$. In view of conservativity

$$\langle \ln f(\boldsymbol{r},t)\rangle + \frac{1}{2}\sigma^2_{\ln f(\boldsymbol{r},t)} = \ln f_0,$$

and the typical realization curve is

$$f^*(\boldsymbol{r},t) = e^{\langle \ln f(\boldsymbol{r},t)\rangle} = f_0 \exp^{-\alpha t}$$

with the Lyapunov characteristic index

$$\alpha = \lim_{t\to\infty} \frac{1}{2t}\sigma^2_{\ln f(\boldsymbol{r},t)} > 0.$$

The problem is to calculate it from the corresponding dynamic equation. Here, as noted above, $\alpha = D$ for a conservative $f(\boldsymbol{r},t)$ (see Eq. (10.71)). Therefore, the characteristic time of cluster structure formation $\alpha t \sim 4$, which is four times greater than the characteristic decay time of the field at almost every point of space. ♦

So, we have demonstrated that if clustering is realized with probability one (i.e., in almost all realizations of a random field), then it arises from very rare events whose probability tends to zero. In this connection, I cannot but note that a point commonly accepted in many works suggests that, for an event to happen, it is required that this event was most probable. For example, in a recent paper [37], a hypothesis of the origin of life from the perspective of physics was suggested as a result of calculations of some probabilities (on the base of random graphs and chains): "*Life can be briefly described as a result of a game process, an interplay between a part of the system and its environment. During the game, this part acquired an ability to remember the probabilities of gains and losses in previous rounds, which gave it a chance to exist in the following ones*". I cannot agree with the idea that the origin of life is a game process. I believe that the origin of life is an event happened with probability one [63].

10.6 Causal Integral Equations

In problems discussed earlier, we succeeded in deriving the closed statistical description in the approximation of the delta-correlated random field due to the fact that every of these problems corresponded to a system of the first-order (in temporal coordinate) differential equations with given initial values at $t = 0$. The solutions to such systems satisfy the dynamic causality condition, which means that the solution at instant t depends only on system parameter fluctuations for preceding times and is independent of fluctuations for consequent times.

However, problems described in terms of integral equations that generally cannot be reduced to a system of differential equations also can satisfy the causality condition.

10.6.1 General Remarks

In this case, the parent stochastic equation is the linear integral equation for Green's function

$$S(\boldsymbol{r},\boldsymbol{r}') = S_0(\boldsymbol{r},\boldsymbol{r}') + \int d\boldsymbol{r}_1 \int d\boldsymbol{r}_2 \int d\boldsymbol{r}_3 S_0(\boldsymbol{r},\boldsymbol{r}_1)\Lambda(\boldsymbol{r}_1,\boldsymbol{r}_2,\boldsymbol{r}_3)f(\boldsymbol{r}_2)S(\boldsymbol{r}_3,\boldsymbol{r}'), \tag{10.78}$$

where $\boldsymbol{r}$ denotes all arguments of functions $S(\boldsymbol{r},\boldsymbol{r}')$ and $f(\boldsymbol{r})$ including the index arguments that assume summation instead of integration. It is assumed here that function $f(\boldsymbol{r})$ is a random field and function $S_0(\boldsymbol{r},\boldsymbol{r}')$ is Green's function for the problem with absent parameter fluctuations, i.e., for $f(\boldsymbol{r}) = 0$. We will assume additionally that quantity $\Lambda(\{\boldsymbol{r}_i\})$ is a function.

The solution to Eq. (10.78) is a functional of field $f(\boldsymbol{r})$, i.e.

$$S(\boldsymbol{r},\boldsymbol{r}') = S[\boldsymbol{r},\boldsymbol{r}'; f(\widetilde{\boldsymbol{r}})]$$

and Eq. (10.78) appears to be equivalent to the functional equation that contains the variational derivative in functional space $\{f(\widetilde{\boldsymbol{r}})\}$ (8.30), page 205

$$\frac{\delta}{\delta f(\boldsymbol{r}_0)} S[\boldsymbol{r},\boldsymbol{r}'; f(\widetilde{\boldsymbol{r}})] = \int d\boldsymbol{r}_1 \int d\boldsymbol{r}_2 S[\boldsymbol{r},\boldsymbol{r}_1; f(\widetilde{\boldsymbol{r}})]\Lambda(\boldsymbol{r}_1,\boldsymbol{r}_0,\boldsymbol{r}_2)S[\boldsymbol{r}_2,\boldsymbol{r}'; f(\widetilde{\boldsymbol{r}})] \tag{10.79}$$

and satisfies the initial value

$$S[\boldsymbol{r},\boldsymbol{r}'; f(\widetilde{\boldsymbol{r}})]_{f=0} = S_0(\boldsymbol{r},\boldsymbol{r}').$$

Now, we separate the temporal coordinate t in Eq. (10.78), i.e., rewrite it in the form

$$\begin{aligned} S(\boldsymbol{r},t;\boldsymbol{r}',t') &= S_0(\boldsymbol{r},t;\boldsymbol{r}',t') \\ &+ \int d\boldsymbol{r}_1 \int d\boldsymbol{r}_2 \int d\boldsymbol{r}_3 \int d\tau S_0(\boldsymbol{r},t;\boldsymbol{r}_1,\tau)\Lambda(\boldsymbol{r}_1,\boldsymbol{r}_2,\boldsymbol{r}_3)f(\boldsymbol{r}_2,\tau)S(\boldsymbol{r}_3,\tau;\boldsymbol{r}',t'). \end{aligned} \tag{10.80}$$

In what follows, we will assume that

$$S_0(\boldsymbol{r},t;\boldsymbol{r}',t') = g(\boldsymbol{r},t;\boldsymbol{r}',t')\theta(t-t'),$$

where $\theta(t)$ is the Heaviside step function . In this case, the solution to Eq. (10.80) also has the form

$$S(\boldsymbol{r},t;\boldsymbol{r}',t') = G(\boldsymbol{r},t;\boldsymbol{r}',t')\theta(t-t'),$$

where function $G(\boldsymbol{r},t;\boldsymbol{r}',t')$ is described by the causal (in time) integral equation

$$G(\boldsymbol{r},t;\boldsymbol{r}',t') = g(\boldsymbol{r},t;\boldsymbol{r}',t')$$
$$+\int d\boldsymbol{r}_1 \int d\boldsymbol{r}_2 \int d\boldsymbol{r}_3 \int\limits_{t'}^{t} d\tau g(\boldsymbol{r},t;\boldsymbol{r}_1,\tau)\Lambda(\boldsymbol{r}_1,\boldsymbol{r}_2,\boldsymbol{r}_3)f(\boldsymbol{r}_2,\tau)G(\boldsymbol{r}_3,\tau;\boldsymbol{r}',t'). \tag{10.81}$$

Function $G(\boldsymbol{r},t;\boldsymbol{r}',t')$ is a functional of field $f(\widetilde{\boldsymbol{r}},\tau)$, i.e.,

$$G(\boldsymbol{r},t;\boldsymbol{r}',t') = G[\boldsymbol{r},t;\boldsymbol{r}',t';f(\widetilde{\boldsymbol{r}},\tau)].$$

Consequently, the dynamic causality condition has the form

$$\frac{\delta}{\delta f(\boldsymbol{r}_0,\tau)} G(\boldsymbol{r},t;\boldsymbol{r}',t') = 0 \quad \text{for} \quad \tau < t' \quad \text{and} \quad \tau > t.$$

In this case, the variational derivative is given, in view of Eq. (10.79), by the expression

$$\frac{\delta}{\delta f(\boldsymbol{r}_0,t_0)} G(\boldsymbol{r},t;\boldsymbol{r}',t') = \int d\boldsymbol{r}_1 \int d\boldsymbol{r}_2 G(\boldsymbol{r},t;\boldsymbol{r}_1,t_0)\Lambda(\boldsymbol{r}_1,\boldsymbol{r}_0,\boldsymbol{r}_2)G(\boldsymbol{r}_2,t_0;\boldsymbol{r}',t'), \tag{10.82}$$

from which follows that

$$\frac{\delta}{\delta f(\boldsymbol{r}_0,t-0)} G(\boldsymbol{r},t;\boldsymbol{r}',t') = \int d\boldsymbol{r}_1 \int d\boldsymbol{r}_2 g(\boldsymbol{r},t;\boldsymbol{r}_1,t)\Lambda(\boldsymbol{r}_1,\boldsymbol{r}_0,\boldsymbol{r}_2)G(\boldsymbol{r}_2,t;\boldsymbol{r}',t'). \tag{10.83}$$

10.6.2 Statistical Averaging

Let now random field $f(\boldsymbol{r},t)$ is the Gaussian random field whose average value is zero. In this case, assignment of the correlation function

$$B(\boldsymbol{r},t;\boldsymbol{r}',t') = \langle f(\boldsymbol{r},t)f(\boldsymbol{r}',t')\rangle .$$

describes all statistical characteristics of the field. For statistically homogeneous and stationary random field $f(\boldsymbol{r},t)$, we have

$$B(\boldsymbol{r},t;\boldsymbol{r}',t') = B(\boldsymbol{r}-\boldsymbol{r}',t-t').$$

Averaging Eq. (10.81) over an ensemble of realizations of field $f(\boldsymbol{r},t)$, we obtain the equation

$$\langle G(\boldsymbol{r},t;\boldsymbol{r}',t')\rangle = g(\boldsymbol{r},t;\boldsymbol{r}',t')$$
$$+\int d\boldsymbol{r}_1\int d\boldsymbol{r}_2\int d\boldsymbol{r}_3\int\limits_{t'}^{t} d\tau g(\boldsymbol{r},t;\boldsymbol{r}_1,\tau)\Lambda(\boldsymbol{r}_1,\boldsymbol{r}_2,\boldsymbol{r}_3)\left\langle f(\boldsymbol{r}_2,\tau)G(\boldsymbol{r}_3,\tau;\boldsymbol{r}',t')\right\rangle. \quad (10.84)$$

Using the Furutsu–Novikov formula (10.10), page 307 to split the correlation in the right-hand side of Eq. (10.84), we obtain

$$\langle G(\boldsymbol{r},t;\boldsymbol{r}',t')\rangle = g(\boldsymbol{r},t;\boldsymbol{r}',t') + \int d\boldsymbol{r}_1\int d\boldsymbol{r}_2\int d\boldsymbol{r}_3\int\limits_{t'}^{t} d\tau g(\boldsymbol{r},t;\boldsymbol{r}_1,\tau)\Lambda(\boldsymbol{r}_1,\boldsymbol{r}_2,\boldsymbol{r}_3)$$
$$\times\int d\boldsymbol{r}_0\int\limits_{t'}^{\tau} dt_0 B(\boldsymbol{r}_2-\boldsymbol{r}_0,\tau-t_0)\left\langle\frac{\delta}{\delta f(\boldsymbol{r}_0,t_0)}G(\boldsymbol{r}_3,\tau;\boldsymbol{r}',t')\right\rangle. \quad (10.85)$$

If we use Eq. (10.82) for the variational derivative, we obtain that Eq. (10.85) assumes the form of the equality

$$\langle G(\boldsymbol{r},t;\boldsymbol{r}',t')\rangle = g(\boldsymbol{r},t;\boldsymbol{r}',t') + \int d\boldsymbol{r}_1\int d\boldsymbol{r}_2\int d\boldsymbol{r}_3\int\limits_{t'}^{t} d\tau g(\boldsymbol{r},t;\boldsymbol{r}_1,\tau)$$
$$\times\Lambda(\boldsymbol{r}_1,\boldsymbol{r}_2,\boldsymbol{r}_3)\int d\boldsymbol{r}_0\int\limits_{t'}^{\tau} dt_0 B(\boldsymbol{r}_2-\boldsymbol{r}_0,\tau-t_0)$$
$$\times\int d\boldsymbol{r}'\int d\boldsymbol{r}"\,\langle G(\boldsymbol{r}_3,\tau;\boldsymbol{r}',t_0)\Lambda(\boldsymbol{r}',\boldsymbol{r}_0,\boldsymbol{r}")G(\boldsymbol{r}",t_0;\boldsymbol{r}',t')\rangle. \quad (10.86)$$

Now, the correlation function of field $G(\boldsymbol{r},t;\boldsymbol{r}',t')$ appears in the right-hand side of Eq. (10.86).

If we tend the temporal correlation radius of random field $f(\boldsymbol{r},t)$ to zero, $\tau_0\to 0$, then Eq. (10.86) is simplified and assumes, for $t\gg\tau_0$, the form of the closed integral equation

$$\langle G(\boldsymbol{r},t;\boldsymbol{r}',t')\rangle = g(\boldsymbol{r},t;\boldsymbol{r}',t')\int d\boldsymbol{r}_1\int d\boldsymbol{r}_2\int d\boldsymbol{r}_3\int\limits_{t'}^{t} d\tau g(\boldsymbol{r},t;\boldsymbol{r}_1,\tau)\Lambda(\boldsymbol{r}_1,\boldsymbol{r}_2,\boldsymbol{r}_3)$$
$$\times\int d\boldsymbol{r}_0 F(\boldsymbol{r}_2-\boldsymbol{r}_0)\int d\boldsymbol{r}'\int d\boldsymbol{r}' g(\boldsymbol{r}_3,\tau;\boldsymbol{r}',\tau)\Lambda(\boldsymbol{r}',\boldsymbol{r}_0,\boldsymbol{r}')\,\langle G(\boldsymbol{r}',\tau;\boldsymbol{r}',t')\rangle,$$

where $F(\boldsymbol{r}) = \int_0^\infty dt B(\boldsymbol{r},t)$. This result is equivalent to the introduction of the effective correlation function of random field $f(\boldsymbol{r},t)$ in Eq. (10.85)

$$B(\boldsymbol{r},t) = 2F(\boldsymbol{r})\delta(t),\quad F(\boldsymbol{r}) = \int\limits_0^\infty dt B(\boldsymbol{r},t)$$

and the use of Eq. (10.83) instead of (10.82), which just corresponds to the delta-correlated approximation for random field $f(\boldsymbol{r}, t)$ in time.

The equation for the correlation function of solution to Eq. (10.81) can be derived in a similar way. For short, we illustrate this derivation by the simplest example of the one-dimensional causal equation ($t > t'$)

$$G(t;t') = g(t;t') + \Lambda \int_{t'}^{t} d\tau g(t;\tau) z(\tau) G(\tau;t'), \tag{10.87}$$

where we assume that $z(t)$ is the Gaussian delta-correlated random function with the parameters

$$\langle z(t) \rangle = 0, \quad \langle z(t) z(t') \rangle = 2D\delta(t - t') \quad (D = \sigma_z^2 \tau_0).$$

Averaging then Eq. (10.87) over an ensemble of realizations of random function $z(t)$, we obtain the equation

$$\langle G(t;t') \rangle = g(t;t') + \Lambda \int_{t'}^{t} d\tau g(t;\tau) \langle z(\tau) G(\tau;t') \rangle . \tag{10.88}$$

Taking into account Eq. (10.83) that assumes here the form

$$\frac{\delta}{\delta z(t)} G(t;t') = g(t;t) \Lambda G(t;t'), \tag{10.89}$$

we can rewrite the correlation in the right-hand side of Eq. (10.88) in the from

$$\langle z(\tau) G(\tau;t') \rangle = D \left\langle \frac{\delta}{\delta z(\tau)} G(\tau;t') \right\rangle = \Lambda D g(\tau;\tau) \langle G(\tau;t') \rangle .$$

As a consequence, Eq. (10.88) grades into the closed integral equation for average Green's function

$$\langle G(t;t') \rangle = g(t;t') + \Lambda^2 D \int_{t'}^{t} d\tau g(t;\tau) g(\tau;\tau) \langle G(\tau;t') \rangle , \tag{10.90}$$

which, according to the general-form derivation technique, has the form of the *Dyson equation*

$$\langle G(t;t')\rangle = g(t;t') + \Lambda \int\limits_{t'}^{t} d\tau g(t;\tau) \int\limits_{t'}^{\tau} d\tau' Q(\tau;\tau') \langle G(\tau';t')\rangle ,$$
$$\langle G(t;t')\rangle = g(t;t') + \Lambda \int\limits_{t'}^{t} d\tau \langle G(t;\tau)\rangle \int\limits_{t'}^{\tau} d\tau' Q(\tau;\tau') g(\tau';t'), \quad (10.91)$$

with the *mass function*

$$Q(\tau;\tau') = \Lambda^2 D g(\tau;\tau)\delta(\tau-\tau'). \quad (10.92)$$

Derive now the equation for the correlation function

$$\Gamma(t,t';t_1,t_1') = \langle G(t;t')G^*(t_1;t_1')\rangle \quad (t>t', \quad t_1>t_1'),$$

where $G^*(t;t')$ is complex conjugated Green's function. With this goal in view, we multiply Eq. (10.87) by $G^*(t_1;t_1')$ and average the result over an ensemble of realizations of random function $z(t)$. The result is the equation that can be symbolically represented as

$$\Gamma = g\langle G^*\rangle + \Lambda g\langle zGG^*\rangle . \quad (10.93)$$

Taking into account the Dyson equation (10.91)

$$\langle G\rangle = \{1 + \langle G\rangle Q\}g,$$

we apply operator $\{1 + \langle G\rangle Q\}$ to Eq. (10.93). As a result, we obtain the symbolic-form equation

$$\Gamma = \langle G\rangle\langle G^*\rangle + \langle G\rangle \Lambda \{\langle zGG^*\rangle - Q\Gamma\},$$

which can be represented in common variables in the form

$$\Gamma(t,t';t_1,t_1') = \langle G(t;t')\rangle \langle G^*(t_1;t_1')\rangle$$
$$+ \Lambda D \int\limits_0^t d\tau \langle G(t;\tau)\rangle \left[\left\langle \frac{\delta G(\tau;t')}{\delta z(\tau)} G^*(t_1;t_1') + 2G(\tau;t')\frac{\delta G^*(t_1;t_1')}{\delta z(\tau)}\right\rangle\right]$$
$$- \Lambda^2 D \int\limits_0^t d\tau \langle G(t;\tau)\rangle g(\tau;\tau)\Gamma(\tau,t';t_1,t_1'). \quad (10.94)$$

Deriving Eq. (10.94), we used additionally Eq. (7.61), page 182 for splitting correlators between the Gaussian delta-correlated process $z(t)$ and functionals of this process

$$
\langle z(t')R[t;z(\tau)]\rangle = \begin{cases} D\left\langle \dfrac{\delta}{\delta z(t)} R[t;z(\tau)] \right\rangle & (t'=t, \quad \tau<t), \\ 2D\left\langle \dfrac{\delta}{\delta z(t')} R[t;z(\tau)] \right\rangle & (t'<t, \quad \tau<t). \end{cases}
$$

Taking into account formulas (10.89) and (10.79) of which the latter assumes in our case the form

$$
\frac{\delta}{\delta z(\tau)} G^*(t_1;t_1') = \Lambda G^*(t_1;\tau) G^*(\tau;t_1'),
$$

we can rewrite Eq. (10.94) as

$$
\Gamma(t,t';t_1,t_1') = \langle G(t;t')\rangle \langle G^*(t_1;t_1')\rangle + 2|\Lambda|^2 D \int_0^t d\tau \, \langle G(t;\tau)\rangle \langle G^*(t_1;\tau) G(\tau;t') G^*(\tau;t_1')\rangle . \quad (10.95)
$$

Now, we take into account the fact that function $G^*(t_1;\tau)$ functionally depends on random process $z(\widetilde{\tau})$ for $\widetilde{\tau} \geq \tau$ while functions $G(\tau;t')$ and $G^*(\tau;t_1')$ depend on it for $\widetilde{\tau} \leq \tau$. Consequently, these functions are statistically independent in the case of the delta-correlated process $z(\widetilde{\tau})$, and we can rewrite Eq. (10.95) in the form of the closed equation $(t_1 \geq t)$

$$
\Gamma(t,t';t_1,t_1') = \langle G(t;t')\rangle \langle G^*(t_1;t_1')\rangle + 2|\Lambda|^2 D \int_0^t d\tau \, \langle G(t;\tau)\rangle \langle G^*(t_1;\tau)\rangle \, \Gamma(\tau;t';\tau;t_1'), \quad (10.96)
$$

which corresponds to the *Bete–Salpeter equation* (8.55), page 211 with the *intensity operator kernel*

$$
K(\tau_1,\tau';\tau_2,\tau'') = 2|\Lambda|^2 D \delta(\tau_1-\tau')\delta(\tau_2-\tau'')\delta(\tau_1-\tau_2). \quad (10.97)
$$

Thus, for the one-dimensional causal equation (10.87), the *ladder approximation* appears the exact equality in the case of the delta-correlated process $z(t)$.

Chapter 11
Methods for Solving and Analyzing the Fokker-Planck Equation

The Fokker–Planck equations for the one-time probability density (10.11), page 307 and for the transitional probability density (10.16), page 309 are the partial differential equations of parabolic type, so that we can use methods of the theory of mathematical physics equations to solve them. In this context, the basic methods are such as the method of separation of variables, the Fourier transformation with respect to spatial coordinates, and other integral transformations.

11.1 Integral Transformations

Integral transformations are very practicable for solving the Fokker–Planck equation. Indeed, earlier we mentioned the convenience of the Fourier transformation in Eq. (10.11), page (307) if the tensor of diffusion coefficients $F_{kl}(\boldsymbol{x}, \boldsymbol{x}; t)$ is independent of $\boldsymbol{x}$. Different integral transformations related to eigenfunctions of the diffusion operator

$$\widehat{L} = \frac{\partial^2}{\partial x_k \partial x_l} F_{kl}(\boldsymbol{x}, \boldsymbol{x}; t)$$

can be used in other situations.

For example, in the case of the Legendre operator

$$\widehat{L} = \frac{\partial}{\partial x}(x^2 - 1)\frac{\partial}{\partial x},$$

it is quite natural to use the integral transformation related to the *Legendre functions*. This transformation is called the *Meler–Fock* transform (see, e.g., [12]) and is defined by the formula

V.I. Klyatskin, *Stochastic Equations: Theory and Applications in Acoustics, Hydrodyn., Magnetohydrodyn., and Radiophys., Vol. 1*, Understanding Complex Systems,
DOI: 10.1007/978-3-319-07587-7_11,

$$F(\mu) = \int_1^\infty dx f(x) P_{-1/2+i\mu}(x) \quad (\mu > 0), \tag{11.1}$$

where $P_{-1/2+i\mu}(x)$ is the *complex index Legendre function of the first kind*, which satisfies the equation

$$\frac{d}{dx}(x^2-1)\frac{d}{dx}P_{-1/2+i\mu}(x) = -\left(\mu^2+\frac{1}{4}\right)P_{-1/2+i\mu}(x). \tag{11.2}$$

The inversion of the transform (11.1) has the form

$$f(x) = \int_0^\infty d\mu\, \mu \tanh(\pi\mu) F(\mu) P_{-1/2+i\mu}(x) \quad (1 \le x \le \infty), \tag{11.3}$$

where $F(\mu)$ is given by formula (11.1).

Another integral transformation called the *Kontorovich–Lebedev* transform (see, e.g., [12]), is related to diffusion operator

$$\widehat{L} = \frac{\partial}{\partial x} x^2 \frac{\partial}{\partial x}$$

and has the form

$$F(\tau) = \int_0^\infty dx f(x) K_{i\tau}(x) \quad (\tau > 0), \tag{11.4}$$

where $K_{i\tau}(x)$ is the *imaginary index McDonalds function of the first kind*, which satisfies the equations

$$\begin{aligned}\left(x^2\frac{d^2}{dx^2} + x\frac{d}{dx} - x^2 + \tau^2\right) K_{i\tau}(x) &= 0,\\ \left(\frac{d}{dx}x^2\frac{d}{dx} - x\frac{d}{dx}\right) K_{i\tau}(x) &= \left(x^2 - \tau^2\right) K_{i\tau}(x).\end{aligned} \tag{11.5}$$

The corresponding inversion has the form

$$f(x) = \frac{2}{\pi^2 x} \int_0^\infty d\tau \sinh(\pi\tau) F(\tau) K_{i\tau}(x). \tag{11.6}$$

As a concrete example, consider the Fokker–Planck equation ($x \ge 1$)

$$\frac{\partial}{\partial t}p(x,t|x_0,t_0) = D\frac{\partial}{\partial x}(x^2-1)\frac{\partial}{\partial x}p(x,t|x_0,t_0),$$
$$p(x,t_0|x_0,t_0) = \delta(x-x_0). \tag{11.7}$$

Multiplying Eq. (11.7) by $P_{-1/2+i\mu}(x)$, integrating the result over x from 1 to ∞, and introducing function

$$p(t,\mu) = \int\limits_1^\infty dx p(x,t|x_0,t_0)P_{-1/2+i\mu}(x),$$

we obtain the equation

$$\frac{\partial}{\partial t}p(t,\mu) = D\int\limits_1^\infty dx P_{-1/2+i\mu}(x)\frac{\partial}{\partial x}(x^2-1)\frac{\partial}{\partial x}p(x,t|x_0,t_0) \tag{11.8}$$

with the initial value

$$p(t_0,\mu) = P_{-1/2+i\mu}(x_0). \tag{11.9}$$

Integrating two times by parts in the right-hand side of Eq. (11.8) and using the differential Legendre equation for function $P_{-1/2+i\mu}(x)$ (11.2), we obtain the ordinary differential equation in $p(t,\mu)$

$$\frac{d}{dt}p(t,\mu) = -D\left(\mu^2+\frac{1}{4}\right)p(t,\mu),$$

whose solution satisfying the initial value (11.9) has the form

$$p(t,\mu) = P_{-1/2+i\mu}(x_0)e^{-D\left(\mu^2+\frac{1}{4}\right)(t-t_0)}.$$

Using now inversion (11.3), we obtain the solution to Eq. (11.7) in terms of the Meler–Fock integral

$$p(x,t|x_0,t_0) = \int\limits_0^\infty d\mu\,\mu\tanh(\pi\mu)e^{-D\left(\mu^2+\frac{1}{4}\right)(t-t_0)}P_{-1/2+i\mu}(x)P_{-1/2+i\mu}(x_0). \tag{11.10}$$

If $x_0 = 1$ at the initial instant $t_0 = 0$, we obtain the expression

$$P(x,t) = \int\limits_0^\infty d\mu\,\mu\tanh(\pi\mu)e^{-D\left(\mu^2+\frac{1}{4}\right)(t-t_0)}P_{-1/2+i\mu}(x) \tag{11.11}$$

corresponding to the solution of the Fokker–Planck equation for the one-time probability density (11.7) with the initial value

$$P(x,0) = \delta(x-1).$$

11.2 Steady-State Solutions of the Fokker–Planck Equation

In previous sections, we discussed the general methods of solving the Fokker–Planck equation for both transition and one-time probability densities. However, the problem on the one-time probability density can have peculiarities related to possible existence of the steady-state solution; in a number of cases, such a solution can be obtained immediately. The steady-state solution, if it exists, is independent of initial values and is the solution of the Fokker–Planck equation in the limit $t \to \infty$.

There are two classes of problems for which the steady-state solution of the Fokker–Planck equation can be easily found. These classes deal with one-dimensional differential equations and with the Hamiltonian systems of equations. Consider them in greater detail.

11.2.1 One-Dimensional Nonlinear Differential Equation

The one-dimensional nonlinear systems are described by the stochastic equation

$$\frac{d}{dt}x(t) = f(x) + z(t)g(x), \quad x(0) = x_0, \tag{11.12}$$

where $z(t)$ is, as earlier, the Gaussian delta-correlated process with the parameters

$$\langle z(t)\rangle = 0, \quad \langle z(t)z(t')\rangle = 2D\delta(t-t') \quad (D = \sigma_z^2\tau_0).$$

The corresponding Fokker–Planck equation has the form

$$\left(\frac{\partial}{\partial t} + \frac{\partial}{\partial x}f(x)\right)P(x,t) = D\frac{\partial}{\partial x}g(x)\frac{\partial}{\partial x}g(x)P(x,t). \tag{11.13}$$

The steady-state probability distribution $P(x)$, if it exists, satisfies the equation

$$f(x)P(x) = Dg(x)\frac{d}{dx}g(x)P(x) \tag{11.14}$$

(we assume that $P(x)$ is distributed over the whole space, i.e., for $-\infty < x < \infty$) whose solution is as follows

$$P(x) = \frac{C}{|g(x)|}\exp\left\{\frac{1}{D}\int dx\frac{f(x)}{g^2(x)}\right\}, \tag{11.15}$$

where constant C is determined from the normalization condition

$$\int_{-\infty}^{\infty} dx P(x) = 1.$$

In the special case of the Langevin equation (10.32), page 315

$$f(x) = -\lambda x,\ g(x) = 1,$$

Eq. (11.15) grades into the Gaussian probability distribution

$$P(x) = \sqrt{\frac{\lambda}{2\pi D}} \exp\left\{-\frac{\lambda}{2D}x^2\right\}. \tag{11.16}$$

11.2.2 Hamiltonian Systems

Another type of dynamic systems that allow obtaining the steady-state probability distribution is described by the Hamiltonian system with linear friction and external random force

$$\begin{aligned} \frac{d}{dt}\boldsymbol{r}_i(t) &= \frac{\partial}{\partial \boldsymbol{p}_i} H(\{\boldsymbol{r}_i\}, \{\boldsymbol{p}_i\}), \\ \frac{d}{dt}\boldsymbol{p}_i(t) &= -\frac{\partial}{\partial \boldsymbol{r}_i} H(\{\boldsymbol{r}_i\}, \{\boldsymbol{p}_i\}) - \lambda \boldsymbol{p}_i + \boldsymbol{f}_i(t), \end{aligned} \tag{11.17}$$

where $i = 1, 2, \cdots, N$,

$$H(\{\boldsymbol{r}_i\}, \{\boldsymbol{p}_i\}) = \frac{\boldsymbol{p}_i^2}{2} + U(\boldsymbol{r}_1, \cdots, \boldsymbol{r}_N)$$

is the Hamiltonian; $\{r_i\}, \{p_i\}$ denotes collections of all quantities $r(t)$ and $p(t)$, i.e., $\{r_i\} = \{r_1, \cdots, r_N\}$, $\{p_i\} = \{p_1, \cdots, p_N\}$; λ is a constant coefficient (friction), and random forces $f_i(t)$ are the Gaussian delta-correlated random vector functions with the correlation tensor

$$\left\langle f_i^{\alpha}(t) f_j^{\beta}(t')\right\rangle = 2D\delta_{ij}\delta_{\alpha\beta}\delta(t-t'), \quad D = \sigma_f^2 \tau_0. \tag{11.18}$$

Here, α and β are the vector indices.

System of equations (11.17) describes the Brownian motion of a system of N interacting particles. The corresponding indicator function $\varphi(\{r_i\}, \{p_i\}, t)$ satisfies the equation

$$\frac{\partial}{\partial t}\varphi(\{\boldsymbol{r}_i\}, \{\boldsymbol{p}_i\}, t) + \sum_{k=1}^{N} \{H, \varphi\}_{(k)} - \lambda \sum_{k=1}^{N} \frac{\partial}{\partial \boldsymbol{p}_k}\{\boldsymbol{p}_k \varphi\} = \sum_{k=1}^{N} \frac{\partial}{\partial \boldsymbol{p}_k}\{\boldsymbol{f}_k(t)\varphi\},$$

which is an extension of Eq. (3.12), page 98 and, hence, the Fokker–Planck equation for the joint probability density of the solution to system (11.17) has the form

$$\frac{\partial}{\partial t}P(\{\boldsymbol{r}_i\},\{\boldsymbol{p}_i\},t)+\sum_{k=1}^{N}\{H,P(\{\boldsymbol{r}_i\},\{\boldsymbol{p}_i\},t)\}_{(k)}$$
$$-\lambda\sum_{k=1}^{N}\frac{\partial}{\partial \boldsymbol{p}_k}\{\boldsymbol{p}_k P(\{\boldsymbol{r}_i\},\{\boldsymbol{p}_i\},t)\}=D\sum_{k=1}^{N}\frac{\partial^2}{\partial \boldsymbol{p}_k^2}P(\{\boldsymbol{r}_i\},\{\boldsymbol{p}_i\},t), \quad (11.19)$$

where

$$\{\varphi,\psi\}_{(k)}=\frac{\partial\varphi}{\partial \boldsymbol{p}_k}\frac{\partial\psi}{\partial \boldsymbol{r}_k}-\frac{\partial\psi}{\partial \boldsymbol{p}_k}\frac{\partial\varphi}{\partial \boldsymbol{r}_k}$$

is the Poisson bracket for the k-th particle.

One can easily check that the steady-state solution to Eq. (11.19) is the *canonical Gibbs distribution*

$$P(\{\boldsymbol{r}_i\},\{\boldsymbol{p}_i\})=C\exp\left\{-\frac{\lambda}{D}H(\{\boldsymbol{r}_i\},\{\boldsymbol{p}_i\})\right\}. \quad (11.20)$$

The specificity of this distribution consists in the Gaussian behavior with respect to momenta and statistical independence of particle coordinates and momenta.

Integrating Eq. (11.20) over all $\boldsymbol{r}$, we can obtain the *Maxwell distribution* that describes velocity fluctuations of the Brownian particles. The case $U(\boldsymbol{r}_1,\cdots,\boldsymbol{r}_N)=0$ corresponds to the Brownian motion of a system of free particles (11.16).

If we integrate probability distribution (11.20) over momenta (velocities), we obtain the *Boltzmann distribution* of particle coordinates

$$P(\{\boldsymbol{r}_i\})=C\exp\left\{-\frac{\lambda}{D}U(\{\boldsymbol{r}_i\})\right\}. \quad (11.21)$$

In the case of sufficiently strong friction, the equilibrium distribution (11.20) is formed in two stages. First, the Gaussian momentum distribution (the Maxwell distribution) is formed relatively quickly and then, the spatial distribution (the Boltzmann distribution) is formed at much slower rate. The latter stage is described by the Fokker–Planck equation

$$\frac{\partial}{\partial t}P(\{\boldsymbol{r}_i\},t)=\frac{1}{\lambda}\sum_{k=1}^{N}\frac{\partial}{\partial \boldsymbol{r}_k}\left[\left(\frac{\partial U(\{\boldsymbol{r}_i\})}{\partial \boldsymbol{r}_k}+\frac{D}{\lambda}\frac{\partial}{\partial \boldsymbol{r}_k}\right)P(\{\boldsymbol{r}_i\},t)\right], \quad (11.22)$$

which is usually called the *Einstein–Smolukhovsky equation* . Derivation of Eq. (11.22) from the Fokker–Planck equation (11.19) is called the *Kramers problem* (see, e.g., [131] and the corresponding discussion in Sect. 5.4.1, where dynamics of particles under a random force is considered as an example). Note that Eq. (11.22) statistically corresponds to the stochastic equation

$$\frac{d}{dt}\boldsymbol{r}_i(t) = -\frac{1}{\lambda}\frac{\partial}{\partial \boldsymbol{r}_i}U(\{\boldsymbol{r}_i\}) + \frac{1}{\lambda}\boldsymbol{f}_i(t), \tag{11.23}$$

which, nevertheless, cannot be considered as the limit of Eq. (11.17) for $\lambda \to \infty$.

In the one-dimensional case, Eqs. (11.17) are simplified and assume the form of the system of two equations

$$\frac{d}{dt}x(t) = y(t), \quad \frac{d}{dt}y(t) = -\frac{\partial}{\partial x}U(x) - \lambda y(t) + f(t). \tag{11.24}$$

The corresponding steady-state probability distribution has the form

$$P(x,y) = C\exp\left\{-\frac{\lambda}{D}H(x,y)\right\}, \quad H(x,y) = \frac{y^2}{2} + U(x). \tag{11.25}$$

11.2.3 Systems of Hydrodynamic Type

In Sect. 1.1.3, page 13, we considered general dynamics of simplest hydrodynamic-type systems (HTS). Now, we consider these systems in terms of the statistical description.

Hydrodynamic-type systems with linear friction are described by the dynamic equations

$$\frac{d}{dt}v_i(t) = F_i(\boldsymbol{v}) - \lambda^{(i)}v_i(t) \qquad (i = 1, \cdots, N), \tag{11.26}$$

where $\lambda^{(i)}$ is the friction coefficient of the i-th component of the N-dimensional vector $\boldsymbol{v}$ [1] and $F_i(\boldsymbol{v})$ is the function quadratic in $\boldsymbol{v}$ and having the following properties:

(a) $v_iF_i(\boldsymbol{v}) = 0$,

energy conservation holds at $\lambda^{(i)} = 0$: $\dfrac{d}{dt}E(t) = 0$, $E(t) = \dfrac{\boldsymbol{v}_i^2(t)}{2}$;

(b) $\dfrac{\partial}{\partial v_i}F_i(\boldsymbol{v}) = 0$,

[1] In the general case, the dissipative term in Eq. (11.26) has the form $\lambda_{ik}v_k$. However, we can always choose the coordinate system in which two positively defined quadratic forms—energy $E = v_i^2/2$ and dissipation $\lambda_{ik}v_iv_k$—have the diagonal representation. The form of Eq. (11.26) assumes the use of namely such a coordinate system.

conditions of the Liouville theorem are satisfied at $\lambda^{(i)} = 0$, and this equality is the equation of incompressibility in the phase space.

Equilibrium Thermal Fluctuations in the Hydrodynamic-Type Systems

Here, we dwell on a class of phenomena closely related to the Brownian motion; namely, we dwell on equilibrium thermal fluctuations in solids.

Microscopic equations describe the behavior of physical systems only in terms of spatial scales large in comparison with the molecule free path in the medium and temporal scales great in comparison with the time between molecule collisions. This means that macroscopic equations adequately describe the behavior of systems only on average. However, in view of molecule thermal motion, macroscopic variables are in general terms stochastic variables, and a complete macroscopic theory must describe not only system behavior on average, but also the fluctuations around the average.

Such a description can be performed in terms of macroscopic variables by supplementing the corresponding macroscopic equations with the 'external forces' specified as the Gaussian random fields delta-correlated in time (this approach is closely related to the *fluctuation–dissipation theorem*). See [81, 84, 87] for the corresponding correlation theories of equilibrium thermal fluctuations in electrodynamics, hydrodynamics, and viscoelastic media.

In HTS, equilibrium thermal fluctuations are described by Eq. (11.26) supplemented with external forces $f_i(t)$

$$\frac{d}{dt}v_i(t) = F_i(\boldsymbol{v}) - \lambda^{(i)}v_i(t) + f_i(t) \qquad (i = 1, \cdots, N). \tag{11.27}$$

External forces are assumed to be the Gaussian random functions delta-correlated in time with the correlation tensor of the form

$$\langle f_i(t)f_j(t')\rangle = 2\delta_{ij}\sigma^2_{(i)}\delta(t-t') \qquad (\langle \boldsymbol{f}(t)\rangle = 0). \tag{11.28}$$

Note that one can consider Eqs. (11.27) as the system describing the Brownian motion in HTS. Such a treatment assumes that coefficients $\lambda^{(i)}$ are certain effective friction coefficients. For example, at $N = 3$, system (11.27) describes (in the velocity space) the rotary Brownian motion of a solid in a medium, and quantities $\lambda^{(i)}v_i(t)$ play the role of the corresponding resistance forces. The probability density of solution $\boldsymbol{v}(t)$ to Eqs. (11.27), i.e., function $P(\boldsymbol{v},t) = \langle\delta(\boldsymbol{v}(t)-\boldsymbol{v})\rangle$, satisfies the Fokker–Planck equation

$$\frac{\partial}{\partial t}P(\boldsymbol{v},t) = -\frac{\partial}{\partial v_i}\left\{[F_i(\boldsymbol{v}) - \lambda^{(i)}v_i]P(\boldsymbol{v},t)\right\} + \sigma^2_{(i)}\frac{\partial^2}{\partial v_i^2}P(\boldsymbol{v},t). \tag{11.29}$$

The steady-state, initial data-independent solution to Eq. (11.29) must behave like the Maxwell distribution that corresponds to the uniform distribution of energy over the degrees of freedom

$$P(\boldsymbol{v}) = C \exp\left\{-\frac{v_i^2}{2kT}\right\}, \tag{11.30}$$

where k is the Boltzmann constant and T is the equilibrium temperature in the system.

Substituting Eq. (11.30) in Eq. (11.29), we obtain the expression for $\sigma^2_{(i)}$

$$\sigma^2_{(i)} = \lambda^{(i)} kT \tag{11.31}$$

called the *Einstein formula*. Here, we used that, in view of conditions (a) and (b),

$$\frac{\partial}{\partial v_i}\left\{F_i(\boldsymbol{v})P(\boldsymbol{v})\right\} = 0.$$

Thus, Eqs. (11.28) and (11.31) completely determine statistics of external forces, and the nonlinear term in Eq. (11.27) plays no role for one-time fluctuations of $\boldsymbol{v}(t)$. Namely this feature constitutes the subject matter of the fluctuation–dissipation theorem (see [84, 87]) in the context of HTS.

Note that Eqs. (11.27) with the correlation relationships (11.28) describe not only equilibrium thermal fluctuations in HTS, but also the interaction of small-scale motions (such as microturbulence, for example) with the motions of larger scales. If such an HTS allows the description in terms of phenomenological equations (11.27) with $\lambda^{(i)} = \lambda = \text{const}$, $\sigma^2_{(i)} = \sigma^2 = \text{const}$, the steady-state probability distribution of $\boldsymbol{v}$ will have the form similar to distribution (11.30):

$$P(\boldsymbol{v}) = C \exp\left\{-\frac{\lambda}{2\sigma^2} v_i^2\right\}. \tag{11.32}$$

Noises in Hydrodynamic-Type Systems under the Action of a Regular Force

We illustrate statistical description of HTS by the example of the simplest system with three degrees of freedom (S_3), which is described by system of equations (1.25), page 15:

$$\begin{aligned}
\frac{d}{dt}v_0(t) &= v_2^2(t) - v_1^2(t) - v_0(t) + R + f_0(t),\\
\frac{d}{dt}v_1(t) &= v_0(t)v_1(t) - v_1(t) + f_1(t),\\
\frac{d}{dt}v_2(t) &= -v_0(t)v_2(t) - v_2(t) + f_2(t).
\end{aligned} \tag{11.33}$$

We assume that $f_i(t)$ are the Gaussian random forces delta-correlated in time with the correlation tensor

$$\langle f_i(t) f_j(t')\rangle = 2\delta_{ij}\sigma^2\delta(\tau).$$

In the absence of external forces, this system is equivalent to the Euler equations of the dynamics of a gyroscope with isotropic friction, which is excited by a constant moment of force relative to the unstable axis. The steady-state solution to this system depends on parameter R (an analog to the Reynolds number), and the critical value of this parameter is $R_{\text{cr}} = 1$.

For $R < 1$, it has the stable steady-state solution

$$v_{1\,\text{s-s}} = v_{\text{s-s}} = 0, \quad v_{0\,\text{s-s}} = R. \tag{11.34}$$

For $R > 1$, this solution becomes unstable with respect to small disturbances of parameters, and new steady-state regime is formed:

$$v_{0\,\text{s-s}} = 1, \quad v_{2\,\text{s-s}} = 0, \quad v_{1\,\text{s-s}} = \pm\sqrt{R-1}. \tag{11.35}$$

Here, we have an element of randomness because quantity $v_{1\,\text{s-s}}$ can be either positive or negative, depending on the amplitude of small disturbances.

As was mentioned earlier, the steady-state probability distribution at $R = 0$ has the form

$$P(\boldsymbol{v}) = C\,\exp\left\{-\frac{v_i^2}{2\sigma^2}\right\}. \tag{11.36}$$

Let now $R \neq 0$ and $R < 1$. In this case, we can easily obtain that fluctuations of components relative to their steady-state values ($\widetilde{v}_i = v_i - v_{i\,\text{s-s}}$) satisfy the system of equations

$$\begin{aligned}
\frac{d}{dt}\widetilde{v}_0(t) &= \widetilde{v}_2^2(t) - \widetilde{v}_1^2(t) - \widetilde{v}_0(t) + f_0(t),\\
\frac{d}{dt}\widetilde{v}_1(t) &= \widetilde{v}_0(t)\widetilde{v}_1(t) - (1-R)\widetilde{v}_1(t) + f_1(t),\\
\frac{d}{dt}\widetilde{v}_2(t) &= -\widetilde{v}_0(t)\widetilde{v}_2(t) - (1+R)\widetilde{v}_2(t) + f_2(t).
\end{aligned} \tag{11.37}$$

Statistical characteristics of the solution to system (11.37) can be determined using the perturbation theory with respect to small parameter σ^2. The second moments of components $\widetilde{v}_i$ will then be described by the linearized system of equations (11.37); the second moment being determined, mean values can be obtained by averaging system (11.37) directly. The corresponding steady-state variances of fluctuations $\widetilde{v}_i$ have the form [14, 24, 46]

$$\langle \widetilde{v}_0^2(t)\rangle = \sigma^2, \quad \langle \widetilde{v}_1^2(t)\rangle = \frac{\sigma^2}{1-R}, \quad \langle \widetilde{v}_2^2(t)\rangle = \frac{\sigma^2}{1+R}. \tag{11.38}$$

Note that expressions (11.38) hold for $R \ll 1$. With increasing R, both intensity and temporal correlation radius of component $\widetilde{v}_1$ increase, while the intensity of quantity $\widetilde{v}_2$ decreases. In this process, maximum fluctuations of quantity $\widetilde{v}_1$ occur when the dynamic system gets over the critical regime.

Consider now the case of $R > 1$. The steady-state probability distribution of component v_1 has two maxima near $v_1 = \pm\sqrt{R-1}$ (they correspond to the stable steady-state positions) and a minimum $v_1 = 0$ that corresponds to the unstable position. This probability distribution is formed by ensemble averaging over realizations of random forces $f_i(t)$. In a single realization, the system arrives with a probability of 1/2 at one of the stable positions corresponding to distribution maxima. In this case, averaging over time or ensemble averaging over the forces that bring the system to this state will form the probability distribution near the maximum, and we can determine statistical characteristics of the solution using the perturbation theory in parameter σ^2.

Let the system arrives at the state corresponding to the stable steady-state position $v_{2\,\text{s-s}} = 0$, $v_{1\,\text{s-s}} = \sqrt{R-1}$. Then, fluctuations relative to this state will satisfy the system of equations

$$\begin{aligned}
\frac{d}{dt}\widetilde{v}_0(t) &= \widetilde{v}_2^2(t) - \widetilde{v}_1^2(t) - 2\sqrt{R-1}\widetilde{v}_1(t) - \widetilde{v}_0(t) + f_0(t),\\
\frac{d}{dt}\widetilde{v}_1(t) &= \widetilde{v}_0(t)\widetilde{v}_1(t) + \sqrt{R-1}\widetilde{v}_0(t) + f_1(t),\\
\frac{d}{dt}\widetilde{v}_2(t) &= -\widetilde{v}_0(t)\widetilde{v}_2(t) - 2\widetilde{v}_2(t) + f_2(t).
\end{aligned} \tag{11.39}$$

For $R \gg 1$, the second moments of fluctuations $\widetilde{v}_i$ are obtained from the linearized system (11.39) and mean values are then determined by direct averaging of Eqs. (11.39). Using this procedure, we obtain that, for $R \gg 1$, the steady-state variances and correlation coefficients of triplet components have the form

$$\begin{aligned}
&\langle\widetilde{v}_0(t)\widetilde{v}_2(t)\rangle = \langle\widetilde{v}_1(t)\widetilde{v}_2(t)\rangle = 0 \quad \langle\widetilde{v}_0(t)\widetilde{v}_1(t)\rangle = -\frac{\sigma^2}{\sqrt{R}}\\
&\langle\widetilde{v}_0^2(t)\rangle = 3\sigma^2, \quad \langle\widetilde{v}_1^2(t)\rangle = \frac{3}{2}\sigma^2, \quad \langle\widetilde{v}_2^2(t)\rangle = \frac{\sigma^2}{2}.
\end{aligned} \tag{11.40}$$

As we mentioned earlier, these statistical characteristics correspond either to averaging over time, or to ensemble averaging over realizations of forces $f_i(t)$ that bring the system to the steady-state position. It should be noted that if the system has arrived at one of the most probable states under the action of certain realization of forces $f_i(t)$, it will be transferred into another most probable state after the lapse of certain time T (the greater T, the less σ^2 and the greater R) in view of availability of sufficiently great values of $f_i(t)$. This process simulated for a realization of random forces is shown in Fig. 1.5 on page 16.

Consider now fluctuations of components $v_i(t)$ at critical regime (i.e., at $R = 1$). Equations for the fluctuations relative to the state $v_{1\,\mathrm{CT}} = v_{2\,\mathrm{CT}} = 0$, $v_{0\,\mathrm{CT}} = R$ can be obtained by setting $R = 1$ either in Eq. (11.37), or in Eq. (11.39). In this case, fluctuations of component $\widetilde{v}_2(t)$ can be described in terms of the linearized equation ($\sigma^2 \ll 1$), and the corresponding steady-state variance is given by the formula

$$\left\langle \widetilde{v}_2^2(t) \right\rangle = \frac{\sigma^2}{2}.$$

Components $\widetilde{v}_0(t)$ and $\widetilde{v}_1(t)$ are described by the nonlinear system of equations

$$\begin{aligned} \frac{d}{dt}\widetilde{v}_0(t) &= -\widetilde{v}_1^2(t) - \widetilde{v}_0(t) + f_0(t), \\ \frac{d}{dt}\widetilde{v}_1(t) &= \widetilde{v}_0(t)\widetilde{v}_1(t) + f_1(t), \end{aligned} \tag{11.41}$$

and no linearization is allowed here.

Averaging system (11.41), we obtain the following relationships for steady-state values of fluctuations

$$\langle \widetilde{v}_0(t)\widetilde{v}_1(t) \rangle = 0, \quad \langle \widetilde{v}_0(t) \rangle = -\left\langle \widetilde{v}_1^2(t) \right\rangle.$$

As we noted earlier, intensity of fluctuations $\widetilde{v}_1$ increases in the critical regime and, consequently, quantity $\langle \widetilde{v}_0(t) \rangle$ increases. At the same time, the variance of fluctuation $\widetilde{v}_0(t)$ remains nearly intact. Indeed, as follows from Eqs. (11.41), the variance of steady-state fluctuation $\widetilde{v}_0(t)$ is given by the formula

$$\left\langle \widetilde{v}_0^2(t) \right\rangle = 2\sigma^2.$$

It becomes clear that fluctuations $\widetilde{v}_0(t)$ and $\widetilde{v}_1(t)$ can be estimated from the simplified system

$$\begin{aligned} \frac{d}{dt}\widetilde{v}_0(t) &= -\widetilde{v}_1^2(t) - \widetilde{v}_0(t) + f_0(t), \\ \frac{d}{dt}\widetilde{v}_1(t) &= \langle \widetilde{v}_0(t) \rangle \widetilde{v}_1(t) + f_1(t) \end{aligned}$$

obtained by the replacement of quantity $\widetilde{v}_0(t)$ in the second equation of system (11.41) by $\langle \widetilde{v}_0(t) \rangle$.

Using this system, we obtain that steady-state fluctuations $\widetilde{v}_0(t)$ and $\widetilde{v}_1(t)$ satisfy the expressions

$$\left\langle \widetilde{v}_1^2(t) \right\rangle = -\langle \widetilde{v}_0(t) \rangle = \sigma.$$

In a similar way, we obtain the estimate of the temporal correlation radius of quantity $\widetilde{v}_1(t)$:

$$\tau \sim 1/\sigma.$$

Earlier, we considered the behavior of a triplet under the assumption that random forces act on all variables. Let us see what will be changed if only one unstable component $v_0(t)$ is affected by the random force (we assume that the force acting on component $v_0(t)$ has regular (R) and random components). In this case, the dynamic system assumes the form ($R > 1$)

$$\begin{aligned} \frac{d}{dt}v_0(t) &= -v_1^2(t) - v_0(t) + R + f(t), \\ \frac{d}{dt}v_1(t) &= v_0(t)v_1(t) - v_1(t). \end{aligned} \tag{11.42}$$

In system (11.42), we omitted the terms related to component $v_2(t)$, because one can easily see that it will not be excited in the problem at hand.

Represent component $v_0(t)$ as $v_0(t) = 1 + \widetilde{v}_0(t)$. The system of equations (11.42) assumes then the form

$$\begin{aligned} \frac{d}{dt}\widetilde{v}_0(t) &= -v_1^2(t) - \widetilde{v}_0(t) + (R-1) + f(t), \\ \frac{d}{dt}v_1(t) &= \widetilde{v}_0(t)v_1(t), \end{aligned} \tag{11.43}$$

from which follows that temporal evolution of component $v_1(t)$ depends on its initial value.

If $v_1(0) > 0$, then $v_1(t) > 0$ too. In this case, we can represent $v_1(t)$ in the form

$$v_1(t) = e^{\varphi(t)}$$

and rewrite the system of equations (11.43) in the Hamiltonian form (11.24)

$$\frac{d}{dt}\widetilde{v}_0(t) = -\frac{\partial U(\varphi)}{\partial \varphi} - \widetilde{v}_0(t) + f(t), \quad \frac{d}{dt}\varphi(t) = \widetilde{v}_0(t), \tag{11.44}$$

where $U(\varphi) = \frac{1}{2}e^{2\varphi} - (R-1)\,\varphi$. Here, variable $\varphi(t)$ plays the role of particle's coordinate and variable $v_0(t)$ plays the role of particle's velocity.

The solid line in Fig. 11.1 shows the behavior of function $U(\varphi)$. At point $\varphi_0 = \ln\sqrt{R-1}$, this function has a minimum

$$U(\varphi_0) = \frac{1}{2}(R-1)\,[1 - \ln(R-1)]$$

corresponding to the stable equilibrium state $v_1 = \sqrt{R-1}$. Thus, the steady-state probability distribution of $\varphi(t)$ and $\widetilde{v}_0(t)$ is similar to the Gibbs distribution (11.25)

$$P(\widetilde{v}_0, \varphi) = C \exp\left\{-\frac{1}{D}H(\widetilde{v}_0, \varphi)\right\}, \quad H(\widetilde{v}_0, \varphi) = \frac{\widetilde{v}_0^2}{2} + U(\varphi). \tag{11.45}$$

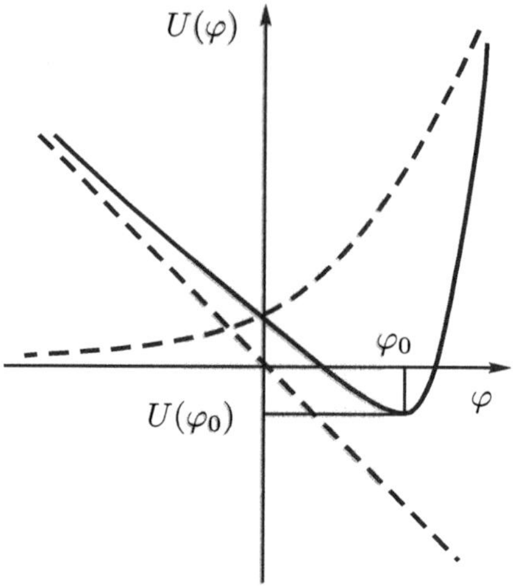

Fig. 11.1 Potential function $U(\varphi)$. The dashed lines show curve $U(\varphi) = \frac{1}{2}\exp\{2\varphi\}$ and straight line $U(\varphi) = -(R-1)\varphi$.

From Eq. (11.45) follows that, for $R > 1$, the steady-state probability distribution is composed of two independent steady-state distributions, of which the distribution of component $v_0(t)$ of system (11.42) is the Gaussian distribution

$$P(v_0) = \frac{1}{\sqrt{2\pi D}} \exp\left\{-\frac{(v_0-1)^2}{2D}\right\}, \tag{11.46}$$

and the distribution of quantity $\varphi(t)$ is the non-Gaussian distribution. If we turn back to variable $v_1(t)$, we obtain the corresponding steady-state probability distribution in the form

$$P(v_1) = \text{const}\, v_1^{\frac{R-1}{D}-1} \exp\left\{-\frac{v_1^2}{2D}\right\}. \tag{11.47}$$

As may be seen from Eq. (11.47), no steady-state probability distribution exists for component $v_1(t)$ in critical regime($R = 1$), which contrasts with the above case when random forces acted on all components of the triplet.

Earlier, we mentioned that two stages are characteristic of the formation of distribution (11.45). First, the Maxwellian distribution with respect to v_0 (11.46) is rapidly formed and the distribution with respect to φ is then formed at a much slower rate. The second stage is described by the Einstein–Smolukhovsky equation (11.22) that, in terms of variable v_1, assumes the form

$$\frac{\partial}{\partial t}P(v_1,t) = \frac{\partial}{\partial v_1}\left\{v_1\left[v_1^2-(R-1)\right] + \sigma^2 v_1 \frac{\partial}{\partial v_1} v_1\right\} P(v_1,t). \tag{11.48}$$

This equation is the Fokker–Planck equation for the stochastic dynamic equation

$$\frac{d}{dt}v_1(t) = -v_1(t)\left[v_1^2(t) - (R-1)\right] + v_1(t)f(t).$$

At critical regime $R = 1$, this equation assumes the form

$$\frac{d}{dt}v_1(t) = -v_1^3(t) + v_1(t)f(t),$$

from which follows that, despite strong nonlinear friction, its solution is stochastically unstable due to the specific form of the random term. Stochastic stabilization occurs, as we saw earlier, due to inclusion of random forces into the equations for other components.

11.3 Boundary-Value Problems for the Fokker-Planck Equation (Hopping Phenomenon)

The Fokker–Planck equations are the partial differential equations and they generally require boundary conditions whose particular form depends on the problem under consideration. One can proceed from both forward and backward Fokker–Planck equations, which are equivalent. Consider several examples.

11.3.1 Hopping Phenomenon in Regular Systems

Consider the nonlinear oscillator with friction described by the equation

$$\frac{d^2}{dt^2}x(t) + \lambda\frac{d}{dt}x(t) + \omega_0^2 x(t) + \beta x^3(t) = f(t) \quad (\beta,\, \lambda > 0) \tag{11.49}$$

and assume that random force $f(t)$ is the delta-correlated random function with the parameters

$$\langle f(t)\rangle = 0, \quad \langle f(t)f(t')\rangle = 2D\delta(t-t') \quad (D = \sigma_f^2\tau_0).$$

At $\lambda = 0$ and $f(t) = 0$, this equation is called the *Duffing equation* .

We can rewrite Eq. (11.49) in the standard form of the Hamiltonian system in functions $x(t)$ and $v(t) = \dfrac{d}{dt}x(t)$,

$$\frac{d}{dt}x(t) = \frac{\partial}{\partial v}H(x,v), \quad \frac{d}{dt}v(t) = -\frac{\partial}{\partial x}H(x,v) - \lambda v + f(t),$$

where

$$H(x,v)=\frac{v^2}{2}+U(x),\quad U(x)=\frac{\omega_0^2x^2}{2}+\beta\frac{x^4}{4}$$

is the Hamiltonian.

According to Eq. (11.25), the steady-state solution to the corresponding Fokker–Planck equation has the form

$$P(x,v)=C\exp\left\{-\frac{\lambda}{D}H(x,v)\right\}. \tag{11.50}$$

It is clear that this distribution is the product of two independent distributions, of which one — the steady-state probability distribution of quantity $v(t)$ — is the Gaussian distribution and the other — the steady-state probability distribution of quantity $x(t)$ — is the non-Gaussian distribution. Integrating Eq. (11.50) over v, we obtain the steady-state probability distribution of $x(t)$

$$P(x)=C\exp\left\{-\frac{\lambda}{D}\left(\frac{\omega_0^2x^2}{2}+\beta\frac{x^4}{4}\right)\right\}.$$

This distribution is maximum at the stable equilibrium point $x=0$.

Consider now the equation

$$\frac{d^2}{dt^2}x(t)+\lambda\frac{d}{dt}x(t)-\omega_0^2x(t)+\beta x^3(t)=f(t)\quad(\beta,\,\lambda>0). \tag{11.51}$$

In this case again, the steady-state probability distribution has the form (11.50), where now

$$H(x,v)=\frac{v^2}{2}+U(x),\quad U(x)=-\frac{\omega_0^2x^2}{2}+\beta\frac{x^4}{4}.$$

The steady-state probability distribution of $x(t)$ assumes now the form

$$P(x)=C\exp\left\{-\frac{\lambda}{D}\left(-\frac{\omega_0^2x^2}{2}+\beta\frac{x^4}{4}\right)\right\} \tag{11.52}$$

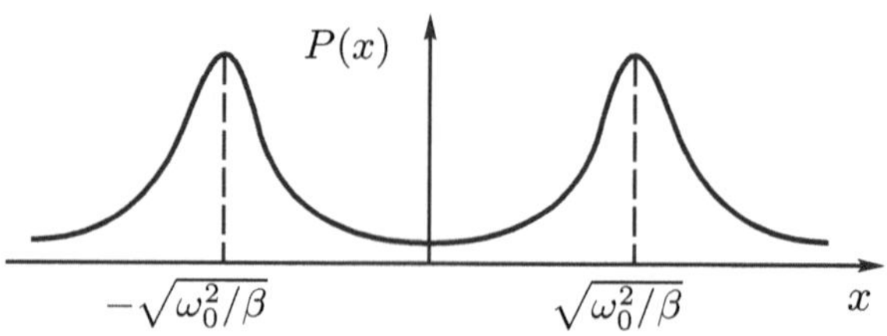

Fig. 11.2 Probability distribution (11.52)

and has maxima at points

$$x = \pm\sqrt{\omega_0^2/\beta}$$

and a minimum at point $x = 0$; the maxima correspond to the stable equilibrium points of problem (11.51) for $f(t) = 0$ and the minimum, to the instable equilibrium point. Figure 11.2 shows the behavior of the probability distribution (11.52).

As we mentioned earlier, the formation of distribution (11.52) is described by the Einstein–Smolukhovsky equation (11.22), which has in this case the form

$$\frac{\partial}{\partial t}P(x,t) = \frac{1}{\lambda}\frac{\partial}{\partial x}\left(\frac{\partial U(x)}{\partial x}P(x,t)\right) + \frac{1}{\lambda^2}\frac{\partial^2}{\partial x^2}P(x,t). \tag{11.53}$$

This equation is statistically equivalent to the dynamic equation

$$\frac{d}{dt}x(t) = -\frac{1}{\lambda}\frac{\partial U(x)}{\partial x} + \frac{1}{\lambda}f(t). \tag{11.54}$$

Probability distribution (11.52) corresponds to averaging over an ensemble of realizations of random process $f(t)$. If we deal with a single realization, the system arrives at one of states corresponding to the distribution maxima with a probability of 1/2. In this case, averaging over time will form the probability distribution around the maximum position. However, after a lapse of certain time T (the longer, the smaller D), the system will be transferred in the vicinity of the other maximum due to the fact that function $f(t)$ can assume sufficiently large values. For this reason, temporal averaging will form probability distribution (11.52) only if averaging time $t \gg T$.

Introducing dimensionless coordinate $x \to \sqrt{\dfrac{\omega_0^2}{\beta}}x$ and time $t \to \dfrac{\lambda}{\omega_0^2}t$, we can rewrite Eq. (11.53) in the form

$$\frac{\partial}{\partial t}P(x,t) = \frac{\partial}{\partial x}\left(\frac{\partial U(x)}{\partial x}P(x,t)\right) + \mu\frac{\partial^2}{\partial x^2}P(x,t), \tag{11.55}$$

where

$$\mu = \frac{\beta D}{\lambda\omega_0^4}, \quad U(x) = -\frac{x^2}{2} + \frac{x^4}{4}.$$

In this case, the equivalent stochastic equation (11.54) assumes the form of Eq. (1.17), page 10

$$\frac{d}{dt}x(t) = -\frac{\partial U(x)}{\partial x} + f(t). \tag{11.56}$$

Estimate the time required for the system to switch from a most probable state $x = -1$ to the other $x = 1$.

Let the system described by stochastic equation (11.56) was at a point from the interval (a, b) at instant t_0. The corresponding probability for the system to leave this interval

$$G(t; x_0, t_0) = 1 - \int_a^b dx p(x, t|x_0, t_0)$$

satisfies Eq. (10.21) following from the backward Fokker–Planck equation (10.20), page 310, i.e., the equation

$$\frac{\partial}{\partial t_0} G(t; x_0, t_0) = \frac{\partial U(x_0)}{\partial x_0} \frac{\partial}{\partial x_0} G(t; x_0, t_0) - \mu \frac{\partial^2}{\partial x_0^2} G(t; x_0, t_0)$$

with the boundary conditions

$$G(t; x_0, t) = 0, \quad G(t; a, t_0) = G(t; b, t_0) = 1.$$

Taking into account the fact that $G(t; x_0, t_0) = G(t - t_0; x_0)$ in our problem, we can denote $(t - t_0) = \tau$ and rewrite the boundary-value problem in the form

$$\begin{gathered} \frac{\partial}{\partial \tau} G(\tau; x_0) = \frac{\partial U(x_0)}{\partial x_0} \frac{\partial}{\partial x_0} G(\tau; x_0) - \mu \frac{\partial^2}{\partial x_0^2} G(\tau; x_0), \\ G(0; x_0) = 0, \, G(\tau; a,) = G(\tau; b) = 1 \quad \left(\lim_{\tau \to \infty} G(\tau; x_0) = 0 \right). \end{gathered} \tag{11.57}$$

From Eq. (11.57), one can easily see that average time required for the system to leave interval (a, b)

$$T(x_0) = \int_0^\infty d\tau \, \tau \frac{\partial G(\tau; x_0)}{\partial \tau}$$

satisfies the boundary-value problem

$$\mu \frac{d^2 T(x_0)}{d x_0^2} - \frac{dU(x_0)}{dx_0} \frac{dT(x_0)}{dx_0} = -1, \quad T(a) = T(b) = 0. \tag{11.58}$$

Equation (11.58) can be easily solved, and we obtain that the average time required for the system under random force to switch its state from $x_0 = -1$ to $x_0 = 1$ (this time is usually called the *Kramers time*) is given by the expression

$$T = \frac{1}{\mu}\int\limits_{-1}^{1} d\xi \int\limits_{-\infty}^{\xi} d\eta \exp\left\{\frac{1}{\mu}\left[U(\xi) - U(\eta)\right]\right\} = \frac{C(\mu)}{\mu}\int\limits_{0}^{1} d\xi \exp\left\{\frac{1}{\mu}U(\xi)\right\}, \tag{11.59}$$

where $C(\mu) = \int\limits_{-\infty}^{\infty} d\xi \exp\left\{\frac{1}{\mu}U(\xi)\right\}$. For $\mu \ll 1$, we obtain

$$T \approx \sqrt{2}\pi \exp\left\{\frac{1}{4\mu}\right\},$$

i.e., the average switching time increases exponentially with decreasing the intensity of fluctuations of the force.

Remark 11.1. ***Stochastic Resonance.***

In addition to the Duffing stochastic equation (11.51), a great attention is given recently to the equation

$$\frac{d^2}{dt^2}x(t) + \lambda\frac{d}{dt}x(t) - \omega_0^2 x(t) + \beta x^3(t) = f(t) + A\cos\omega_0 t \quad (\beta,\, \lambda > 0)$$

and, in particular, to the effect of an additional (except the noise) periodic impact on the statistical characteristics of the solution to Eq. (11.51) (see, e.g., reviews [5] and [40]). In this case, there sometimes occurs the phenomenon commonly called the stochastic resonance. In the context of this problem, the physical meaning of the term 'resonance' differs from the generally accepted one. Here, it reflects the fact that the response of the nonlinear stochastic oscillator on an external action appears a non-monotonous (i.e., 'resonance') function of the intensity of stochastic noise $f(t)$. In the case of the above problem, such a stochastic resonance occurs if the periodic signal frequency ω_0 coincides with the frequency of system switching between two stable states $\omega \sim 1/T$, which is called the *Kramers frequency* . ♦

11.3.2 Hopping Phenomena in Singular Systems

Consider now the singular stochastic problem described by Eq. (1.31), page 20. We rewrite this equation in the form ($\lambda = 1$)

$$\frac{d}{dt}x(t) = -x^2(t) + f(t), \quad x(0) = x_0, \tag{11.60}$$

where we assume as earlier that random process $f(t)$ is the Gaussian delta-correlated process with the parameters

$$\langle f(t)\rangle = 0, \quad \langle f(t)f(t')\rangle = 2D\delta(t - t') \quad (D = \sigma_f^2\tau_0).$$

In the absence of fluctuations, the solution to Eq. (11.60) has the form

$$x(t) = \frac{1}{t - t_0}, \quad t_0 = -\frac{1}{x_0}.$$

If $x_0 > 0$, the solution monotonously tends to zero. But if $x_0 < 0$, the solution arrives at the infinite value within a finite time t_0.

The solution of the statistical problem (11.60) is described by the forward and backward Fokker–Planck equations ($t - t_0 = \tau$)

$$\begin{aligned} \frac{\partial}{\partial \tau} p(x,\tau|x_0) &= \frac{\partial}{\partial x} x^2 p(x,\tau|x_0) + D\frac{\partial^2}{\partial x^2} p(x,\tau|x_0), \\ \frac{\partial}{\partial \tau} p(x,\tau|x_0) &= -x_0^2 \frac{\partial}{\partial x_0} p(x,\tau|x_0) + D\frac{\partial^2}{\partial x_0^2} p(x,\tau|x_0). \end{aligned} \tag{11.61}$$

Note that respective dimensions of quantities $x, p(x,\tau|x_0)$ and D are

$$[x] = \tau^{-1}, \quad [D] = \tau^{-3}, \quad [p] = \tau.$$

Consequently, we can reduce Eq. (11.61) to the following dimensionless form

$$\begin{aligned} \frac{\partial}{\partial \tau} p(x,\tau|x_0) &= \frac{\partial}{\partial x} x^2 p(x,\tau|x_0) + \frac{\partial^2}{\partial x^2} p(x,\tau|x_0), \\ \frac{\partial}{\partial \tau} p(x,\tau|x_0) &= -x_0^2 \frac{\partial}{\partial x_0} p(x,\tau|x_0) + \frac{\partial^2}{\partial x_0^2} p(x,\tau|x_0). \end{aligned} \tag{11.62}$$

Now, we must formulate boundary conditions to Eq. (11.62). Two types of boundary conditions are of the first-hand interest.

Boundary conditions of the first type correspond to the assumption that curve $x(t)$ stops at point t_0 where it becomes equal to $-\infty$. This means that probability flux density

$$J(\tau, x) = x^2 p(x,\tau|x_0) + \frac{\partial}{\partial x} p(x,\tau|x_0) \tag{11.63}$$

must vanish for $x \to \infty$, i.e.,

$$\begin{aligned} J(\tau, x) &\to 0, \quad \text{for} \quad x \to \infty; \\ p(x,\tau|x_0) &\to 0, \quad \text{for} \quad x \to -\infty. \end{aligned}$$

In this case, quantity

$$G(\tau|x_0) = \int_{-\infty}^{\infty} dx\, p(x,\tau|x_0) \neq 1$$

is the probability of the event that function $x(t)$ remain finite along the whole axis $(-\infty, \infty)$; in other words, this quantity is the probability of the absence of singular point at instant t:

$$G(\tau|x_0) = P(t < t_0).$$

Consequently, the probability of the appearance of singular point at instant t is given by the equality

$$P(t > t_0) = 1 - \int_{-\infty}^{\infty} dx p(x, \tau|x_0),$$

and the corresponding probability density

$$p(\tau|x_0) = \frac{\partial}{\partial t} P(t > t_0) = -\frac{\partial}{\partial \tau} \int_{-\infty}^{\infty} dx p(x, \tau|x_0) \tag{11.64}$$

satisfies the equation

$$\frac{\partial}{\partial \tau} p(\tau|x_0) = -x_0^2 \frac{\partial}{\partial x_0} p(\tau|x_0) + \frac{\partial^2}{\partial x_0^2} p(\tau|x_0), \qquad \lim_{\tau \to 0,\, \tau \to \infty} p(\tau|x_0) \to 0 \tag{11.65}$$

following from the backward Fokker–Planck equation (11.62).

Estimate the average time

$$\langle T(x_0) \rangle = \int_{0}^{\infty} \tau d\tau p(\tau|x_0)$$

during which the system switches from state x_0 to state $(-\infty)$. This time is described by the equation following from Eq. (11.65)

$$-1 = -x_0^2 \frac{d}{dx_0} \langle T(x_0) \rangle + \frac{d^2}{dx_0^2} \langle T(x_0) \rangle \tag{11.66}$$

the boundary conditions to which are formulated as $\langle T(x_0) \rangle \to 0$ for $x_0 \to -\infty$ and finiteness of $\langle T(x_0) \rangle$ for $x_0 \to \infty$. This equation can be easily integrated, and the result has the form

$$\langle T(x_0) \rangle = \int_{-\infty}^{x_0} d\xi \int_{\xi}^{\infty} d\eta \exp\left\{ \frac{1}{3} \left(\xi^3 - \eta^3 \right) \right\}. \tag{11.67}$$

From Eq. (11.67), we obtain for the average time of switching between two singular points ($x_0 \to \infty$)

$$\langle T(\infty)\rangle = \sqrt{\pi}\frac{12^{1/6}}{3}\Gamma\left(\frac{1}{6}\right) \approx 4.976.$$

Additionally, we note that quantity

$$\langle T(0)\rangle = \frac{2}{3}\langle T(\infty)\rangle$$

is the average time of switching from state $x_0 = 0$ to state $x_0 = -\infty$.

Drastically different boundary conditions appear under the assumption that function $x(t)$ is discontinuous and defined for all times t. If we assume additionally that function value $-\infty$ at instant $t \to t_0 - 0$ is immediately followed by value ∞ at instant $t \to t_0 + 0$, the boundary condition to Eq. (11.62) will be the condition of continuity of probability density flux (11.63), i.e., the condition

$$J(\tau, x)|_{x=-\infty} = J(\tau, x)|_{x=+\infty}.$$

In this case, the steady-state probability density exists and is independent of x_0,

$$P(x) = J\int\limits_{-\infty}^{x} d\xi \exp\left\{\frac{1}{3}\left(\xi^3 - x^3\right)\right\}, \tag{11.68}$$

where

$$J = \frac{1}{\langle T(\infty)\rangle}$$

is the steady-state probability flux density .

From (11.68) follows the asymptotic formula

$$P(x) \approx \frac{1}{\langle T(\infty)\rangle x^2} \tag{11.69}$$

for great x. This asymptotic is formed by discontinuities of function $x(t)$. Indeed, function $x(t)$ behaves near the discontinuity as

$$x(t) = \frac{1}{t - t_k},$$

and the effect of randomness appears insignificant. In this case, we have for sufficiently great t ($t \gg \langle T(\infty)\rangle$) and x

$$p(x, t|x_0) = \sum_{k=0}^{\infty}\left\langle \delta\left(x - \frac{1}{t - t_k}\right)\right\rangle = \frac{1}{x^2}\sum_{k=0}^{\infty}\langle\delta(t - t_k)\rangle$$

$$= \frac{1}{2\pi x^2}\int\limits_{-\infty}^{\infty} d\omega e^{-i\omega t}\sum_{k=0}^{\infty}\left\langle e^{i\omega t_k}\right\rangle = \frac{1}{2\pi x^2}\int\limits_{-\infty}^{\infty} d\omega e^{-i\omega t}\frac{\Phi_0(\omega)}{1 - \Phi(\omega)},$$

where $\Phi_0(\omega) = \left\langle e^{i\omega t_0} \right\rangle$ is the characteristic function of the first singular point, and $\Phi(\omega) = \left\langle e^{i\omega T} \right\rangle$ is the characteristic function of the temporal interval between the singularities. As a result, for $t \to \infty$, we obtain the asymptotic expression

$$P(x) = -\frac{1}{2\pi i x^2 \langle T(\infty) \rangle} \int\limits_{-\infty}^{\infty} d\omega e^{-i\omega t} \frac{1}{\omega + i0} = \frac{J}{x^2},$$

coincident with Eq. (11.69).

11.4 Asymptotic and Approximate Methods of Solving the Fokker-Plank Equation

If parameter fluctuations of the dynamic system are sufficiently small, the Fokker–Planck equation can be analyzed using different asymptotic and approximate methods. Consider in more detail three methods used in statistical analysis most actively.

11.4.1 Asymptotic Expansion

First of all, one can formulate some convergence method with respect to small parameters related to fluctuating quantities. This is the standard procedure for partial differential equations in which the small parameter appears as a factor of the highest derivative. The schematic of such a method is as follows (see, e.g., [21]).

Rewrite the Fokker–Planck equation in the form

$$\begin{aligned} \frac{\partial}{\partial t} P(\boldsymbol{x}, t) + A(\boldsymbol{x}, t; \varepsilon) P(\boldsymbol{x}, t) + B_i(\boldsymbol{x}, t; \varepsilon) \frac{\partial}{\partial x_i} P(\boldsymbol{x}, t) \\ = \varepsilon^2 D_{ij}(\boldsymbol{x}, t; \varepsilon) \frac{\partial^2}{\partial x_i \partial x_j} P(\boldsymbol{x}, t), \quad P(\boldsymbol{x}, 0) = p_0(\boldsymbol{x}), \quad (11.70) \end{aligned}$$

where we introduced parameter ε^2 that characterizes the intensity of fluctuations of dynamic system parameters. Representing the solution to Eq. (11.70) in the form

$$P(\boldsymbol{x}, t) = C(\varepsilon) \exp\left\{ -\frac{1}{\varepsilon^2} \phi(\boldsymbol{x}, t; \varepsilon) \right\}, \quad (11.71)$$

we obtain the nonlinear equation for function $\phi(\boldsymbol{x}, t; \varepsilon)$

$$\begin{aligned}\frac{\partial}{\partial t}\phi(\boldsymbol{x},t;\varepsilon) - \varepsilon^2 A(\boldsymbol{x},t;\varepsilon) + B_i(\boldsymbol{x},t;\varepsilon)\frac{\partial}{\partial x_i}\phi(\boldsymbol{x},t;\varepsilon) \\ - \varepsilon^2 D_{ij}(\boldsymbol{x},t;\varepsilon)\frac{\partial^2}{\partial x_i \partial x_j}\phi(\boldsymbol{x},t;\varepsilon) \\ + D_{ij}(\boldsymbol{x},t;\varepsilon)\left(\frac{\partial}{\partial x_i}\phi(\boldsymbol{x},t;\varepsilon)\right)\left(\frac{\partial}{\partial x_j}\phi(\boldsymbol{x},t;\varepsilon)\right) = 0, \quad (11.72)\end{aligned}$$

whose solution can be sought in the form of the series in powers of ε^2

$$\phi(\boldsymbol{x},t;\varepsilon) = \phi_0(\boldsymbol{x},t) + \varepsilon^2\phi_1(\boldsymbol{x},t) + \cdots .$$

To derive the convergence method for Eq. (11.72), we substitute this expansion in Eq. (11.72), expand the equation coefficients in series in ε^2, and group the terms with the corresponding powers of ε^2. In particular, for function $\phi_0(\boldsymbol{x},t)$, we obtain the equation

$$\left(\frac{\partial}{\partial t} + B_i(\boldsymbol{x},t;0)\frac{\partial}{\partial x_i}\right)\phi_0(\boldsymbol{x},t) + D_{ij}(\boldsymbol{x},t;0)\left(\frac{\partial}{\partial x_i}\phi_0(\boldsymbol{x},t)\right)\left(\frac{\partial}{\partial x_j}\phi_0(\boldsymbol{x},t)\right) = 0,$$

which is the first-order partial differential equation and can be solved by the method of characteristics, for example. Function $\phi_0(\boldsymbol{x},t)$ is the first term of the convergence method series; it describes the main singularity of the Fokker–Planck equation. The next term $\phi_1(\boldsymbol{x},t)$ describes the preexponential factor, and constant $C(\varepsilon^2)$ in (11.71) can be obtained from the behavior of the solution to Eq. (11.70) for $t \to 0$ and the corresponding initial value.

This convergence method holds only for a finite-duration initial stage of evolution and fails in the limit $t \to \infty$. To analyze Eq. (11.70) in this limit, this equation is usually rearranged to the form containing the self-adjoint operator with respect to spatial variables, which has the discrete spectrum.

Consider now two approximate methods of solving the Fokker–Planck equation.

11.4.2 Method of Cumulant Expansions

The first method is called the method of cumulant expansions [99]. If we perform the Fourier transform of the Fokker–Planck equation (10.11), page 307 with respect to spatial variables $\boldsymbol{x}$, i.e., turn from the probability density of the solution to stochastic equations (10.1), page 305 to the characteristic function

$$\Phi(\boldsymbol{\lambda},t) = \left\langle e^{i\boldsymbol{\lambda}\boldsymbol{x}(t)}\right\rangle = e^{\Theta(\boldsymbol{\lambda},t)} \quad (11.73)$$

and expand this function in the Taylor series in powers of $\boldsymbol{\lambda}$, we obtain that the expansion coefficients (i.e., one-time cumulants of random process $\boldsymbol{x}(t)$) satisfy the infinite system of nonlinear equations. The method of cumulant expansions considers this system neglecting all higher-order cumulants beginning from some certain order (if this order is three, we arrive at the Gaussian approximation, if four, the excess approximation, and so on). The retained cumulants satisfy the closed nonlinear system of ordinary differential equations whose solution determines the time-dependent behavior of cumulants. Note that monograph [99] suggests the general approach for deriving these equations directly from stochastic equations (10.1), without considering the Fokker–Planck equation (10.11), page 307 or the equation for the characteristic function. A disadvantage of this method consists in the fact that the neglect of the infinite number of cumulants, as is known, impairs the probability distribution. In particular, such impaired distribution appears negative in certain regions of spatial variables. Nevertheless, examples show that the method of cumulant expansions adequately describes time-dependent behavior of certain cumulants for a wide class of problems. It seems that this class of problems is limited to the problems for which statistical characteristics of the solution are analytic functions with respect to the intensity of random actions. Most likely, this method will fail for problems characterized by the nonanalytic behavior with respect to this parameter (such as problems on the escape of the trajectory of a system out of certain spatial region and the problem on the arrival at a given boundary).

11.4.3 Method of Fast Oscillation Averaging

Another approximate method widely used in the context of stochastic oscillating systems is called the method of averaging over fast parameters. For example, let a stochastic system is described by the dynamic equations

$$\begin{aligned}\frac{d}{dt}x(t) &= A(x,\widetilde{\phi}) + z(t)B(x,\widetilde{\phi}),\\ \frac{d}{dt}\phi(t) &= C(x,\widetilde{\phi}) + z(t)D(x,\widetilde{\phi}),\end{aligned} \tag{11.74}$$

where

$$\widetilde{\phi}(t) = \omega_0 t + \phi(t),$$

functions $A(x,\widetilde{\phi})$, $B(x,\widetilde{\phi})$, $C(x,\widetilde{\phi})$, and $D(x,\widetilde{\phi})$ are the periodic functions of variable $\widetilde{\phi}$, and $z(t)$ is the Gaussian delta-correlated process with the parameters

$$\langle z(t)\rangle = 0, \quad \langle z(t)z(t')\rangle = 2D\delta(t-t'), \quad D = \sigma^2\tau_0.$$

Variables $x(t)$ and $\phi(t)$ can mean the vector module and phase, respectively. The Fokker–Planck equation corresponding to system of equations (11.74) has the form

$$\frac{\partial}{\partial t}P(x,\phi,t) = -\frac{\partial}{\partial x}A(x,\widetilde{\phi})P(x,\phi,t) - \frac{\partial}{\partial \phi}C(x,\widetilde{\phi})P(x,\phi,t) + D\left[\frac{\partial}{\partial x}B(x,\widetilde{\phi}) + \frac{\partial}{\partial \phi}D(x,\widetilde{\phi})\right]^2 P(x,\phi,t). \quad (11.75)$$

Commonly, Eq. (11.75) is very complicated to immediately analyze the joint probability density. We rewrite this equation in the form

$$\begin{aligned}\frac{\partial}{\partial t}P(x,\phi,t) &= -\frac{\partial}{\partial x}A(x,\widetilde{\phi})P(x,\phi,t) - \frac{\partial}{\partial \phi}C(x,\widetilde{\phi})P(x,\phi,t)\\ &- D\frac{\partial}{\partial x}\left(\frac{\partial B^2(x,\widetilde{\phi})}{2\partial x} + \frac{\partial B(x,\widetilde{\phi})}{\partial \widetilde{\phi}}D(x,\widetilde{\phi})\right)P(x,\phi,t)\\ &- D\frac{\partial}{\partial \phi}\left(\frac{\partial D(x,\widetilde{\phi})}{\partial x}B(x,\widetilde{\phi}) + \frac{\partial D^2(x,\widetilde{\phi})}{2\partial \widetilde{\phi}})\right)P(x,\phi,t)\\ &+ D\left\{\frac{\partial^2}{\partial x^2}B^2(x,\widetilde{\phi}) + 2\frac{\partial^2}{\partial x\partial \phi}B(x,\widetilde{\phi})D(x,\widetilde{\phi}) + \frac{\partial^2}{\partial \phi^2}D^2(x,\widetilde{\phi})\right\}P(x,\phi,t).\end{aligned} \quad (11.76)$$

Now, we assume that functions $A(x,\widetilde{\phi})$ and $C(x,\widetilde{\phi})$ are sufficiently small and fluctuation intensity of process $z(t)$ is also small. In this case, statistical characteristics of system of equations (11.74) only slightly vary during times $\sim 1/\omega_0$. To study these small variations (accumulated effects), we can average Eq. (11.76) over the period of all oscillating functions. Assuming that function $P(x,\phi,t)$ remains intact under averaging, we obtain the equation

$$\begin{aligned}\frac{\partial}{\partial t}\overline{P(x,\phi,t)} &= -\frac{\partial}{\partial x}\overline{A(x,\widetilde{\phi})}\,\overline{P(x,\phi,t)} - \frac{\partial}{\partial \phi}\overline{C(x,\widetilde{\phi})}\,\overline{P(x,\phi,t)}\\ &- D\left\{\frac{\partial}{\partial x}\left(\overline{\frac{\partial B^2(x,\widetilde{\phi})}{2\partial x}} + \overline{\frac{\partial B(x,\widetilde{\phi})}{\partial \widetilde{\phi}}D(x,\widetilde{\phi})}\right) + \frac{\partial}{\partial \phi}\overline{\frac{\partial D(x,\widetilde{\phi})}{\partial x}B(x,\widetilde{\phi})}\right\}\overline{P(x,\phi,t)}\\ &+ D\left\{\frac{\partial^2}{\partial x^2}\overline{B^2(x,\widetilde{\phi})} + 2\frac{\partial^2}{\partial x\partial \phi}\overline{B(x,\widetilde{\phi})D(x,\widetilde{\phi})} + \frac{\partial^2}{\partial \phi^2}\overline{D^2(x,\widetilde{\phi})}\right\}\overline{P(x,\phi,t)},\end{aligned} \quad (11.77)$$

where the overbar denotes quantities averaged over the oscillation period.

Integrating Eq. (11.77) over ϕ, we obtain the Fokker–Planck equation for function $\overline{P(x,t)}$

$$\begin{aligned}\frac{\partial}{\partial t}\overline{P(x,t)} &= -\frac{\partial}{\partial x}\overline{A(x,\widetilde{\phi})}\,\overline{P(x,t)} + D\frac{\partial}{\partial x}\left(\overline{\frac{\partial B^2(x,\widetilde{\phi})}{2\partial x}} + \overline{\frac{\partial B(x,\widetilde{\phi})}{\partial \widetilde{\phi}}D(x,\widetilde{\phi})}\right)\overline{P(x,t)}\\ &+ D\frac{\partial}{\partial x}\overline{B^2(x,\widetilde{\phi})}\frac{\partial}{\partial x}\overline{P(x,t)}.\end{aligned} \quad (11.78)$$

Note that quantity $x(t)$ appears to be the one-dimensional Markovian random process in this approximation.

If we assume that $\overline{B(x,\widetilde{\phi})D(x,\widetilde{\phi})} = \overline{\frac{\partial D(x,\widetilde{\phi})}{\partial x}B(x,\widetilde{\phi})} = 0$, and $\overline{C(x,\widetilde{\phi})} =$ const, $\overline{D^2(x,\widetilde{\phi})} =$ const in Eq. (11.77), then processes $x(t)$ and $\phi(t)$ become statistically independent, and process $\phi(t)$ becomes the Markovian Gaussian process whose variance is the linear increasing function of time t. This means that probability distribution of quantity $\phi(t)$ on segment $[0, 2\pi]$ becomes uniform for large t (at $\overline{C(x,\widetilde{\phi})} = 0$).

As an illustration of using the above technique, we consider the problems on the stochastic parametric resonance, diffusion of passive tracer in a random hydrodynamic flow, and noises in hydrodynamic flows near the instability threshold.

Stochastic Parametric Resonance

Consider the stochastic second-order equation equivalent to system of the first-order equations (8.180), page 249

$$\frac{d}{dt}x(t) = y(t), \quad \frac{d}{dt}y(t) = -2\gamma y(t) - \omega_0^2[1 + z(t)]x(t). \tag{11.79}$$

In Chapter 5, page 244, we discussed the general approach to this problem in the case of the delta-correlated fluctuations of frequency. Here, we will assume that $z(t)$ is the Gaussian random process with the parameters

$$\langle z(t)\rangle = 0, \quad \langle z(t)z(t')\rangle = 2\sigma^2\tau_0\delta(t-t').$$

Replace functions $x(t)$ and $y(t)$ with the variables — oscillation amplitude $A(t)$ and phase $\phi(t)$ — defined by the formulas

$$x(t) = A(t)\sin\left(\omega_0 t + \phi(t)\right), \quad y(t) = \omega_0 A(t)\cos\left(\omega_0 t + \phi(t)\right). \tag{11.80}$$

Substituting Eq. (11.80) in system of equations (11.79), we obtain the system of equations in functions $A(t)$ and $\phi(t)$

$$\begin{aligned} \frac{d}{dt}A(t) &= -2\gamma A(t)\cos^2\psi(t) - \frac{\omega_0}{2}z(t)A(t)\sin\left(2\psi(t)\right), \\ \frac{d}{dt}\phi(t) &= 2\gamma\sin\left(2\psi(t)\right) + \omega_0 z(t)\sin^2\psi(t), \end{aligned} \tag{11.81}$$

where $\psi(t) = \omega_0 t + \phi(t)$. Representing amplitude $A(t)$ as $A(t) = e^{u(t)}$, we can rewrite system (11.81) in the form

$$
\begin{aligned}
\frac{d}{dt}u(t) &= -2\gamma\cos^2\psi(t) - \frac{\omega_0}{2}z(t)\sin\left(2\psi(t)\right), \\
\frac{d}{dt}\phi(t) &= \gamma\sin\left(2\psi(t)\right) + \omega_0 z(t)\sin^2\psi(t).
\end{aligned}
\tag{11.82}
$$

Consider now the joint probability density of the solution to system of equations (11.81) $P(u,\phi,t) = \langle\varphi(u,\phi,t)\rangle$, where the indicator function

$$
\varphi(u,\phi,t) = \delta\left(u(t)-u\right)\delta\left(\phi(t)-\phi\right)
$$

satisfies the Liouville equation

$$
\begin{aligned}
\frac{\partial}{\partial t}\varphi(u,\phi,t) &= \gamma\left\{2\frac{\partial}{\partial u}\cos^2\psi(t) - \frac{\partial}{\partial\phi}\sin\left(2\psi(t)\right)\right\}\varphi(u,\phi,t) \\
&+ z(t)\omega_0\left\{\frac{1}{2}\frac{\partial}{\partial u}\sin\left(2\psi(t)\right) - \frac{\partial}{\partial\phi}\sin^2\psi(t)\right\}\varphi(u,\phi,t).
\end{aligned}
\tag{11.83}
$$

Averaging now Eq. (11.83) over an ensemble of realizations of random delta-correlated process $z(t)$, using the Furutsu–Novikov formula that assumes in this case the form

$$
\langle z(t)\varphi(u,\phi,t)\rangle = \sigma^2\tau_0\left\langle\frac{\delta}{\delta z(t-0)}\varphi(u,\phi,t)\right\rangle,
$$

and the equality

$$
\frac{\delta}{\delta z(t-0)}\varphi(u,\phi,t) = \omega_0\left\{\frac{1}{2}\frac{\partial}{\partial u}\sin\left(2\psi(t)\right) - \frac{\partial}{\partial\phi}\sin^2\psi(t)\right\}\varphi(u,\phi,t)
$$

following from Eq. (11.83), we obtain the Fokker–Planck equation for the probability density

$$
\begin{aligned}
\frac{\partial}{\partial t}P(u,\phi,t) &= \gamma\left\{2\frac{\partial}{\partial u}\cos^2\psi(t) - \frac{\partial}{\partial\phi}\sin\left(2\psi(t)\right)\right\}P(u,\phi,t) \\
&+ D\left\{\frac{1}{2}\frac{\partial}{\partial u}\sin\left(2\psi(t)\right) - \frac{\partial}{\partial\phi}\sin^2\psi(t)\right\}^2 P(u,\phi,t),
\end{aligned}
$$

where $D = \sigma^2\tau_0\omega_0^2$. This equation can be rewritten in the form

$$\frac{\partial}{\partial t}P(u,\phi,t) = \gamma\left\{2\frac{\partial}{\partial u}\cos^2\psi(t) - \frac{\partial}{\partial\phi}\sin\left(2\psi(t)\right)\right\}P(u,\phi,t)$$

$$+ D\left\{\frac{\partial}{\partial u}\cos\left(2\psi(t)\right)\sin^2\psi(t) - 2\frac{\partial}{\partial\phi}\sin^3\psi(t)\cos\psi(t)\right\}P(u,\phi,t)$$

$$+ D\left\{\frac{1}{4}\frac{\partial^2}{\partial u^2}\sin^2\left(2\psi(t)\right) - \frac{\partial^2}{\partial u\partial\phi}\sin\left(2\psi(t)\right)\sin^2\psi(t) + \frac{\partial^2}{\partial\phi^2}\sin^4\psi(t)\right\}$$

$$\times P(u,\phi,t). \quad (11.84)$$

Assuming that absorption parameter γ is small in comparison with oscillation frequency ω_0 ($\gamma \ll \omega_0$), we can average Eq. (11.84) over oscillation period $T = 2\pi/\omega_0$ (the assumption that statistical characteristics only slightly vary during times $\sim T$ allows us to average solely trigonometric functions appeared in the right-hand side of Eq. (11.84)) to obtain the equation for the averaged (i.e., describing slow variations of statistical characteristics) probability density

$$\frac{\partial}{\partial t}\overline{P(u,\phi,t)} = \gamma\frac{\partial}{\partial u}\overline{P(u,\phi,t)} - \frac{D}{4}\frac{\partial}{\partial u}\overline{P(u,\phi,t)} + \frac{D}{8}\frac{\partial^2}{\partial u^2}\overline{P(u,\phi,t)} + \frac{3D}{8}\frac{\partial^2}{\partial\phi^2}\overline{P(u,\phi,t)} \quad (11.85)$$

with the initial value

$$\overline{P(u,\phi,0)} = \delta(u-u_0)\delta(\phi-\phi_0).$$

For example, in the case of initial values $u_0 = 0$, $\phi_0 = 0$ corresponding to $x(0) = 0$, $y(0) = \omega_0$, from Eq. (11.85) follows that statistical characteristics of amplitude and phase (averaged over the oscillation period) are statistically independent and the corresponding probability densities are the Gaussian densities,

$$\overline{P(u,t)} = \frac{1}{\sqrt{2\pi\sigma_u^2(t)}}\exp\left\{-\frac{\left(u-\langle u(t)\rangle\right)^2}{2\sigma_u^2(t)}\right\},$$
$$\overline{P(\phi,t)} = \frac{1}{\sqrt{2\pi\sigma_\phi^2(t)}}\exp\left\{-\frac{\left(\phi-\phi_0\right)^2}{2\sigma_\phi^2(t)}\right\}, \quad (11.86)$$

where

$$\langle u(t)\rangle = u_0 - \gamma t + \frac{D}{4}t, \quad \sigma_u^2(t) = \frac{D}{4}t, \quad \langle\phi(t)\rangle = \phi_0, \quad \sigma_\phi^2(t) = \frac{3D}{4}t.$$

As an example of using the above expressions, consider expressions for $\langle x(t)\rangle$ and $\langle x^2(t)\rangle$ corresponding to initial values $u_0 = 0$ and $\phi_0 = 0$.

For the average value, we have the expression

$$\langle x(t)\rangle = \langle A(t)\rangle \langle \sin(\omega_0 t + \phi(t))\rangle = \frac{1}{2i}\left\langle e^{u(t)}\right\rangle \left\langle e^{i\omega_0 t + i\phi(t)} - e^{-i\omega_0 t - i\phi(t)}\right\rangle$$
$$= \exp\left\{\langle u(t)\rangle + \frac{1}{2}\sigma_u^2(t) - \frac{1}{2}\sigma_\phi^2(t)\right\}\sin(\omega_0 t) = e^{-\gamma t}\sin(\omega_0 t) \quad (11.87)$$

coinciding with the problem solution in the case of absent fluctuations.

For quantity $\langle x^2(t)\rangle$, we obtain the expression

$$\langle x^2(t)\rangle = \left\langle e^{2u(t)}\right\rangle \langle \sin^2(\omega_0 t + \phi(t))\rangle = \frac{1}{2}\left\langle e^{2u(t)}\right\rangle \{1 - \langle \cos 2(\omega_0 t + \phi(t))\rangle\}$$
$$= \frac{1}{2}e^{2\langle u(t)\rangle + 2\sigma_u^2(t)}\left\{1 - e^{-2\sigma_\phi^2(t)}\cos(\omega_0 t)\right\} = \frac{1}{2}e^{(D-2\gamma)t}\left\{1 - e^{-3Dt/2}\cos(2\omega_0 t)\right\} \quad (11.88)$$

that coincides (in the absence of absorption) with Eq. (8.179), page 248 to terms of order $D/\omega_0 \ll 1$, and statistical parametric excitation of the system occurs if the condition $D > 2\gamma$ is satisfied.

As was mentioned earlier, the random amplitude has the lognormal probability distribution; consequently, its moment functions are given by the expression

$$\langle A^n(t)\rangle = \left\langle e^{nu(t)}\right\rangle = A_0^n \exp\left\{-n\gamma t + \frac{1}{8}n(n+2)Dt\right\}. \quad (11.89)$$

Under the condition $8\gamma < (n+2)D$, stochastic dynamic system (11.79) is statistically excited beginning from the moment function of order n. Nevertheless, the typical realization curve of the random amplitude has the form

$$A^*(t) = A_0 e^{-\left(\gamma - \frac{D}{4}\right)t},$$

and, under sufficiently weak absorption , namely if

$$1 < 4\frac{\gamma}{D} < 1 + \frac{1}{2}n,$$

which is the case if n is sufficiently great, the typical realization curve decreases exponentially with time, whereas all moment functions of random amplitude $A(t)$ of order n and higher are exponentially increasing functions of time. This means that statistics of random amplitude $A(t)$ is formed by high spikes above the exponentially decreasing typical realization curve, which is a consequence of the fact that random amplitude $A(t)$ is the lognormal quantity.

Diffusion of Passive Tracer in a Random Hydrodynamic Flow

In Chapter 1, we gave examples of the formation of cluster structure of the tracer field in random velocity field $\boldsymbol{u}(\boldsymbol{r},t) = \boldsymbol{v}(t)f(\boldsymbol{r})$, where $f(\boldsymbol{r})$ is the deterministic function and $\boldsymbol{v}(t)$ is the vector random Gaussian process. In this case, the field of tracer density satisfies Eq. (3.16), page 100. In terms of the model under consideration, this equation assumes the form

$$\left(\frac{\partial}{\partial t}+\boldsymbol{v}(t)\frac{\partial}{\partial \boldsymbol{r}}f(\boldsymbol{r})\right)\rho(\boldsymbol{r},t)=0,\quad \rho(\boldsymbol{r},0)=\rho_0(\boldsymbol{r}).$$

We will assume that $\boldsymbol{v}(t)$ is the Gaussian random stationary vector process with the parameters

$$\langle\boldsymbol{v}(t)\rangle=0,\quad B_{ij}(t-t')=\langle v_i(t)v_j(t')\rangle\quad \left(B_v(0)=\langle\boldsymbol{v}^2(t)\rangle\right).$$

In the approximation of delta-correlated process $\boldsymbol{v}(t)$, we have

$$B_{ij}(t-t')=2\sigma^2\delta_{ij}\tau_0\delta(t-t')\quad\left(\sigma^2\delta_{ij}\tau_0=\int\limits_0^\infty d\tau B_{ij}(\tau)\right),\tag{11.90}$$

where σ^2 is the variance of the velocity field and τ_0 is the temporal correlation radius of field $\boldsymbol{v}$.

In the Eulerian representation, the indicator function

$$\varphi(\boldsymbol{r},t;\rho)=\delta(\rho(\boldsymbol{r},t)-\rho)$$

satisfies the Liouville equation (3.20), page 101; in the case under consideration, this equation assumes the form

$$\left(\frac{\partial}{\partial t}+\boldsymbol{v}(t)f(\boldsymbol{r})\frac{\partial}{\partial \boldsymbol{r}}\right)\varphi(\boldsymbol{r},t;\rho)=\boldsymbol{v}(t)\frac{\partial f(\boldsymbol{r})}{\partial \boldsymbol{r}}\frac{\partial}{\partial\rho}\rho\varphi(\boldsymbol{r},t;\rho),$$

$$\varphi(\boldsymbol{r},0;\rho)=\delta(\rho_0(\boldsymbol{r})-\rho).$$

We rewrite this equation in the form

$$\frac{\partial}{\partial t}\varphi(\boldsymbol{r},t;\rho)=-\boldsymbol{v}(t)\left\{\frac{\partial}{\partial \boldsymbol{r}}f(\boldsymbol{r})-\frac{\partial f(\boldsymbol{r})}{\partial \boldsymbol{r}}\left(1+\frac{\partial}{\partial\rho}\rho\right)\right\}\varphi(\boldsymbol{r},t;\rho).\tag{11.91}$$

Averaging Eq. (11.91) over an ensemble of random process $\boldsymbol{v}(t)$, we obtain the equation for the probability density

$$\frac{\partial}{\partial t}P(\boldsymbol{r},t;\rho)=\sigma^2\tau_0\left\{\frac{\partial^2}{\partial \boldsymbol{r}^2}\,f^2(\boldsymbol{r})-\left(3+2\frac{\partial}{\partial\rho}\rho\right)\frac{\partial}{\partial \boldsymbol{r}}f(\boldsymbol{r})\frac{\partial f(\boldsymbol{r})}{\partial \boldsymbol{r}}+\right.$$

$$\left.+f(\boldsymbol{r})\frac{\partial^2 f(\boldsymbol{r})}{\partial \boldsymbol{r}^2}\left(1+\frac{\partial}{\partial\rho}\rho\right)+\frac{\partial f(\boldsymbol{r})}{\partial \boldsymbol{r}}\frac{\partial f(\boldsymbol{r})}{\partial \boldsymbol{r}}\left(1+\frac{\partial}{\partial\rho}\rho\right)^2\right\}P(\boldsymbol{r},t;\rho).\tag{11.92}$$

If characteristic scale of function $f(\boldsymbol{r})$ variation with $\boldsymbol{r}$ is $\widetilde{k}^{-1}$ and these variations show nearly periodic behavior ('fast variations'), then we can additionally average Eq. (11.92) over this scale to obtain the equation for 'slow' spatial variations

$$\frac{\partial}{\partial t}P(\boldsymbol{r},t;\rho) = \sigma^2\tau_0\left\{\overline{f^2(\boldsymbol{r})}\Delta + \overline{\left(\frac{\partial f(\boldsymbol{r})}{\partial \boldsymbol{r}}\right)^2}\frac{\partial^2}{\partial\rho^2}\rho^2\right\}P(\boldsymbol{r},t;\rho). \tag{11.93}$$

Note that Eq. (11.93) coincides with Eq. (4.44), Vol. 2, page 39, which means that, in the context of the one-time statistical characteristics, the velocity field model considered here is statistically equivalent to the model of the Gaussian delta-correlated field $\boldsymbol{u}(\boldsymbol{r},t)$. As a consequence, this model of the velocity field fluctuations must also result in clustering of tracer if $\boldsymbol{v}(t)\partial f(\boldsymbol{r})/\partial\boldsymbol{r} \neq 0$, which was observed in simulations for the simplest function $f(\boldsymbol{r}) = \sin 2(\boldsymbol{kr})$ (see the corresponding results on page 37).

Equation (11.93) becomes simpler if we consider this model using the x-axis directed along vector $\boldsymbol{k}$ and assume that the initial distribution of density is independent of $\boldsymbol{r}$ ($\rho_0(\boldsymbol{r}_0) = \rho_0$):

$$\frac{\partial}{\partial t}P(t;\rho) = 2k^2\sigma^2\tau_0\frac{\partial^2}{\partial\rho^2}\rho^2 P(t;\rho). \tag{11.94}$$

The solution of this equation corresponds to the logarithmically normal probability distribution with the moment functions of the form

$$\langle\rho^n(\boldsymbol{r},t)\rangle = \rho_0^n e^{2n(n-1)k^2\sigma^2\tau_0 t},$$

and, in accordance with Eq. (5.22), page 132, the typical realization curve is the exponentially decaying curve

$$\rho^*(\boldsymbol{r},t) = \rho_0 e^{-2k^2\sigma^2\tau_0 t}.$$

In these conditions, the statistical field of density can be represented in the form of Eq. (1.68), page (36)

$$\rho(\boldsymbol{r},t)/\rho_0 = \frac{1}{e^{T(t)}\cos^2(kx) + e^{-T(t)}\sin^2(kx)},$$

where

$$T(t) = 2k\int\limits_0^t d\tau v_x(\tau)$$

is the 'random time'. Consequently, quantity $\rho(\boldsymbol{r},t)/\rho_0$ averaged over fast spatial variables is equal to unity,

$$\overline{\rho(\boldsymbol{r},t)/\rho_0} = 1,$$

and is independent of random time $T(t)$.

In a similar way, we can obtain for example that

$$\overline{\left(\rho(\boldsymbol{r},t)/\rho_0\right)^2} = \frac{1}{2}\left(e^{T(t)} + e^{-T(t)}\right).$$

In the case of the Gaussian delta-correlated random process $v_x(t)$ with parameters (11.90), we have

$$\left\langle \overline{\left(\rho(\boldsymbol{r},t)/\rho_0\right)^2} \right\rangle = \left\langle e^{T(t)} \right\rangle = \exp\left\{\frac{1}{2}\left\langle T^2(t)\right\rangle\right\} = e^{4k^2\sigma^2\tau_0 t},$$

which agrees with the logarithmically normal probability distribution of field $\rho(\boldsymbol{r},t)$ (11.94).

Noises in Hydrodynamic Flows Near the Instability Threshold

In actuality, physical processes described by macroscopic hydrodynamic equations occur against a background of processes characterized by smaller scales (noises), which require statistical description. Among these processes are, for example, such as molecular noises in microscopic hydrodynamics, microturbulence in respect of large-scale motions, and the effect of rejected terms in the finite-dimensional approximation of hydrodynamic equations. It appears that the effect of such noises can be statistically described within the framework of macroscopic variables, by supplementing the corresponding microscopic equations with random 'external forces' (sources) with specified statistical characteristics.

A characteristic feature of such formulation of the problem consists in the existence of small parameter σ^2, which is the ratio of noise energy to the kinetic energy of macroscopic motion. The smallness of this parameter drastically simplifies the problem in a number of cases, because it allows using a perturbation theory to describe the fluctuations in a hydrodynamic flow far from the critical regime. The perturbation theory fails only in the region, where the type of flow changes (i.e., in the region, where a bifurcation occurs in the solution of the corresponding dynamic problem). In this region, one must consider the complete nonlinear problem, which is very difficult in the general case. Moreover, the issue of the nonlinear stability of flows is difficult and poorly known even in the absence of fluctuations. For this reason, it is natural to limit ourselves to the consideration of simple models with a finite number of degrees of freedom [44].

The simplest model is the Landau scheme of instability and secondary flows origination [80] (see also [83, 112]). According to this scheme, near the instability threshold, the velocity field is representable in the form (the one-dimensional instability)

$$u(x,t) = A(t)\varphi(x) + A^*(t)\varphi^*(x), \tag{11.95}$$

where, in the linear statement of the problem, $A(t) = \exp\{\gamma t - i\omega t\}$ ($\gamma < 0$ corresponds to the undercritical regime, $\gamma = 0$ to the critical regime, and $\gamma > 0$ to the overcritical regime) and $\varphi(x)$ is the spatial eigenfunction of the corresponding boundary-value problem. In addition, $|\gamma| \ll |\omega|$ near the critical regime. According to Landau, the behavior of flow for $\gamma > 0$ can be considered on the basis of the equation

$$\frac{d}{dt}\widetilde{I}(t) = 2\gamma\widetilde{I}(t) - \delta\widetilde{I}^2(t) \tag{11.96}$$

in slowly varying quantity $\widetilde{I}(t) = \widetilde{A(t)A(t)^*}$. This equation is called the *Landau equation*; for $\delta > 0$ it determines the steady-state secondary flow $\widetilde{I} = 2\gamma/\delta$. Here, we performed averaging over times $t \sim 1/\omega$.

In principle, hydrodynamic equations allow obtaining an equation for amplitude $A(t)$ (with allowance for random noises such as, in particular, thermal fluctuations [2]) in the form

$$\frac{d}{dt}A(t) = \{\cdots\} + f(t), \quad \frac{d}{dt}A^*(t) = \{\cdots\}^* + f^*(t), \tag{11.97}$$

where $\{\cdots\}$ denotes the terms of the third degree in A and A^*, asterisk denotes the complex conjugated quantity, and $f(t)$ is the corresponding 'external force' acting on the instable mode. Within the framework of model statement of the problem, we can consider forces $f(t)$ and $f^*(t)$ as the zero-mean Gaussian random forces characterized by the correlation functions of the form

$$\begin{gathered}\langle f(t)f(t')\rangle = 2\sigma_1^2 h\delta(t-t'), \quad \langle f^*(t)f^*(t')\rangle = 2\sigma_1^2 h^*\delta(t-t'), \\ \langle f(t)f^*(t')\rangle = 2\sigma_0^2\delta(t-t'),\end{gathered} \tag{11.98}$$

where h is the complex constant appeared due to separation of random force $f(t)$ acting on the instable mode. Delta-correlated property is conditioned here by the fact that the correlation radius of the force is much less than the characteristic times of the corresponding dynamic problem.

Introduce probability density of quantity $I(t) = A(t)A(t)^*$:

$$P(I,t) = \langle\delta\left(I(t) - I\right)\rangle .$$

In view of the relationships

---

[2] The problem on thermal noises in simple hydrodynamic-type system was considered in Sect. 11.2.3, page 348.

$$\frac{\delta A(t)}{\delta f(t-0)} = \frac{\delta A^*(t)}{\delta^* f(t-0)} = 1, \quad \frac{\delta A(t)}{\delta f^*(t-0)} = \frac{\delta A^*(t)}{\delta f(t-0)} = 0,$$

we obtain that probability density $P(I,t)$ satisfies the equation

$$\frac{\partial}{\partial t}P(I,t) = -\frac{\partial}{\partial I}\left\langle\left[A^*(t)\{\cdots\} + A(t)\{\cdots\}^*\right]\delta\left(I(t)-I\right)\right\rangle$$
$$+ 2\sigma_0^2\frac{\partial}{\partial I}I\frac{\partial}{\partial I}P(I,t) + \sigma_1^2\frac{\partial^2}{\partial I^2}\left\langle\left[hA^*(t)A^*(t) + h^*A(t)A(t)\right]\delta\left(I(t)-I\right)\right\rangle. \tag{11.99}$$

Averaging Eq. (11.99) over times $t \sim 1/\omega$ and taking into account the Landau equation (11.96), we obtain the Fokker–Planck equation for the probability density $P(I,t)$ of the slowly varying quantity $I(t)$ in the following form

$$\frac{\partial}{\partial t}P(I,t) = -\frac{\partial}{\partial I}\left\{\left(2\gamma I - \delta I^2\right)P(I,t)\right\} + 2\sigma_0^2\frac{\partial}{\partial I}I\frac{\partial}{\partial I}P(I,t). \tag{11.100}$$

The steady-state solution of Eq. (11.100) has the form

$$P(I) = C\exp\left\{\frac{1}{2\sigma_0^2}\left(2\gamma I - \frac{\delta I^2}{2}\right)\right\},$$

from which follows that fluctuations strongly increase in the critical regime ($\gamma = 0$) because $\langle I\rangle \sim \sigma_0$ in this regime, in contrast to $\langle I\rangle \sim \sigma_0^2$ in the undercritical region ($\gamma < 0$).

Relative to the temporal correlation $\langle A(t)A^*(t')\rangle$, the model under consideration tells nothing, because this quantity depends on the particular form of the terms denoted $\{\cdots\}$ in Eqs. (11.97).

Chapter 12
Diffusion and Higher Approximations

12.1 General Remarks

Applicability of the approximation of delta-correlated random field $\boldsymbol{f}(\boldsymbol{x},t)$ (i.e., applicability of the Fokker–Planck equation) is restricted by the smallness of the temporal correlation radius τ_0 of random field $\boldsymbol{f}(\boldsymbol{x},t)$ with respect to all temporal scales of the problem under consideration. The effect of the finite-valued temporal correlation radius of random field $\boldsymbol{f}(\boldsymbol{x},t)$ can be considered within the framework of the diffusion approximation (see, e.g., [51, 52]). The diffusion approximation appears to be more obvious and physical than the formal mathematical derivation of the approximation of delta-correlated random field. This approximation also holds for sufficiently weak parameter fluctuations of the stochastic dynamic system and allows describing new physical effects caused by the finite-valued temporal correlation radius of random parameters, rather than only obtaining the applicability range of the delta-correlated approximation. The diffusion approximation assumes that the effect of random actions is insignificant during temporal scales about τ_0, i.e., the system behaves during these times as a free system.

Again, let vector function $\boldsymbol{x}(t)$ satisfies the dynamic equation (10.1), page 305

$$\frac{d}{dt}\boldsymbol{x}(t) = \boldsymbol{v}(\boldsymbol{x},t) + \boldsymbol{f}(\boldsymbol{x},t), \;\; \boldsymbol{x}(t_0) = \boldsymbol{x}_0, \tag{12.1}$$

where $\boldsymbol{v}(\boldsymbol{x},t)$ is the deterministic vector function and $\boldsymbol{f}(\boldsymbol{x},t)$ is the random statistically homogeneous and stationary Gaussian vector field with the statistical characteristics

$$\langle f(\boldsymbol{x},t)\rangle = 0, \quad B_{ij}(\boldsymbol{x},t;\boldsymbol{x}',t') = B_{ij}(\boldsymbol{x}-\boldsymbol{x}',t-t') = \langle f_i(\boldsymbol{x},t)f_j(\boldsymbol{x}',t')\rangle .$$

Introduce the indicator function

$$\varphi(\boldsymbol{x},t) = \delta(\boldsymbol{x}(t)-\boldsymbol{x}), \tag{12.2}$$

V.I. Klyatskin, *Stochastic Equations: Theory and Applications in Acoustics, Hydrodyn., Magnetohydrodyn., and Radiophys., Vol. 1*, Understanding Complex Systems,
DOI: 10.1007/978-3-319-07587-7_12, © Springer International Publishing Switzerland 2015

where $\boldsymbol{x}(t)$ is the solution to Eq. (12.1) satisfying the Liouville equation (10.6), page 306

$$\left(\frac{\partial}{\partial t}+\frac{\partial}{\partial \boldsymbol{x}}\boldsymbol{v}(\boldsymbol{x},t)\right)\varphi(\boldsymbol{x},t)=-\frac{\partial}{\partial \boldsymbol{x}}\boldsymbol{f}(\boldsymbol{x},t)\varphi(\boldsymbol{x},t). \tag{12.3}$$

As earlier, we obtain the equation for the probability density of the solution to Eq. (12.1)

$$P(\boldsymbol{x}(t)=\langle\varphi(\boldsymbol{x},t)\rangle=\langle\delta(\boldsymbol{x}(t)-\boldsymbol{x})\rangle$$

by averaging Eq. (12.3) over an ensemble of realizations of field $\boldsymbol{f}(\boldsymbol{x},t)$

$$\begin{gathered}\left(\frac{\partial}{\partial t}+\frac{\partial}{\partial \boldsymbol{x}}\boldsymbol{v}(\boldsymbol{x},t)\right)P(\boldsymbol{x},t)=-\frac{\partial}{\partial \boldsymbol{x}}\left\langle\boldsymbol{f}(\boldsymbol{x},t)\varphi(\boldsymbol{x},t)\right\rangle,\\ P(\boldsymbol{x},t_0)=\delta(\boldsymbol{x}-\boldsymbol{x}_0).\end{gathered} \tag{12.4}$$

Using the Furutsu–Novikov formula (10.10), page 307

$$\langle f_k(\boldsymbol{x},t)R[t;\boldsymbol{f}(\boldsymbol{y},\tau)]\rangle=\int d\boldsymbol{x}'\int dt' B_{kl}(\boldsymbol{x},t;\boldsymbol{x}',t')\left\langle\frac{\delta}{\delta f_l(\boldsymbol{x}',t')}R[t;\boldsymbol{f}(\boldsymbol{y},\tau)]\right\rangle$$

valid for the correlation between the Gaussian random field $\boldsymbol{f}(\boldsymbol{x},t)$ and arbitrary functional $R[t;\boldsymbol{f}(\boldsymbol{y},\tau)]$ of this field, we can rewrite Eq. (12.4) in the form

$$\left(\frac{\partial}{\partial t}+\frac{\partial}{\partial \boldsymbol{x}}\boldsymbol{v}(\boldsymbol{x},t)\right)P(\boldsymbol{x},t)=-\frac{\partial}{\partial x_i}\int d\boldsymbol{x}'\int\limits_{t_0}^{t} dt' B_{ij}(\boldsymbol{x},t;\boldsymbol{x}',t')\left\langle\frac{\delta}{\delta f_j(\boldsymbol{x}',t')}\varphi(\boldsymbol{x},t)\right\rangle. \tag{12.5}$$

12.2 Diffusion Approximation

In the diffusion approximation, Eq. (12.5) is the exact equation, and the variational derivative and indicator function satisfy, within temporal scales of about temporal correlation radius τ_0 of random field $\boldsymbol{f}(\boldsymbol{x},t)$, the system of dynamic equations

$$\begin{gathered}\frac{\partial}{\partial t}\frac{\delta\varphi(\boldsymbol{x},t)}{\delta f_i(\boldsymbol{x}',t')}=-\frac{\partial}{\partial \boldsymbol{x}}\left\{\boldsymbol{v}(\boldsymbol{x},t)\frac{\delta\varphi(\boldsymbol{x},t)}{\delta f_i(\boldsymbol{x}',t')}\right\},\\ \left.\frac{\delta\varphi(\boldsymbol{x},t)}{\delta f_i(\boldsymbol{x}',t')}\right|_{t=t'}=-\frac{\partial}{\partial x_i}\left\{\delta(\boldsymbol{x}-\boldsymbol{x}')\varphi(\boldsymbol{x},t')\right\},\\ \frac{\partial}{\partial t}\varphi(\boldsymbol{x},t)=-\frac{\partial}{\partial \boldsymbol{x}}\left\{\boldsymbol{v}(\boldsymbol{x},t)\varphi(\boldsymbol{x},t)\right\},\quad \varphi(\boldsymbol{x},t)|_{t=t'}=\varphi(\boldsymbol{x},t').\end{gathered} \tag{12.6}$$

The solution to problem (12.5), (12.6) holds for all times t. In this case, the solution $\boldsymbol{x}(t)$ to problem (12.1) cannot be considered as the Markovian vector random process because its multi-time probability density cannot be factorized in terms of the transition probability density. However, in asymptotic limit $t \gg \tau_0$, the diffusion-approximation solution to the initial dynamic system (12.1) will be the Markovian random process, and the corresponding conditions of applicability are formulated as smallness of all statistical effects within temporal scales of about temporal correlation radius τ_0.

12.2.1 Dynamics of a Particle

The use of the diffusion approximation in concrete physical problems will be discussed in Part 4. Here, we illustrate this approximation by the example of the dynamics of a particle with linear friction under random forces, which is described by the stochastic system (1.14), page 9

$$\frac{d}{dt}\boldsymbol{r}(t) = \boldsymbol{v}(t), \quad \frac{d}{dt}\boldsymbol{v}(t) = -\lambda\boldsymbol{v}(t) + \boldsymbol{f}(\boldsymbol{r},t),$$
$$\boldsymbol{r}(0) = \boldsymbol{r}_0, \quad \boldsymbol{v}(0) = \boldsymbol{v}_0. \tag{12.7}$$

Introduce the indicator function for particle position and velocity

$$\varphi(\boldsymbol{r},\boldsymbol{v},t) = \delta(\boldsymbol{r}(t)-\boldsymbol{r})\delta(\boldsymbol{v}(t)-\boldsymbol{v}).$$

This function satisfies the stochastic Liouville equation

$$\left(\frac{\partial}{\partial t} + \boldsymbol{v}\frac{\partial}{\partial \boldsymbol{r}} - \lambda\frac{\partial}{\partial \boldsymbol{v}}\boldsymbol{v}\right)\varphi(\boldsymbol{r},\boldsymbol{v},t) = -\boldsymbol{f}(\boldsymbol{r},t)\frac{\partial}{\partial \boldsymbol{v}}\varphi(\boldsymbol{r},\boldsymbol{v},t). \tag{12.8}$$

Averaging Eq. (12.8) over an ensemble of realizations of field $\boldsymbol{f}(\boldsymbol{r},t)$, we obtain that the one-time probability density

$$P(\boldsymbol{r},\boldsymbol{v},t) = \langle\delta(\boldsymbol{r}(t)-\boldsymbol{r})\delta(\boldsymbol{v}(t)-\boldsymbol{v})\rangle$$

satisfies the equation

$$\left(\frac{\partial}{\partial t} + \boldsymbol{v}\frac{\partial}{\partial \boldsymbol{r}} - \lambda\frac{\partial}{\partial \boldsymbol{v}}\boldsymbol{v}\right)P(\boldsymbol{r},\boldsymbol{v},t) = -\frac{\partial}{\partial \boldsymbol{v}}\left\langle\boldsymbol{f}(\boldsymbol{r},t)\varphi(\boldsymbol{r},\boldsymbol{v},t)\right\rangle. \tag{12.9}$$

Taking into account the Furutsu–Novikov formula (10.10), page 307 we can rewrite this equation in the form

$$\left(\frac{\partial}{\partial t} + \boldsymbol{v}\frac{\partial}{\partial \boldsymbol{r}} - \lambda\frac{\partial}{\partial \boldsymbol{v}}\boldsymbol{v}\right)P(\boldsymbol{r},\boldsymbol{v},t) = -\frac{\partial}{\partial v_i}\int d\boldsymbol{r}'\int_0^t dt' B_{ij}(\boldsymbol{r}-\boldsymbol{r}',t-t')\left\langle\frac{\delta\varphi(\boldsymbol{r},\boldsymbol{v},t)}{\delta f_j(\boldsymbol{r}',t')}\right\rangle. \tag{12.10}$$

If we express the variational derivative in the right-hand side of Eq. (12.10) in terms of the variational derivatives of functions $\boldsymbol{r}(t)$ and $\boldsymbol{v}(t)$, the equation will assume the form

$$\left(\frac{\partial}{\partial t}+\boldsymbol{v}\frac{\partial}{\partial \boldsymbol{r}}-\lambda\frac{\partial}{\partial \boldsymbol{v}}\boldsymbol{v}\right)P(\boldsymbol{r},\boldsymbol{v},t)=\frac{\partial}{\partial v_i}\int d\boldsymbol{r}'\int\limits_0^t dt' B_{ij}(\boldsymbol{r}-\boldsymbol{r}',t-t')$$
$$\times\left\langle\left[\frac{\partial}{\partial r_k}\frac{\delta r_k(t)}{\delta f_j(\boldsymbol{r}',t')}+\frac{\partial}{\partial v_k}\frac{\delta v_k(t)}{\delta f_j(\boldsymbol{r}',t')}\right]\varphi(\boldsymbol{r},\boldsymbol{v},t)\right\rangle. \quad (12.11)$$

The variational derivatives of functions $\boldsymbol{r}(t)$ and $\boldsymbol{v}(t)$ appeared in Eq. (12.11) satisfy the system of equations following from Eq. (12.7) for $t' < t$,

$$\begin{aligned}\frac{d}{dt}\frac{\delta r_k(t)}{\delta f_j(\boldsymbol{r}',t')}&=\frac{\delta v_k(t)}{\delta f_j(\boldsymbol{r}',t')},\\ \frac{d}{dt}\frac{\delta v_k(t)}{\delta f_j(\boldsymbol{r}',t')}&=-\lambda\frac{\delta v_k(t)}{\delta f_j(\boldsymbol{r}',t')}+\frac{\partial f_k(\boldsymbol{r},t)}{\partial r_l}\frac{\delta r_l(t)}{\delta f_j(\boldsymbol{r}',t')}\end{aligned} \quad (12.12)$$

with the initial values

$$\left.\frac{\delta r_k(t)}{\delta f_j(\boldsymbol{r}',t')}\right|_{t-t'}=0,\quad \left.\frac{\delta v_k(t)}{\delta f_j(\boldsymbol{r}',t')}\right|_{t=t'}=\delta_{kj}\delta(\boldsymbol{r}(t')-\boldsymbol{r}'). \quad (12.13)$$

The integral with respect to time in the right-hand side of Eq. (12.11) depends mainly on the behavior of variational derivatives within the temporal interval $t-t' \sim \tau_0$. Assuming that the effect of random forces is insignificant within this temporal scale, we can omit the last term in the second equation of system (12.12) to obtain the deterministic system of equations

$$\frac{d}{dt}\frac{\delta r_k(t)}{\delta f_j(\boldsymbol{r}',t')}=\frac{\delta v_k(t)}{\delta f_j(\boldsymbol{r}',t')},\quad \frac{d}{dt}\frac{\delta v_k(t)}{\delta f_j(\boldsymbol{r}',t')}=-\lambda\frac{\delta v_k(t)}{\delta f_j(\boldsymbol{r}',t')}. \quad (12.14)$$

Nevertheless, the initial values (12.13) to this system remain random because $\boldsymbol{r}(t')$ is the stochastic function.

The solution to system (12.14) with initial values (12.13) is given by the formulas

$$\begin{aligned}\frac{\delta v_k(t)}{\delta f_j(\boldsymbol{r}',t')}&=\delta_{kj}e^{-\lambda(t-t')}\delta(\boldsymbol{r}(t')-\boldsymbol{r}'),\\ \frac{\delta r_k(t)}{\delta f_j(\boldsymbol{r}',t')}&=\frac{1}{\lambda}\delta_{kj}\left[1-e^{-\lambda(t-t')}\right]\delta(\boldsymbol{r}(t')-\boldsymbol{r}').\end{aligned} \quad (12.15)$$

Assuming further that the dynamics of the particle is also only slightly affected by random forces, we can express function $\boldsymbol{r}(t')$ in Eq. (12.15) through function $\boldsymbol{r}(t)$ that satisfies the simplified system of equations (12.7)

$$\frac{d}{dt}\boldsymbol{r}(t)=\boldsymbol{v}(t),\quad \frac{d}{dt}\boldsymbol{v}(t)=-\lambda\boldsymbol{v}(t) \quad (12.16)$$

with the initial values

$$\boldsymbol{r}(t)|_{t=t'} = \boldsymbol{r}(t'), \quad \boldsymbol{v}(t)|_{t=t'} = \boldsymbol{v}(t'), \tag{12.17}$$

from which follows that

$$\boldsymbol{r}(t') = \boldsymbol{r}(t) - \frac{1}{\lambda}\boldsymbol{v}\left[e^{\lambda(t-t')} - 1\right], \quad \boldsymbol{v}(t') = e^{\lambda(t-t')}\boldsymbol{v}(t). \tag{12.18}$$

The above simplifying procedures of transition from Eqs. (12.12), (12.13) to Eqs. (12.15) and from Eqs. (12.7) to Eqs. (12.18) form the basis of the diffusion approximation in the context of the problem under consideration.

Using now Eqs. (12.15) and (12.18), we can rewrite Eq. (12.11) in the closed form

$$\begin{aligned}\left(\frac{\partial}{\partial t} + \boldsymbol{v}\frac{\partial}{\partial \boldsymbol{r}} - \lambda\frac{\partial}{\partial \boldsymbol{v}}\boldsymbol{v}\right) P(\boldsymbol{r},\boldsymbol{v},t) &= -\frac{\partial}{\partial v_i}\int d\boldsymbol{r}' \int_0^t d\tau B_{ij}(\boldsymbol{r}-\boldsymbol{r}',\tau) \\ &\times \left\{\frac{\partial}{\partial r_j}\frac{1}{\lambda}\left[1-e^{-\lambda\tau}\right] + \frac{\partial}{\partial v_j}e^{-\lambda\tau}\right\}\delta\left(\boldsymbol{r}-\boldsymbol{r}'-\frac{1}{\lambda}\left[e^{\lambda\tau}-1\right]\boldsymbol{v}\right)P(\boldsymbol{r},\boldsymbol{v},t).\end{aligned} \tag{12.19}$$

The operator in braces commutes with the delta-function, and integration over $\boldsymbol{r}'$ results in the equation

$$\left(\frac{\partial}{\partial t} + \boldsymbol{v}\frac{\partial}{\partial \boldsymbol{r}} - \lambda\frac{\partial}{\partial \boldsymbol{v}}\boldsymbol{v}\right) P(\boldsymbol{r},\boldsymbol{v},t) = \frac{\partial}{\partial v_i}\left\{D_{ij}^{(1)}(\boldsymbol{v},t)\frac{\partial}{\partial v_j} + D_{ij}^{(2)}(\boldsymbol{v},t)\frac{\partial}{\partial r_j}\right\}P(\boldsymbol{r},\boldsymbol{v},t), \tag{12.20}$$

where we introduced the diffusion coefficients

$$\begin{aligned} D_{ij}^{(1)}(\boldsymbol{v},t) &= \int_0^t d\tau e^{-\lambda\tau} B_{ij}\left(\frac{1}{\lambda}\left[e^{\lambda\tau}-1\right]\boldsymbol{v},\tau\right), \\ D_{ij}^{(2)}(\boldsymbol{v},t) &= \frac{1}{\lambda}\int_0^t d\tau \left[1-e^{-\lambda\tau}\right] B_{ij}\left(\frac{1}{\lambda}\left[e^{\lambda\tau}-1\right]\boldsymbol{v},\tau\right). \end{aligned} \tag{12.21}$$

Equation (12.20) for the one-time probability density is correct even for times $t < \tau_0$. In this case, the solution $\{\boldsymbol{r}(t), \boldsymbol{v}(t)\}$ to problem (12.7) will not be the Markovian vector process because its multi-time probability density cannot be factorized in terms of the transition probability density. Nevertheless, it will be the Markovian random process in asymptotic limit $t \gg \tau_0$. In this limit, we can replace the upper limits of integrals in Eqs. (12.21) with infinity. This replacement results in the Fokker–Planck equation for the one-time probability density

$$\left(\frac{\partial}{\partial t} + \boldsymbol{v}\frac{\partial}{\partial \boldsymbol{r}} - \lambda\frac{\partial}{\partial \boldsymbol{v}}\boldsymbol{v}\right) P(\boldsymbol{r},\boldsymbol{v},t) = \frac{\partial}{\partial v_i}\left\{D_{ij}^{(1)}(\boldsymbol{v})\frac{\partial}{\partial v_j} + D_{ij}^{(2)}(\boldsymbol{v})\frac{\partial}{\partial r_j}\right\}P(\boldsymbol{r},\boldsymbol{v},t), \tag{12.22}$$

with the diffusion coefficients

$$D_{ij}^{(1)}(\boldsymbol{v}) = \int\limits_0^\infty d\tau e^{-\lambda\tau} B_{ij}\left(\frac{1}{\lambda}\left[e^{\lambda\tau}-1\right]\boldsymbol{v},\tau\right),$$
$$D_{ij}^{(2)}(\boldsymbol{v}) = \frac{1}{\lambda}\int\limits_0^\infty d\tau \left[1-e^{-\lambda\tau}\right] B_{ij}\left(\frac{1}{\lambda}\left[e^{\lambda\tau}-1\right]\boldsymbol{v},\tau\right). \tag{12.23}$$

Note that the approximation of delta-correlated random field corresponds to Eq. (12.20) with the diffusion coefficients

$$D_{ij}^{(1)}(\boldsymbol{v}) = \int\limits_0^\infty d\tau B_{ij}(0,\tau), \quad D_{ij}^{(2)}(\boldsymbol{v}) = 0.$$

Integrating Eq. (12.22) over $\boldsymbol{r}$, we arrive at the Fokker–Planck equation for the one-time probability density of particle velocity

$$\left(\frac{\partial}{\partial t} - \lambda\frac{\partial}{\partial \boldsymbol{v}}\boldsymbol{v}\right) P(\boldsymbol{v},t) = \frac{\partial}{\partial v_i} D_{ij}^{(1)}(\boldsymbol{v}) \frac{\partial}{\partial v_j} P(\boldsymbol{v},t).$$

The steady-state probability density corresponding to the limit process $t \to \infty$ satisfies the equation

$$-\lambda\frac{\partial}{\partial \boldsymbol{v}}\boldsymbol{v}P(\boldsymbol{v}) = \frac{\partial}{\partial v_i} D_{ij}^{(1)}(\boldsymbol{v}) \frac{\partial}{\partial v_j} P(\boldsymbol{v}), \tag{12.24}$$

whose solution essentially depends on the behavior of the diffusion coefficient, i.e., on the correlation function of random vector field $\boldsymbol{f}(\boldsymbol{r},t)$. For example, if we consider the one-dimensional case and specify the correlation function $B_f(x,t)$ in the form

$$B_f(x,t) = \sigma_f^2 \exp\left\{-\frac{|x|}{l_0} - \frac{|t|}{\tau_0}\right\},$$

where l_0 and τ_0 are the spatial and temporal correlation radii, respectively, then we obtain that, for a sufficiently small friction ($\lambda\tau_0 \ll 1$), the solution to Eq. (12.24) has the form [51]

$$P(v) = C\exp\left\{-\frac{\lambda v^2}{2\sigma_f^2\tau_0}\left[1+\frac{2}{3}\frac{|v|\tau_0}{l_0}\right]\right\}. \tag{12.25}$$

For small particle velocity $|v|\tau_0 \ll l_0$, probability distribution (12.25) grades into the Gaussian distribution corresponding to the approximation of the delta-correlated (in time) random field $f(x,t)$. However, in the opposite limiting case $|v|\tau_0 \gg l_0$, probability distribution (12.25) decreases significantly faster than in the case of the approximation of the delta-correlated (in time) random field $f(x,t)$, namely,

$$P(v) = C\exp\left\{-\frac{\lambda v^2|v|}{3\sigma_f^2 l_0}\right\}, \tag{12.26}$$

which corresponds to the diffusion coefficient decreasing according to the law $D^{(1)} \sim 1/|v|$ for great particle velocities. Physically, it means that the effect of random force $f(x,t)$ on faster particles is significantly smaller than on slower ones.

Thus, the diffusion approximation lifts the basic restriction on smallness of the temporal correlation radius τ_0 remaining within the framework of the Markovian process.

12.3 Higher Approximations

In some cases, the diffusion coefficients can vanish in both approximation of the delta-correlated random field and diffusion approximation. Such a situation occurs, for example, when a particle moves in fast random wavefields of velocity [74] (see also [132–134]).

In this case, particle diffusion is described by the equation

$$\frac{d}{dt}\boldsymbol{r}(t) = \boldsymbol{u}(\boldsymbol{r},t), \qquad \boldsymbol{r}(0) = \boldsymbol{r}_0, \tag{12.27}$$

where $\boldsymbol{u}(\boldsymbol{r},t)$ is the statistically homogeneous and stationary random wave vector field such that $\langle \boldsymbol{u}(\boldsymbol{r},t)\rangle = 0$ and the correlation tensor has the form

$$B_{ij}(\boldsymbol{r},t) = \int d\boldsymbol{k}F_{ij}(\boldsymbol{k})\cos\{\boldsymbol{k}\boldsymbol{r}-\omega(\boldsymbol{k})t\}. \tag{12.28}$$

The spectral function $F_{ij}(\boldsymbol{k})$ is such that $\int d\boldsymbol{k}F_{ii}(\boldsymbol{k}) = \sigma_{\boldsymbol{u}}^2$ and $\omega = \omega(\boldsymbol{k}) > 0$ is the equation of the dispersion curve for wave motions. For conventional wave motions, the spectral functions of the velocity satisfies the condition $\Phi_{ij}(0) = 0$, where

$$\Phi_{ij}(\omega) = \int d\boldsymbol{k}F_{ij}(\boldsymbol{k})\delta[\omega - \omega(\boldsymbol{k})],$$

so that the tensor diffusion coefficient in the Fokker–Planck equation vanishes, i.e.,

$$D_{ij} = \int_0^\infty B_{ij}(0,t)dt = 0.$$

The same diffusion coefficient appears in the diffusion approximation for $t \gg \tau_0$, where τ_0 is the temporal correlation radius of the velocity field. Consequently, both approximation of the delta-correlated field of velocity and diffusion approximation give no finite result, and one needs to take into account higher-order terms. The general procedure of constructing the theory of successive approximations was developed in paper [73]. Here, we will adhere to work [74].

Let the maximum of spectral function $F_{ij}(\boldsymbol{k})$ corresponds to a certain wave number k_m, and the maximum of spectral function $\Phi_{ij}(\omega)$, to frequency ω_m. The corresponding spatial and temporal scales are $l = 1/k_m$ and $\tau_0 = 1/\omega_m$. Quantity $\varepsilon = \sigma_{\boldsymbol{u}}\tau_0/l$ appears usually small for actual wavefields and can be used as the basic small parameter of the problem, i.e., $\varepsilon \ll 1$. This case will be considered in Volume 2 in more detail.

Introduce now new field $\widetilde{\boldsymbol{u}}(\boldsymbol{r},t)$ characterized by unit variance and such that $\boldsymbol{u}(\boldsymbol{r},t) = \sigma_{\boldsymbol{u}}\widetilde{\boldsymbol{u}}(\boldsymbol{r},t)$, where $\sigma_{\boldsymbol{u}}^2 = B_{ii}(0,0)$ is the variance of velocities and $\widetilde{\boldsymbol{u}}(\boldsymbol{r},t)$ is the statistically homogeneous and stationary random wave field with mean value $\langle \widetilde{\boldsymbol{u}}(\boldsymbol{r},t)\rangle = 0$ and correlation tensor

$$\langle \widetilde{u}_i(\boldsymbol{r},t)\widetilde{u}_j(\boldsymbol{r}',t')\rangle = B_{ij}(\boldsymbol{r}-\boldsymbol{r}',t-t') \quad (B_{ii}(0,0) = 1)\,.$$

Rewrite Eq. (12.27) in the form

$$\frac{d}{dt}\boldsymbol{r}(t) = \sigma_{\boldsymbol{u}}\widetilde{\boldsymbol{u}}(\boldsymbol{r},t), \quad \boldsymbol{r}(0) = \boldsymbol{r}_0. \tag{12.29}$$

Introduce designations for the indicator function of the particle coordinate,

$$\varphi(\boldsymbol{r},t) = \delta\left(\boldsymbol{r}(t) - \boldsymbol{r}\right), \tag{12.30}$$

and the first and second variational derivatives that will be used later for calculating statistical averages

$$\begin{aligned}
\frac{\delta\varphi(\boldsymbol{r},t)}{\delta\widetilde{u}_i(\boldsymbol{r}',t')} &= \sigma_{\boldsymbol{u}} S_i(\boldsymbol{r},t;\boldsymbol{r}',t'),\\
\frac{\delta^2\varphi(\boldsymbol{r},t)}{\delta\widetilde{u}_i(\boldsymbol{r}',t')\delta\widetilde{u}_j(\boldsymbol{r}'',t'')} &= \sigma_{\boldsymbol{u}}^2 S_{ij}(\boldsymbol{r},t;\boldsymbol{r}',t';\boldsymbol{r}'',t'').
\end{aligned} \tag{12.31}$$

The indicator function satisfies the stochastic Liouville equation

$$\frac{\partial}{\partial t}\varphi(\boldsymbol{r},t) = -\sigma_{\boldsymbol{u}}\frac{\partial}{\partial r_k}\left\{\widetilde{u}_k(\boldsymbol{r},t)\varphi(\boldsymbol{r},t)\right\}, \quad \varphi(\boldsymbol{r},0) = \delta(\boldsymbol{r}-\boldsymbol{r}_0) \tag{12.32}$$

that can be rewritten in the integral form

$$\varphi(\boldsymbol{r},t) = \delta(\boldsymbol{r}-\boldsymbol{r}_0) - \sigma_{\boldsymbol{u}}\frac{\partial}{\partial r_k}\int\limits_0^t d\tau\, \widetilde{u}_k(\boldsymbol{r},\tau)\varphi(\boldsymbol{r},\tau). \tag{12.33}$$

For the first variational derivative (12.31) (recall that it takes on nonzero values only for $t \geq t'$), we obtain the stochastic integral equation

$$S_i(\boldsymbol{r},t;\boldsymbol{r}',t') = \widehat{L}_i(\boldsymbol{r},\boldsymbol{r}')\varphi(\boldsymbol{r},t')\theta(t-t') - \sigma_{\boldsymbol{u}}\frac{\partial}{\partial r_k}\int\limits_{t'}^{t} d\tau\, \widetilde{u}_k(\boldsymbol{r},\tau)S_i(\boldsymbol{r},\tau;\boldsymbol{r}',t'), \tag{12.34}$$

where operator $\widehat{L}_j(\boldsymbol{r},\boldsymbol{r}')$ acts on function $f(\boldsymbol{r})$ according to the formula

$$\widehat{L}_i(\boldsymbol{r},\boldsymbol{r}')f(\boldsymbol{r}) = -\frac{\partial}{\partial r_i}\left\{\delta(\boldsymbol{r}-\boldsymbol{r}')f(\boldsymbol{r})\right\}. \tag{12.35}$$

The stochastic integral equation for the second variational derivative is obtained similarly:

$$\begin{aligned} S_{ij}(\boldsymbol{r},t;\boldsymbol{r}',t';\boldsymbol{r}'',t'') &= \widehat{L}_i(\boldsymbol{r},\boldsymbol{r}')S_j(\boldsymbol{r},t';\boldsymbol{r}'',t'')\theta(t-t')\theta(t'-t'') \\ &+ \widehat{L}_j(\boldsymbol{r},\boldsymbol{r}'')S_i(\boldsymbol{r},t'';\boldsymbol{r}',t')\theta(t-t'')\theta(t''-t') \\ &- \sigma_{\boldsymbol{u}}\frac{\partial}{\partial r_l}\int\limits_{\max\{t',t''\}}^{t} d\tau\, \widetilde{u}_l(\boldsymbol{r},\tau)S_{ij}(\boldsymbol{r},\tau;\boldsymbol{r}',t';\boldsymbol{r}'',t''). \end{aligned} \tag{12.36}$$

If we now average Eq. (12.32) over an ensemble of realizations of field $\{\widetilde{\boldsymbol{u}}(\boldsymbol{r},t)\}$ and take into account the Furutsu–Novikov formula (10.10), page 307, we obtain the equation for the probability density of particle position $P(\boldsymbol{r},t) = \langle\varphi(\boldsymbol{r},t)\rangle$:

$$\begin{gathered} \frac{\partial}{\partial t}P(\boldsymbol{r},t) = -\sigma_{\boldsymbol{u}}^2\frac{\partial}{\partial r_k}\int d\boldsymbol{r}'\int\limits_0^t dt' B_{ki}(\boldsymbol{r}-\boldsymbol{r}',t-t')\left\langle S_i(\boldsymbol{r},t;\boldsymbol{r}',t')\right\rangle, \\ P(\boldsymbol{r},0) = \delta(\boldsymbol{r}-\boldsymbol{r}_0). \end{gathered} \tag{12.37}$$

Integrating Eq. (12.37) with respect to time over the interval (t_1,t), where $t_1 < t$, we obtain the equality

$$P(\boldsymbol{r},t) - P(\boldsymbol{r},t_1) = -\sigma_{\boldsymbol{u}}^2\frac{\partial}{\partial r_l}\int\limits_{t_1}^{t} d\tau\int d\boldsymbol{r}''\int\limits_0^{\tau} dt'' B_{lj}(\boldsymbol{r}-\boldsymbol{r}'',\tau-t'')\left\langle S_j(\boldsymbol{r},\tau;\boldsymbol{r}'',t'')\right\rangle. \tag{12.38}$$

Average now Eq. (12.34) over an ensemble of realizations of field $\{\widetilde{u}_k(\boldsymbol{r},t)\}$. Then we obtain the equation for quantity $\langle S_i(\boldsymbol{r},t;\boldsymbol{r}',t')\rangle$:

$$\begin{aligned} \left\langle S_i(\boldsymbol{r},t;\boldsymbol{r}',t')\right\rangle &= \widehat{L}_i(\boldsymbol{r},\boldsymbol{r}')P(\boldsymbol{r},t')\theta(t-t') \\ &- \sigma_{\boldsymbol{u}}^2\frac{\partial}{\partial r_l}\int\limits_{t'}^{t} d\tau\int d\boldsymbol{r}''\int\limits_0^{t} dt'' B_{lj}(\boldsymbol{r}-\boldsymbol{r}'',\tau-t'')\left\langle S_{ij}(\boldsymbol{r},\tau;\boldsymbol{r}',t';\boldsymbol{r}'',t'')\right\rangle. \end{aligned} \tag{12.39}$$

We approximate function $\langle S_{ij}(\boldsymbol{r},t;\boldsymbol{r}',t'';\boldsymbol{r}'',t)\rangle$ by the expression

$$\langle S_{ij}(\boldsymbol{r},t;\boldsymbol{r}',t'';\boldsymbol{r}'',t)\rangle = \widehat{L}_i(\boldsymbol{r},\boldsymbol{r}')\,\langle S_j(\boldsymbol{r},t';\boldsymbol{r}'',t'')\rangle\,\theta(t-t')\theta(t'-t'') + \widehat{L}_j(\boldsymbol{r},\boldsymbol{r}'')\,\langle S_i(\boldsymbol{r},t'';\boldsymbol{r}',t')\rangle\,\theta(t-t'')\theta(t''-t') \quad (12.40)$$

that corresponds to the neglect of the third-order variational derivatives in Eq. (12.36). Using this approximation and Eq. (12.38), we can rewrite Eq. (12.39) in the form of the closed integral equation:

$$\begin{aligned}\langle S_i(\boldsymbol{r},t;\boldsymbol{r}',t')\rangle &= \widehat{L}_i(\boldsymbol{r},\boldsymbol{r}')P(\boldsymbol{r},t)\theta(t-t')\\
&+ \sigma_{\boldsymbol{u}}^2\widehat{L}_i(\boldsymbol{r},\boldsymbol{r}')\frac{\partial}{\partial r_l}\int\limits_{t'}^{t} d\tau\int d\boldsymbol{r}''\int\limits_{0}^{\tau} dt''B_{lj}(\boldsymbol{r}-\boldsymbol{r}'',\tau-t'')\,\langle S_j(\boldsymbol{r},\tau;\boldsymbol{r}'',t'')\rangle\\
&- \sigma_{\boldsymbol{u}}^2\frac{\partial}{\partial r_l}\int\limits_{t'}^{t} d\tau\int d\boldsymbol{r}''\int\limits_{0}^{t'} dt''B_{lj}(\boldsymbol{r}-\boldsymbol{r}'',\tau-t'')\widehat{L}_i(\boldsymbol{r},\boldsymbol{r}')\,\langle S_j(\boldsymbol{r},t';\boldsymbol{r}'',t'')\rangle\\
&- \sigma_{\boldsymbol{u}}^2\frac{\partial}{\partial r_l}\int\limits_{t'}^{t} d\tau\int d\boldsymbol{r}''\int\limits_{t'}^{\tau} dt''B_{lj}(\boldsymbol{r}-\boldsymbol{r}'',\tau-t'')\widehat{L}_j(\boldsymbol{r},\boldsymbol{r}'')\,\langle S_i(\boldsymbol{r},t'';\boldsymbol{r}',t')\rangle\,.\end{aligned} \quad (12.41)$$

Solving Eq. (12.41) in $\langle S_i(\boldsymbol{r},t;\boldsymbol{r}',t')\rangle$ by the convergence method with respect to parameter $\sigma_{\boldsymbol{u}}^2$ to terms proportional to $\sigma_{\boldsymbol{u}}^2$ (in this case, temporal arguments t_i of functions $P(\boldsymbol{r},t_i)$ can be replaced with t), we have

$$\begin{aligned}\langle S_i(\boldsymbol{r},t;\boldsymbol{r}',t')\rangle = \Big\{&\widehat{L}_i(\boldsymbol{r},\boldsymbol{r}')\\
&+ \sigma_{\boldsymbol{u}}^2\widehat{L}_i(\boldsymbol{r},\boldsymbol{r}')\frac{\partial}{\partial r_l}\int\limits_{t'}^{t} d\tau\int d\boldsymbol{r}''\int\limits_{0}^{\tau} dt''B_{lj}(\boldsymbol{r}-\boldsymbol{r}'',t'')\widehat{L}_j(\boldsymbol{r},\boldsymbol{r}'')\\
&- \sigma_{\boldsymbol{u}}^2\frac{\partial}{\partial r_l}\int\limits_{t'}^{t} d\tau\int d\boldsymbol{r}''\int\limits_{0}^{t'} dt''B_{lj}(\boldsymbol{r}-\boldsymbol{r}'',\tau-t'')\widehat{L}_i(\boldsymbol{r},\boldsymbol{r}')\widehat{L}_j(\boldsymbol{r},\boldsymbol{r}'')\\
&- \sigma_{\boldsymbol{u}}^2\frac{\partial}{\partial r_l}\int\limits_{t'}^{t} d\tau\int d\boldsymbol{r}''\int\limits_{t'}^{\tau} dt''B_{lj}(\boldsymbol{r}-\boldsymbol{r}'',\tau-t'')\widehat{L}_j(\boldsymbol{r},\boldsymbol{r}'')\,\widehat{L}_i(\boldsymbol{r},\boldsymbol{r}')\Big\}\,P(\boldsymbol{r},t).\end{aligned} \quad (12.42)$$

Substituting Eq. (12.42) in Eq. (12.37) and integrating the result over all spatial variables, we arrive at the equation of the third order in $\boldsymbol{r}$ (where the terms containing the first-order derivative and proportional to $\sigma_{\boldsymbol{u}}^4$ can be omitted)

$$
\begin{aligned}
\frac{\partial}{\partial t}P(\boldsymbol{r},t) = &-\sigma_{\boldsymbol{u}}^2 \int\limits_0^t dt' \frac{\partial B_{ki}(0,t')}{\partial r_i}\frac{\partial}{\partial r_k}P(\boldsymbol{r},t) + \sigma_{\boldsymbol{u}}^2 \int\limits_0^t dt' B_{ki}(0,t')\frac{\partial^2}{\partial r_k \partial r_i}P(\boldsymbol{r},t) \\
&+ \sigma_{\boldsymbol{u}}^4 \frac{\partial^2}{\partial r_k \partial r_l} \int\limits_0^t dt' B_{ki}(0,t-t') \int\limits_{t'}^t d\tau \int\limits_0^{t'} dt'' \frac{\partial^2 B_{lj}(0,\tau-t'')}{\partial r_i \partial r_j} P(\boldsymbol{r},t) \\
&+ \sigma_{\boldsymbol{u}}^4 \frac{\partial^2}{\partial r_k \partial r_l} \int\limits_0^t dt' \frac{\partial^2 B_{ki}(0,t-t')}{\partial r_i \partial r_j} \int\limits_{t'}^t d\tau \int\limits_{t'}^{\tau} dt'' B_{lj}(0,\tau-t'') P(\boldsymbol{r},t) \\
&+ \sigma_{\boldsymbol{u}}^4 \frac{\partial^2}{\partial r_k \partial r_j} \int\limits_0^t dt' \frac{\partial B_{ki}(0,t-t')}{\partial r_l} \int\limits_{t'}^t d\tau \int\limits_0^{t'} dt'' \frac{\partial B_{lj}(0,\tau-t'')}{\partial r_i} P(\boldsymbol{r},t) \\
&+ \sigma_{\boldsymbol{u}}^4 \frac{\partial^2}{\partial r_k \partial r_i} \int\limits_0^t dt' \frac{\partial B_{ki}(0,t-t')}{\partial r_l} \int\limits_{t'}^t d\tau \int\limits_0^{\tau} dt'' \frac{\partial B_{lj}(0,t'')}{\partial r_j} P(\boldsymbol{r},t) \\
&+ \sigma_{\boldsymbol{u}}^4 \frac{\partial^2}{\partial r_k \partial r_j} \int\limits_0^t dt' \frac{\partial^2 B_{ki}(0,t-t')}{\partial r_i \partial r_l} \int\limits_{t'}^t d\tau \int\limits_0^{\tau} dt'' B_{lj}(0,t'') P(\boldsymbol{r},t) \\
&+ \sigma_{\boldsymbol{u}}^4 \frac{\partial^2}{\partial r_k \partial r_i} \int\limits_0^t dt' \frac{\partial^2 B_{ki}(0,t-t')}{\partial r_l \partial rj} \int\limits_{t'}^t d\tau \int\limits_{t'}^{\tau} dt'' B_{lj}(0,\tau-t'') P(\boldsymbol{r},t) \\
&- \sigma_{\boldsymbol{u}}^4 \frac{\partial^3}{\partial r_k \partial r_l \partial r_j} \int\limits_0^t dt' B_{ki}(0,t-t') \int\limits_{t'}^t d\tau \int\limits_0^{t'} dt'' \frac{\partial B_{lj}(0,\tau-t'')}{\partial r_i} P(\boldsymbol{r},t) \\
&- \sigma_{\boldsymbol{u}}^4 \frac{\partial^3}{\partial r_k \partial r_l \partial r_i} \int\limits_0^t dt' \frac{\partial B_{ki}(0,t-t')}{\partial r_j} \int\limits_{t'}^t d\tau \int\limits_{t'}^{\tau} dt'' B_{lj}(0,\tau-t'') P(\boldsymbol{r},t) \\
&- \sigma_{\boldsymbol{u}}^4 \frac{\partial^3}{\partial r_k \partial r_i \partial r_j} \int\limits_0^t dt' \frac{\partial B_{ki}(0,t-t')}{\partial r_l} \int\limits_{t'}^t d\tau \int\limits_0^{\tau} dt'' B_{lj}(0,t'') P(\boldsymbol{r},t). \qquad (12.43)
\end{aligned}
$$

Generally, Eq. (12.43) cannot be considered as the equation for probability density because the solution can be negative in the region of small values. Nevertheless, the solution of Eq. (12.43) adequately describe statistical moments and, in this sense, Eq. (12.43) generalizes the Fokker–Planck equation. Using now velocity field in the spectral representation (12.28) and taking into account properties of this representation, we can calculate integrals in the coefficients of Eq. (12.43) and obtain the equation for great times ($t \gg \tau_0$) in the form

$$
\begin{aligned}
\frac{\partial}{\partial t}P(\boldsymbol{r},t) &= -\sigma_{\boldsymbol{u}}^2 \int \frac{d\boldsymbol{k}}{\omega(\boldsymbol{k})} k_i F_{ki}(\boldsymbol{k}) \frac{\partial}{\partial r_k} P(\boldsymbol{r},t) \\
&+ \sigma_{\boldsymbol{u}}^4 \frac{\pi}{2} \int d\boldsymbol{k}_1 \int \frac{d\boldsymbol{k}_2}{\omega_2^2} k_{1l} k_{1j} F_{ki}(\boldsymbol{k}_1) F_{lj}(\boldsymbol{k}_2) \delta\left(\omega_1 - \omega_2\right) \frac{\partial^2}{\partial r_k \partial r_i} P(\boldsymbol{r},t) \\
&+ \sigma_{\boldsymbol{u}}^4 \frac{\pi}{2} \int d\boldsymbol{k}_1 \int \frac{d\boldsymbol{k}_2}{\omega_2^2} k_{1l} k_{2i} F_{ki}(\boldsymbol{k}_1) F_{lj}(\boldsymbol{k}_2) \delta\left(\omega_1 - \omega_2\right) \frac{\partial^2}{\partial r_k \partial r_j} P(\boldsymbol{r},t),
\end{aligned}
\tag{12.44}
$$

where $\omega_1 = \omega(\boldsymbol{k}_1)$ and $\omega_2 = \omega(\boldsymbol{k}_2)$.

Equation (12.44) is now the Fokker–Planck equation and describes the probability density of the position of a particle transferred by the statistically homogeneous Gaussian wave velocity field.

In the case of isotropic fluctuations of field $\widetilde{\boldsymbol{u}}(\boldsymbol{r},t)$, Eq. (12.44) is simplified and assumes the form

$$
\frac{\partial}{\partial t}P(\boldsymbol{r},t) = D\frac{\partial^2}{\partial \boldsymbol{r}^2}P(\boldsymbol{r},t)
$$

corresponding to the Gaussian random vector process $\boldsymbol{r}(t)$ with mean value $\langle \boldsymbol{r}(t) \rangle = \boldsymbol{r}_0$ and variance

$$
\sigma_{\boldsymbol{r}}^2(t) = \left\langle \left(\boldsymbol{r}(t) - \boldsymbol{r}_0\right)^2 \right\rangle = 2dD\,t,
$$

where d is the dimension of space and

$$
D = \sigma_{\boldsymbol{u}}^4 \frac{\pi}{2d} \int d\boldsymbol{k}_1 \int \frac{d\boldsymbol{k}_2}{\omega_2^2} k_{1l} k_{1j} F_{ii}(\boldsymbol{k}_1) F_{lj}(\boldsymbol{k}_2) \delta(\omega_1 - \omega_2)
$$

is the diffusion coefficient. In this case, the structure of the wave field spectral tensor has the form

$$
F_{ki}(\boldsymbol{k}) = F^{\mathrm{s}}(k)\left(\delta_{ik} - \frac{k_i k_k}{\boldsymbol{k}^2}\right) + F^{\mathrm{p}}(k)\frac{k_i k_k}{\boldsymbol{k}^2},
$$

where $F^{\mathrm{s}}(k)$ and $F^{\mathrm{p}}(k)$ are the solenoidal and potential components, respectively and $\omega(\boldsymbol{k}) \equiv \omega(k)$; as a result, the diffusion coefficient is given by the expression

$$
\begin{aligned}
D &= \sigma_{\boldsymbol{u}}^4 \frac{\pi}{2d} \int \frac{d\boldsymbol{k}_1}{\omega_1^2} k_1^2 F_{ii}(\boldsymbol{k}_1) \int d\boldsymbol{k}_2 F_{ll}(\boldsymbol{k}_2) \delta(\omega_1 - \omega_2) \\
&= \sigma_{\boldsymbol{u}}^4 \frac{\pi}{2d} \int \frac{d\boldsymbol{k}_1}{\omega_1^2} k_1^2 \left[F^{\mathrm{s}}(k_1)\left(N-1\right) + F^{\mathrm{p}}(k_1)\right]^2 \int d\boldsymbol{k}_2 \delta(\omega_1 - \omega_2).
\end{aligned}
$$

In the case of anisotropic medium, vector process $\boldsymbol{r}(t)$ shows spatial asymmetry and the corresponding mean value and variance are given by the expressions

$$\langle r_m(t) \rangle = r_{0m} + t\sigma_{\boldsymbol{u}}^2 \int \frac{d\boldsymbol{k}}{\omega(\boldsymbol{k})} k_i F_{mi}(\boldsymbol{k}),$$

$$\sigma_{\boldsymbol{r}}^2(t) = \left\langle \boldsymbol{r}^2(t) - \langle \boldsymbol{r}(t) \rangle^2 \right\rangle = t\sigma_{\boldsymbol{u}}^4 \pi \int d\boldsymbol{k}_1 \int \frac{d\boldsymbol{k}_2}{\omega_2^2} k_{1l} k_{1j} F_{ii}(\boldsymbol{k}_1) F_{lj}(\boldsymbol{k}_2) \delta(\omega_1 - \omega_2)$$
$$+ t\sigma_{\boldsymbol{u}}^4 \pi \int d\boldsymbol{k}_1 \int \frac{d\boldsymbol{k}_2}{\omega_2^2} k_{1l} k_{2i} F_{ki}(\boldsymbol{k}_1) F_{lk}(\boldsymbol{k}_2) \delta\left(\omega_1 - \omega_2\right).$$

The diffusion coefficient appears to be proportional to the square of the velocity field variance, rather than simply to the velocity field variance. This fact follows from the absence of wave–particle resonances in this problem, which decreases the variance of the velocity of random particle drift by an order of magnitude.

Chapter 13
Some Other Approximate Approaches to the Problems of Statistical Hydrodynamics

13.1 Quasi-Elastic Properties of Isotropic and Stationary Noncompressible Turbulent Media

In Part 2, we analyzed statistics of solutions to the nonlinear equations of hydrodynamics using the rigorous approach based on deriving and investigating the exact variational differential equations for characteristic functionals of nonlinear random fields. However, this approach encounters severe difficulties caused by the lack of development of the theory of variational differential equations. For this reason, many researchers prefer to proceed from more habitual partial differential equations for different moment functions of fields of interest. The nonlinearity of the input dynamic equations governing random fields results in the appearance of higher moment functions of fields of interest in the equations governing any moment function. As a result, even determination of average field or correlation functions requires, in the strict sense, solving an infinite system of linked equations.

Thus, the main problem of this approach consists in cutting the mentioned system of equations by means of one or another physical hypothesis. The most known example of such a hypothesis is the *Millionshchikov hypothesis* according to which the higher moment functions of even orders are expressed in terms of the lower ones by the laws of the Gaussian statistics. The disadvantage of such approaches consists in the fact that the validity of hypothesis usually cannot be proved; moreover, cutting the system of equations may often yield physically contradictory results, namely, energy spectra of turbulence can appear negative for certain wave numbers. Nevertheless, these approximate approaches provide a deeper insight into physical mechanisms of forming the statistics of strongly nonlinear random fields and make it possible to derive quantitative expressions for fields' correlation functions and spectra. It seems that the Millionshchikov hypothesis provides correct spectra of the developed turbulence in viscous interval [112].

V.I. Klyatskin, *Stochastic Equations: Theory and Applications in Acoustics, Hydrodyn., Magnetohydrodyn., and Radiophys., Vol. 1*, Understanding Complex Systems,
DOI: 10.1007/978-3-319-07587-7_13, © Springer International Publishing Switzerland 2015

We emphasize additionally that the mentioned approximate equations reveal many nontrivial effects inherent in nonlinear random fields and having no analogs in the behavior of deterministic fields and waves. Here, we illustrate these methods of analysis by the example of an interesting physical effect, which consists in the fact that average flows of noncompressible fluids superimposed on the background of developed turbulent pulsations acquire quasi-elastic wave properties. This effect was first mentioned by Moffatt [111] who studied the reaction of turbulence on variations of the transverse gradient of average velocity. Moffatt noted that turbulent medium behaves in some sense like an elastic medium; namely, variations of average velocity profile of a plane-parallel flow satisfy the wave equation.

Let $\boldsymbol{u}^{\mathrm{T}}(\boldsymbol{r},t)$ is the random velocity field of developed turbulence stationary in time and homogeneous and isotropic in space (we assume that $\left\langle \boldsymbol{u}^{\mathrm{T}}(\boldsymbol{r},t)\right\rangle = 0$). In actuality, regular average flows are imposed on turbulent pulsations of fluid. We denote $\boldsymbol{U}(\boldsymbol{r},t)$ the velocity field of these regular flows. Because the turbulent pulsations and average flows interact nonlinearly, we can represent the total velocity field in the form

$$\boldsymbol{u}(\boldsymbol{r},t) = \boldsymbol{U}(\boldsymbol{r},t) + \boldsymbol{u}^{\mathrm{T}}(\boldsymbol{r},t) + \boldsymbol{u}'(\boldsymbol{r},t), \tag{13.1}$$

Here, $\boldsymbol{u}^{\mathrm{T}}(\boldsymbol{r},t)$ is the unperturbed turbulent field in the absence of average flows and $\boldsymbol{u}'(\boldsymbol{r},t)$ is the turbulent field disturbance caused by the interaction with the regular flow (we assume that $\langle \boldsymbol{u}'(\boldsymbol{r},t)\rangle = 0$).

Investigation of nonlinear interactions between regular flows and turbulent pulsations (and the analysis of turbulent pulsations $\boldsymbol{u}^{\mathrm{T}}(\boldsymbol{r},t)$ by themselves, though) is very difficult in the general case. However, in the case of weak average flows, i.e., under the condition that

$$\boldsymbol{U}^2(\boldsymbol{r},t) \ll 2T(t),$$

where $T(t) = \dfrac{1}{2}\left\langle \left[\boldsymbol{u}^{\mathrm{T}}(\boldsymbol{r},t)\right]^2\right\rangle$ is the average density of turbulent energy pulsation, we can discuss the effect of turbulence on the average flow evolution in sufficient detail by considering the linear approximation in small disturbances of fields $\boldsymbol{U}(\boldsymbol{r},t)$ and $\boldsymbol{u}'(\boldsymbol{r},t)$ and assuming known the spectrum of turbulent pulsations $\boldsymbol{u}^{\mathrm{T}}(\boldsymbol{r},t)$.

The total velocity field (13.1) and its turbulent component $\boldsymbol{u}^{\mathrm{T}}(\boldsymbol{r},t)$ satisfy the Navier–Stokes equation (1.164), page 68. Hence, substituting Eq. (13.1) in Eq. (1.164) and linearizing the equations in average field $\boldsymbol{U}(\boldsymbol{r},t)$ and fluctuation component $\boldsymbol{u}'(\boldsymbol{r},t)$, we arrive at the approximate system of equations

$$\frac{\partial}{\partial t}U_i(\boldsymbol{r},t) + \frac{\partial}{\partial r_k}T_{ik}(\boldsymbol{r},t) = -\frac{\partial}{\partial r_i}P(\boldsymbol{r},t), \tag{13.2}$$

$$\frac{\partial}{\partial t}u_i'(\boldsymbol{r},t)+\frac{\partial}{\partial r_k}\left[u_i(\boldsymbol{r},t)u_k(\boldsymbol{r},t)-\langle u_i(\boldsymbol{r},t)u_k(\boldsymbol{r},t)\rangle\right]$$
$$+u_l^{\mathrm{T}}(\boldsymbol{r},t)\frac{\partial U_i(\boldsymbol{r},t)}{\partial r_l}+U_l(\boldsymbol{r},t)\frac{\partial u_i^{\mathrm{T}}(\boldsymbol{r},t)}{\partial r_l}=-\frac{\partial}{\partial r_i}p'(\boldsymbol{r},t), \quad (13.3)$$

where $T_{ik}(\boldsymbol{r},t)=\langle u_i(\boldsymbol{r},t)u_k(\boldsymbol{r},t)\rangle$ is the *Reynolds stress tensor* and $P(\boldsymbol{r},t)$ and $p'(\boldsymbol{r},t)$ are the average and perturbed turbulent pressure components, respectively. Here and below, we neglect the effect of viscosity on dynamics and statistics of perturbed fields $\boldsymbol{U}(\boldsymbol{r},t)$ and $\boldsymbol{u}'(\boldsymbol{r},t)$. In addition, we will bear in mind that fields $\boldsymbol{U}(\boldsymbol{r},t)$ and $\boldsymbol{u}'(\boldsymbol{r},t)$ satisfy the incompressibility condition (1.164)

$$\boldsymbol{\nabla}\boldsymbol{U}(\boldsymbol{r},t)=0, \quad \boldsymbol{\nabla}\boldsymbol{u}'(\boldsymbol{r},t)=0. \quad (13.4)$$

Being combined with Eq. (1.164), page 68 for turbulent pulsations $\boldsymbol{u}^{\mathrm{T}}(\boldsymbol{r},t)$, Eq. (13.3) yields the following equation for the Reynolds stress tensor $T_{ik}(\boldsymbol{r},t)$

$$\frac{\partial}{\partial t}T_{ik}(\boldsymbol{r},t)+\frac{\partial}{\partial r_l}\langle u_i(\boldsymbol{r},t)u_k(\boldsymbol{r},t)u_l(\boldsymbol{r},t)\rangle$$
$$+\left\langle u_k^{\mathrm{T}}(\boldsymbol{r},t)u_l^{\mathrm{T}}(\boldsymbol{r},t)\right\rangle\frac{\partial U_i(\boldsymbol{r},t)}{\partial r_l}+\left\langle u_i^{\mathrm{T}}(\boldsymbol{r},t)u_l^{\mathrm{T}}(\boldsymbol{r},t)\right\rangle\frac{\partial U_k(\boldsymbol{r},t)}{\partial r_l}$$
$$+U_l(\boldsymbol{r},t)\frac{\partial\left\langle u_i^{\mathrm{T}}(\boldsymbol{r},t)u_k^{\mathrm{T}}(\boldsymbol{r},t)\right\rangle}{\partial r_l}-\left\langle\left[u_k^{\mathrm{T}}(\boldsymbol{r},t)\frac{\partial}{\partial r_i}p'(\boldsymbol{r},t)+u_i^{\mathrm{T}}(\boldsymbol{r},t)\frac{\partial}{\partial r_k}p'(\boldsymbol{r},t)\right]\right\rangle. \quad (13.5)$$

Here,

$$u_i(\boldsymbol{r},t)u_k(\boldsymbol{r},t)u_l(\boldsymbol{r},t)=u_i^{\mathrm{T}}(\boldsymbol{r},t)u_k^{\mathrm{T}}(\boldsymbol{r},t)u_l'(\boldsymbol{r},t)+\cdots.$$

Pressure pulsations p' in this equations can be expressed in terms of perturbed velocities $\boldsymbol{U}(\boldsymbol{r},t)$ and $\boldsymbol{u}'(\boldsymbol{r},t)$,

$$p'(\boldsymbol{r},t)=-\Delta^{-1}(\boldsymbol{r},\boldsymbol{r}')\left\{\frac{\partial^2}{\partial r_m'\partial r_n'}\left[u_m(\boldsymbol{r}',t)u_n(\boldsymbol{r}',t)-\langle u_m(\boldsymbol{r}',t)u_n(\boldsymbol{r}',t)\rangle\right]\right.$$
$$\left.+2\frac{\partial u_l^{\mathrm{T}}(\boldsymbol{r}',t)}{\partial r_m'}\frac{\partial U_m(\boldsymbol{r}',t)}{\partial r_l'}\right\},$$

where $\Delta^{-1}(\boldsymbol{r},\boldsymbol{r}')$ is the integral operator inverse to the Laplace operator. Equations (13.2), (13.5), and the expression for pressure $p'(\boldsymbol{r},t)$ form the system of equations in average field $\boldsymbol{U}(\boldsymbol{r},t)$ and stress tensor $T_{ik}(\boldsymbol{r},t)$. These equations are not closed because they depend on higher-order velocity correlators like the triple correlator $\left\langle u_i^{\mathrm{T}}(\boldsymbol{r},t)u_k^{\mathrm{T}}(\boldsymbol{r},t)u_l'(\boldsymbol{r},t)\right\rangle$. Assuming that such correlators only slightly affect the dynamics of perturbations and taking into account conditions (13.4) and statistical homogeneity of field $\boldsymbol{u}^{\mathrm{T}}(\boldsymbol{r},t)$, we reduce the system of equations (13.2), (13.5) to the form $\left(\tau_i(\boldsymbol{r},t)=\dfrac{\partial}{\partial r_k}T_{ik}(\boldsymbol{r},t)\right)$

$$\frac{\partial}{\partial t}U_i(\boldsymbol{r},t)+\tau_i(\boldsymbol{r},t)=-\frac{\partial}{\partial r_i}P(\boldsymbol{r},t), \tag{13.6}$$

$$\frac{\partial}{\partial t}\tau_i(\boldsymbol{r},t)+\left\langle u_k^{\mathrm{T}}(\boldsymbol{r},t)u_l^{\mathrm{T}}(\boldsymbol{r},t)\right\rangle\frac{\partial^2 U_i(\boldsymbol{r},t)}{\partial r_l\partial r_k}$$

$$=2\frac{\partial}{\partial r_k}\left[\left\langle u_k^{\mathrm{T}}(\boldsymbol{r},t)\frac{\partial}{\partial r_i}\Delta^{-1}(\boldsymbol{r},\boldsymbol{r}')\left(\frac{\partial u_l^{\mathrm{T}}(\boldsymbol{r}',t)}{\partial r'_m}\frac{\partial U_m(\boldsymbol{r}',t)}{\partial r'_l}\right)\right\rangle\right.$$

$$\left.+\left\langle u_i^{\mathrm{T}}(\boldsymbol{r},t)\frac{\partial}{\partial r_k}\Delta^{-1}(\boldsymbol{r},\boldsymbol{r}')\left(\frac{\partial u_l^{\mathrm{T}}(\boldsymbol{r}',t)}{\partial r'_m}\frac{\partial U_m(\boldsymbol{r}',t)}{\partial r'_l}\right)\right\rangle\right]. \tag{13.7}$$

This system of equations is closed in $\boldsymbol{U}(\boldsymbol{r},t)$. The coefficients in the left-hand side of Eq. (13.7) and the integral operator in the right-hand side can be expressed in terms of the correlation tensor of vortex field of unperturbed turbulence $\boldsymbol{u}^{\mathrm{T}}(\boldsymbol{r},t)$

$$\left\langle u_i^{\mathrm{T}}(\boldsymbol{r},t)u_j^{\mathrm{T}}(\boldsymbol{r}',t)\right\rangle=\int d\boldsymbol{q}\varPhi_{ij}(\boldsymbol{q})e^{i\boldsymbol{q}(\boldsymbol{r}-\boldsymbol{r}')}. \tag{13.8}$$

In view of the fact that field $\boldsymbol{u}^{\mathrm{T}}(\boldsymbol{r},t)$ is the solenoidal field, we have

$$\varPhi_{ij}(\boldsymbol{q})=\Delta_{ij}(\boldsymbol{q})F(q),$$

where

$$\Delta_{ij}(\boldsymbol{q})=\delta_{ij}-\frac{q_iq_j}{\boldsymbol{q}^2}.$$

We will assume the energy spectrum of turbulent pulsations $F(q)$ known. From Eq. (13.8) follows in particular that the energy of turbulent pulsations is expressed through the energy spectrum by the relationship

$$T=\frac{1}{2}\left\langle\left[\boldsymbol{u}^{\mathrm{T}}(\boldsymbol{r},t)\right]^2\right\rangle=4\pi\int\limits_0^\infty dq\,q^2F(q),$$

and coefficients in the left-hand side of Eq. (13.7) are given by the formula

$$\left\langle u_k^{\mathrm{T}}(\boldsymbol{r},t)u_l^{\mathrm{T}}(\boldsymbol{r},t)\right\rangle=\frac{2}{3}T\delta_{kl}. \tag{13.9}$$

Finally, using Eq. (13.8), the right-hand side of Eq. (13.7) can be represented in the form

$$-\int d\boldsymbol{q} e^{i\boldsymbol{q}\boldsymbol{r}} U_m(\boldsymbol{q},t) g_{im}(\boldsymbol{q}), \tag{13.10}$$

where $\boldsymbol{U}(\boldsymbol{q},t)$ is the spatial Fourier transform of average velocity field $\boldsymbol{U}(\boldsymbol{r},t)$ and tensor $g_{im}(\boldsymbol{q})$ is given by the formula

$$g_{im}(\boldsymbol{p}) = 2\int d\boldsymbol{q}\frac{q_m p_l p_k}{(\boldsymbol{q}+\boldsymbol{p})^2}\left[(q_i+p_i)\,\Delta_{kl}(\boldsymbol{q}) + (q_k+p_k)\,\Delta_{il}(\boldsymbol{q})\right]F(q). \tag{13.11}$$

From the obvious fact that tensor $g_{im}(\boldsymbol{p})$ is invariant relative rotations in space follows that it must have the form

$$g_{im}(\boldsymbol{p}) = A(p)\delta_{im} + B(p)\frac{p_i p_m}{\boldsymbol{p}^2}. \tag{13.12}$$

Moreover, the term proportional to $B(p)$ disappears in Eq. (13.12) in view of Eq. (13.4) (it expresses the property of incompressibility of average field $\boldsymbol{U}(\boldsymbol{r},t)$) and the identity $\boldsymbol{p}\boldsymbol{U}(\boldsymbol{p},t) = 0$ following from Eq. (13.4), so that the right-hand side (13.10) of Eq. (13.7) assumes the form

$$-\int d\boldsymbol{q} e^{i\boldsymbol{q}\boldsymbol{r}} U_i(\boldsymbol{q},t) A(q).$$

Substituting this expression in the right-hand side of Eq. (13.7) and taking into account Eq. (13.9), we rewrite Eqs. (13.6), (13.7) in the form

$$\begin{aligned} &\frac{\partial}{\partial t}U_i(\boldsymbol{r},t) + \tau_i(\boldsymbol{r},t) = -\frac{\partial}{\partial r_i}P(\boldsymbol{r},t), \\ &\frac{\partial}{\partial t}\tau_i(\boldsymbol{r},t) + \frac{2}{3}T\Delta U_i(\boldsymbol{r},t) = -A(-i\boldsymbol{\nabla})U_i(\boldsymbol{r},t). \end{aligned} \tag{13.13}$$

The kernel of the integral operator appeared in the second equation can be obtained from comparison of Eq. (13.11) with (13.12). The result is as follows

$$A(p) = \int d\boldsymbol{q}\frac{F(q)}{(q^2+p^2)}\left[p^2 - \frac{(\boldsymbol{q}\boldsymbol{p})^2}{q^2}\right]\left[q^2 - (\boldsymbol{q}\boldsymbol{p}) - 2\frac{(\boldsymbol{q}\boldsymbol{p})^2}{p^2}\right]. \tag{13.14}$$

We can perform integration in Eq. (13.14) over angular coordinate to obtain the expression

$$A(p) = 2\pi p^2\int_0^\infty dq\,F(q)\left\{\frac{2q^2}{3p^2} - \frac{q^4-p^4}{4p^4} + \frac{(q^2-p^2)^3}{8qp^5}\ln\left|\frac{q+p}{q-p}\right|\right\}. \tag{13.15}$$

We note additionally that Eq. (13.13) with allowance for Eq. (13.4) yields the identity $\Delta P(\boldsymbol{r},t) = 0$, so that $P(\boldsymbol{r},t) = P_0 = \text{const}$. Then, eliminating quantity $\tau_i(\boldsymbol{r},t)$ from system of equations (13.11) and (13.4), we arrive at a single equation in the vector field of average velocity $\boldsymbol{U}(\boldsymbol{r},t)$ of flow

$$\frac{\partial^2}{\partial t^2}\boldsymbol{U}(\boldsymbol{r},t) - \left[\frac{2}{3}T\Delta + A(-i\boldsymbol{\nabla})\right]\boldsymbol{U}(\boldsymbol{r},t) = 0. \tag{13.16}$$

The corresponding dispersion equation

$$\omega^2(p) = \frac{2}{3}Tp^2 - A(p)$$

can be rewritten in the form

$$\omega^2(p) = 2\pi p^2 \int\limits_0^\infty dq\, F(q) f\left(\frac{q}{p}\right),$$

where

$$f(x) = \frac{2-x^2}{3} + \frac{x^4-1}{8} - \frac{\left(x^2-1\right)^3}{16x}\ln\left|\frac{1+x}{1-x}\right|.$$

Function $f(x)$ has the following asymptotics

$$f(x) = \begin{cases} 2/3, & x \ll 1, \\ 4/15 & x \gg 1. \end{cases}$$

Moreover, the inequalities

$$\frac{2}{3} \geq f(x) \geq \frac{4}{15}$$

hold for arbitrary x.

Thus, the time-dependent development of disturbances in average flow is governed by the hyperbolic equation (13.16). As a result, the turbulent medium possesses certain quasi-elastic properties; namely, disturbances diffuse in the turbulent medium as transverse waves showing dispersion property. The phase and group velocities of these waves vary in the limits

$$\sqrt{\frac{2}{3}T} \geq c \geq \sqrt{\frac{4}{15}T}.$$

13.2 Sound Radiation by Vortex Motions

In the previous section, we considered the physical effect immediately related to statistical averaging of the nonlinear system of hydrodynamical equations in the case of an incompressible flow. Here, we consider an effect related to weakly compressible media; namely, we consider the problem on sound radiation by a weakly compressible medium. This problem corresponds to the inclusion of random fields in the linearized equations of hydrodynamics. Note that, within the framework of linear equations, parametric action of turbulent medium yields the equations of acoustics with random refractive index. We dealt with such problems in Chapter 13, Vol. 2

Turbulent motion of a flow in certain finite spatial region excites acoustic waves outside this region. In the case of a weakly compressible flow, this sound field is such as if it were generated by the static distribution of acoustic quadrupoles whose instantaneous power per unit volume is given by the relationship

$$T_{ij}(\boldsymbol{r},t) = \rho_0 v_i^{\mathrm{T}}(\boldsymbol{r},t) v_j^{\mathrm{T}}(\boldsymbol{r},t),$$

where ρ_0 is the average density and $v_i^{\mathrm{T}}(\boldsymbol{r},t)$ are the components of the velocity of flow in turbulent region V inside which the turbulence is assumed homogeneous and isotropic in space and stationary in time (we use the coordinate system in which the whole of the flow is quiescent). Turbulent motions cause the fluctuating waves of density $\rho(\boldsymbol{r},t)$ that satisfy the wave equation

$$\left(\frac{\partial^2}{\partial t^2} - c_0^2 \Delta\right) \rho(\boldsymbol{r},t) = \frac{\partial^2}{\partial r_i \partial r_j} T_{ij}(\boldsymbol{r},t), \tag{13.17}$$

where c_0 is the sound velocity in the homogeneous portion of the medium, i.e., outside the region of turbulent motions.

The solution to this equation has the form of the retarded solution

$$\rho(\boldsymbol{r},t) = \frac{1}{4\pi c_0^2} \frac{\partial^2}{\partial r_i \partial r_j} \int\limits_V d\boldsymbol{y} T_{ij}\left(\boldsymbol{y}, t - \frac{|\boldsymbol{r}-\boldsymbol{y}|}{c_0}\right).$$

For distances significantly exceeding linear sizes of turbulent region V, this solution can be reduced to the asymptotic expression

$$\rho(\boldsymbol{r},t) = \frac{1}{4\pi c_0^4} \frac{r_i r_j}{x^3} \int\limits_V d\boldsymbol{y} \ddot{T}_{ij}\left(\boldsymbol{y}, t - \frac{|\boldsymbol{r}-\boldsymbol{y}|}{c_0}\right), \tag{13.18}$$

where $\ddot{T}_{ij}(\boldsymbol{y},t) = \dfrac{\partial^2}{\partial t^2} T_{ij}(\boldsymbol{y},t)$. Correspondingly, average energy flux density of acoustic wave excited by turbulent motions $q(\boldsymbol{r},t) = \dfrac{c_0^3 \left\langle \rho^2(\boldsymbol{r},t)\right\rangle}{\rho_0}$ is given by the expression

$$q(\boldsymbol{r},t) = \frac{1}{16\pi^2 c_0^5} \frac{r_i r_j r_k r_l}{r^6} \int\limits_V d\boldsymbol{y} \int\limits_V d\boldsymbol{z} \left\langle \ddot{T}_{ij}\left(\boldsymbol{y}, t - \frac{|\boldsymbol{r}-\boldsymbol{y}|}{c_0}\right) \ddot{T}_{kl}\left(\boldsymbol{z}, t - \frac{|\boldsymbol{r}-\boldsymbol{z}|}{c_0}\right)\right\rangle. \tag{13.19}$$

Further calculations require the knowledge of the space–time correlators of the velocity field. The current theory of strong turbulence is incapable of yielding the corresponding relationships. For this reason, investigators limit themselves to various plausible hypotheses that allow complete calculations to be performed (see, e.g., [89–91, 112]). In particular, it was shown that, with allowance for incompressibility of flow in volume V and the use of the Kolmogorov–Obukhov hypothesis and a number of simplifying hypotheses for splitting space-time correlators, from Eq. (13.19) follows that both average energy flux density and acoustic power are proportional to $\sim M^5$, where $M = \dfrac{\sqrt{\langle \boldsymbol{v}^2(\boldsymbol{r},t)\rangle}}{c_0}$ is the *Mach number* (significantly smaller than unity). We note that this result can be explained purely hydrodynamically, by analyzing vortex interactions in weakly compressible medium [41]. The simplest sound-radiating vortex systems are the pair of vortex lines (radiating cylindrical waves) and the pair of vortex rings (radiating spherical waves).

13.2.1 Sound Radiation by Vortex Lines

Consider two parallel vortex lines separated by distance $2h$ and characterized by equal intensities $\kappa = \dfrac{1}{2}\pi\xi\sigma$, where ξ is the vorticity (the size of the vortex uniformly distributed over the area of infinitely small section σ), so that the circulation about each of vortex line is $\Gamma = 2\pi\kappa$. We will call these vortex lines simply vortices. In a noncompressible flow, these vortices revolve with angular velocity $\omega = \dfrac{\kappa}{2h^2}$ around the center of the line connecting these vortices (see, fig. 13.1a and e.g., [109]).

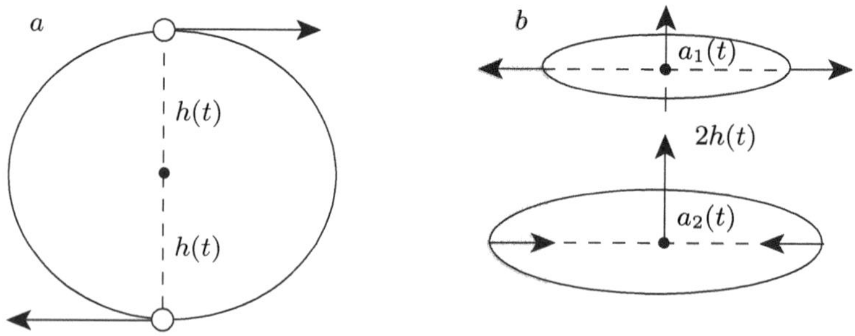

Fig. 13.1 Schematics of dynamics of a) vortex lines and (b) vortex rings

Select the coordinate system with the origin at a fixed point and z-axis along the vortex line. In this coordinate system, velocity potential $\varphi_0(\boldsymbol{r},t)$ ($\boldsymbol{v}_0(\boldsymbol{r},t) = -\mathrm{Re}\boldsymbol{\nabla}\varphi_0(\boldsymbol{r},t)$) and squared velocity $\boldsymbol{v}_0^2(\boldsymbol{r},t)$ assume the forms

$$\varphi_0(\boldsymbol{r},t) = ik\ln\left[r^2 e^{2i\theta} - h^2 e^{2i\omega t}\right], \quad \boldsymbol{v}_0^2(\boldsymbol{r},t) = \frac{4\kappa^2 r^2}{r^4 + h^4 - 2r^2h^2\cos 2(\omega t - \theta)}. \tag{13.20}$$

Here $re^{i\theta}$ is the radius-vector of the observation point.

According to the Bernoulli equation, local velocity pulsations described by Eq. (13.20) must produce the corresponding pressure pulsations; in the case of weakly compressible medium, namely, under the condition that $M \ll 1$, where $M = \dfrac{\kappa}{2hc_0}$ is the Mach number and c_0 is the sound velocity, these pressure pulsations will propagate at large distances as sound waves.

Encircle the origin with a circle of radius R such that $h \ll R \ll \lambda$, where λ is the sound wavelength. This is possible because

$$\frac{\lambda}{h} = \frac{\pi}{M} \gg 1 \quad \text{for} \quad M \ll 1.$$

In region $r < R$, dynamics of flow approximately coincides with the dynamics of incompressible flow. In other words, the dynamics of flow is described by Eqs. (13.20) in this region.

In region $r > R$, equations of motion coincide with the standard equations of acoustics (see, e.g., [83])

$$\left(\frac{\partial^2}{\partial t^2} - c_0^2 \Delta\right) \varphi(\boldsymbol{r}, t) = 0, \quad p'(\boldsymbol{r}, t) = \rho_0 \frac{\partial}{\partial t} \varphi(\boldsymbol{r}, t). \tag{13.21}$$

Here, $\varphi(\boldsymbol{r}, t)$ is the potential of velocity in the acoustic wave, $p'(\boldsymbol{r}, t)$ is the pressure in the wave, and ρ_0 is the density of the medium.

Taking into account the fact that velocity $\boldsymbol{v}_0(\boldsymbol{r}, t)$ depends on t and θ only in combination $2(\omega t - \theta)$, we will seek the solution to Eq. (13.21) in the form

$$\varphi(\boldsymbol{r}, t) = f(r) e^{2i(\omega t - \theta)}, \tag{13.22}$$

Substituting Eq. (13.22) in Eq. (13.21) and solving the resulting equation in function $f(r)$ with allowance for the radiation condition, we obtain

$$\varphi(\boldsymbol{r}, t) = A H_2^2 (2\omega r / c_0)\, e^{2i(\omega t - \theta)}, \tag{13.23}$$

where $H_2^2(z)$ is the Hankel function of the second kind and A is a constant.

For $r \gg c_0 / 2\omega$ we have the standard divergent cylindrical wave with wavelength $\lambda = \pi c_0 / \omega$

$$\varphi(\boldsymbol{r}, t) = A \sqrt{\frac{c}{\pi \omega r}} \exp\{2i(\omega t - \omega r / c_0 - \theta - 3\pi/8)\}.$$

In the opposite case $r \ll \lambda$, we obtain

$$\varphi(\boldsymbol{r}, t) = i \frac{A}{\pi} \left(\frac{c}{\omega r}\right)^2 e^{2i(\omega t - \theta)}.$$

Potential $\varphi(\boldsymbol{r}, t)$ in region $h \ll r \ll \lambda$ must coincide with the oscillating portion of potential $\varphi_0(\boldsymbol{r}, t)$ (see, e.g., [83]), i.e., with

$$\varphi_0^{(1)}(\boldsymbol{r},t) = -i\kappa\frac{h^2}{r^2}e^{2i(\omega t-\theta)}.$$

This condition yields the following expression for constant A

$$A = -\pi\kappa M^2 = -\frac{\pi\kappa^3}{4h^2c_0^2}.$$

Consequently, potential $\varphi(\boldsymbol{r},t)$ in the wave zone assumes the form

$$\varphi(\boldsymbol{r},t) = -\kappa M^{3/2}\sqrt{\frac{\pi h}{r}}\exp\{2i(\omega t - \omega r/c_0 - \theta - 3\pi/8)\}. \tag{13.24}$$

The pressure in sound wave can be determined from potential (13.24) using Eq. (13.21). The result is as follows

$$p'(\boldsymbol{r},t) = -2i\kappa\omega\rho_0 M^{3/2}\sqrt{\frac{\pi h}{r}}\exp\{2i(\omega t - \omega r/c_0 - \theta - 3\pi/8)\}.$$

The sound intensity (energy radiated per unit time) can be obtained by integrating along the circle of radius $R \gg \lambda$

$$I = \frac{c_0}{\rho_0}\oint dl\,\left\langle |p'(r,t)|^2\right\rangle = 2\pi^2\rho_0 M^4\frac{\kappa^3}{h^2}. \tag{13.25}$$

The radiated energy must coincide with the interaction energy of vortices located in region $r < R$. Total energy in region $r < R$ is

$$E = \frac{\rho_0}{2}\int dS\boldsymbol{v}_0^2(\boldsymbol{r},t). \tag{13.26}$$

Substituting Eq. (13.20) in (13.26) and discarding infinite terms corresponding to the energy of motion of vortices themselves (we assume vortices the point vortices), we obtain the interaction energy in the form

$$E_1 = 4\pi\kappa^2\rho_0\ln(R/h). \tag{13.27}$$

The interaction energy can vary only at the expense of varying the distance between vortices ($h = h(t)$), because the circulation remains intact due to the fact that we consider nonviscous medium.

Differentiating Eq. (13.27) with respect to time, we obtain energy variation rate

$$I(t) = -4\pi\rho_0\frac{\kappa^2}{h(t)}\frac{d}{dt}h(t), \tag{13.28}$$

which is just transferred into the energy of acoustic waves. Using Eq. (13.25) and (13.28), we obtain that the distance between vortices satisfies the equation

$$\frac{d}{dt}h(t) = \frac{\pi\kappa M^4}{2h(t)}. \tag{13.29}$$

Integrating Eq. (13.29) with allowance for the fact that

$$M = M(t) = \frac{\kappa}{2h(t)c_0},$$

we obtain

$$h(t) = h_0 \left[1 + 6\pi M_0^4 \omega_0 t\right]^{1/6}.$$

Thus, the intensity of radiated sound is proportional to M^4, $I \sim M^4$. It is obvious that, in the case of statistically distributed system of vortex line pairs, this estimate remains valid for certain portion of the plane.

13.2.2 Sound Radiation by Vortex Rings

In an incompressible flow, a vortex ring of intensity κ causes the flow to move with a velocity

$$\boldsymbol{v}(\boldsymbol{r},t) = \frac{\kappa}{2}a\int_0^{2\pi} d\phi \frac{\boldsymbol{s}\times\boldsymbol{r}}{r^3},$$

which follows from the *Biot–Savart law* (see, e.g. [109]). Here, $\boldsymbol{s}$ is the unit vector tangent to the vortex ring (it is directed along the vortex vector), a is the radius of the ring, and $\boldsymbol{r}$ is the vector specifying observation point position relative to points lying on the ring.

In the cylindrical coordinate system with origin at the center of ring and z-axis directed along ring axis, we have

$$v_R = \frac{\kappa}{2}a\int_0^{2\pi} d\phi \frac{\cos\phi}{r^3}, \quad v_\theta = 0, \quad v_z = \frac{\kappa}{2}a\int_0^{2\pi} d\phi \frac{a - R\cos\phi}{r^3}, \tag{13.30}$$

where

$$r = \left(R^2 + z^2 + a^2 - 2Ra\cos\phi\right)^{1/2}.$$

Here (R, θ, z) are the coordinates of the radius-vector of the observation point.

Let now we have two vortex rings of equal intensities and equal radii a_0 at distance $2h_0$. In this case, the front ring will increase in size, while the rear ring will decrease and pursue the front ring (fig. 13.1*b*). At certain instant, it will penetrate through the front ring, and the rings switch places. This phenomenon is called the *game of vortex rings* . In a weakly compressible flow, these motions of rings produce local regions of compression and rarefaction; these regions propagate in the medium and, for large distances assume the form of spherical acoustic waves. To determine the structure of radiated sound, we must know relative motions of rings in an incompressible flow.

Let rings have radii $a_1(t)$ and $a_2(t)$ and are separated by distance $2h(t)$ at certain instant t. The rates of variations of ring radii are equal to radial velocities that rings induce at each other, and the rate of variation of the distance between the rings is equal to the difference of the z-components of velocities induced by the rings. Consequently, we have

$$
\begin{aligned}
\frac{d}{dt}a_1(t) &= 2\kappa a_2(t)h(t)\int\limits_0^\pi d\phi\frac{\cos\phi}{|a_1(t)-a_2(t)|^3},\\
\frac{d}{dt}a_2(t) &= -2\kappa a_1(t)h(t)\int\limits_0^\pi d\phi\frac{\cos\phi}{|a_1(t)-a_2(t)|^3},\\
\frac{d}{dt}h(t) &= -\frac{\kappa}{2}\left[a_1^2(t)-a_2^2(t)\right]\int\limits_0^\pi d\phi\frac{1}{|a_1(t)-a_2(t)|^3},
\end{aligned}
\tag{13.31}
$$

where

$$
|a_1(t)-a_2(t)| = \left[a_1^2(t)+a_2^2(t)+4h^2(t)-2a_1(t)a_2(t)\cos\phi\right]^{1/2}.
$$

Equations (13.31) should be solved with the initial conditions at $t=0$

$$
a_1(0)=a_2(0)=a_0, \quad h(0)=h_0.
$$

The first pair of equations immediately yields the relationship between $a_1(t)$ and $a_2(t)$; namely,

$$
a_1^2(t)+a_2^2(t)=2a_0^2. \tag{13.32}
$$

This relationship shows that the moment of inertia of rings relative to the z -axis is conserved.

The integrals in the right-hand side of Eq. (13.31) can be expressed in terms of elliptic functions.

If rings are far from each other ($\gamma = h_0/a_0 \gg 1$), they interact only slightly, and we can assume that, in the first approximation, they move independently with the velocities determined by the ring areas. In the other limiting case $\gamma \ll 1$ (namely this case will be considered in what follows), the rings interact actively. The integrals in the right-hand side of Eq. (13.31) are mainly contributed by the neighborhood of point $\phi = 0$. For this reason, we can replace $\cos\phi$ in the numerators of the first pair of equations by unity. Thus we obtain the second integral of motion

$$
4h^2(t)+\left[a_1(t)-a_2(t)\right]^2 = 4h_0^2. \tag{13.33}
$$

Integral (13.33) means that the distance between points on different rings at the same polar angle is the conserved quantity.

In view of the existence of integrals (13.32) and (13.33), we can reduce system (13.31) to a single equation in variable $\theta(t)$ determined by the equalities

$$a_1(t) = \sqrt{2}a_0 \cos\left(\frac{\pi}{4} - \gamma \sin\theta(t)\right),\ a_2(t) = \sqrt{2}a_0 \sin\left(\frac{\pi}{4} - \gamma \sin\theta(t)\right),$$

$$h(t) = h_0 \cos\theta(t). \tag{13.34}$$

With this definition, conservation laws (13.32) and (13.33) are satisfied automatically (in the first order with respect to γ). Substituting Eq. (13.34) in Eq. (13.31), expanding the result in the series, and calculating the integral, we obtain

$$\theta(t) = \frac{\kappa}{2h_0^2}t.$$

Consequently, the ring radii and the distance between rings are given by the expression

$$a_1(t) = a_0\left(1 + \gamma\sin(\omega t)\right), \quad a_2(t) = a_0\left(1 - \gamma\sin(\omega t)\right),$$

$$h(t) = h_0\cos(\omega t), \tag{13.35}$$

where

$$\omega = \frac{\kappa}{2h_0^2}, \tag{13.36}$$

as in the above case of vortex lines.

We note that the infinitely thin rings move with an infinite velocity. However, actual vortex rings move with a finite velocity significantly smaller than the sound velocity, but the dynamics of relative motion of rings only slightly differs from that we just obtained. The fact that angular velocity (13.36) coincides with the corresponding angular velocity in the case of vortex lines shows that the points lying on rings at equal polar angles revolve relative the center point of the line connecting them at the rotational speed coinciding with the rotational speed of vortex lines located at the same distance and having the same intensity as the rings.

To study the structure of the sound radiated by the system of rings, we need to know the velocity field for large distances from the system. We associate the coordinate system with the point located at the center of the segment connecting the centers of the rings. The velocity of flow outside the rings is given by the formula

$$\boldsymbol{v}(\boldsymbol{r},t) = \frac{\kappa}{2}a_1(t)\int_0^{2\pi} d\phi \frac{\boldsymbol{s}_1\times\boldsymbol{r}_1}{r_1^3} + \frac{\kappa}{2}a_2(t)\int_0^{2\pi} d\phi \frac{\boldsymbol{s}_2\times\boldsymbol{r}_2}{r_2^3}. \tag{13.37}$$

Here, $\boldsymbol{s}_1$ and $\boldsymbol{s}_2$ are the unit vectors tangent to the vortex rings and $\boldsymbol{r}_1$ and $\boldsymbol{r}_2$ are the vectors specifying observation point position relative to points lying on the rings. For large distances from the rings, the oscillating parts of the velocity have the form (in the cylindrical coordinates)

$$v_R^{(1)}(t) = \frac{3\kappa}{4}\frac{R\left(4z^2 - R^2\right)}{(R^2+z^2)^{3/2}} h\left(a_1^2 - a_2^2\right), \quad v_\theta^{(1)}(t) = 0,$$
$$v_z^{(1)}(t) = \frac{3\kappa}{4}\frac{z\left(2z^2 - 3R^2\right)}{(R^2+z^2)^{3/2}} h\left(a_1^2 - a_2^2\right). \tag{13.38}$$

Introducing potential by the formula $\boldsymbol{v}^{(1)} = -\boldsymbol{\nabla}\varphi^{(1)}$, we obtain

$$\varphi^{(1)}(t) = -\frac{\kappa}{4}\frac{R^2 - 2z^2}{(R^2+z^2)^{5/2}} h\left(a_1^2 - a_2^2\right). \tag{13.39}$$

Substituting Eq. (13.35) in Eq. (13.39) and changing to spherical coordinates, we can rewrite Eq. (13.39) in the complex form

$$\varphi^{(1)}(t) = i\frac{\kappa}{2} a_0^3 \gamma^2 \frac{1 - 3\cos^2\theta}{r^3} e^{2i\omega t}. \tag{13.40}$$

Now, we will proceed similar to the case of two vortex lines. We encircle the origin by a sphere of radius L such that $a_0 \ll L \ll \lambda$, where λ is the wavelength of radiated sound waves. We will use the fact that the motion of flow inside the sphere approximately coincides with the motion of incompressible flow. Outside the sphere, the equations of motion will have the form of Eq. (13.21) with the only difference that now Δ is the Laplace operator in the three-dimensional space.

Represent potential φ in the form

$$\varphi(r,\theta) = f(r,\theta) e^{2i\omega t}.$$

Substituting this expression in Eq. (13.21) and taking into account that we must obtain divergent spherical waves for $r \gg \lambda$, we obtain $f(r,\theta)$ in the form

$$f(r,\theta) = \sum_{n=0}^{\infty} \frac{A_n}{\sqrt{r}} H_{n+1/2}^{(2)}\left(\frac{2\omega r}{c_0}\right) P_n(\cos\theta),$$

where $H_{n+1/2}^{(2)}(z)$ is the Hankel function of the second kind and $P_n(z)$ is the Legendre polynomial . Comparing potential $\varphi(r,\theta)$ for $r \ll \lambda$ with $\varphi^{(1)}(r,\theta)$, we obtain that $n = 2$ and

$$A_2 = \frac{2\sqrt{\pi}}{3}\frac{\kappa\gamma^2 a_0^3 \omega^{5/2}}{c_0^{5/2}}.$$

Consequently, potential and pressure in the wave zone have the forms

$$\varphi(r,\theta) = \frac{2\kappa}{3}\left(\frac{\kappa}{2h_0c_0}\right)^2 \frac{a_0}{r}\left(1 - 3\cos^2\theta\right)\exp\left\{2i\left(\omega t - \frac{\omega r}{c_0} - \frac{3\pi}{4}\right)\right\},$$

$$p'(r,\theta) = \frac{4i\kappa}{3}\left(\frac{\kappa}{2h_0c_0}\right)^2 \frac{a_0\omega}{r}\rho_0\left(1 - 3\cos^2\theta\right)\exp\left\{2i\left(\omega t - \frac{\omega r}{c_0} - \frac{3\pi}{4}\right)\right\},$$

so that angular distribution of the energy radiated per unit time is

$$I(\theta) = \frac{8\pi}{9}\frac{\kappa^3 a_0^2}{h_0^3}\left(\frac{\kappa}{2h_0c_0}\right)^5 \rho_0\left(1 - 3\cos^2\theta\right)^2.$$

Vortex rings radiate energy as a quadrupole; the main portion of energy is radiated in a cone about z-axis with corner angle 106° in both positive and negative z-directions. Integrating over θ, we obtain that total energy radiated per unit time is given by the expression

$$I = \frac{64\pi}{45}\frac{\kappa^3 a_0^2}{h_0^3}\left(\frac{\kappa}{2h_0c_0}\right)^5 \rho_0.$$

Thus, intensity of sound radiated by a pair of vortex rings is proportional to M^5. In the case of statistically distributed system of pairs of vortex rings, this proportionality will remain valid for certain portion of space, which agrees with estimates obtained in [89–91].

References

1. Adler, R.J.: The Geometry of Random Fields. Wiley, New York (1991)
2. Ambartsumyan, V.A.: Diffuse reflection of light by a foggy medium. Comptes Rendus (Doklady) de l'USSR 38(8), 229–232 (1943)
3. Ambartsumyan, V.A.: On the problem of diffuse reflection of light. Journal of Physics USSR 8(1), 65 (1944)
4. Ambartsumyan, V.A.: On the principle of invariance and its some applications. In: Mnatsakanyan, M.A., Pickichyan, H.V. (eds.) Principle of Invariance and its Applications, pp. 9–18. Armenian SSR Acad. of Sciences, Yerevan (1989)
5. Anishchenko, V.S., Neiman, A.B., Moss, F., Schimansky-Geier, L.: Stochastic resonance: noise enhanced order. Physics–Uspekhi 42(1) (1999)
6. Aref, H.: The development of chaotic advection. Physics of Fluids 14, 1315–1325 (2002)
7. Batchelor, G.K.: Theory of Homogeneous Turbulence. Cambridge Univ. Press, London (1953)
8. Bellman, R., Wing, G.M.: An Introduction to Invariant Imbedding. Classics in Applied Mathematics, vol. 8. SIAM, Philadelphia (1992)
9. Bourret, R.C., Frish, U., Pouquet, A.: Brownian motion of harmonical oscillator with stochastic frequency. Physica A 65(2), 303–320 (1973)
10. Casti, J., Kalaba, R.: Imbedding Methods in Applied Mathematics. Addison-Wesley, Reading (1973)
11. Cramer, H., Leadbetter, M.R.: Stationary and Related Stochastic Processes. Wiley, New York (1967)
12. Ditkin, V.A., Prudnikov, A.P.: Integral Transforms and Operation Calculus. Pergamon Press, Oxford (1965)
13. Dolzhansky, F.V.: Fundamentals of Geophysical Hydrodynamics. In: Encyclopedia of Mathematical Sciences. Mathematical Physics IV, vol. 103. Springer, Heidelberg (2013)
14. Dolzhansky, F.V., Klyatskin, V.I., Obukhov, A.M., Chusov, M.A.: Nonlinear Hydrodynamic Type Systems. Nauka, Moscow (1974) (in Russian)
15. Donsker, M.D.: On functional space integrals. In: Proc. Conf. Theory and Appl. of Analysis in Funct. Space, vol. 17. The M.I.I. Press (1964)
16. Er-El, J., Peskin, R.L.: Relative diffusion of constant-level balloons in the Southern Hemisphere. J. Atmos. Sci. 38(10), 2264–2274 (1981)
17. Feller, W.: An Introduction to Probability Theory and its Applications, vol. I, 3rd edn., vol. I, 2nd edn. Wiley, New York (1968)

18. Flatté, S.M., Wang, G.-Y., Martin, J.: Irradiance variance of optical waves through atmospheric turbulence by numerical simulation and comparison with experiment. JOSA A10, 2363–2370 (1993)
19. Flatté, S.M., Bracher, C., Wang, G.Y.: Probability - density functions of irradiance for waves in atmospheric turbulence calculated by numerical simulation. JOSA A11, 2080–2092 (1994)
20. Furutsu, K.: On the statistical theory of electromagnetic waves in a fluctuating medium. J. Res. NBS D-67, 303 (1963)
21. Gardiner, C.W.: Handbook of Stochastic Methods for Physics, Chemistry and the Natural Sciences. Springer, Berlin (1985)
22. Gel'fand, I.M., Shilov, G.E.: Generalized Functions, vol. 1. Academic Press, New York (1964)
23. Gledzer, E.B., Dolzhanskii, F.V., Obukhov, A.M.: Systems of hydrodynamical type and their applications. Nauka, Moscow (1981) (in Russian)
24. Glukhovsky, A.B., Klyatskin, V.I.: Stochastic 'noise' in elementary nonlinear fluid-dynamic systems. Izvestiya, Atmospheric and Oceanic Physics 9(7), 381–386 (1973)
25. Gluhovsky, A., Klyatskin, V.I.: On dynamics of flipover phenomena in simple hydrodynamic models. Doklady, Earth Science Sections 237, 18–20 (1977)
26. Golberg, M.A.: Invariant imbedding and Riccati transformations. Appl. Math. and Com. 1(1), 1–24 (1975)
27. Gryanik, N.V., Klyatskin, V.I.: Contribution to the statistical theory of wave localization in a two-layer medium. JETP 84(6), 1106–1113 (1997a)
28. Gryanik, N.V., Klyatskin, V.I.: Localization of Rossby waves under the influence of random cylindrical topography (two-layer model). Izvestiya, Atmospheric and Oceanic Physics 33(6), 669–678 (1997b)
29. Gurbatov, S., Malakhov, A., Saichev, A.: Nonlinear Random Waves and Turbulence in Nondispersive Media: Waves, Rays and Particles. Manchester University Press, Manchester (1991)
30. Gurvich, A.S., Kallistratova, M.A., Martvel', F.E.: An investigation of strong fluctuations of light intensity in a turbulent medium at a small wave parameter. Radiophys. & Quantum Electron. 20(7), 705–714 (1977)
31. Guzev, M.A., Klyatskin, V.I.: Approximation of the parabolic equation and the wavefield of a point source in a layered random medium. Waves in Random Media 1(4), 275–286 (1991)
32. Guzev, M.A., Klyatskin, V.I., Popov, G.V.: Phase fluctuations and localization length in layered randomly inhomogeneous media. Waves in Random Media 2(2), 117–123 (1992)
33. Hopf, E.: Statistical hydromechanics and functional calculus. J. Ration. Mech. Anal. 1, 87–123 (1952)
34. Hopf, E.: On the application of functional calculus to the statistical theory of turbulence. In: Proc. Symp. Appl. Math., vol. 7, pp. 41–50. Am. Math. Society (1957)
35. Hopf, E.: Remarks on the functional-analytic approach to turbulence. In: Proc. Symp. Appl. Math., vol. 13, pp. 157–163. Am. Math. Society (1962)
36. Isichenko, M.B.: Percolation, statistical topography, and transport in random media. Rev. Modern Phys. 64(4), 961–1043 (1992)
37. Ivanittskii, G.R.: 21st century: what is life from the perspective of physics? Phys. Usp. 53, 327–356 (2010)

38. Kagiwada, H.H., Kalaba, R.: Integral Equations Via Imbedding Methods. Addison-Wesley, Reading (1974)
39. Kharif, C., Pelinovskyy, E., Slyunaen, A.: Rogue Waves in the Ocean. Springer, Berlin (2009)
40. Klimontovich, Y.L.: What are stochastic filtration and stochastic resonance? Physics–Uspekhi 42(1) (1999)
41. Klyatskin, V.I.: Sound radiation from vortex system. Izvestiya AN SSSR, Mekhanika Zhidkosti i Gaza (6), 87–92 (1966)
42. Klyatskin, V.I.: Space-time description of stationary and homogeneous turbulence. Fluid Dynamics 6(4), 655–661 (1971)
43. Klyatskin, V.I.: To the nonlinear theory of stability of periodic flow. J. Appl. Math. and Mech., 36(2) (1972)
44. Klyatskin, V.I.: Noise in the hydrodynamic flow near the instability threshold. Radiophys. & Quantum Electron. 17(4), 476–481 (1974)
45. Klyatskin, V.I.: Statistical theory of light propagation in a medium having random inhomogeneities. Radiophys. & Quantum Electron. 18(1), 47–51 (1975a)
46. Klyatskin, V.I.: Statistical Description of Dynamical Systems with Fluctuating Parameters. Nauka, Moscow (1975b) (in Russian)
47. Klyatskin, V.I.: Dynamic systems with parameter fluctuations as telegraphic-process type. Radiophys. & Quantum Electron. 20(4), 382–392 (1977)
48. Klyatskin, V.I.: Stochastic Equations and Waves in Randomly Inhomogeneous Media. Nauka, Moscow (1980) (in Russian)
49. Klyatskin, V.I.: Ondes et Équations Stochastiques dans les milieus Aléatoirement non Homogènes. Les éditions de Physique, Besançon-Cedex (1985) (in French)
50. Klyatskin, V.I.: The Imbedding Method in Wave Propagation Theory. Nauka, Moscow (1986) (in Russian)
51. Klyatskin, V.I.: Approximations by delta-correlated random processes and diffusion approximation in stochastic problems. In: Kohler, W., White, B.S. (eds.) Mathematics of Random Media. Lectures in Appl. Math, vol. 27, pp. 447–476. AMS, Providence RI (1991)
52. Klyatskin, V.I.: The imbedding method in statistical boundary-value wave problems. In: Wolf, E. (ed.) Progress in Optics, vol. XXXIII, pp. 1–128. North-Holland, Amsterdam (1994a)
53. Klyatskin, V.I.: Statistical description of the diffusion of a passive tracer in a random velocity field. Physics–Uspekhi 37(5), 501–513 (1994b)
54. Klyatskin, V.I.: Localization of Rossby waves over a random cylindrical topography of the ocean bottom. Atmospheric and Oceanic Physics 32(6), 757–765 (1996)
55. Klyatskin, V.I.: Diffusion and Clustering of Passive Tracers in Random Hydrodynamic Flows. Fizmatlit, Moscow (2005) (in Russian)
56. Klyatskin, V.I.: Stochastic equations (Theory and Applications in Acoustics, Hydrodynamics, and Radiophysics). Basic Concepts, Exact Results, and Asymptotic Approximations, vol. 1. Fizmatlit, Moscow (2008) (in Russian)
57. Klyatskin, V.I.: Stochastic equations (Theory and Applications in Acoustics, Hydrodynamics, and Radiophysics). Coherent Phenomena in Stochastic Dynamic Systems, vol. 2. Fizmatlit, Moscow (2008) (in Russian)
58. Klyatskin, V.I.: Dynamic stochastic systems, typical realization curve, and Lyapunov's exponents. Atmospheric and Oceanic Physics 44(1), 18–32 (2008a)

59. Klyatskin, V.I.: Statistical topography and Lyapunov's exponents in dynamic stochastic systems. Physics-Uspekhi 51(4), 395–407 (2008b)
60. Klyatskin, V.I.: Modern methods for the statistical description of dynamical stochastic sestems. Physics-Uspekhi 52(5), 514–519 (2009)
61. Klyatskin, V.I.: Lectures on Dynamics of Stochastic Systems. Elsevier, Boston (2011)
62. Klyatskin, V.I.: Studies on Dynamics of Stochastic Systems. URSS, Moscow (2012) (in Russian)
63. Klyatskin, V.I.: Spatial structures can form in stochastic dynamic systems due to near-zero-probability events: (comment on "21st century: what is life from the perspective of physics?"). Physics-Uspekhi 55(11), 1152–1154 (2012)
64. Klyatskin, V.I.: On the criterion of stochastic structure formation in random media. In: Proceedings of the 4th International Interdisciplinary Chaos Symposium, pp. 69–73. Springer, Berlin (2013)
65. Klyatskin, V.I.: Clastering of Random Positive Fiele as a Law of Nature. Teor. Math. Phys. 176(3), 1252–1256 (2013)
66. Klyatskin, V.I.: On the Statistical Theory of Spatial Structure Formation in Random Media. Russ. Journ. Math. Phys. 20(3), 295–314 (2013)
67. Klyatskin, V.I., Chkhetiani, O.G.: On the Diffusion and Clustering of a magnetic Field in Random Velocity Fields. JETP 109(2), 345–356 (2009)
68. Klyatskin, V.I., Elperin, T.: Clustering of the Low-Inertia Particle Number Density Field in Random Divergence-Free Hydrodynamic Flows. JETP 95(2), 328–340 (2002a)
69. Klyatskin, V.I., Elperin, T.: Diffusion of Low-Inertia Particles in a Field of Random Forces and the Kramers Problem. Atmospheric and Oceanic Physics 38(6), 725–731 (2002b)
70. Klyatskin, V.I., Koshel', K.V.: The simplest example of the development of a cluster-structured passive tracer field in random flows. Physics–Uspekhi 43(7), 717–723 (2000)
71. Klyatskin, V.I., Saichev, A.I.: Statistical and dynamical localization of plane waves in randomly layered media. Soviet Physics Usp. 35(3), 231–247 (1992)
72. Klyatskin, V.I., Saichev, A.I.: Statistical theory of the diffusion of a passive tracer in a random velocity field. JETP 84(4), 716–724 (1997)
73. Klyatskin, V.I., Tatarskii, V.I.: New method of successive approximations in the peoblem of wave propagation in a medium with large-scale inhomogeneities. Radiophys. & Quantum Electron. 14(9), 1100–1111 (1971)
74. Klyatskin, V.I., Yakushkin, I.G.: Stochastic transport in random wave fields. JETP 91(4), 736–747 (2000)
75. Klyatskin, V.I., Gryanik, N.V., Gurarie, D.: Localization of Rossby waves under the influence of random topography (two-layer model). Wave Motion 28(4), 333–352 (1998)
76. Koshel', K.V., Aleksandrova, O.V.: Some resullts of numerical modeling of passive scalar diffusion in randon velocity field. Izvestiya, Atmospheric and Oceanic Physics 35(5), 578–588 (1999)
77. Kravtsov, Y.A.: 'Geometrical' depolarization of light in a turbulent atmosphere. Radiophys. & Quantum Electron. 13(2), 217–220 (1970)
78. Kuzovlev, Y.E., Bochkov, G.N.: Operator methods of analysing stochastic non-gaussian processes and systems. Radiophys. & Quantum Electron. 20(10), 1036–1044 (1977)
79. Lamb, H.: Hydrodinamics, 6th edn. Dover Publications, New York (1932)

80. Landau, L.D.: To the problem of turbulence. Dokl. Akad. Nauk SSSR 44(6), 339 (1944)
81. Landau, L.D., Lifshitz, E.M.: On hydrodynamic fluctuations. JETP 32(3) (1957)
82. Landau, L.D., Lifshitz, E.M.: Course of Theoretical Physics. Electrodynamics of Continuous Media, vol. 8. Butterworth-Heinemann, Oxford (1984)
83. Landau, L.D., Lifshitz, E.M.: Fluid Mechanics, 2nd edn. Pergamon Press, London (1987)
84. Levin, M.L., Rytov, S.M.: Theory of Steady-state Thermal Fluctuations in Electrodynamics. Nauka, Moscow (1967) (in Russian)
85. Leontovich, M.A.: On one method of solving some problems about propagation along the Earth surface. Izvestiya AN SSSR, Physika 8, 16 (1944)
86. Leontovich, M., Fock, V.A.: The solution of the electromagnetic wave propagation problem along the Earth surface by the parabolic equation method. Zh. Éxper. Teor. Fiz. 16, 557–573 (1946)
87. Lifshits, E.M., Pitaevskii, L.P.: Statistical Physics, Pt. 2. Elsevier Science & Technology, UK (1980)
88. Lifshits, I.M., Gredeskul, S.A., Pastur, L.A.: Introduction to the Theory of Disordered Solids. John Wiley, New York (1988)
89. Lighthill, J.: On sound generated aerodynamically. 1. General theory. Proc. Roy. Soc. A.211(1107), 564–587 (1952)
90. Lighthill, J.: On sound generated aerodynamicaly. II. Turbulence as a source of sound. Proc. Roy. Soc. A.222(1148), 1–32 (1954)
91. Lighthill, J.: Sound generated aerodynamicaly. Proc. Roy. Soc. A.267(1329), 147–182 (1962)
92. Longuet-Higgins, M.S.: Statistical properties of a moving wave-form. Proc. Cambridge Philos. Soc. 52, 234–245 (1956)
93. Longuet-Higgins, M.S.: On the velocity of the maxima in a moving wave-form. Proc. Cambridge Philos. Soc. 53, 230–233 (1957a)
94. Longuet-Higgins, M.S.: The statistical analysis of a random, moving surface. Philos. Trans. R. Soc., London, Ser. A249(966), 321–387 (1957b)
95. Longuet-Higgins, M.S.: Statistical properties of an isotropic random surface. Philos. Trans. R. Soc., London, Ser. A250(975), 157–174 (1957c)
96. Longuet-Higgins, M.S.: The statistical geometry of random surfaces. In: Hydrodynamic Stability. In: Proc. 13th Symp. Appl. Math., pp. 105–142. Am. Math. Soc. (1962)
97. Lorenz, E.N.: Maximum simplification of the dynamic equations. Tellus 12(3), 243–254 (1960)
98. Lorenz, E.N.: Deterministic nonperiodic flow. J. Atmos. Sci. 20(3), 130–141 (1963)
99. Malakhov, A.N.: Kumulant Analysis of Non-Gaussian Processes and Their Transforms. Sov. Radio, Moscow (1978) (in Russian)
100. Maxey, M.R.: The Gravitational Settling of Aerosol Particles in Homogeneous turbulence and random flow field. J. Fluid Mech. 174, 441–465 (1987)
101. McLaughlin, J.B., Martin, P.C.: Transition to turbulence in a statically stressed fluid system. Phys. Rev. A12(1), 186–203 (1975)
102. Mehlig, B., Wilkinson, M.: Coagulation by random velocity fields as a Kramers Problem. Phys. Rev. Letters 92(25), 250602-1–250602-4 (2004)
103. Meshalkin, L.D., Sinai, Y.G.: Investigation of the stability of a stationary solution of a system of equations for the plane movement of an incompressible viscous flow. J. Appl. Math. and Mech. 25(6), 1700–1705 (1961)

104. Mesinger, F.: Behavior of a very large number of constant-volume trajectories. J. Atmos. Sci. 22(5), 479–492 (1965)
105. Mesinger, F.: Numerical integration of the primitive equations with a floating set of computation points: experiments with a barotropic global model. Monthly Weather Review 99(1), 15–29 (1971)
106. Mesinger, F., Mintz, Y.: Numerical simulation of the 1970–1971 Eole experiment. Numerical simulation of weather and climate. Technical Report No. 4, Dep. Meteorology. Univ. of California, Los Angeles (1970a)
107. Mesinger, F., Mintz, Y.: Numerical simulation of the clustering of the constant-volume balloons in the global domain. Numerical simulation of weather and climate. Technical Report No. 5, Dep. Meteorology. Univ. of California, Los Angeles (1970b)
108. Mikhailov, A.S., Uporov, I.V.: Critical phenomena in media with breeding, decay, and diffusion. Sov. Phys. Usp. 27(9), 695–714 (1984)
109. Milne-Thomson, L.M.: Theoretical Hydrodynamics, 5th edn. Dover Publication, Inc., New York (1996)
110. Mirabel', A.P., Monin, A.S.: On statistical properties of admixture mixing by two-dimensional turbulence. Izvestiya, Atmospheric and Oceanic Physics 19(9), 681–687 (1983)
111. Moffatt, H.K.: The interaction of turbulence with strong wind shear. In: Yaglom, A.M., Tatarskii, V.I. (eds.) Atmospheric Turbulence and Radio Wave Propagation, pp. 139–154. Nauka, Moscow (1967)
112. Monin, A.S., Yaglom, A.M.: Statistical Fluid Mechanics. MIT Press, Cambridge (1980)
113. Nalbandyan, O.G.: Possible approximations of the gaussian random process in some nonlinear stochastic problems. Radiophys. & Quantum Electron. 20(5) (1977)
114. Novikov, E.A.: Functionals and the random force method in turbulence theory. Sov. Phys. JETP 20(5), 1290–1294 (1965)
115. Obukhov, A.M.: On integral invariants in the hydrodynamic-type systems. Dokl. AN SSSR 184(2), 309–312 (1969)
116. Obukhov, A.M.: On invariant characteristics of systems of fluid mechanical type. Fluid Dynam. Trans. 5(2) (1971)
117. Obukhov, A.M.: On the problem of nonlinear interaction in fluid dynamics. Gerlands Beitr. Geophys. 82(4), 282–290 (1973)
118. Obukhov, A.M.: The Kolmogorov flow and its laboratory modeling. Uspekhi Mat. Nauk 38(4) (232), 101–111 (1983)
119. Obukhov, A.M.: Turbulence and Dynamics of Atmosphere. Leningrad, Gidrometizdat (1988) (in Russian)
120. Pedlosky, J.: Geophysical Fluid Dynamics. Springer, New York (1982)
121. Pelinovskii, E.N.: Spectral analysis of simple waves. Radiophys. & Quantum Electron. 19(3), 262–270 (1976)
122. Rytov, S.M., Kravtsov, Y.A., Tatarskii, V.I.: Principles of Statistical Radiophysics, pp. 1–4. Springer, Berlin (1987-1989)
123. Saichev, A.I., Woyczynski, W.A.: Probability distributions of passive tracers in randomly moving media. In: Molchanov, S.A., Woyczynski, W.A. (eds.) Stochastic Models in Geosystems. IMA Volumes in Mathematics and its Applications, vol. 85, pp. 359–400. Springer, New York (1996)
124. Shapiro, V.E., Loginov, V.M.: 'Formulae of differentiation' and their use for solving stochastic equations. Physica 91A, 563–574 (1978)

125. Shapiro, V.E., Loginov, V.M.: Dynamical Systems Under Random Influences. Nauka, Novosibirsk (1983) (in Russian)
126. Stratonovich, R.L.: Topics in the Theory of Random Noise, vol. 1, 2. Gordon and Breach, New York (1963, 1967)
127. Swerling, P.: Statistical properties of the contours of random surfaces. I.R.E. Tranc. Inf. Theory IT-8, 315–321 (1962)
128. Tatarskii, V.I.: Depolarization of light by turbulent atmospheric inhomogeneties. Radiophys. & Quantum Electron. 10(12), 987–988 (1967)
129. Tatarskii, V.I.: Light propagation in a medium with random index refraction inhomogeneities in the Markov process approximation. Sov. Phys. JETP 29(6), 1133–1138 (1969)
130. Tatarskii, V.I.: The Effects of the Turbulent Atmosphere on Wave Propagation. National Technical Information Service. Springfield, Va (1977)
131. Uhlenbeck, G.E.: The fundamental problems of statistical mechanics. Uspekhi Fiz. Nauk 103(2), 275–318 (1971); Soviet Phys., Uspekhi
132. Weichman, P.B., Glazman, R.E.: Turbulent fluctuation and transport of passive scalar by random wave fields. Phys. Rev. Letters 83(24), 5011–5018 (1999)
133. Weichman, P.B., Glazman, R.E.: Passive scalar transport by travelling wave fields. J. Fluid Mech. 420, 147–200 (2000)
134. Weichman, P.B., Glazman, R.E.: Spatial variations of a passive tracer in a random wave field. J. Fluid Mech. 453, 263–287 (2002)
135. Whitham, G.B.: Linear and Nonlinear Waves. John Wiley & Sons Inc., New York (1974)
136. Wilkinson, M., Mehlig, B.: Caustics in turbulent aerosols. Europhys. Letters 71(2), 186–192 (2005)
137. Yaroshchuk, I.O.: Numerical Modeling of the Plane Wave Propagation in Randomly Layered Linear and Nonlinear Media. Ph.D dissertation, Pacific Oceanological Institute, Vladivostok (1986)
138. Yudovich, V.I.: Example of the birth of secondary steady-state or periodic flow at stability failure of the laminar flow of viscous noncompressible fluid. J. Appl. Math. and Mech. 29(3), 527–544 (1965)
139. Zel'dovich, Y.B., Molchanov, S.A., Ruzmaikin, A.A., Sokolov, D.D.: Intermittency of passive fields in random media. Sov. Phys. JETP 62, 1188–1194 (1985)
140. Zel'dovich, Y.B., Molchanov, S.A., Ruzmaikin, A.A., Sokolov, D.D.: Intermittency in random media. Sov. Phys. Usp. 30, 353–369 (1987)
141. Ziman, J.M.: Models of Disorder, The Theoretical Physics of Homogeneously Disordered Systems. Cambr. Univ. Press, Cambridge (1979)
142. Zirbel, C.L., Çinlar, E.: Mass transport by Brownian motion. In: Molchanov, S.A., Woyczynski, W.A. (eds.) Stochastic Models in Geosystems. IMA Volumes in Mathematics and its Applications, vol. 85, pp. 459–492. Springer, New York (1996)

Index

MIX
Papier aus verantwortungsvollen Quellen
Paper from responsible sources
FSC® C105338

If you have any concerns about our products,
you can contact us on
ProductSafety@springernature.com

In case Publisher is established outside the EU,
the EU authorized representative is:
Springer Nature Customer Service Center GmbH
Europaplatz 3, 69115 Heidelberg, Germany

Printed by Libri Plureos GmbH
in Hamburg, Germany